FUNDAMENTALS OF EARTHQUAKE ENGINEERING

FUNDAMENTALS OF EARTHQUAKE ENGINEERING

Amr S. Elnashai
Department of Civil and Environmental Engineering, University of Illinois, USA

and

Luigi Di Sarno
Department of Structural Analysis and Design, University of Sannio, Benvenuto, Italy

WILEY
A John Wiley & Sons, Ltd, Publication

This edition first published 2008

Registered office

John Wiley & Sons Ltd, The Atrium, Southern Gate, Chichester, West Sussex, PO19 8SQ, United Kingdom

For details of our global editorial offices, for customer services and for information about how to apply for permission to reuse the copyright material in this book please see our website at www.wiley.com.

Library of Congress Cataloging-in-Publication Data

Elnashai, Amr S.
Fundamentals of earthquake engineering / Amr S. Elnashai and Luigi Di Sarno.
p. cm.
Includes bibliographical references and index.
ISBN 978-0-470-02483-6 (Hbk)   1. Earthquake engineering.   I. Di Sarno, Luigi.   II. Title.
TA654.6.E485   2008
624.1'762–dc22

2008033265

ISBN: 978-0-470-02483-6 (Hbk)

A catalogue record for this book is available from the British Library.

Set in 9 on 11pt Times by SNP Best-set Typesetter Ltd., Hong Kong
Printed in England by Antony Rowe Ltd, Chippenham, Wilts.

Contents

About the Authors

Professor Amr Elnashai

Professor Amr Elnashai is Bill and Elaine Hall Endowed Professor at the Civil and Environmental Engineering Department, University of Illinois at Urbana-Champaign. He is Director of the National Science Foundation (NSF) multi-institution multi-disciplinary Mid-America Earthquake Center. He is also Director of the NSF Network for Earthquake Engineering Simulation (NEES) Facility at Illinois. Amr obtained his MSc and PhD from Imperial College, University of London, UK. Before joining the University of Illinois in June 2001, Amr was Professor and Head of Section at Imperial College. He has been Visiting Professor at the University of Surrey since 1997. Other visiting appointments include the University of Tokyo, the University of Southern California and the European School for Advanced Studies in Reduction of Seismic Risk, Italy, where he serves on the Board of Directors since its founding in 2000. Amr is a Fellow of the Royal Academy of Engineering in the United Kingdom (UK-equivalent of the NAE), Fellow of the American Society of Civil Engineers and the UK Institution of Structural Engineers.

He is founder and co-editor of the *Journal of Earthquake Engineering*, editorial board member of several other journals, a member of the drafting panel of the European design code, and past senior Vice-President of the European Association of Earthquake Engineering. He is the winner of the Imperial College Unwin Prize for the best PhD thesis in Civil and Mechanical Engineering (1984), the Oscar Faber Medal for best paper in the Institution of Structural Engineering, and two best paper medals from the International Association of Tall Buildings, Los Angeles. He is the administrative and technical team builder and director of both the MAE Center and NEES@UIUC Simulation Laboratory, at Illinois.

Amr is President of the Asia-Pacific Network of Centers of Earthquake Engineering Research (ANCER), a member of the FIB Seismic Design Commission Working Groups and two Applied Technology Council (ATC, USA) technical committees. He founded the Japan–UK Seismic Risk Forum in 1995 and served as its director until 2004. He leads a FEMA project for impact assessment for the eight central US states, was advisor to the UK Department of the Environment, advisor to the Civil Defense Agency of Italy, and review panel member for the Italian Ministry of Research and the New Zealand and Canadian Science Research Councils.

Amr's technical interests are multi-resolution distributed analytical simulations, network analysis, large-scale hybrid testing, and field investigations of the response of complex networks and structures to extreme loads, on which he has more than 250 research publications, including over 110 refereed journal papers, many conference, keynote and prestige lectures (including the Nathan Newmark Distinguished Lecture), research reports, books and book chapters, magazine articles, and field investigation reports. Amr has successfully supervised 29 PhD and over 100 Masters Theses. Many of his students hold significant positions in industry, academia and government in over 12 countries. He has a well-funded research group, with a large portfolio of projects from private industry, state agencies, federal agencies, and international government and private entities. Amr taught many different subjects both at Illinois and at Imperial College. He is recognized as an effective teacher and has been on the 'incomplete list of teachers considered excellent by their students' twice at UIUC.

He has contributed to major projects for a number of international companies and other agencies such as the World Bank, GlaxoWellcome (currently GSK), Shell International, AstraZeneca, Minorco, British Nuclear Fuels, UK Nuclear Installations Inspectorate, Mott MacDonald, BAA, Alstom Power, the Greek, Indonesian and Turkish Governments, and the National Geographic Society. He is currently working on large projects for the Federal Emergency Management Agency (FEMA), State Emergency Management Agencies, Istanbul Municipality, US AID, Governments of Pakistan and Indonesia, among others. Amr enjoys scuba-diving and holds several certificates from the British Sub-Aqua Club and the US Professional Association of Diving Instructors. He also enjoys reading on history, the history of painting and film-making.

Dr Luigi Di Sarno

Dr. Luigi Di Sarno is Assistant Professor in Earthquake Engineering at the University of Sannio (Benevento), and holds the position of Research Associate at the Department of Structural Engineering (DIST), University of Naples, Federico II in Italy. He graduated cum laude in Structural Engineering from the University of Naples, Federico II. He then obtained two MSc degrees in Earthquake Engineering and Structural Steel Design from Imperial College, London. In 2001 Dr. Di Sarno obtained his PhD from University of Salerno in Italy and moved to the University of Illinois at Urbana Champaign in 2002 where he worked as a Post-doctoral Research Associate. He has been Visiting Professor at the Mid-America Earthquake Center at Illinois since 2004. His research interests are seismic analysis and design of steel, reinforced concrete and composite structures, and the response of tall buildings to extreme loads, on which he has written more than 60 research publications, including over 15 refereed journal papers, many conference papers, research reports, book chapters and field investigation reports. Dr. Di Sarno continues to work with the active research group at the University of Naples, with a large portfolio of projects from private industry, state agencies, and international government and private entities. He taught several courses at Naples, Benevento and the Mid-America Earthquake Center. He is currently working on large projects funded by the Italian State Emergency Management Agency (DPC) and the Italian Ministry of Education and Research, amongst others. Dr. Di Sarno enjoys reading on history, science and art. He also enjoys playing tennis and swimming.

Foreword

Congratulations to both authors! A new approach for instruction in Earthquake Engineering has been developed. This package provides a new and powerful technique for teaching – it incorporates a book, worked problems and comprehensive instructional slides available on the web site. It has undergone numerous prior trials at the graduate level as the text was being refined.

The book, in impeccable English, along with the virtual material, is something to behold. 'Intense' is my short description of this book and accompanying material, crafted for careful study by the student, so much so that the instructor is going to have to be reasonably up-to-date in the field in order to use it comfortably. The writer would have loved to have had a book like this when he was teaching Earthquake Engineering.

The text has four main chapters and two appendices. The four main chapters centre on (a) Earthquake Characteristics, (b) Response of Structures, (c) Earthquake Input Motions and (d) Response Evaluation, with two valuable appendices dealing with Structural Configurations and Systems for Effective Earthquake Resistance, and Damage to Structures. The presentation, based on stiffness, strength and ductility concepts, comprises a new and powerful way of visualizing many aspects of the inelastic behaviour that occurs in structures subjected to earthquake excitation.

The book is written so as to be appropriate for international use and sale. The text is supplemented by numerous references, enabling the instructor to pick and choose sections of interest, and to point thereafter to sources of additional information. It is not burdened by massive reference to current codes and standards in the world. Unlike most other texts in the field, after studying this book, the students should be in a position to enter practice and adapt their newly acquired education to the use of regional seismic codes and guidelines with ease, as well as topics not covered in codes. Equally importantly, students who study this book will understand the bases for the design provisions.

Finally, this work has application not only in instruction, but also in research. Again, the authors are to be congratulated on developing a valuable work of broad usefulness in the field of earthquake engineering.

William J. Hall
Professor Emeritus of Civil Engineering
University of Illinois at Urbana-Champaign

Preface and Acknowledgements

This book forms one part of a complete system for university teaching and learning the fundamentals of earthquake engineering at the graduate level. The other components are the slide sets, the solved examples, including the comprehensive project, and a free copy of the computer program Zeus-NL, which are available on the book web site. The book is cast in a framework with three key components, namely (i) earthquake causes and effects are traced from Source to Society; (ii) structural response under earthquake motion is characterized primarily by the varying and inter-related values of stiffness, strength and ductility; and (iii) all structural response characteristics are presented on the material, section, member, sub-assemblage and structural system levels. The four chapters of the book cover an overview of earthquake causes and effects, structural response characteristics, features and representations of strong ground motion, and modelling and analysis of structural systems, including design and assessment response quantities. The slide sets follow closely the contents of the book, while being a succinct summary of the main issues addressed in the text. The slide sets are intended for use by professors in the lecture room, and should be made available to the students only at the end of each chapter. They are designed to be also a capping revision tool for students. The solved examples are comprehensive and address all the important and intricate sub-topics treated in the four chapters of the book. The comprehensive project is used to provide an integration framework for the various components of the earthquake source, path, site, and structural features that affect the actions and deformations required for seismic design. The three teaching and learning components of (i) the book, (ii) the slide sets and (iii) the solved examples are inseparable. Their use in unison has been tested and proven in a top-tier university teaching environment for a number of years.

We have written this book whilst attending to our day jobs. We have not taken a summer off, or gone on sabbatical leave. It has therefore been difficult to extract ourselves from the immediate and more pressing priorities of ongoing academic and personal responsibilities. That authoring the book took four years has been somewhat frustrating. The extended period has however resulted in an improved text through the feedback of end-users, mainly graduate students of exceptional talent at the University of Illinois. Our first thanks therefore go to our students who endured the experimental material they were subjected to and who provided absolutely essential feedback. We are also grateful for a number of world-class researchers and teachers who voluntarily reviewed the book and provided some heart-warming praise alongside some scathing criticism. These are, in alphabetical order, Nicholas Ambraseys, Emeritus Professor at Imperial College; Mihail Garevski, Professor and Director, Institute of Seismology and Earthquake Engineering, University of Skopje 'Kiril and Methodius'; Ahmed Ghobarah, Professor at McMaster University; William Hall, Emeritus Professor at the University of Illinois; and Sashi Kunnath, Professor at University of California-Irvine. Many other colleagues have read parts of chapters and commented on various aspects of the book, the set of slides and the worked examples. Finally our thanks go to six anonymous reviewers who were contacted by Wiley to assess the book proposal, and to all Wiley staff who have been invariably supportive and patient over the four years.

Amr S. Elnashai
Luigi Di Sarno

Introduction

Context, Framework and Scope

Earthquakes are one of the most devastating natural hazards that cause great loss of life and livelihood. On average, 10,000 people die each year due to earthquakes, while annual economic losses are in the billions of dollars and often constitute a large percentage of the gross national product of the country affected.

Over the past few decades, earthquake engineering has developed as a branch of engineering concerned with the estimation of earthquake consequences and the mitigation of these consequences. It has become an interdisciplinary subject involving seismologists, structural and geotechnical engineers, architects, urban planners, information technologists and social scientists. This interdisciplinary feature renders the subject both exciting and complex, requiring its practitioners to keep abreast of a wide range of rapidly evolving disciplines. In the past few years, the earthquake engineering community has been reassessing its procedures, in the wake of devastating earthquakes which caused extensive damage, loss of life and property (e.g. Northridge, California, 17 January 1994; $30 billion and 60 dead; Hyogo-ken Nanbu, Japan, 17 January 1995; $150 billion and 6,000 dead).

The aim of this book is to serve as an introduction to and an overview of the latest structural earthquake engineering. The book deals with aspects of geology, engineering seismology and geotechnical engineering that are of service to the earthquake structural engineering educator, practitioner and researcher. It frames earthquake structural engineering within a framework of balance between 'Demand' and 'Supply' (requirements imposed on the system versus its available capacity for action and deformation resistance).

In a system-integrated framework, referred to as 'From Source-to-Society', where 'Source' describes the focal mechanisms of earthquakes, and 'Society' describes the compendium of effects on complex societal systems, this book presents information pertinent to the evaluation of actions and deformations imposed by earthquakes on structural systems. It is therefore a 'Source-to-Structure' text. Source parameters, path and site characteristics are presented at a level of detail sufficient for the structural earthquake engineer to understand the effect of geophysical and seismological features on strong ground-motion characteristics pertinent to the evaluation of the response of structures. Structural response characteristics are reviewed and presented in a new framework of three quantities: stiffness, strength and ductility, which map onto the three most important limit states of serviceability, structural damage control and collapse prevention. This three-parameter approach also matches well with the consequential objectives of reducing down time, controlling repair costs and protecting life. By virtue of the fact that the text places strong emphasis on the varying values of stiffness, strength and ductility as a function of the available deformation capacity, it blends seamlessly with deformation-based design concepts and multi-limit state design, recently referred to as performance-based design. The book stops where design codes start, at the stage of full and detailed evaluation of elastic and inelastic actions and deformations to which structures are likely to be subjected. Emphasis is placed on buildings and bridges, and material treatment is constrained to steel and concrete. The scope of the book is depicted in the figure below.

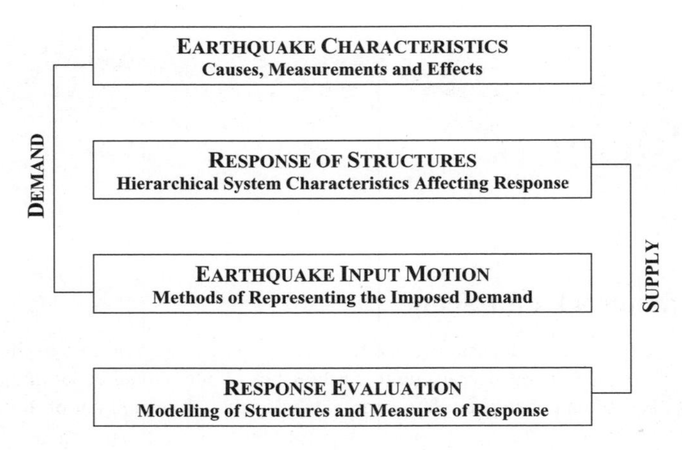

Scope of the book

Chapter 1 belongs to the Demand sub-topic and is a standard exposé of the geological, seismological and earth sciences aspects pertinent to structural earthquake engineering. It concludes with two sections; one on earthquake damage, bolstered by a detailed Appendix of pictures of damaged buildings and bridges categorized according to the cause of failure. The last section is on earthquake losses and includes global statistics, as well as description of the various aspects of impact of earthquakes on communities in a regional context.

Chapter 2, which belongs to the Supply or Capacity sub-topic, establishes a new framework of understanding structural response and relating milestones of such a response to (i) probability of occurrence of earthquakes and (ii) structural and societal limit states. Viewing the response of structures in the light of three fundamental parameters, namely Stiffness, Strength and Ductility, and their implications on system performance opens the door to a new relationship between measured quantities, limit states and consequences, as described in Table 2.1. The two most important 'implications' of stiffness, strength and ductility are overstrength and damping. The latter two parameters have a significant effect on earthquake response and are therefore addressed in detail. All five response quantities of (1) Stiffness, (2) Strength, (3) Ductility, (4) Overstrength and (5) Damping are related to one another and presented in a strictly hierarchical framework of the five levels of the hierarchy, namely (i) material, (ii) section, (iii) member, (iv) connection and (v) system. Finally, principles of capacity design are demonstrated numerically and their use to improve structural response is emphasized.

Chapter 3 brings the readers back to description of the Demand sub-topic and delves into a detailed description of the input motion in an ascending order of complexity. It starts with point estimates of peak ground parameters, followed by simplified, detailed and inelastic spectra. Evaluation of the required response modification factors, or the demand response modification factors, is given prominence in this chapter, to contrast the capacity response modification factors addressed in Chapter 2. The chapter concludes with selection and scaling of acceleration time histories, as well as a discussion of the significance of duration on response of inelastic structures.

Chapter 4 concludes the Supply sub-topic by discussing important aspects of analytically representing the structure and the significance or otherwise of some modelling details. The chapter is presented in a manner consistent with Chapter 2 in terms of dealing with modelling of materials, sections, members, connections, sub-assemblages and systems. The final section of Chapter 4 presents expected and important outcomes from analytical modelling for use in assessment of the adequacy of the structure under consideration, as well as conventional design forces and displacements. The chapter also includes a brief review of methods of quasi-dynamic and dynamic analysis pertinent to earthquake response evaluation.

Use Scenarios

Postgraduate Educators and Students

As discussed in the preceding section, the book was written with the university professor in mind as one of the main users, alongside students attending a graduate course. It therefore includes a large number of work assignments and additional worked examples, provided on the book web site. Most importantly, summary slides are also provided on the book web site. The slides are intended to be used in the classroom, and also to be used in final revision by students. The book and the slides have been used in teaching the postgraduate level course in earthquake engineering at the University of Illinois at Urbana-Champaign for a number of years, and are therefore successfully tested in a leading university environment. Parts of the book were also used in teaching short courses on a number of occasions in different countries. For the earthquake engineering professor, the whole book is recommended for postgraduate courses, with the exception of methods of analysis (Section 4.5 in Chapter 4) which are typically taught in structural dynamics courses that should be a prerequisite to this course.

Researchers

The book is also useful to researchers who have studied earthquake engineering in a more traditional context, where strength and direct assessment for design were employed, as opposed to the integrated strength-deformation and capacity assessment for design approach presented in this book. Moreover, structural earthquake engineering researchers will find Chapter 3 of particular interest because it bridges the conventional barriers between engineering seismology and earthquake engineering, and brings the concepts from the former in a palatable form to the latter. From the long experience of working with structural earthquake engineers, Chapter 3 is recommended as an essential read prior to undertaking research, even for individuals who have attended traditional earthquake engineering courses. Researchers from related fields, such as geotechnical earthquake engineering or structural control, may find Chapter 2 of value, since it heightens their awareness of the fundamental requirements of earthquake response of structures and the intricate relationship between stiffness, strength, ductility, overstrength and damping.

Practitioners

Practising engineers with long and relatively modern experience in earthquake-resistant design in high-seismicity regions will find the book on the whole easy to read and rather basic. They may however appreciate the presentation of fundamental response parameters and may find their connection to the structural and societal limit states refreshing and insightful. They may also benefit from the modelling notes of Chapter 4, since use is made of concepts of finite element representation in a specifically earthquake engineering context. Many experienced structural earthquake engineering practitioners will find Chapter 3 on input motion useful and practical. The chapter will aid them in selection of appropriate characterization of ground shaking. The book as a whole, especially Chapters 3 and 4 is highly recommended for practising engineers with limited or no experience in earthquake engineering.

Abbreviations

AI = Arias Intensity
AIJ = Architectural Institute of Japan
ASCII = American Standard Code for Information Interchange
ATC = Applied Technology Council
BF = Braced Frame
CBF = Concentrically Braced Frame
CEB = Comité Euro-international du Beton
CEUS = Central and Eastern United States
COSMOS = Consortium of Organisations for Strong-Motion Observation Systems
COV = Coefficient Of Variation
CP = Collapse Prevention
CQC = Complete Quadratic Combination
CSMIP = California Strong-Motion Instrumentation Program
CSUN = California State University Northridge
CTBUH = Council on Tall Building and Urban Habitat
CUE = Conference on Usage of Earthquakes
DC = Damage Control
DL = Dead Load
EBF = Eccentrically Braced Frame
EERI = Earthquake Engineering Research Institute
ELF = Equivalent Lateral Force
EPM = Elastic-Plastic Model
EPP = Elastic Perfectly-Plastic
EMS = European Modified Scale
EQ = Earthquake
FE = Finite Element
FEMA = Federal Emergency Management Agency
FRP = Fibre-Reinforced Plastic
FW = Frame-Wall structure
GNP = Gross National Product
HF = Hybrid Frame
HPGA = Horizontal Peak Ground Acceleration
ICSMD = Imperial College Strong-Motion Databank
ID = Inter-storey Drift
IDA = Incremental Dynamic Analysis
IF = Irregular Frame
JMA = Japanese Meteorological Agency

KBF = Knee-Braced Frame
K-NET = Kyoshin Net
LEM = Linear Elastic Model
LENLH = Linear Elastic-plastic with Non-Linear Hardening
LEPP = Linear Elastic-Perfectly Plastic
LESH = Linear Elastic-plastic with Strain Hardening
LL = Live Load
LQ = Love wave
LR = Rayleigh wave
LRH = Linear Response History
LS = Limit State
MCS = Mercalli-Cancani-Seiberg
MDOF = Multi-Degree-Of-Freedom
MM = Modified Mercalli
MP = Menegotto-Pinto model
MRF = Moment-Resisting Frame
MSK = Medvedev-Sponheuer-Karnik
NGA = New Generation Attenuation
NLEM = Non-Linear Elastic Model
NRH = Non-linear Response History
NSP = Non-linear Static Pushover
OBF = Outrigger-Braced Frame
PA = Pushover Analysis
PGA = Peak Ground Acceleration
PGD = Peak Ground Displacement
PGV = Peak Ground Velocity
PEER = Pacific Earthquake Engineering Research Center
PL = Performance Level
RC = Reinforced Concrete
RO = Ramberg-Osgood model
RF = Regular Frame
RSA = Response Spectrum Analysis
SCWB = Strong Column-Weak Beam
SDOF = Single-Degree-Of-Freedom
SH = Shear Horizontal
SI = Spectral Intensity
SL = Serviceability Limit
SPEAR = Seismic Performance Assessment and Rehabilitation
SRSS = Square Root of the Sum of Squares
SV = Shear Vertical
SW = Structural Wall
TS = Tube System
URM = Unreinforced masonry
USA = United States of America
USEE = Utility Software for Earthquake Engineering
USSR = Union of Soviet Socialist Republics
VPGA = Vertical Peak Ground Acceleration
WCSB = Weak Column-Strong Beam.

Symbols

Symbols defined in the text that are used only once, and those which are clearly defined in a relevant figure or table, are in general not listed herein.

A_v = effective shear area
C_M = centre of mass
C_R = centre of rigidity
d = distance from the earthquake source
E = Young's modulus
E_0 = initial Young's modulus (at the origin)
E_t = tangent Young's modulus
f_c = concrete compression strength
f_t = concrete tensile strength
f_u = steel ultimate strength
f_y = steel yield strength
G = shear modulus
G_b = shear modulus of the bedrock
g = acceleration of gravity
H = total height
H_{eff} = effective height
h = height
I = intensity
= moment of inertia
I_i = Modified Mercalli intensity of the ith isoseismal
I_{JMA} = intensity in the Japanese Meteorological Agency (JMA) scale
I_{max} = maximum intensity
I_{MM} = intensity in the Modified Mercalli (MM) scale
I_0 = epicentral intensity
J = torsional moment of inertia
K = stiffness
K_s = secant stiffness
K_t = tangent stiffness
K_0 = initial stiffness (at origin)
K = connection rotational stiffness
k_{eff} = effective stiffness
k_f = flexural stiffness
k_s = shear stiffness

L_p = plastic hinge length
L_w = wall length
M = magnitude
= bending moment
m_b = body wave magnitude
M_{eff} = effective mass
M_L = local (or Richter) magnitude
M_{JMA} = Japanese Meteorological Agency (JMA) magnitude
m_r = rotational mass
M_S = surface wave magnitude
m_t = translational mass
M_w = moment magnitude
N = axial load
q = force reduction factor
R = focal distance
= force reduction factor
r_i = radius of the equivalent area enclosed in the *i*th isoseismal
S_a = spectral acceleration
S_d = spectral displacement
SI_H = Housner's spectral intensity
SI_M = Matsumura's spectral intensity
S_v = spectral velocity
T = period of vibration
T_h = hardening period
T_R = return period
T_S = site fundamental period of vibration
$T_{S,n}$ = site period of vibration relative to the nth mode
T_y = yield period
t_r = reference time period
V_{base} = global base shear
V_e = elastic shear
V_i = storey shear
V_y = yield shear
V_d = design base shear
V_u = ultimate shear
v_{LQ} = velocity of Love waves
v_{LR} = velocity of Rayleigh waves
v_P = velocity of P-waves
v_S = velocity of S-waves
α_s = shear span ratio
Γ_i = modal participation factor for the *i*th mode
γ_D, γ_E, γ_L = load factors
γ_I = importance factor
Δ = global lateral displacement
Δ_y = global yield lateral displacement
Δ_u = global ultimate lateral displacement
δ = lateral displacement
δ_i = storey lateral displacement
δ_{top} = top lateral displacement
δ_u = ultimate lateral displacement

δ_y = yield lateral displacement
ε = strain
ε_c = concrete strain
ε_{cu} = concrete crushing strain
ε_u = ultimate strain
ε_y = yield strain
θ = rotation
θ_p = plastic rotation
θ_u = ultimate rotation
θ_y = yield rotation
μ = ductility
μ_a = available ductility
μ_d = ductility demand
μ_Δ = global displacement ductility
μ_δ = displacement ductility
μ_ε = material ductility
μ_θ = rotation ductility
μ_χ = curvature ductility
ν = Poisson's ratio
ξ = damping
ξ_{eff} = effective damping
ξ_{eq} = equivalent damping
ρ = density
σ = normal stress
σ_y = yielding normal stress
χ = curvature
χ_u = ultimate curvature
χ_y = yield curvature
Ψ = combination coefficient
Ω_d = observed overstrength
Ω_i = inherent overstrength
ω = natural circular frequency
ω_i = circular frequency relative to the *i*th mode

1

Earthquake Characteristics

1.1 Causes of Earthquakes

1.1.1 Plate Tectonics Theory

An earthquake is manifested as ground shaking caused by the sudden release of energy in the Earth's crust. This energy may originate from different sources, such as dislocations of the crust, volcanic eruptions, or even by man-made explosions or the collapse of underground cavities, such as mines or karsts. Thus, while earthquakes are defined as natural disturbances, different types of earthquake exist: fault rupture-induced, volcanic, mining-induced and large reservoir-induced. Richter (1958) has provided a list of major earth disturbances recorded by seismographs as shown in Figure 1.1. Tectonic earthquakes are of particular interest to the structural engineers, and further discussion will therefore focus on the latter type of ground disturbance.

Earthquake occurrence may be explained by the theory of large-scale tectonic processes, referred to as 'plate tectonics'. The theory of plate tectonics derives from the theory of continental drift and sea-floor spreading. Understanding the relationship between geophysics, the geology of a particular region and seismic activity began only at the end of the nineteenthth century (Udias, 1999). Earthquakes are now recognized to be the symptoms of active tectonic movements (Scholz, 1990). This is confirmed by the observation that intense seismic activity occurs predominantly on known plate boundaries as shown in Figure 1.2.

Plates are large and stable rigid rock slabs with a thickness of about 100 km, forming the crust or lithosphere and part of the upper mantle of the Earth. The crust is the outer rock layer with an internal complex geological structure and a non-uniform thickness of 25–60 km under continents and 4–6 km under oceans. The mantle is the portion of the Earth's interior below the crust, extending from a depth of about 30 km to about 2,900 km; it consists of dense silicate rocks. The lithosphere moves differentially on the underlying asthenosphere, which is a softer warmer layer around 400 km thick at a depth of about 50 km in the upper mantle. It is characterized by plastic or viscous flow. The horizontal movement of the lithosphere is caused by convection currents in the mantle; the velocity of the movement is about 1 to 10 cm/year. Current plate movement can be tracked directly by means of reliable space-based geodetic measurements, such as very long baseline interferometry, satellite laser ranging and global positioning systems.

Large tectonic forces take place at the plate edges due to the relative movement of the lithosphere–asthenosphere complex. These forces instigate physical and chemical changes and affect the geology of the adjoining plates. However, only the lithosphere has the strength and the brittle behaviour to fracture, thus causing an earthquake.

Fundamentals of Earthquake Engineering Amr S. Elnashai and Luigi Di Sarno

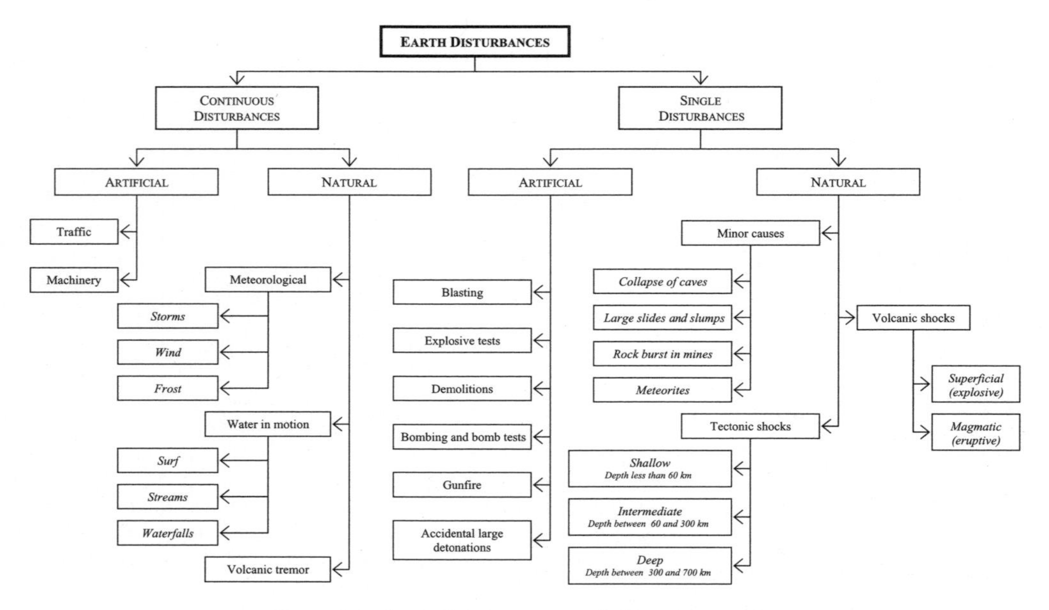

Figure 1.1 Earth disturbances recorded by seismographs

According to the theory of continental drift, the lithosphere is divided into 15 rigid plates, including continental and oceanic crusts. The plate boundaries, where earthquakes frequently occur, are also called 'seismic belts' (Kanai, 1983). The Circum-Pacific and Eurasian (or Alpine) belts are the most seismically active. The former connects New Zealand, New Guinea, the Philippines, Japan, the Aleutians, the west coast of North America and the west coast of South America. The 1994 Northridge (California) and the 1995 Kobe (Japan) earthquakes occurred along the Circum-Pacific belt. The Eurasian belt links the northern part of the Mediterranean Sea, Central Asia, the southern part of the Himalayas and Indonesia. The Indian Ocean earthquake of 26 December 2004 and the Kashmir earthquake of 8 October 2005 were generated by the active Eurasian belt.

The principal types of plate boundaries can be grouped as follows (Figure 1.3):

(i) *Divergent or rift zones*: Plates separate themselves from one another and either an effusion of magma occurs or the lithosphere diverges from the interior of the Earth. Rifts are distinct from mid-ocean ridges, where new oceanic crust and lithosphere is created by sea-floor spreading. Conversely, in rifts no crust or lithosphere is produced. If rifting continues, eventually a mid-ocean ridge may form, marking a divergent boundary between two tectonic plates. The Mid-Atlantic ridge is an example of a divergent plate boundary. An example of rift can be found in the middle of the Gulf of Corinth, in Greece. However, the Earth's surface area does not change with time and hence the creation of new lithosphere is balanced by the destruction at another location of an equivalent amount of rock crust, as described below.

(ii) *Convergent* or *subduction zones*: Adjacent plates converge and collide. A subduction process carries the slab-like plate, known as the 'under-thrusting plate', into a dipping zone, also referred to as the 'Wadati–Benioff zone', as far downward as 650–700 km into the Earth's interior. Two types of convergent zones exist: oceanic and continental lithosphere convergent boundaries. The first type occurs when two plates consisting of oceanic lithosphere collide. Oceanic rock is mafic, and heavy compared to continental rock; therefore, it sinks easily and is destroyed in a subduc-

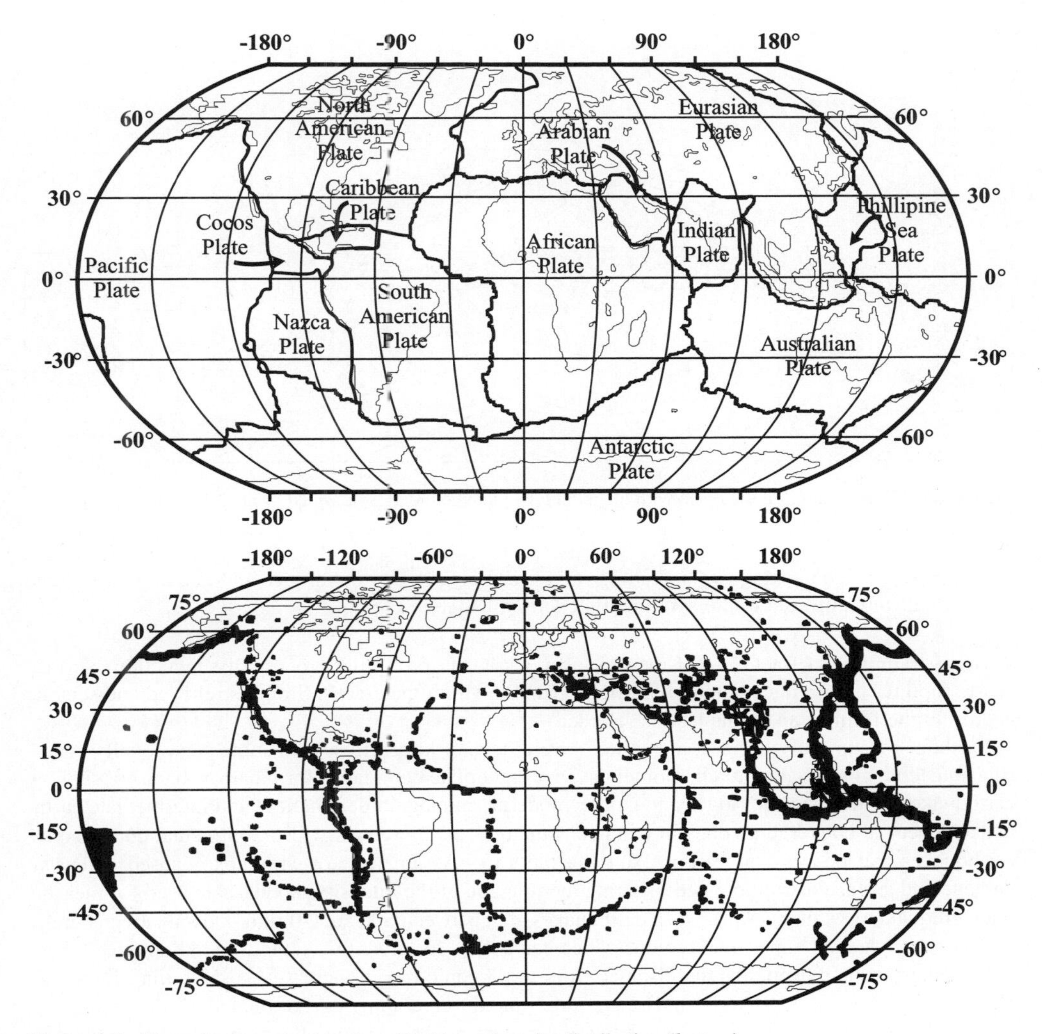

Figure 1.2 Tectonic plates (*top*) and worldwide earthquake distribution (*bottom*)

tion zone. The second type of convergent boundary occurs when both grinding plates consist of continental lithosphere. Continents are composed of lightweight rock and hence do not subduct. However, in this case the seismicity is extended over a wider area. The Circum-Pacific and Eurasian belts are examples of oceanic and continental lithosphere convergent boundaries, respectively.

(iii) *Transform zones* or *transcurrent horizontal slip*: Two plates glide past one another but without creating new lithosphere or subducting old lithosphere. Transform faults can be found either in continental or oceanic lithosphere. They can offset mid-ocean ridges, subduction zones or both. Boundaries of transcurrent horizontal slip can connect either divergent and convergent zones or two convergent zones. The San Andreas Fault in California is an example of a transform

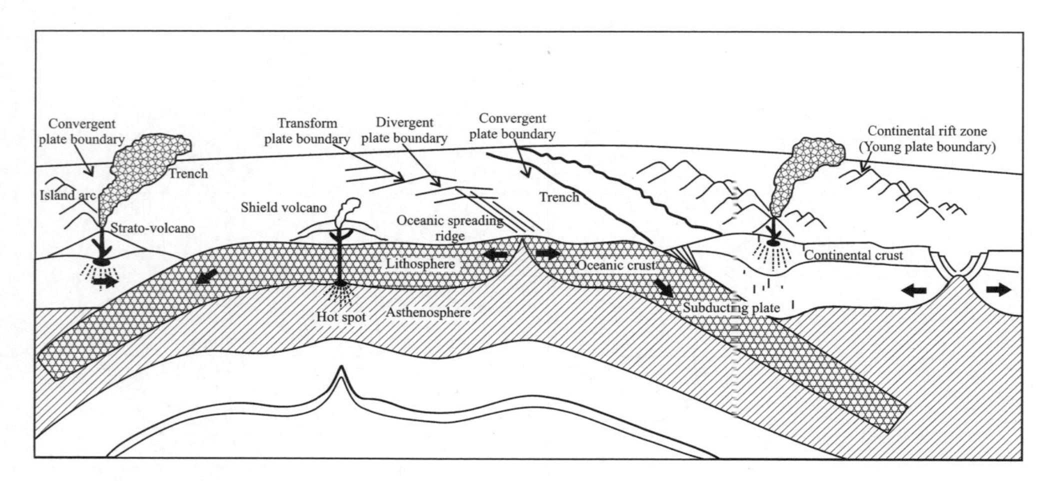

Figure 1.3 Cross-section of the Earth with the main type plate boundaries (*adapted from* U.S. Geological Survey)

boundary connecting two spreading ridges, namely the North America and Pacific plates in the Gulf of California to the south and the Gorda Ridge in the north

High straining and fracturing of the crustal rocks is caused by the process of subduction. Surface brittle ruptures are produced along with frictional slip within the cracks. Strain is relieved and seismic energy in the form of an earthquake is released.

Earthquakes normally occur at a depth of several tens of kilometres, with some occasionally occurring at a depth of several hundred kilometres. Divergent plate boundaries form narrow bands of shallow earthquakes at mid-oceanic ridges and can be moderate in magnitude. Shallow and intermediate earthquakes occur at convergent zones in bands of hundreds of kilometres wide. Continental convergence earthquakes can be very large. For example, the 1897 Assam (India) earthquake caused extensive damage and surface disruption, necessitating the upgrade of the intensity model scale used for measuring earthquakes (Richter, 1958). Deep earthquakes, e.g. between 300 and 700 km in depth, are generally located in subduction zones over regions which can extend for more than a thousand kilometres. These earthquakes become deeper as the distance from the oceanic trench increases as shown in Figure 1.4. However, the seismic Wadati–Benioff zones are limited to the upper part of the subduction zones, i.e. about 700 km deep. Beyond this depth, either the plates are absorbed into the mantle or their properties are altered and the release of seismic energy is inhibited. Shallow earthquakes with large magnitude can occur along transform faults. For example, Guatemala City was almost destroyed during the devastating 1976 earthquake, which occurred on the Motagua Fault. The latter constitutes the transform boundary between two subduction zones, located respectively off the Pacific Coast of Central America and the Leeward and Windward Islands in the Atlantic Ocean.

Plate tectonic theory provides a simple and general geological explanation for plate boundary or inter-plate earthquakes, which contribute 95% of worldwide seismic energy release. It is, however, to be noted that earthquakes are not confined to plate boundaries. Local small magnitude intra-plate earthquakes, which may occur virtually anywhere, can cause considerable damage. Several examples of such events exist and the devastating effects are well documented (e.g. Scholz, 1990; Bolt, 1999, among others). The Newcastle (Australia) earthquake of 28 December 1989 caused about 30 deaths and $750 million in economic loss. The Dahshour (Egypt) earthquake of 12 October 1992 caused damage estimated at $150 million and more than 600 fatalities. In the USA, three of the largest intra-plate earthquakes in modern record occurred in the mid-continent in 1811 and 1812. They caused

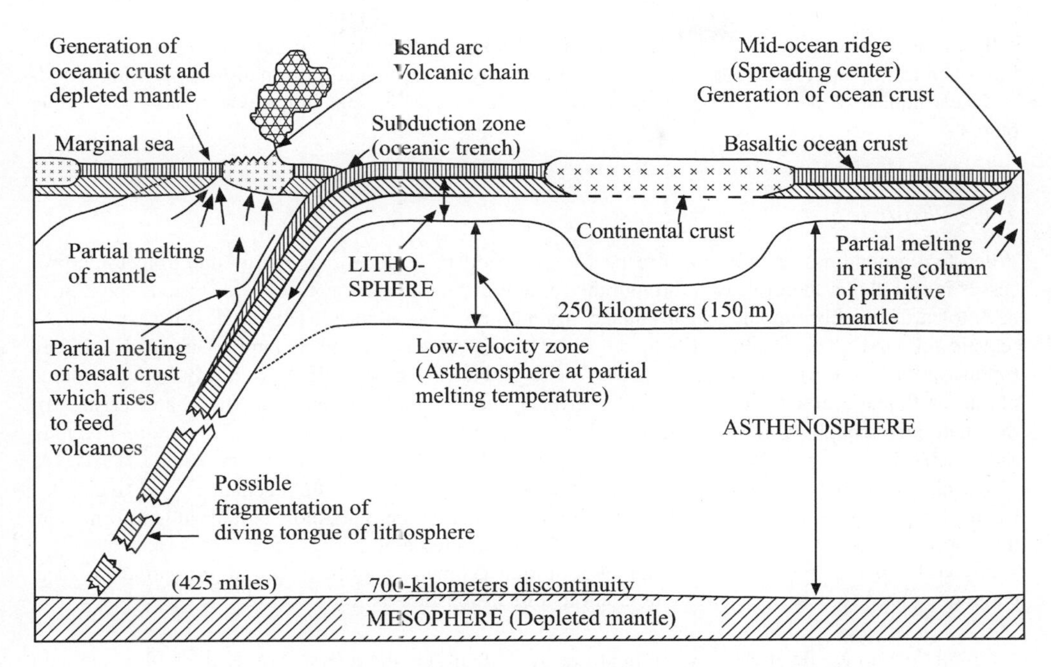

Figure 1.4 Tectonic mechanisms at plate boundaries (*after* Dewey, 1972)

Table 1.1 Classification of tectonic earthquakes (*after* Scholz, 1990).

Earthquake (type)	Slip rate (v) (mm/year)	Recurrence time (year)
Inter-plate	$v > 10$	~100
Intra-plate (*plate boundary related*)	$0.1 \leq v \leq 10$	10^2–10^4
Intra-plate (*mid-plate*)	$v < 0.1$	$>10^4$

significant ground effects in the New Madrid area of Missouri and were felt as far away as New England and Canada. From a tectonic standpoint, the occurrence of intra-plate earthquakes shows that the lithosphere is not rigid and internal fractures can take place; the latter are, however, difficult to predict. The genesis of this seismic activity is attributed either to the geological complexity of the lithosphere or anomalies in its temperature and strength. Stress build-ups at the edges may be transmitted across the plates and are released locally in weak zones of the crust. It has been shown that intra-plate events exhibit much higher stress drops than their inter-plate counterparts, the difference being a factor five (Scholz *et al.*, 1986). Intra-plate and inter-plate earthquakes can be distinguished quantitatively on the basis of the slip rate of their faults and the recurrence time (Scholz, 1990) as outlined in Table 1.1. For example, the Kashmir earthquake of 8 October 2005 is associated with the known subduction zone of an active fault where the Eurasian and the Indian plates are colliding and moving northward at a rate of 40 mm/year (Durrani *et al.*, 2005). The data collected for the Kashmir earthquake correspond to the figures given in Table 1.1 for slip rate and recurrence time of a typical inter-plate seismic event.

Intra-plate earthquakes generally fall into two groups: plate boundary-related and mid-plate. The former take place either in broad bands near plate edges and are tectonically linked to them or in diffuse plate boundaries. Examples of such earthquakes have occurred inland in Japan, and are linked tectonically to the Pacific–Eurasian plate. In contrast, mid-plate earthquakes are not related to plate edges. Inter- and intra-plate crustal movements are continuously occurring and information concerning world-wide earthquake activity can be found at several Internet sites, e.g. http://www.usgs.gov, among others.

1.1.2 Faulting

When two groundmasses move with respect to one another, elastic strain energy due to tectonic processes is stored and then released through the rupture of the interface zone. The distorted blocks snap back towards equilibrium and an earthquake ground motion is produced. This process is referred to as 'elastic rebound'. The resulting fracture in the Earth's crust is termed a 'fault'. During the sudden rupture of the brittle crustal rock, seismic waves are generated. These waves travel away from the source of the earthquake along the Earth's outer layers. Their velocity depends on the characteristics of the material through which they travel. Further details on types of seismic waves are given in Section 1.1.3.

The characteristics of earthquake ground motions are affected by the slip mechanism of active faults. Figure 1.5 provides two examples of significant active faults: the San Andreas fault in California and the Corinth Canal fault in Greece, with about 70 m exposure height.

Active faults may be classified on the basis of their geometry and the direction of relative slip. The parameters used to describe fault motion and its dimensions are as follows:

(i) *Azimuth* (ϕ): the angle between the trace of the fault, i.e. the intersection of the fault plane with the horizontal, and the northerly direction ($0° \le \phi \le 360°$). The angle is measured so that the fault plane dips to the right-hand side;
(ii) *Dip* (δ): the angle between the fault and the horizontal plane ($0° \le \delta \le 90°$);
(iii) *Slip or rake* (λ): the angle between the direction of relative displacement and the horizontal direction ($-180° \le \lambda \le 180°$). It is measured on the fault plane;
(iv) *Relative displacement* (Δu): the distance travelled by a point on either side of the fault plane. If Δu varies along the fault plane, its mean value is generally used;
(v) *Area* (S): surface area of the highly stressed region within the fault plane.

Figure 1.5 Active faults: San Andreas in California (*left*) (*courtesy of* National Information Service for Earthquake Engineering, University of California, Berkeley) and the Corinth Canal in Greece (*right*)

The orientation of fault motion is defined by the three angles ϕ, δ and λ, and its dimensions are given by its area S as displayed in Figure 1.6; the fault slip is measured by the relative displacement Δu.

Several fault mechanisms exist depending on how the plates move with respect to one another (Housner, 1973). The most common mechanisms of earthquake sources are described below (Figure 1.7):

(i) *Dip-slip faults*: One block moves vertically with respect to the other. If the block underlying the fault plane or 'footwall' moves up the dip and away from the block overhanging the fault plane, or 'hanging wall', normal faults are obtained. Tensile forces cause the shearing failure of normal faults. In turn, when the hanging wall moves upward in relation to the footwall, the faults are reversed; compressive forces cause the failure. Thrust faults are reverse faults characterized by a very small dip. Mid-oceanic ridge earthquakes are due chiefly to normal faults. The 1971 San Fernando earthquake in California was caused by rupture of a reverse fault. Earthquakes along the Circum-Pacific seismic belt are caused by thrust faults;

(ii) *Strike-slip faults*: The adjacent blocks move horizontally past one another. Strike-slip can be right-lateral or left-lateral, depending on the sense of the relative motion of the blocks for an observer located on one side of the fault line. The slip takes place along an essentially vertical fault plane and can be caused by either compression or tension stresses. They are typical of transform zones. An example of strike-slip occurred in the 1906 San Francisco earthquake on the San Andreas Fault. The latter is characterized by large strike-slip deformations when earthquakes occur (see, for example, also Figure 1.5): part of coastal California is sliding to the northwest relative to the rest of North America – Los Angeles is slowly moving towards San Francisco.

Several faults exhibit combinations of strike-slip and dip-slip movements; the latter are termed 'oblique slip'. Oblique slips can be either normal or reverse and right- or left-lateral. The above fault mechanisms can be defined in mathematical terms through the values of the dip δ and the slip or rake λ. For example, strike-slip faults show $\delta = 90°$ and $\lambda = 0°$. The slip angle λ is negative for normal faults and positive for reverse faults; for $\delta > 0°$ the fault plane is inclined and can exhibit either horizontal ($\lambda = \pm 180°$ and $0°$) or vertical ($\lambda = \pm 90°$) motion. For other λ-values, the relative displacement

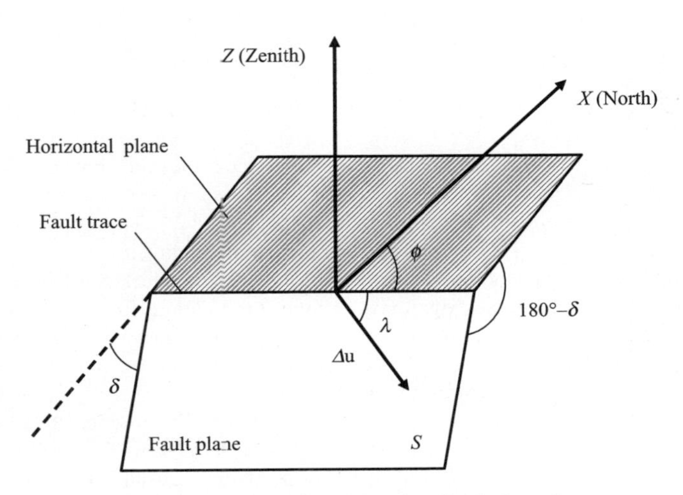

Figure 1.6 Parameters used to describe fault motion

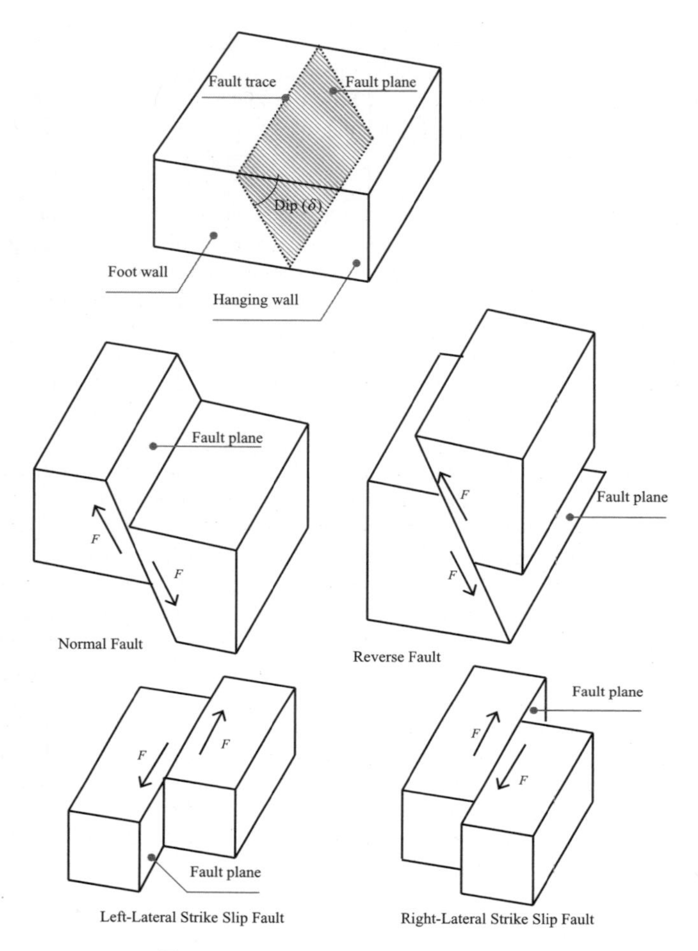

Figure 1.7 Fundamental fault mechanisms

has both vertical and horizontal components; the latter can be of normal or reverse type according to the algebraic sign of the angle λ.

The 'focus' or 'hypocentre' of an earthquake is the point under the surface where the rupture is said to have originated. The projection of the focus on the surface is termed 'epicentre'. The reduction of the focus to a point is the point-source approximation (Mallet, 1862). This approximation is used to define the hypocentral parameters. However, the parameters that define the focus are similar to those that describe the fault fracture and motion. Foci are located by geographical coordinates, namely latitude and longitude, the focal depth and the origin or occurrence time. Figure 1.8 provides a pictorial depiction of the source parameters, namely epicentral distance, hypocentral or focal distance, and focal depth. Earthquakes are generated by sudden fault slips of brittle rocky blocks, starting at the focus depth and observed at a site located at the epicentral distance.

Most earthquakes have focal depths in the range of 5–15 km, while intermediate events have foci at about 20–50 km and deep earthquakes occur at 300–700 km underground. The three types are also referred to as shallow, intermediate and deep focus, respectively. Crustal earthquakes normally have depths of about 30 km or less. For example, in Central California the majority of earthquakes have focal

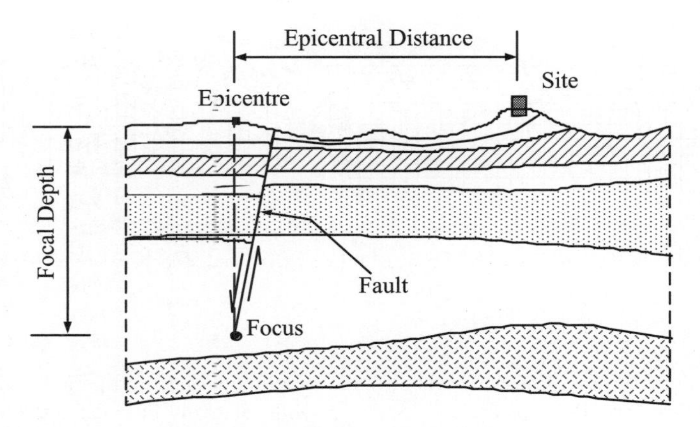

Figure 1.8 Definition of source parameters

depths in the upper 5–10 km. Some intermediate- and deep-focus earthquakes are located in Romania, the Aegean Sea and under Spain.

The above discussion highlights one of the difficulties encountered in characterizing earthquake parameters, namely the definition of the source. From Figure 1.8, it is clear that the source is not a single point, hence the 'distance from the source' required for engineering seismology applications, especially in attenuation relationships as discussed in Section 3.3, is ill-defined. This has led researchers to propose treatments for point, line and area sources (Kasahara, 1981). It is therefore important to exercise caution in using relationships based on source-site measurements, especially for near-field (with respect to site) and large magnitude events. A demonstration of this is the values of ground acceleration measured in the Adana–Ceyhan (Turkey) earthquake of 26 June 1998. Two seismological recording stations, at Ceyhan and Karatas, were located at distances of 32 km and 36 km from the epicentre, respectively. Whereas the peak acceleration in Ceyhan was 0.27 g, that at Karatas was 0.03 g. The observed anomaly may be explained by considering the point of initiation and propagation of the fault rupture or 'directivity', which is presented in Section 1.3.1, possibly travelling towards Ceyhan and away from Karatas.

Problem 1.1

Determine the source mechanism of faults with a dip $\delta = 60°$ and rake $\lambda = 45°$. Comment on the results.

1.1.3 Seismic Waves

Fault ruptures cause brittle fractures of the Earth's crust and dissipate up to 10% of the total plate-tectonic energy in the form of seismic waves. Earthquake shaking is generated by two types of elastic seismic waves: body and surface waves. The shaking felt is generally a combination of these waves, especially at small distances from the source or 'near-field'.

Body waves travel through the Earth's interior layers. They include longitudinal or primary waves (also known as 'P-waves') and transverse or secondary waves (also called 'S-waves'). P- and S-waves are also termed 'preliminary tremors' because in most earthquakes they are felt first (Kanai, 1983). P-waves cause alternate push (or compression) and pull (or tension) in the rock as shown in Figure 1.9. Thus, as the waves propagate, the medium expands and contracts, while keeping the same form. They

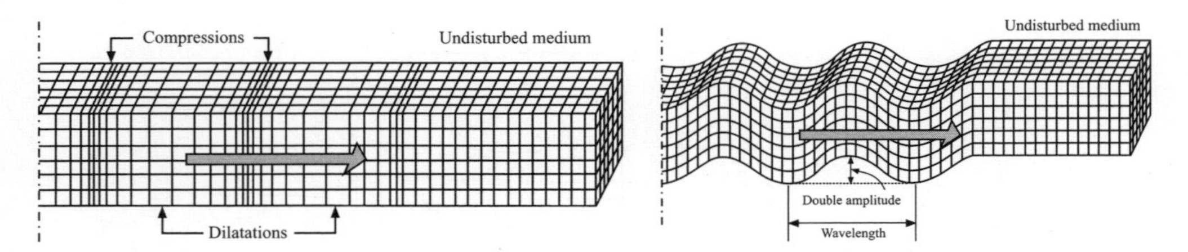

Figure 1.9 Travel path mechanisms of body waves: primary (*left*) and secondary waves (*right*) (*adapted from* Bolt, 1999)

exhibit similar properties to sound waves, show small amplitudes and short periods, and can be transmitted in the atmosphere. P-waves are seismic waves with relatively little damage potential. S-wave propagation, by contrast, causes vertical and horizontal side-to-side motion. Such waves introduce shear stresses in the rock along their paths as displayed in Figure 1.9 and are thus also defined as 'shear waves'. Their motion can be separated into horizontal (SH) and vertical (SV) components, both of which can cause significant damage, as illustrated in Sections 1.4.1 and 1.4.2 as well as in Appendix B. Shear waves are analogous to electromagnetic waves, show large amplitudes and long periods, and cannot propagate in fluids.

Body waves (P and S) were named after their arrival time as measured by seismographs at observation sites. P-waves travel faster, at speeds between 1.5 and 8 kilometres per second while S-waves are slower, usually travelling at 50% to 60% of the speed of P-waves. The actual speed of body waves depends upon the density and elastic properties of the rock and soil through which they pass.

Body waves may be described by Navier's equation for an infinite, homogeneous, isotropic, elastic medium in the absence of body forces (e.g. Udias, 1999). The propagation velocities of P- and S-waves within an isotropic elastic medium with density ρ, denoted as v_P and v_S respectively, are as follows:

$$v_P = \sqrt{\frac{E(1-\nu)}{\rho(1+\nu)(1-2\nu)}} \tag{1.1.1}$$

$$v_S = \sqrt{\frac{E}{2\rho(1+\nu)}} \tag{1.1.2}$$

in which ν is Poisson's ratio and E is Young's modulus of the elastic medium.

The ratio of P- and S-wave velocities is as follows:

$$\frac{v_S}{v_P} = \sqrt{\frac{1-2\nu}{2(1-\nu)}} \tag{1.2.1}$$

and for ν-values characterizing ordinary soil types, i.e. with ν ranging between 0.30 and 0.50:

$$0 \le v_S \le 0.53\, v_P \tag{1.2.2}$$

Equations (1.2.1) and (1.2.2) can be employed along with wave traces of seismogram records to locate earthquakes in time and space. For shallow earthquakes, the effects of the Earth's curvature can be ignored and hence a planar model is used for the propagation of body waves. Assuming homogenous soil profiles between earthquake foci and observation sites, the focal distance Δx is linearly dependent on the time-lag Δt between the P- and S-waves as follows:

Table 1.2 Velocity of primary (P) and secondary (S) waves in Earth's layer.

Layer (type)	Depth (km)	P-waves (km/s)	S-waves (km/s)
Crust	10–30	6.57	3.82
	40	8.12	4.42
Upper mantle	220	8.06	4.35
	400	9.13	5.22
	670	10.75	5.95
Lower mantle	1,200	11.78	6.52
	2,885	13.72	7.26
	2,890	8.06	0.00
Outer core	3,800	9.31	0.00
	5,150	10.36	0.00
Inner core	5,155	11.03	3.50
	6,371	11.26	3.67

$$\Delta x = \frac{v_P \, v_S}{v_P - v_S} \, \Delta t \tag{1.3.1}$$

thus, if the wave velocities v_P and v_S are known, the distance Δx is readily evaluated. Velocities of P- and S-waves in the Earth's interior layers are given in Table 1.2. For a quick evaluation, Omori's formula may also be used (Kanai, 1983):

$$\Delta x \approx 7.42 \, \Delta t \tag{1.3.2}$$

with Δx and Δt expressed in kilometres and seconds, respectively. Equation (1.3.2) assumes that body wave velocities are almost constant within a limited area. A comparison between the coefficient '7.42' used by Omori in equation (1.3.2), the coefficients that are computed by using the first term on the right-hand side in equation (1.3.1), and the values of v_P and v_S given in Table 1.2 is provided in Figure 1.10. It is proposed to make use of a step-function to take into consideration the variability of the body wave velocities in the Earth's interior. The suggested coefficients for equation (1.3.2) are 9.43 and 13.88, for depths below and above 300 km, respectively.

The procedure to locate an earthquake epicentre and origin time, i.e. time of initiating of fault rupture, is as follows:

(a) Obtain seismogram records for a given observation site.
(b) Select the arrival time of the body waves on the record traces.
(c) Compute the time delay Δt in the arrival of P- and S-waves.
(d) Subtract the travel time Δt from the arrival time at the observation site to obtain the origin time.
(e) Use equations (1.3.1) or (1.3.2) to evaluate the distance Δx between the seismic station and the epicentre. The use of either equations (1.3.1) or (1.3.2) depends on the data available for the soil profile and approximation accepted.
(f) Draw a circle on a map around the station location (or centre) with a radius equal to Δx. The curve plotted shows a series of possible locations for the earthquake epicentre.
(g) Repeat steps (a) to (f) for a second seismic station. A new circle is drawn; the latter intersects the circle of the first station at two points.
(h) Repeat steps (a) to (f) for a third seismic station. It identifies which of the two previous possible points is acceptable and corresponds to the earthquake source.

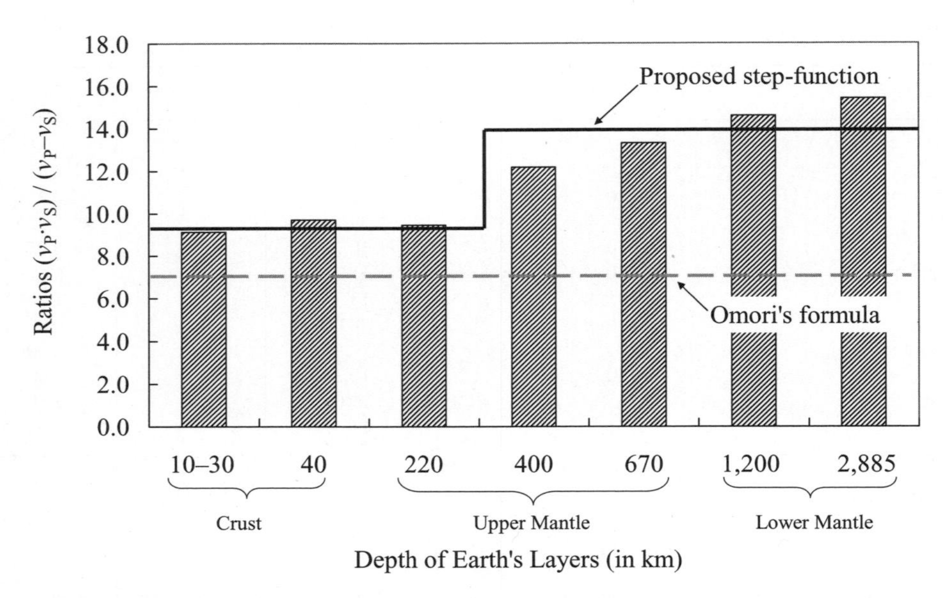

Figure 1.10 Comparison between ratios of body wave velocities in equations (1.3.1) and (1.3.2)

Errors are common in the above graphical method; hence, the procedure becomes more accurate with the increase in the number of measuring stations. In which case, the intersection will correspond to a small area containing the epicentre. In recent times, computer-based techniques have been employed to enhance the accuracy in evaluating earthquake epicentral locations (e.g. Lee *et al.*, 2003).

Equations (1.3.1) and (1.3.2) may be employed to derive travel–time curves, i.e. plots of the time seismic waves take to propagate from the earthquake source to each seismograph station or 'observation site', as a function of the horizontal distance. The use of these curves is twofold: estimating the Earth's internal structure and seismic prospecting (extensively used for underground structures). In particular, travel–time curves for earthquakes observed worldwide have shown that S-waves cannot travel deeper than 2,900 km (reference is also made to Table 1.2). At this depth, the medium has no rigidity and hence only P-waves can propagate through it.

Surface waves propagate across the outer layers of the Earth's crust. They are generated by constructive interference of body waves travelling parallel to the ground surface and various underlying boundaries. Surface waves include Love (indicated as 'L- or LQ-waves') and Rayleigh (indicated as 'R- or LR-waves') waves. These waves induce generally large displacements and hence are also called 'principal motion' (Kanai, 1983). They are most distinct at distances further away from the earthquake source. Surface waves are most prominent in shallow earthquakes while body waves are equally well represented in earthquakes at all depths. Because of their long duration, surface waves are likely to cause severe damage to structural systems during earthquakes.

LQ-waves are generated by constructive interference of SH body waves and hence cannot travel across fluids. Their motion is horizontal and perpendicular to the direction of their propagation, which is parallel to the Earth's surface as illustrated pictorially in Figure 1.11. LQ-waves have large amplitudes and long periods. LQ-waves of long period (60–300 seconds) are also called 'G-waves', after Gutenberg (Richter, 1958). For these periods, the waves travel with a velocity of about 4.0 km/sec and are pulse-like.

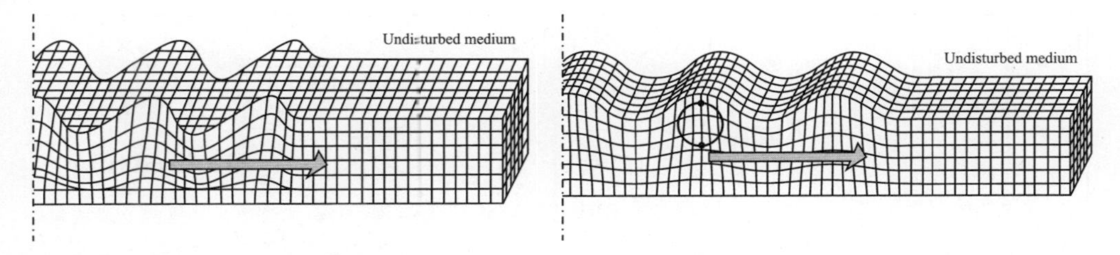

Figure 1.11 Travel path mechanisms of surface waves: Love (*left*) and Rayleigh waves (*right*) (*adapted from* Bolt, 1999)

LR-waves are caused by constructive interference of body waves, such as P and SV. As they pass by, particles of soil move in the form of a retrograde ellipse whose long axis is perpendicular to the Earth's surface (Figure 1.11). R-waves exhibit very large amplitude and regular waveforms.

LR-waves are slower than S-waves. As an approximation, it may be assumed that the velocity of LR-waves v_{LR} is given by the equation (Bolt, 1999):

$$v_{LR} \approx 0.92\, v_S \tag{1.4}$$

For a layered solid, LQ-wave velocity v_{LQ} generally obeys the following relationship:

$$v_{S1} < v_{LQ} < v_{S2} \tag{1.5}$$

with v_{S1} and v_{S2} as the velocities of S-waves in the surface and deeper layers, respectively.

Surface waves are slower than body waves and LQ-waves are generally faster than LR-waves. Moreover, the amplitudes of P- and S-waves show amplitudes linearly decreasing with the increase in distance x, while the amplitude of surface waves attenuates in inverse proportion to the square root of distance x. S-waves damp more rapidly than P-waves; attenuations increase with the wave frequencies. Amplitude attenuation is caused by the viscosity of the Earth's crust; seismic waves also change in form during their travel paths for the same reason (Kanai, 1983). Amplitudes and periods are of great importance because they influence the energy content of seismic waves as discussed in Section 1.2.

Body waves are reflected and refracted at interfaces between different layers of rock according to Snell's law of refraction. When reflection and refraction occur, part of the energy of one type is transformed in the other. Regardless of whether the incident wave is P or S, the reflected and refracted waves, also termed 'multiple phase waves', each consists of P- and S-waves, such as PP, SS, PS and SP. Their name indicates the travel path and mode of propagation (Reiter, 1990). For example, SP starts as S and then continues as P. The phenomenon known as the 'Moho bounce' is due to the simultaneous arrival at the surface of direct S-waves and S-waves reflected by the so-called 'Mohorovicic discontinuity' – or 'Moho' in short – at the boundary between the crust and the underlying mantle in the internal structure of the Earth. The latter discontinuity may be responsible for significant strong motions leading to damage far from the source as illustrated in Section 1.2.1.

Multiple phase waves do not possess significant damage potential. However, when P- and S-waves reach the ground surface, they are reflected back. As a result, waves move upwards and downwards. Such reflections may lead to significant local amplification of the shaking at the surface. It has been shown that seismic waves are influenced by soil conditions and local topography (e.g. Kramer, 1996), as further discussed in Section 1.3.2.

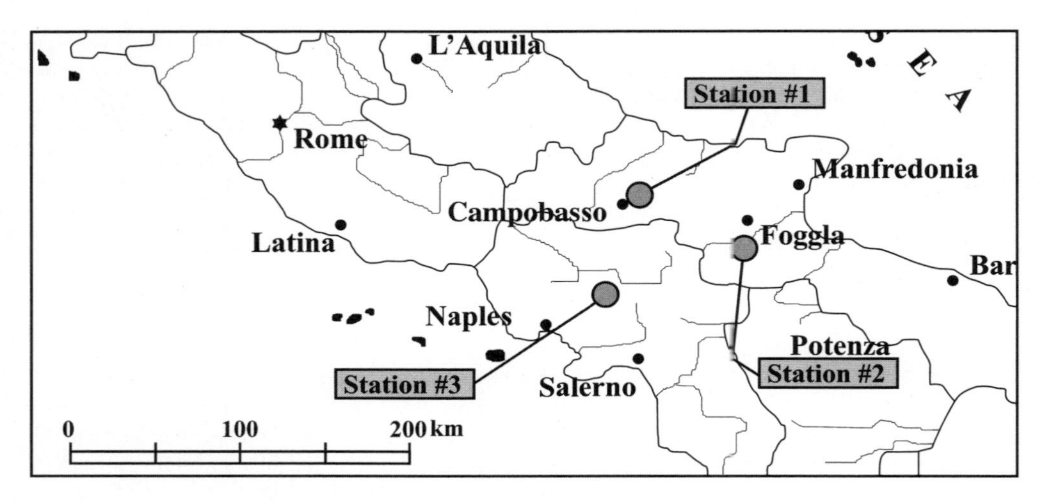

Figure 1.12 Map with the location of the seismological stations

A final point worth noting about the various types of seismic waves is the likelihood of rotatory vibrations, also referred to as 'progressive waves', at ground surface. These waves occur in addition to translational oscillations and are generated either when a plane wave is incident obliquely to the ground surface or when surface waves are present. Progressive waves may excite rocking and torsional vibrations especially in high-rise structures (Okamoto, 1984). Rotatory earthquake motions are complex and not yet fully understood. They are subject to active research.

Problem 1.2

Locate and mark on the map provided in Figure 1.12 the epicentre of an earthquake that was recorded in Italy by three observation sites with a time delay between P- and S-waves of 5.0, 7.5 and 6.0 seconds, respectively. The body wave velocities are 8.5 km/sec and 4.30 km/sec; it is up to the reader to determine which of these values refer to P- and S-waves. Compare the results obtained by equation (1.3.1) with those estimated from equation (1.3.2).

1.2 Measuring Earthquakes

Earthquake size is expressed in several ways. Qualitative or non-instrumental and quantitative or instrumental measurements exist; the latter can be either based on regional calibrations or applicable worldwide. Non-instrumental measurements are of great importance for pre-instrumental events and are hence essential in the compilation of historical earthquake catalogues for purposes of hazard analysis. For earthquakes that have been instrumentally recorded, qualitative scales are complementary to the instrumental data. The assessment and use of historical records is not straightforward and may lead to incorrect results due to inevitable biases (Ambraseys and Finkel, 1986). Moreover, the observation period during which data are employed to determine future projections is an issue of great importance. For example, recent studies (Ambraseys, 2006) indicate that for three active regions around the world, limiting the catalogues used in hazard analysis to a short period of time may grossly overestimate or underestimate the ensuing hazard. The over- and underestimation is a function of whether the observation period was an exceptionally quiescent or energetic epoch. Seismograms recorded at different epicentral distances are employed to determine origin time, epicentre, focal depth and type of faulting – as discussed in Sections 1.1.2 and 1.1.3 – as well as to estimate the energy released during an earthquake. Descriptive methods can also be used to establish earthquake-induced damage and its spatial distribution. In so doing, intensity, magnitude and relevant scales are utilized; these are outlined below.

1.2.1 Intensity

Intensity is a non-instrumental perceptibility measure of damage to structures, ground surface effects, e.g. fractures, cracks and landslides illustrated in Section 1.4.2, and human reactions to earthquake shaking. It is a descriptive method which has been traditionally used to establish earthquake size, especially for pre-instrumental events. It is a subjective damage evaluation metric because of its qualitative nature, related to population density, familiarity with earthquake and type of constructions.

Discrete scales are used to quantify seismic intensity; the levels are represented by Roman numerals and each degree of intensity provides a qualitative description of earthquake effects. Several intensity scales have been proposed worldwide. Early attempts at classifying earthquake damage by intensity were carried out in Italy and Switzerland around the late 1700s and early 1900s (Kanai, 1983). Some of these scales are still used in Europe (alongside modern scales), the USA and Japan. Some of the most common intensity scales are listed below:

(i) *Mercalli–Cancani–Seiberg* (MCS): 12-level scale used in southern Europe;
(ii) *Modified Mercalli* (MM): 12-level scale proposed in 1931 by Wood and Neumann, who adapted the MCS scale to the California data set. It is used in North America and several other countries;
(iii) *Medvedev–Sponheuer–Karnik* (MSK): 12-level scale developed in Central and Eastern Europe and used in several other countries;
(iv) *European Macroseismic Scale* (EMS): 12-level scale adopted since 1998 in Europe. It is a development of the MM scale;
(v) *Japanese Meteorological Agency* (JMA): 7-level scale used in Japan. It has been revised over the years and has recently been correlated to maximum horizontal acceleration of the ground.

Descriptions of the above intensity scales can be found in several textbooks (Reiter, 1990; Kramer, 1996; Lee *et al.*, 2003, among many others). A comparison between MCS, MM, MSK, EMS and JMA scales is provided in Figure 1.13. Intensity scales may include description of construction quality for structures in the exposed region. For example, the MM-scale specifies different damage levels depending on whether the structural system was poorly built or badly designed (VII), ordinary substantial buildings (VIII) or structures built especially to withstand earthquakes (IX). However, intensity scales do not account for local soil conditions, which may significantly affect the earthquake-induced damage and its distribution. Correlations between earthquake source and path, on the one hand, and intensity measures on the other are therefore highly inaccurate.

Intensity scales are used to plot contour lines of equal intensity or 'isoseismals'. Intensity maps provide approximate distributions of damage and the extent of ground shaking. Maps of local site intensity include reports of all observation sites and whether or not the strong motion was felt. For example, the isoseismal map of the 17 October 1989 Loma Prieta earthquake in California shown in Figure 1.14 locates the epicentre (marked as a star) and provides MM intensities between isoseismals (Roman numerals), and MM intensities at specific cities (Arabic numerals). The MM intensity of VIII

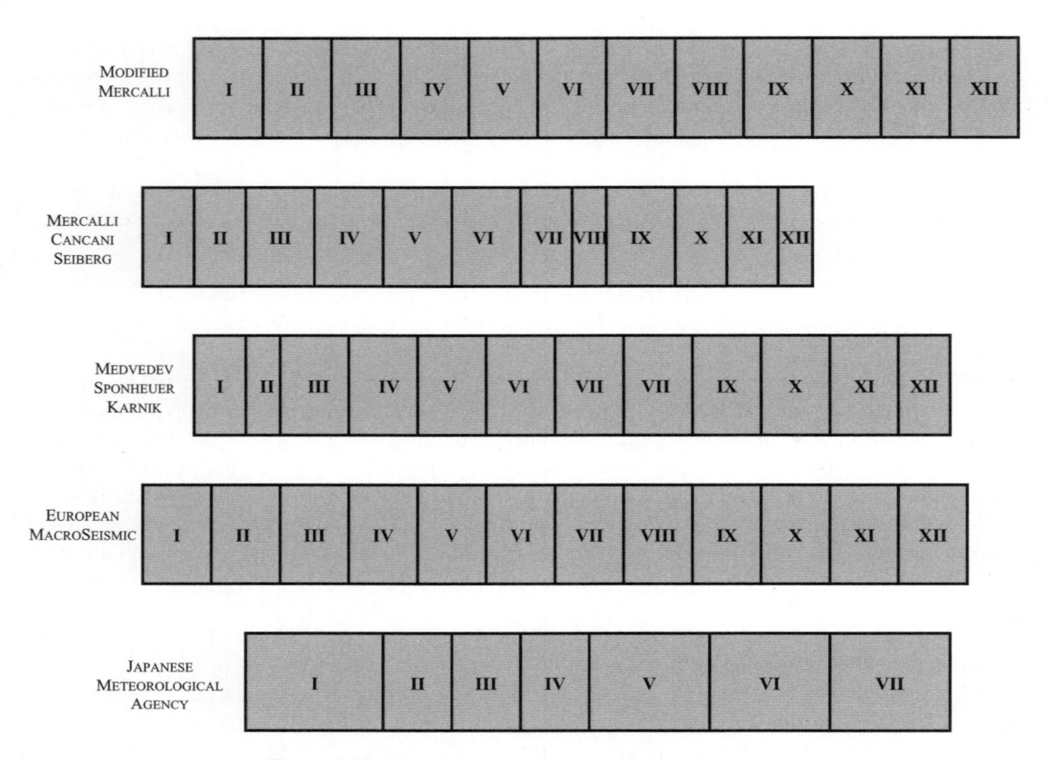

Figure 1.13 Comparison between seismic intensity scales

was assigned to an area of about 50 km long and 25 km wide. Significant ground motions were generated at distances of several tens of kilometres from the earthquake source because of the Moho bounce and the soft soil amplifications, described in Sections 1.1.3 and 1.3.2, respectively.

Anomalous damage distributions may derive from the lack of populated areas in the neighbourhood of the epicentral regions, the depth of soil, local site conditions and directivity effects. Intensity value I_o at the epicentre, or 'epicentral intensity', is equal to the maximum intensity I_{max} felt during ground motion. However, for offshore earthquakes, I_{max} is recorded on the coast and hence does not correspond to I_o.

In some scales, for example JMA, the intensity of earthquakes can also be expressed by the radius R of the felt area (Kanai, 1983). The relationship between R and the earthquake classification is provided in Table 1.3. Epicentral regions in perceptible earthquakes experience ground motions ranked not less than intensity V in the JMA scale.

It has been observed repeatedly that structures in the immediate vicinity of earthquake sources experience very high ground accelerations but sustain little or no damage (e.g. Elnashai *et al.*, 1998). On the other hand, intensity is a measure of the perceptibility of the earthquake and its actual consequential damage. Therefore, relating intensity to peak ground acceleration is, in principle, illogical. However, the necessity of bridging the distance between historical earthquake observations (based mainly on intensity) and code-defined forces (based entirely on peak ground acceleration or displacement) warrants the efforts expended in correlating the two measures. Attenuation relationships correlating intensity and peak ground accelerations, which are presented in Section 3.3, do not reflect parameters influencing earthquake damage potential other than intensity, e.g. site amplification effects and directivity discussed in Sections 1.3.1 and 1.3.2. In addition, source characteristics

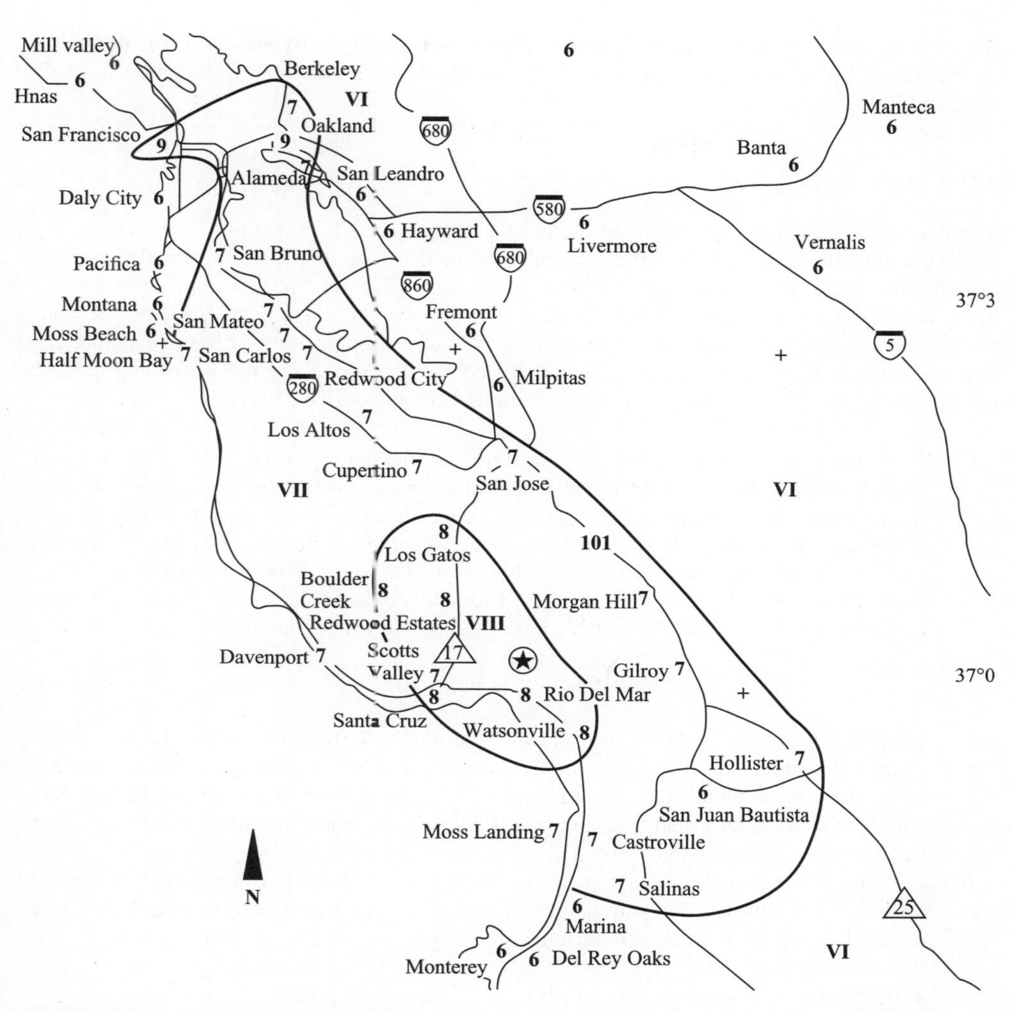

Figure 1.14 Isoseismal map for 1989 Loma Prieta earthquake in California (*after* Plafker and Galloway, 1989)

Table 1.3 Earthquake intensity based on the radius (R) of felt area.

Radius (km)	Earthquake intensity
$R < 100$	Local
$100 < R < 200$	Small region
$200 < R < 300$	Rather conspicuous
$R > 300$	Conspicuous

and mechanisms do not affect intensity scales. The measurement of earthquake size should be based on the amount of energy released at the focus. Therefore, magnitude scales have been defined as presented hereafter.

1.2.2 Magnitude

Magnitude is a quantitative measure of earthquake size and fault dimensions. It is based on the maximum amplitudes of body or surface seismic waves. It is therefore an instrumental, quantitative and objective scale. The first attempts to define magnitude scales were made in Japan by Wadati and in California by Richter in the 1930s. Several scales exist. Many of these scales are frequency-dependent because they measure amplitudes of seismic waves with different properties. Scales related directly to source parameters have also been proposed. These do not depend on specific waves and hence are frequency-independent. The most common magnitude scales are described herein:

(i) *Local* (or *Richter*) *magnitude* (M_L): measures the maximum seismic wave amplitude A (in microns) recorded on standard Wood–Anderson seismographs located at a distance of 100 km from the earthquake epicentre. The standard Wood–Anderson seismograph has a natural period of 0.8 seconds, a critical damping ratio of 0.8 and an amplification factor of 2,800. It amplifies waves with periods between approximately 0.5 and 1.5 seconds i.e. wavelengths of 500 m to 2 km. These waves are of particular interest for earthquake engineers due to their potential to cause damage. Magnitude M_L is related to A by the following relationship:

$$M_L = \log(A) - \log(A_0) \tag{1.6}$$

where A_0 is a calibration factor that depends on distance (Richter 1958). The Richter scale was calibrated assuming that magnitude $M_L = 3$ corresponds to an earthquake at a distance of 100 km with maximum amplitude of A = 1.0 mm. Indeed, $\log A_0 = -3$ for a distance D = 100 km. Earthquakes with M_L greater than 5.5 cause significant damage, while an earthquake of $M_L = 2$ is the smallest event normally felt by people.

(ii) *Body wave magnitude* (m_b): measures the amplitude of P-waves with a period of about 1.0 second, i.e. less than 10-km wavelengths. This scale is suitable for deep earthquakes that have few surface waves. Moreover, m_b can measure distant events, e.g. epicentral distances not less than 600 km. Furthermore, P-waves are not affected by the depth of energy source. Magnitude m_b is related to the amplitude A and period T of P-waves as follows:

$$m_b = \log\left(\frac{A}{T}\right) + \sigma(\Delta) \tag{1.7}$$

in which $\sigma(\Delta)$ is a function of the epicentre distance Δ (in degrees). For example, if $\Delta = 45°$ then $\sigma = 6.80$; other values can be found in the literature (e.g. Udias, 1999).

(iii) *Surface wave magnitude* (M_S): is a measure of the amplitudes of LR-waves with a period of 20 seconds, i.e. wavelength of about 60 km, which are common for very distant earthquakes, e.g. where the epicentre is located at more than 2,000 km. M_S is used for large earthquakes. However, it cannot be used to characterize deep or relatively small, regional earthquakes. This limitation is due to the characteristics of LR-waves as described in Section 1.1.3. The relationship between amplitude A, period T, distance Δ and M_S is given by:

$$M_S = \log\left(\frac{A}{T}\right) + 1.66\log(\Delta) + 3.30 \tag{1.8}$$

where Δ is measured in degrees, the ground-motion amplitude in microns and the period in seconds. Equation (1.8) is applicable for $\Delta > 15°$.

(iv) *Moment magnitude* (M_w): accounts for the mechanism of shear that takes place at earthquake sources. It is not related to any wavelength. As a result, M_w can be used to measure the whole spectrum of ground motions. Moment magnitude is defined as a function of the seismic moment M_0. This measures the extent of deformation at the earthquake source and can be evaluated as follows:

$$M_0 = G A \Delta u \tag{1.9.1}$$

in which G is the shear modulus of the material surrounding the fault, A is the fault rupture area and Δu is the average slip between opposite sides of the fault. The modulus G can be assumed to be 32,000 MPa in the crust and 75,000 MPa in the mantle. M_w is thus given by:

$$M_w = 0.67 \log(M_0) - 10.70 \tag{1.9.2}$$

where M_0 is expressed in ergs.

Richter magnitude M_L exhibits several limitations. It is applicable only to small and shallow earthquakes in California and for epicentral distances less than 600 km. It is, therefore, a regional (or local) scale, while m_b, M_S, and M_w are worldwide scales. The main properties of the above magnitude scales are summarized in Table 1.4. The mathematical definition of magnitude implies that all the above scales have virtually no upper and lower bounds. Notwithstanding, the upper bound is provided by strength of materials in the Earth's crust and the characteristics of the waves measured, while minimum values of magnitude that may be recorded by sensitive seismographs are around –2. As a general guideline, earthquakes with magnitude between 4.5 and 5.5 can be defined as local, while large seismic events generally have a magnitude 5.0 to 7.0. Great earthquakes are those with magnitude larger than 7.0.

Other magnitude scales exist; they are usually based on maximum amplitudes A of certain waves recorded by seismographs. The general correlation between magnitude M and A is as follows (Reiter, 1990):

Table 1.4 Properties of major magnitude scales.

Scale type	Author	Earthquake size	Earthquake depth	Epicentre distance (km)	Reference parameter	Applicability	Saturation
M_L	Richter (1935)	Small	Shallow	<600	Wave amplitude	Regional (California)	✓
m_b	Gutenberg and Richter (1956)	Small-to-medium	Deep	>1,000	Wave amplitude (P-waves)	Worldwide	✓
M_S	Richter and Gutenberg (1936)	Large	Shallow	>2,000	Wave amplitude (LR-waves)	Worldwide	✓
M_w	Kanamori (1977)	All	All	All	Seismic moment	Worldwide	n.a.

Key: n.a. = not applicable; ✓ = saturation occurs.

$$M = \log(A) + f(d, h) + C_S + C_R \tag{1.10}$$

in which the function $f(d,h)$ accounts for epicentral distance d and focal depth h. The coefficients C_S and C_R are station and regional corrections, respectively. They are introduced to account for local and regional effects.

Conversions between different magnitude scales can be performed using simple empirical or semi-empirical relations. For example, the M_{JMA}, which is a long-period measurement adopted by the Japanese Meteorological Agency (JMA), is related to Richter magnitude M_L (Kanai, 1983) by the relationship:

$$M_{JMA} = 2.0\,M_L - 9.7 \tag{1.11}$$

where magnitude M_L is expressed in ergs.

Earthquakes of different size and energy release may have the same magnitude. Typical examples are the 1906 San Francisco (California) and the 1960 Chile earthquakes. Both events showed $M_S = 8.3$. However, the fault rupture area in Chile was about 35 times greater than that observed in California. Different fault rupture lengths correspond to different amounts of energy released; moment magnitude accounts for the extent of fault rupture (Scholz, 1990). The moment magnitude M_w is about 8 for the San Francisco fault while the Chile earthquake has a moment magnitude M_w of 9.5. Magnitude scales do not increase monotonically with earthquake size. This observation is known as 'saturation' and affects all scales that are related to seismic waves of a particular period and wavelength, i.e. frequency-dependent scales. Figure 1.15 shows a comparison between different magnitude scales. Saturation is evident as M_w increases ($M_w > 6.5$). Another magnitude scale, i.e. m_B is included in the plot; m_B is a body wave scale measuring different types of body waves with periods between 1.0 and 10 seconds and is distinct from m_b.

For values of magnitude of about 5.5, scales m_b and M_S coincide; for smaller earthquakes, e.g. $M_w < 5.5$, $m_b > M_S$, while for large magnitude $M_S > m_b$. Thus, surface wave magnitudes underestimate the size of small earthquakes while they overestimate the size of large events. Magnitudes m_b and M_S

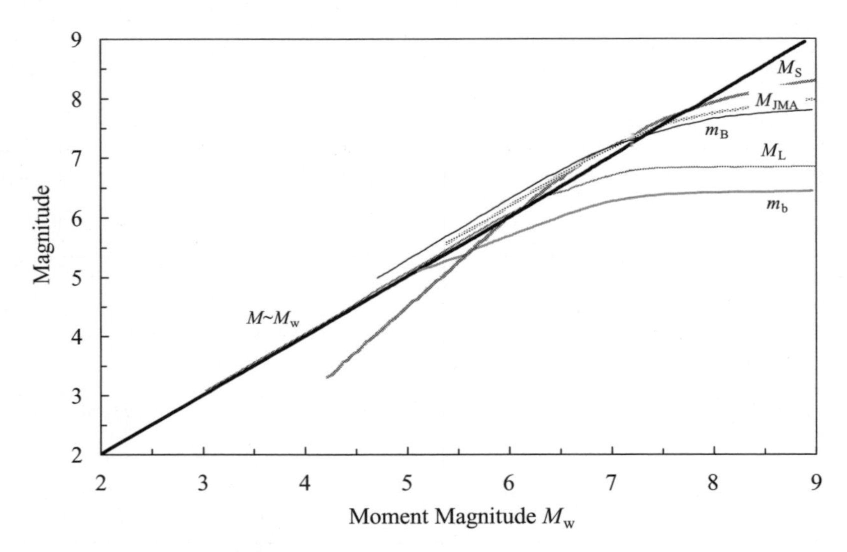

Figure 1.15 Saturation of magnitude scales

saturate at about 6.5 and 8.5, respectively. The Richter scale stops increasing at $M_w = 7.0$. M_w does not suffer from saturation problems in the practical range of magnitude of $2 < M_w < 10$. Therefore, it can be employed for all magnitudes. For shallow earthquakes, Bolt (1999) suggests using M_D, also referred to as 'coda-length magnitude', for magnitudes less than 3, either M_L or m_b for magnitudes between 3 and 7, and M_S for magnitudes between 5 and 7.5. The 1994 Northridge earthquake has been ranked, for example, as 6.4 in the local magnitude scale M_L, 6.6 in M_S and 6.7 in M_w (Broderick *et al.*, 1994). At these magnitudes, the different scales provide similar values, as displayed, for example, in Figure 1.15.

Earthquake magnitude can be used to quantify the amount of energy released during fault ruptures. Energy propagating by seismic waves is proportional to the square root of amplitude–period ratios. Magnitude is proportional to the logarithm of seismic energy E. A semi-empirical relationship between surface wave magnitude M_S and E was formulated by Richter and Gutenberg (Richter, 1958), and is given by:

$$\log(E) = 1.5M_S + 11.8 \tag{1.12}$$

where E is in ergs. As the magnitude increases by one unit, the energy increases by a factor of 31.6 and the difference between two units of magnitude is a factor of 1,000 on energy release. Similarly, m_b and M_S are related to seismic energy E by the following empirical relations:

$$\log(E) = 2.4m_b - 1.3 \tag{1.13.1}$$

$$\log(E) = 1.5M_S + 4.2 \tag{1.13.2}$$

where E is expressed in joules (1 joule = 10^7 ergs). Figure 1.16 indicates the correlation between surface wave magnitude M_S and energy released during earthquakes and other events. The number of earthquakes per year is also provided.

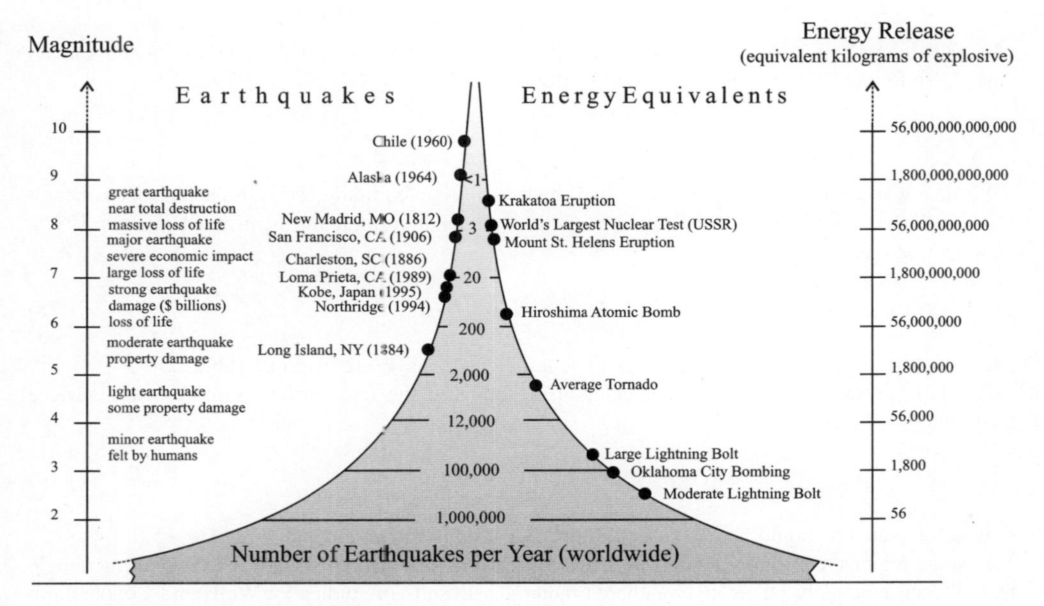

Figure 1.16 Correlation between magnitude and energy release (*adapted from* Bolt, 1999)

Seismic moment M_0 measures the energy E released by fault rupture during earthquakes (Scholz, 1990). The following relationship is applicable to all source mechanisms:

$$E = \frac{\Delta\tau}{2G} M_0 \tag{1.14}$$

where $\Delta\tau$ is the stress drop $\Delta\tau = \tau_1 - \tau_2$, and τ_1 and τ_2 are the shear stresses on the fault before and after brittle fracture occurs, respectively; G is the shear modulus of the material surrounding the fault as also shown in equation (1.9.1). For moderate-to-large earthquakes, the mean values of $\Delta\tau$ are equal to about 6.0 MPa. In the definition of M_w, the stress drop is assumed constant.

Magnitude–moment relationships have been defined empirically for periods less than 20 seconds (Purcaru and Berckhemer, 1978), as below:

$$\log(M_0) = 1.5 M_S + 16.1 \tag{1.15}$$

and body wave magnitude m_b can be related over a wide range to M_S by the following semi-empirical formula proposed by Gutenberg and Richter (Richter, 1958):

$$m_b = 0.63 M_S + 2.5 \tag{1.16}$$

therefore, combining equations (1.15) and (1.16), seismic moment M_0 can be related to body waves m_b and vice versa. Moreover, Figure 1.15 may be used when relationships between M_0 and magnitude scales other than m_b and M_S are sought.

Expressions correlating magnitude scales and fault rupture parameters can be found in the literature (e.g. Tocher, 1958; Housner, 1965; Seed *et al.*, 1969; Krinitzsky, 1974; Mark and Bonilla, 1977). For example, Bonilla *et al.* (1984) computed M_S as a function of the fault rupture length L:

$$M_S(L) = 6.04 + 0.71 \log(L) \tag{1.17.1}$$

where the length is in kilometres. Equation (1.17.1), which is applicable for $M_S > 6.7$, is based on mean values, while the 95th percentile is given as follows:

$$M_S^{0.95} = M_S(L) + 0.52 \tag{1.17.2}$$

Surface wave magnitude M_S has also been related to the maximum observed displacement of fault D. Empirical relationships are provided as a function of the fault rupture mechanism (Slemmons, 1977), as shown below:

$$M_S = a + b \log(D) \tag{1.18}$$

where the displacement D is in metres, while coefficients a and b are given in Table 1.5.

Similarly, Wyss (1979) proposed a relationship between the fault surface rupture S and surface magnitude M_S given by:

$$M_S = 4.15 + \log(S) \tag{1.19}$$

in which the area S should be expressed in km^2. Equation (1.19) is applicable for $M_S > 5.6$.

In some regions, correlations as given above are of little value since many of the important geologic features can be deeply buried by weathered materials. Results of studies by Wells and Coppersmith (1994) are outlined in Table 1.6 for different types of fault mechanisms, i.e. strike-slip, reverse and

Table 1.5 Values of coefficients in equation (1.18).

Fault mechanism	a	b
Normal	6.67	0.75
Reverse	6.79	1.31
Strike-slip	6.97	0.80

Table 1.6 Empirical relationships between moment magnitude M_w, surface rupture length, L (km), subsurface rupture length L' (km), rupture area, A (km^2), maximum D and average $\bar{D}$ surface displacement, in metres (*after* Wells and Coppersmith, 1994).

Fault mechanism	Relationship	σ_{M_W}	Relationship	$\sigma_{\log L,A,D}$	Magnitude range	Length/Width/ Displacement range (km)
Strike-slip	$M_W = 5.16 + 1.12\log L$	0.28	$\log L = 0.74M_W - 3.55$	0.23	5.6 to 8.1	1.3 to 432
Reverse	$M_W = 5.00 + 1.22\log L$	0.28	$\log L = 0.63M_W - 2.86$	0.20	5.4 to 7.4	3.3 to 85
Normal	$M_W = 4.86 + 1.32\log L$	0.34	$\log L = 0.50M_W - 2.01$	0.21	5.2 to 7.3	2.5 to 41
All	$M_W = 5.08 + 1.16\log L$	0.28	$\log L = 0.69M_W - 3.22$	0.22	5.2 to 8.1	1.3 to 432
Strike-slip	$M_W = 4.33 + 1.49\log L'$	0.24	$\log L' = 0.62M_W - 2.57$	0.15	4.8 to 8.1	1.5 to 350
Reverse	$M_W = 4.49 + 1.49\log L'$	0.26	$\log L' = 0.58M_W - 2.42$	0.16	4.8 to 7.6	1.1 to 80
Normal	$M_W = 4.34 + 1.54\log L'$	0.31	$\log L' = 0.50M_W - 1.88$	0.17	5.2 to 7.3	3.8 to 63
All	$M_W = 4.38 + 1.49\log L'$	0.26	$\log L' = 0.59M_W - 2.44$	0.16	4.8 to 8.1	1.1 to 350
Strike-slip	$M_W = 3.98 + 1.02\log A$	0.23	$\log A = 0.90M_W - 3.42$	0.22	4.8 to 7.9	3 to 5,184
Reverse	$M_W = 4.33 + 0.90\log A$	0.25	$\log A = 0.98M_W - 3.99$	0.26	4.8 to 7.6	2.2 to 2,400
Normal	$M_W = 3.93 + 1.02\log A$	0.25	$\log A = 0.82M_W - 2.87$	0.22	5.2 to 7.3	19 to 900
All	$M_W = 4.07 + 0.98\log A$	0.24	$\log A = 0.91M_W - 3.49$	0.24	4.8 to 7.9	2.2 to 5,184
Strike-slip	$M_W = 3.80 + 2.59\log W$	0.45	$\log W = 0.27M_W - 0.76$	0.45	4.8 to 8.1	1.5 to 350
Reverse	$M_W = 4.37 + 1.95\log W$	0.32	$\log W = 0.41M_W - 1.61$	0.32	4.8 to 7.6	1.1 to 80
Normal	$M_W = 4.04 + 2.11\log W$	0.31	$\log W = 0.35M_W - 1.14$	0.31	5.2 to 7.3	3.8 to 63
All	$M_W = 4.06 + 2.25\log W$	0.41	$\log W = 0.32M_W - 1.017$	0.41	4.8 to 8.1	1.5 to 350
Strike-slip	$M_W = 6.81 + 0.78\log D$	0.29	$\log D = 1.03M_W - 7.03$	0.34	5.6 to 8.1	0.01 to 14.6
Reverse*	$M_W = 6.52 + 0.44\log D$	0.52	$\log D = 0.29M_W - 1.84$	0.42	5.4 to 7.4	0.11 to 6.5
Normal	$M_W = 6.61 + 0.71\log D$	0.34	$\log D = 0.89M_W - 5.90$	0.38	5.2 to 7.3	0.06 to 6.1
All	$M_W = 6.69 + 0.74\log D$	0.40	$\log D = 0.82M_W - 5.46$	0.42	5.2 to 8.1	0.01 to 14.6
Strike-slip	$M_W = 7.04 + 0.89\log \bar{D}$	0.28	$\log \bar{D} = 0.90M_W - 6.32$	0.28	5.6 to 8.1	0.05 to 8.0
Reverse*	$M_W = 6.64 + 0.13\log \bar{D}$	0.50	$\log \bar{D} = 0.08M_W - 0.74$	0.38	5.8 to 7.4	0.06 to 1.5
Normal	$M_W = 6.78 + 0.65\log \bar{D}$	0.33	$\log \bar{D} = 0.63M_W - 4.45$	0.33	6.0 to 7.3	0.08 to 2.1
All	$M_W = 6.93 + 0.82\log \bar{D}$	0.39	$\log \bar{D} = 0.69M_W - 4.80$	0.36	5.6 to 8.1	0.05 to 8.0

Key: *Regression relationships are not statistically significant at a 95% probability level.

normal. It was observed that large scatter may characterize the relationship between moment magnitude M_w and surface rupture length L (in km), the subsurface rupture length L' (in km), the rupture area A (in km^2), the downdip rupture width W (in km), the maximum D and the average $\bar{D}$ surface displacement (in metres), especially for reverse-slip earthquakes.

Equations (1.17) to (1.19) and those in Table 1.6 are valid for earthquakes on or closer to tectonic place boundaries (inter-plate earthquakes). For earthquakes distant from plate boundaries (intra-plates events), such as the New Madrid seismic zone, a study by Nuttli (1983) showed that the latter equations may overestimate fault rupture lengths. Average source parameters and relevant magnitude scales are summarized in Table 1.7.

Table 1.7 Average source parameters for mid-plate earthquakes (*after* Nuttli, 1983).

Rupture length (km)	Slip (m)	m_b	M_S	log M_0 (dyne-cm)
2.1	0.01	4.5	3.35	22.2
3.8	0.03	5.0	4.35	23.2
7.0	0.11	5.5	5.35	24.2
13.0	0.34	6.0	6.35	25.2
24.0	1.10	6.5	7.35	26.2
45.0	3.70	7.0	8.32	27.2
58.0	5.80	7.2	8.53	27.6
75.0	9.20	7.4	8.87	28.0
85.0	11.50	7.5	9.00	28.2

Differences between the values predicted by equation (1.17.1) and those provided in Table 1.7 drop as the rupture length increases. For short rupture lengths, e.g. 2 to 5 km, the variations exceed 50%, while for longer fault ruptures, the differences are between 10% and 20%.

1.2.3 Intensity–Magnitude Relationships

Intensity–magnitude relationships are essential for the use of historical earthquakes for which no instrumental records exist. Several simple methods to convert intensity into magnitude have been proposed (e.g. Lee *et al.*, 2003); most of which exhibit large scatter because of the inevitable bias present in the definition of intensity (Ambraseys and Melville, 1982). Gutenberg and Richter (1956) proposed a linear relationship between local magnitude M_L and epicentral intensity I_0 for Southern California, given by:

$$M_L = 0.67 I_0 + 1.00 \tag{1.20}$$

in which the intensity I_0 is expressed in the MM scale. The above equation shows, for example, that the epicentral intensity I_0 of VI corresponds to M_L = 5.02, indicating that the earthquake is likely to cause significant damage.

Street and Turcotte (1977) related m_b magnitude to the intensity I_0 (in the MM scale) as follows:

$$m_b = 0.49 I_0 + 1.66 \tag{1.21}$$

which is useful in converting earthquake data in the central and eastern USA. Equation (1.21) relates an intensity of VI in the MM scale to a magnitude m_b of 4.60, which contradicts the observation that M_L should be systematically lower than m_b for short-period waves, as discussed in Section 1.2.2. This contradiction may be due to different rates of earthquake occurrence in various regions of the USA (Reiter, 1990). It also demonstrates that values obtained from intensity–magnitude relationships should be subject to engineering judgement. Regression analyses carried out on magnitudes predicted by equations (1.20) and (1.21), and values measured for the same events have in many instances indicated poor statistical correlations. For example, correlation coefficients as low as ~0.5 are obtained when comparing earthquakes that occurred between the 1930s and 1970s in Quebec (Canada) and some regions of the USA, such as Illinois and New York (Reiter, 1990). As a result, several other methods have been proposed in an attempt to correlate intensity and magnitude scales. These formulations have been based on different intensity-related parameters, such as the felt area, the area inscribed by intensity IV isoseismals and the fall-off of intensity with distance.

Intensity–magnitude relationships were proposed by Ambraseys (1985, 1989) for European regions as follows:

$$M_S = -1.10 + 0.62 I_i + 1.30 \cdot 10^{-3} r_i + 1.62 \log(r_i) \tag{1.22.1}$$

which is applicable for north-west Europe, and

$$M_S = -0.90 + 0.58 I_i + 1.10 \cdot 10^{-3} r_i + 2.11 \log(r_i) \tag{1.22.2}$$

for the Alpine zone, where I_i is the MM intensity of the ith isoseismal and r_i is the radius of equivalent area enclosed by the ith isoseismal, in kilometres.

Local geological conditions and focal depths can significantly affect the intensity of earthquake ground motion. Semi-empirical formulations accounting for focal depths are available (e.g. Kanai, 1983). Sponheuer (1960) proposed to calculate M from the epicentral intensity I_0 as follows:

$$M_S = 0.66 I_0 + 1.70 \log(h) - 1.40 \tag{1.23}$$

where the focal depth h is in kilometres and the intensity I_0 is in the MM scale.

Attenuation relationships (relationships between a ground-shaking parameter, magnitude, distance and soil condition) for different ground-motion parameters can be derived from intensity and magnitude; they may account for distance, travel path and site effects. The most common attenuation relationships formulated for active seismic regions worldwide are presented in Section 3.3 of Chapter 3.

Problem 1.3

Calculate the surface wave magnitude M_S for an earthquake with I_{MM} of VII in an area that can be approximated by a circle with radius 20 km for a site at the borders of the given isoseismal. This site is located in the Western United States but you may use equation (1.22.1). Compare the ensuing value with the estimations from relationships with other magnitude scales. Calculate the fault surface displacements. Assume that the earthquake mechanism is normal faulting.

1.3 Source-to-Site Effects

The characteristics of seismic waves are altered as they travel from the source to the site of civil engineering works, due to wave dispersion at geological interfaces, damping and changes in the wavefront shape. The latter are referred to as 'distance and travel path effects'. Moreover, local site conditions may affect significantly the amplitude of earthquake ground motions; these are known as 'site effects'. Non-linearity of soil response and topographical effects may also influence ground-motion parameters (Silva, 1988) as shown in Table 1.8. For example during the 26 September 1997 Umbria–Marche (Italy) earthquake, significant site amplification was observed even at large distances from the epicentre (Sano and Pugliese, 1999). Due to the geomorphological conditions in the epicentral area, located in the Apennines, local soil amplifications related both to topographic and basin effects were present. During the long aftershock sequence, a temporary strong-motion array was installed in the area where major damage took place. Some instruments were deployed on different geological and morphological soil conditions in two towns, Cesi and Sellano, to investigate the considerable localization in the observed damage. Field investigations were also carried out to assess the geological profiles across strong-motion sites. The recordings confirmed the importance of site characteristics in the distribution of damage at sites very close to one another. Large amplification at the basin border of the Cesi site and an important three-dimensional effect at the site in Sellano were observed.

Table 1.8 Effects of topographic and subsurface irregularities (*adapted from* Silva, 1988).

Structure	Influencing factors	Effect	Quantitative	Predictability
Surface topography	Sensitive to shape ratio, largest for ratio between 0.2 and 0.6.	Amplification at top of structure, rapid changes in amplitude phase along slopes.	Ranges up to a factor of 30 but generally about 2 to 20.	Poor: generally under-predict size; may be due to ridge–ridge interaction and three-dimensional effects.
Shallow and wide (depth/width <0.25) sediment-filled valleys	Effects most pronounced near edges; largely vertically propagating shear waves away from edges.	Broadband amplification near edges due to generation of surface waves.	One-dimensional models may under-predict at higher frequencies by about 2 near edges.	Good: away from edges one dimension works well, near edges extend one dimension to higher frequencies.
Deep and narrow (depth/width >0.25) sediment-filled valleys	Effects throughout valley width.	Broadband amplification across valley due to whole valley modes.	One-dimensional models may under-predict for a wide bandwidth by about 2 to 4; resonant frequencies shifted from one-dimensional analysis.	Fair: given detailed description of vertical and lateral changes in material properties.

It has been demonstrated that the most important topographical parameter influencing local amplification of ground motion is the steepness of the ridge (Finn, 1991). Displacement amplifications at the crest of a triangular-shaped hill are equal to $2/v$, where v is estimated from the angle formed by the ridges, i.e. $v\,\pi$. Consequently, as the ridge becomes steeper, the displacement amplification increases. Measured amplification at hill crests with respect to the base ranges between 2 and 20. The latter values are higher than those predicted analytically (generally between 2 and 4) because of the significant influence of both ridge-to-ridge interaction and three-dimensional effects, as for example those observed in the town of Sellano during the 1997 Umbria–Marche (Italy) earthquake.

An exhaustive discussion of distance, travel path and site effects from seismological and geotechnical standpoints can be found in Reiter (1990) and Kramer (1996), respectively. Hereafter, directional effects, site amplification, dispersion and incoherence, and their effects on structural response are outlined.

1.3.1 Directional Effects

Earthquakes of small magnitude are frequently generated by sources that may be represented by a point, since the fault rupture extends only a few kilometres. Conversely, for large earthquakes, fault rupture traces can be a few hundred kilometres long. In the latter case, seismic wave radiation is influenced by the source dimensions. Earthquake stress waves propagate in the direction of faulting more intensely than in other directions. This affects the distribution of shaking intensity and hence the distribution of ground-motion parameters and consequently damage distribution. For example, waves propagate away from the fault rupture with different intensity along different directions; this observation is referred to

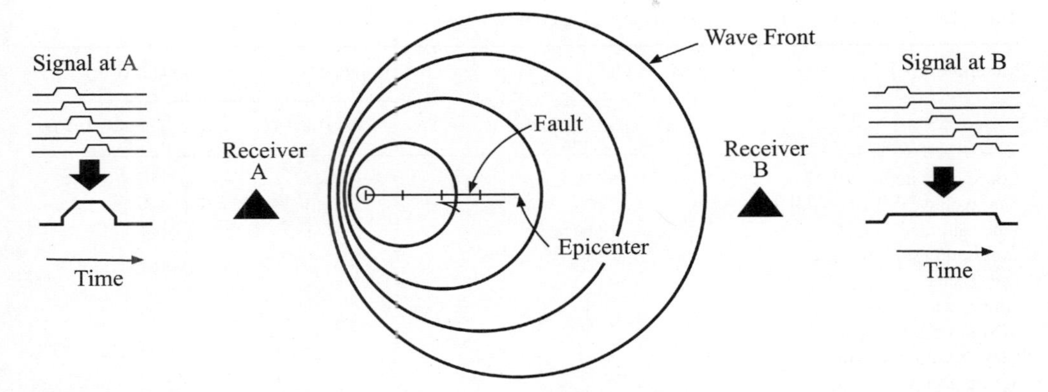

Figure 1.17 Directivity effects on sites towards and away from direction of fault rupture (*adapted from* Singh, 1985)

as 'directivity'. Benioff (1955) and Ben-Menachem (1961) demonstrated that such directivity can lead to azimuthal differences in ground motions. Directivity occurs because fault ruptures are moving wave sources, which travel at a finite velocity along the fault. The engineering implication of such directivity effects is that sites that are equidistant from the source will be subjected to varying degrees of shaking from the same earthquake, thus casting doubt over the concept of distance-based attenuation relationships discussed in Section 3.3. In Figure 1.17, a pictorial representation of directivity effects on ground motions at sites in the direction of, and away from, fault rupture is given. As the fault rupture (or earthquake source) moves away from the epicentre, it generates ground motion from each segment of the breaking fault. The ground motion radiates outward in all directions and the seismic energy propagates through expanding wavefronts.

The over-riding of stress waves or 'constructive interference' results in larger ground-motion magnification with shorter total duration in the direction of rupture propagation. Lower amplitude motions and longer total duration are exhibited in the opposite direction. This effect increases as the velocity of the fault rupture reaches the speed of seismic waves and as the angle between the point of observation (e.g. the recording station and construction site) and the direction of rupture propagation is reduced. Constructive interference, which is in essence a Doppler effect, generates strong pulses of large displacement or 'fling' at nearby sites towards which the fault rupture is progressing (Singh, 1985; Somerville *et al.*, 1997), e.g. towards the left in Figure 1.17. Rupture directivity also causes the polarization of ground motion, i.e. differences between the fault-normal and fault-parallel components of horizontal ground-motion amplitudes (Stewart *et al.*, 2001). This polarization causes more intense shaking in the fault-normal direction than in the fault-parallel direction. Where sufficient information exists, directivity effects should be taken into account in estimating earthquake design parameters. Directivity or focusing of seismic energy caused severe damage to residential buildings and transportation systems in urban areas during the 1994 Northridge and 1995 Kobe earthquakes (Broderick *et al.*, 1994; AIJ, 1995). Damage to structures during past earthquakes is illustrated in detail in Appendix B.

1.3.2 Site Effects

The characteristics of the site affect the frequency and duration of earthquake ground motions. Structures founded on rock will, in general, be subjected to short-period (high frequency) motion, while soft sites result in longer period (low frequency) excitation. The ratio between the period of the site and that of the building is important in estimating the amplification effects; this is known as 'site resonance

Table 1.9 Shear wave velocities for foundation materials (in m/s).

Material (type)	Depth, H (m)		
	$1 < H < 6$	$7 < H < 15$	$H \geq 15$
Loose saturated sand	60	–	–
Sandy clay	100	250	–
Fine saturated sand	110	–	–
Clay/sand mix	140	–	–
Dense sand	160	–	–
Gravel with stone	180	–	–
Medium gravel	200	–	–
Clayey sand with gravel	–	350	–
Medium gravel	–	–	780
Hard sandstone	–	–	1,200

effect'. Resonance is a frequency-dependent phenomenon. The site period T_S for uniform single soil layer on bedrock can be estimated from the relationship:

$$T_S = \frac{4H}{v_S} \tag{1.24.1}$$

where T_S is in seconds. H and v_S are the depth of soil layer (in metres) and soil shear wave velocity (in m/s), respectively. The shear wave velocity v_S of the soil layer is a function of the soil type and the depth of the deposit. The average values given in Table 1.9 may be used with equation (1.24.1); the latter equation provides the natural period of vibration of a single homogeneous soil layer. Periods associated with higher modes can be determined as follows:

$$T_{S,n} = \frac{1}{2n-1} \frac{4H}{v_S} \tag{1.24.2}$$

in which n represents the nth mode of vibration ($n > 1$).

In alluvial surface layers, vibrations are amplified due to multi-reflection effects. The ratio of the amplitude a_g at the ground surface to the amplitude at the lower boundary layer (bedrock) a_b is given by (Okamoto, 1984):

$$\left| \frac{a_g}{a_b} \right| = \left(\cos^2 \frac{\omega H}{v_s} + \alpha^2 \sin^2 \frac{\omega H}{v_s} \right)^{-\frac{1}{2}} \tag{1.24.3}$$

in which ω is the natural circular frequency of the soil layer and α is the wave-propagation impedance given by:

$$\alpha = \frac{\rho_s v_s}{\rho_b v_b} \tag{1.24.4}$$

where ρ and v are the density and velocity of the surface layer (subscript s) and lower layer (subscript b), respectively.

The response of elastic layers of soil of finite depth H and varying shear rigidity G to earthquake ground motions was first investigated analytically by Ambraseys (1959). Auto-frequencies of the overburden were derived when the rigidity of the material G varies with depth. The latter is often encountered in practical applications in comparatively thin superficial weathered layers of soil or in desiccated soils in arid climates. Surface compaction may also produce a decrease in rigidity with depth. It was demonstrated that a good approximation of the periods of vibration can be obtained by considering the rigidity ratio k equal to the mean value $\overline{G}$ of shear modulus at the surface G and at the bedrock G_b and utilizing the following relationship:

$$T_{S,n} = \frac{5.66}{2n-1}\frac{H}{v_S}\frac{k}{\sqrt{1+k^2}} \tag{1.25.1}$$

where n is the nth mode of vibration ($n > 1$), v_s the shear wave velocity near the surface of the layer of height H. The constant of rigidity is given by:

$$k = \sqrt{\frac{\overline{G}}{G_b}} \tag{1.25.2}$$

The expression in equation (1.25.1) holds within less than 6.0% of the true frequencies for small values of the rigidity ratio, i.e. $k \leq 1.5$ to 2.0. Alternatively, for layers of linearly increasing rigidity, the periods of layers of constant rigidity [as per equations (1.24.1) and (1.24.2)] can be reduced through the factors provided in Table 1.10. Periods of vibrations of layers with uniform rigidity are always higher than those corresponding to a layer of linearly increasing rigidity. The listed correction factors are given for the first six modes of vibration and may be used to estimate site periods.

An example of significant site amplifications was observed in the 1985 Mexico City earthquake. On 19 September 1985, an earthquake of magnitude M_S = 8.1 struck the Mexican capital and caused widespread structural damage especially downtown, as shown in damage pictures in Appendix B. More than 10,000 people were killed. Downtown Mexico City is built on sediments from an ancient 40-m-thick soft layer of lake deposits. The average shear wave velocity of the soil layer is about 80 m/s and hence the resonant period T_S computed from equations (1.24.1) and (1.24.2) is about 2.0 seconds (0.5 Hz). Medium-to-high rise buildings with 5 to 15 storeys were particularly susceptible to damage

Table 1.10 Reduction factors (in %) for period of elastic soil layers with uniform rigidity (*after* Ambraseys, 1959).

G_b / G	Mode (n)					
	1st	2nd	3rd	4th	5th	6th
1.00	0.0	0.0	0.0	0.0	0.0	0.0
1.10	3.4	2.5	2.4	2.3	2.1	2.0
1.21	6.6	5.0	4.9	4.6	4.6	4.6
1.32	9.5	7.5	7.1	7.0	6.9	6.9
1.56	15.0	11.7	11.3	11.1	11.1	11.0
1.96	22.0	17.2	17.0	16.7	16.6	16.6
2.25	28.7	20.8	20.3	20.0	20.0	20.0
4.00	41.7	34.6	34.0	33.6	33.5	33.4
9.00	59.1	51.6	50.6	50.4	50.2	50.1
25.00	74.6	68.5	67.3	67.1	66.8	66.8

(e.g. Osteraas and Krawinkler, 1990). These structures exhibit fundamental periods close to the resonant value T_S. Site amplifications also caused several structural collapses during the 1994 Northridge earthquake, in California (Broderick *et al.*, 1994).

It is recommended that the ratio between the building and site periods be as distinct from unity as possible. In estimating the period of the site, assessment of the deep geology, not only the surface soil condition is crucial. Higher vibration modes of the site should be checked with respect to the predominant response periods of the structure under consideration.

The nature of soil response in earthquakes depends on the amplitude and duration of motion. High-amplitude motion tends to cause inelasticity in the soil. Long-duration shaking increases the susceptibility to liquefaction of saturated and partially saturated soils. When the soil responds elastically, the observed motions at the surface are amplified proportional to the input ground motion. On the other hand, for inelastic response, the soil absorbs large amounts of the energy corresponding to large amplitude of ground motions. Therefore, in general, large earthquake vibrations travelling through inelastic media will exhibit lower accelerations (relative to small-magnitude earthquakes) and large displacements, corresponding to long periods. The displacement demand on structural systems is thus increased, especially on medium- and long-period structures, such as high-rise multi-storey buildings and long-span bridges. Long-duration shaking applies a large number of cycles that may cause a significant increase in pore water pressure leading to total loss of cohesion in soils that then turn into a liquid. This is referred to as liquefaction (e.g. Kramer, 1996, among others).

1.3.3 Dispersion and Incoherence

Earthquake ground motion may exhibit spatial variability on regional and local levels. Large-scale effects are described mathematically by attenuation relationships, which are presented in Section 3.3 of Chapter 3. Herein, two strong-motion characteristics associated with local spatial variations, i.e. 'dispersion' and 'incoherence', are discussed primarily from a physical, as opposed to a mathematical, point of view.

Dispersion and incoherence may be caused by several factors. They can be thought of as the result of the combination of three basic effects as shown in Figure 1.18 and summarized below (Abrahamson, 1991):

(i) *Wave passage effect*: represents the time delay in the arrival of seismic waves on the ground surface at different stations or sites. This effect is due to the finite travelling velocity of seismic waves through media (*see* Section 1.1.3);
(ii) *Extended source effect*: number and size of earthquake sources affecting the seismicity at a site may cause delays in the arrival time of waves. This time lag generates different motions at different points;

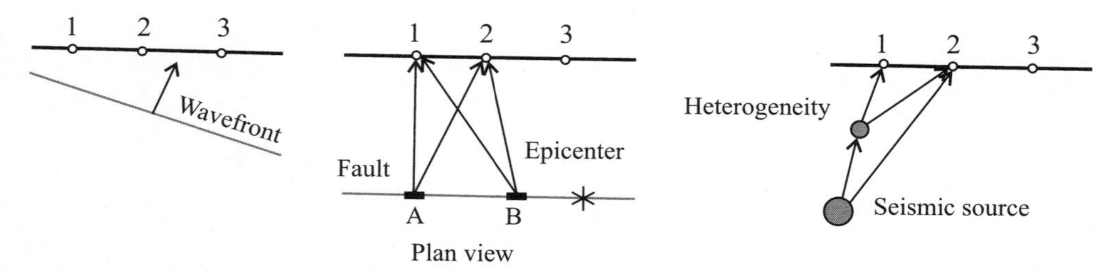

Figure 1.18 Sources of local spatial variability of ground motions: wave passage effect (*left*), extended source effect (*middle*) and ray path effects (*right*)

(iii) *Ray path effect* (or *scattering effect*): caused by reflection and refraction of waves through the soil during their propagation, inhomogeneities of soil layers and other differences in local soil conditions under the various stations.

Spatial variability of earthquakes can be described mathematically either in the time domain (generally by auto-covariance and cross-covariance) or frequency domain (by coherency functions). It is beyond the scope of this chapter to discuss analytical techniques employed to define dispersion and incoherence. The reader may consult one of the textbooks that deal specifically with random vibrations in earthquake engineering (e.g. Manolis and Koliopoulos, 2001, among others). It is noteworthy that ground motions recorded by dense arrays in several regions worldwide, e.g. USA, Japan and Taiwan, have shown coherency decreases with increasing distance between measuring points and increasing frequency of motion (e.g. Clough and Penzien, 1993; Kramer, 1996). The coherency of two ground motions is a measure of correlation of amplitudes and phase angles at different frequencies. Incoherence (or loss of coherence) is strongly frequency-dependent (Luco and Wong, 1986). The coherence factor or absolute value of coherency is a measure of the incoherence. More significant effects are observed at higher frequencies: for frequencies lower than 1.0–2.0 Hz (periods T of 0.5 to 1.0 seconds), the loss of coherence can be ignored (coherence factor is close to 1.0). Coherence starts to decrease significantly for higher frequencies. For frequencies higher than 5 Hz (T less than 0.2 seconds), the coherence factor is reduced by more than 40–50%. Several expressions for smooth coherence functions have been proposed for design purposes (e.g. Harichandran and Vanmarcke, 1986; Luco and Wong, 1986; Abrahamson, 1991; Oliveira *et al.*, 1991; Somerville *et al.*, 1991; Der Kiureghian, 1996). These relationships typically depend on the separation distance and frequency.

Dispersion and incoherence of earthquake ground motions do not generally affect short-span structures, such as buildings, but they may significantly influence the dynamic response of long-span structures, for example medium- to long-span bridges, stadiums and pipelines that extend over considerable distances. Significant spatial variability may often occur whenever the large plan dimensions are combined with irregularities in the soil profile along the travel path. For long distances and rather stiff structures, totally uncorrelated ground motions with appropriate frequency content should be considered. Loss of coherence can be ignored in all the other cases, although time delay should always be accounted for.

Problem 1.4

What is the natural period of a layered soil with medium gravel of depth 40 m? Is it safe to build a multi-storey framed building with fundamental period of vibration equal to 1.5 seconds, as that displayed in Figure 1.19, on a site with the above soil type? Is this site more suitable for a particular type of structure shown in Figure 1.19?

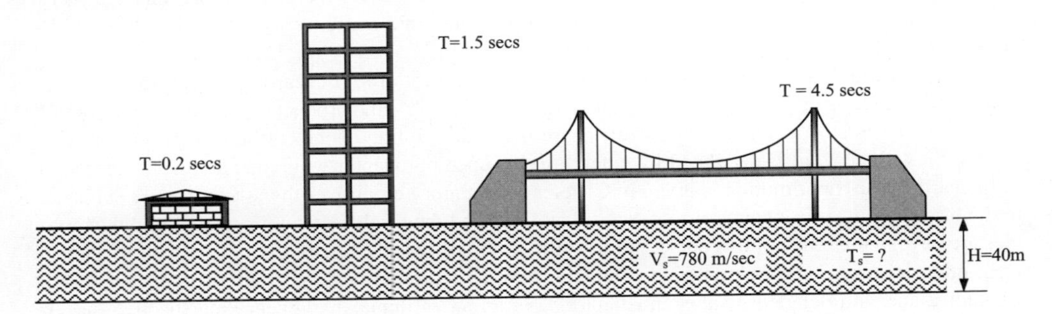

Figure 1.19 Structural systems with different natural periods of vibration

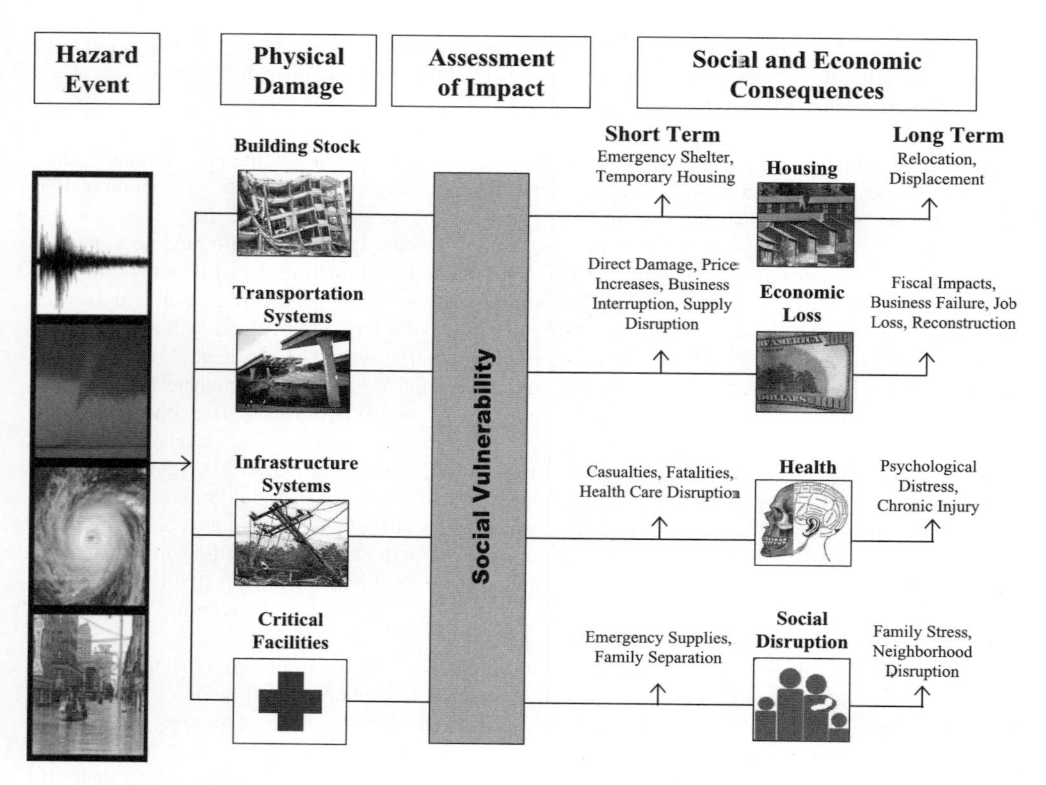

Figure 1.20 Correlation between typical hazard events and social and economic consequences (*courtesy of* Dr. Steve French, Georgia Institute of Technology, USA)

1.4 Effects of Earthquakes

Comprehensive regional earthquake impact assessment requires an interdisciplinary framework that encompasses the definition of the hazard event, physical damage, and social and economic consequences. Such an integrated framework may provide the most credible estimates with associated uncertainty that can stand scientific and political scrutiny. Physical damage should be evaluated for the building stocks, lifeline systems, transportation networks and critical facilities. Short- and long-term effects should be considered in quantifying social and economic consequences. Figure 1.20 provides an overview of causes and effects of natural disasters.

The fundamental components of earthquake loss assessment are (i) hazard, (ii) inventory and (iii) vulnerability or fragility, as depicted in Figure 1.21. Seismic risk is the product of hazard and vulnerability for a unit value of assets. Hazard or exposure is the description of the earthquake ground motion. In this book, the hazard is described in general in this opening chapter while detailed characterization of the earthquake input motion is given in Chapter 3. Inventory comprises the assets that are subjected to the hazard; thus, it is a count of the exposed systems and their value. Inventory issues and technologies are beyond the scope of this book. Vulnerability or fragility is the sensitivity of the assets to damage from intensity of ground shaking. The vulnerability of structural systems is addressed conceptually in Chapter 2 and in a detailed manner in Chapter 4. From an earthquake engineer's perspective, hazard can be quantified but not reduced. Vulnerability can be both evaluated and reduced, by measures of

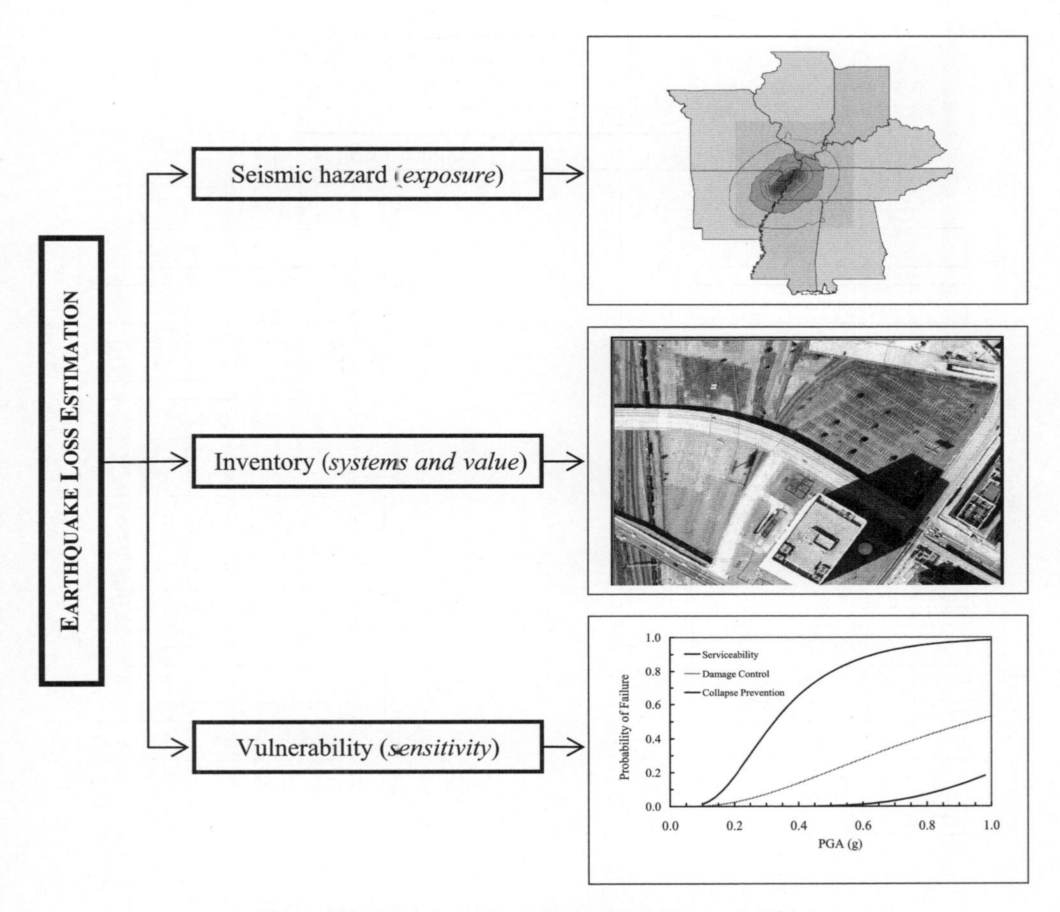

Figure 1.21 Basic components for earthquake loss estimations

retrofitting for example. Vulnerability can also be reduced by other means, such as long-term land-use management and education. Obtaining accurate inventories of exposed assets and their values remains a significant challenge that requires not only technical tools, but also political will and national commitment, especially in regions where private industry holds large inventory data sets that are not in the public domain.

Earthquakes can cause devastating effects in terms of loss of life and livelihood. The destructive potential of earthquakes depends on many factors. The size of an event (expressed by either intensity or magnitude as described in Sections 1.2.1 and 1.2.2), focal depth and epicentral distance, topographical conditions and local geology are important earthquake characteristics. However, the causes of fatalities and extent of damage depend to a great extent on the type of constructions and the density of population present in the area. Earthquakes exact a heavy toll on all aspects of exposed societal systems. They can have several direct and indirect effects as shown in Figure 1.22.

Ground shaking is by far the most important hazard resulting from earthquakes, with some exceptions (e.g. the Asian tsunami of 26 December 2004 with about 280,000 people killed). Structural damage, which is a feature of the primary vertical and lateral load-resisting systems, may vary between light damage and collapse. Non-structural damage consists of the failure or malfunctioning of architectural,

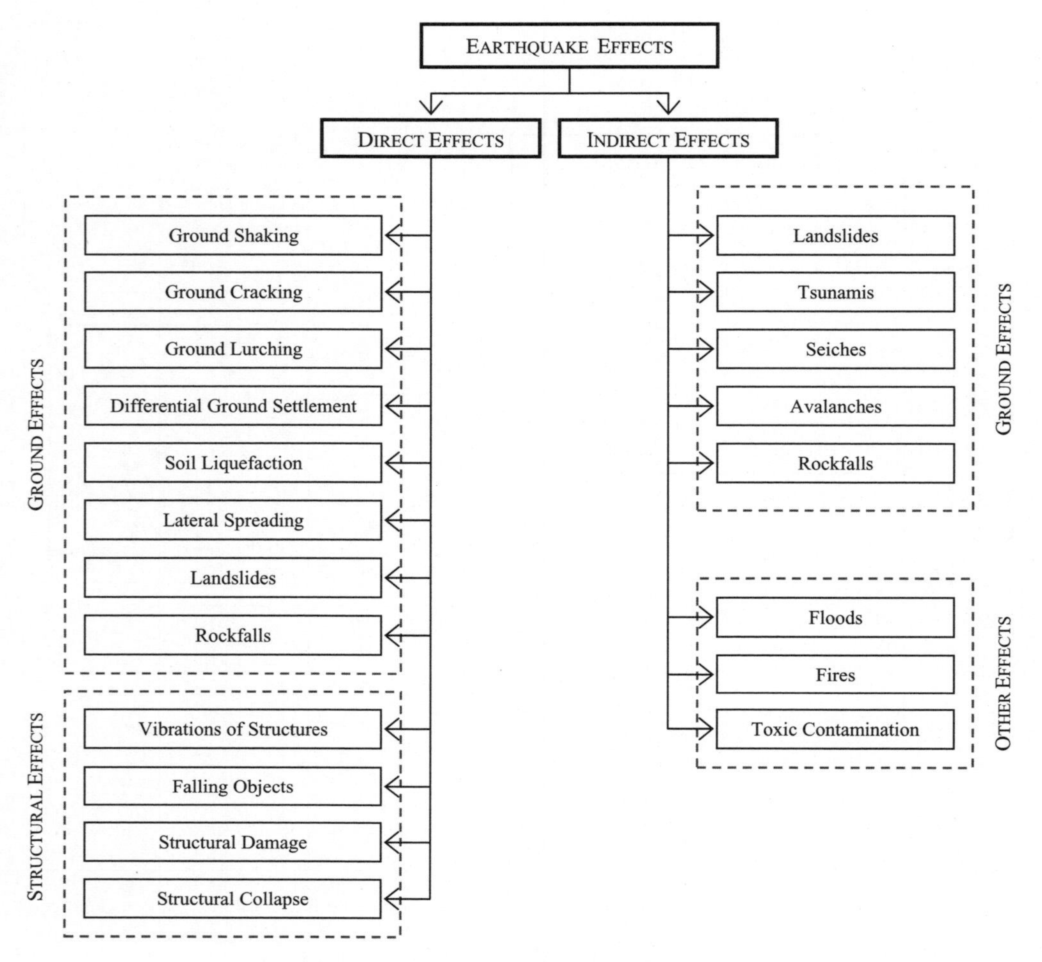

Figure 1.22 Direct and indirect earthquake effects

mechanical and electrical systems, and components within a building. Non-structural damage may lead to large financial losses, as well as pose significant risk to life. Further details on non-structural damage can be found, for example, in ATC (1998) and the reconnaissance reports published in the aftermath of damaging earthquakes.

1.4.1 Damage to Buildings and Lifelines

Extensive structural damage is suffered by buildings, bridges, highways and other lifelines during earthquakes. Seismic vulnerability of structures varies as a function of construction materials and earthquake action-resisting system employed. Typical damage to masonry, reinforced concrete (RC), steel and composite (steel–concrete) buildings is summarized in Table 1.11. Damage is classified under the categories of structural members, connections and systems. It should be noted that in some cases, a pattern of damage is common to different structural members. For example, shear failure may occur

Table 1.11 Typical damage to building structures.

Masonry and reinforced concrete		Steel and composite	
Structural element/ system	Observed damage	Structural element/ system	Observed damage
Beams	Shear failure, concrete cover spalling, reinforcing bar buckling	Beams	Flange and web yielding, local buckling, brittle fracture
Columns	Cracking, crushing, concrete cover spalling, reinforcing bar buckling and pull-out, flexural and shear failure, short-column effect	Columns	Flange yielding, local buckling, brittle fracture, splice failure, member buckling
Connections	Cracking, crushing, reinforcing bar buckling and pull-out, shear failure	Braces	Local and member buckling, brittle fracture
Structural walls and infills	X-shaped cracks, crushing, reinforcing bar buckling, overturning, rocking, sliding	Connections	Yielding, local buckling, brittle fracture, weld cracks, excessive panel deformations, bolt rupture
Foundations	Settlement, reinforcing bar pull-out, rocking, sliding, uplifting	Foundations	Bolt anchorage rupture, weld cracks and fracture, pull-out, excessive base plate deformations
Frames	Soft and weak storeys, excessive residual deformations, distress in diaphragms and connectors, pounding, rocking, uplifting, fall of parapets and brick chimneys	Frames	Soft and weak storeys, excessive residual deformations, distress in diaphragms and connectors, pounding, uplifting

in RC beams and columns. Moreover, local buckling may affect steel beams, columns and braces. Several examples of damage to buildings and bridges are provided in Appendix B, which also contains a detailed discussion of common structural deficiencies observed for steel, concrete and masonry systems. Timber structures have been used extensively especially in Japan, New Zealand and the USA. They include both older non-engineered single-storey family residences and newer two- to three-storey apartment and condominium buildings. Wood-framed buildings are inherently lightweight and flexible; both features are advantageous under earthquake loading conditions (Ambrose and Vergun, 1999). Low- to medium-rise wood buildings, however, have been affected by structural damage during large earthquakes (Bertero, 2000). Observed damage consists of cracking in interior walls and brick chimneys, cracking and collapse of brick veneer on exterior walls. Wooden constructions have often experienced failures similar to those of masonry buildings. Indeed, several partial or total collapses are due to soft and weak storeys, insufficient lateral bracing, and inadequate ties and connections between the components of the building. Inadequate foundation anchorage led to uplifting and sliding in many cases during recent earthquakes in California (e.g. Baker *et al.*, 1989; Andreason and Rose, 1994, among others).

Lifelines are those services that are vital to the health and safety of communities and the functioning of urban and industrial regions. These include electric power, gas, water and wastewater systems. Infrastructures, such as transportation systems (highways and railways), bridges, ports and airports are also classified as lifelines. Damage to lifelines imposes devastating economic effects on the community.

Figure 1.23 Tilting of oxygen tanks *(left)* and brittle fracture of circuit-breaker *(right)* during the 1999 Izmit (Turkey) earthquake (*courtesy of* Dr. Andrew S. Whittaker)

Their seismic performance affects emergency response, short-term and long-term recovery. Broken gas and power lines are serious threats to safety, largely because of risk of fire and explosions. The lack of water also inhibits firefighting efforts. Leaks and rupture of wastewater systems may lead to toxic contamination. For example, during the 1995 Kobe earthquake, the destruction of lifelines and utilities made it impossible for firefighters to reach fires started by broken gas lines (Bukowski and Scawthorn, 1995; Elnashai *et al.*, 1995; Scawthorn *et al.*, 2005). Large sections of the city burned, greatly contributing to the loss of life. Examples of damage to fuel tanks and electrical power systems are displayed in Figure 1.23. Tilting and 'elephant foot' buckling are common failure modes of fluid-holding steel tanks, while brittle fractures are generally observed in substations, which receive and distribute energy to large urban areas. The major causes of outages during past earthquakes were the catastrophic failures of circuit breakers, transformer bushings and disconnected switches at substations. Major damage to lifelines observed during recent earthquakes is summarized in Table 1.12.

The list of types of damage in Table 1.12 is indicative rather than exhaustive, given the variety and complexity of lifeline systems, which are beyond the scope of this book. Several textbooks and manuals that specialize in this subject are available (e.g. Okamoto, 1984; Taylor *et al.*, 1998; Taylor and Van-Marcke, 2002, among others). Reconnaissance reports of damage to lifelines are published by the Earthquake Engineering Research Institute on the Internet (http://www.eeri.org).

1.4.2 Effects on the Ground

Analysis of earthquake-induced damage indicates that ground effects are a serious contributor to damage of the built environment. Local geology and topography influence the travel path and amplification characteristics of seismic waves. For example, natural and artificial unconsolidated foundation materials, such as sediments in river deltas and materials used as landfill, amplify ground motions in comparison to motion measured on consolidated sediments or bedrock. The thickness of unconsolidated soil also affects the ground shaking, as discussed in Section 1.3.2. Quasi-resonance between the underlying soil layers and the structures has led to increased damage during past earthquakes as presented in preceding sections of this book. Ground motions may be amplified by sedimentary layers with various thicknesses and degrees of consolidation.

In addition to direct shaking effects, earthquakes may lead to several forms of ground failure which cause damage to the built environment. For example, the more than $200 million in property losses and a substantial number of deaths in the 1964 Alaska earthquake (M_S = 8.6) were due to earthquake-induced ground failures. Similarly, soil effects were clear in the 1971 San Fernando and the 1989 Loma Prieta earthquakes in California. In particular, many apartment buildings in the Marina District of San

Table 1.12 Typical damage to lifelines.

Highways and railways	Gas and electric power	Water and waste systems	Communication systems
Bending and shear failure of reinforced concrete piers.	Cracks and ruptures in the network	Breakage of pipelines and leakages in the network	Damage to electronic switching systems
Local and overall buckling of steel and composite piers. Brittle fracture of welded components.	Brittle fracture to porcelain components in high-voltage transmission stations and substations	Sloshing and suction damage in metal storage tanks	Damage to phone lines
Pounding and unseating at hinge seats and deck supports.	Damage to switching systems, cranes and tanks in power plants	Elephant foot and shell buckling in metal tanks	Damage to telephone system buildings
Cracks, large gaps and settlements in pavements of highways.	Disruptions of electric power supply	Cracks and leaks in concrete basins	Malfunctioning of computer networks
Rails bending or rupture and train derails.	Fires and explosions due to gas leaks	Malfunctioning of process equipment associated with ground settlement or rocking	Malfunctioning and collapse of transmission towers

Francisco suffered damage because of soil liquefaction. Geological and geotechnical aspects of earthquakes are beyond the scope of this book. A detailed treatment of geotechnical earthquake engineering may be found in Kramer (1996). Failure modes that are of primary concern for structural earthquake engineering are summarized below. Effects of water waves, such as tsunamis (or sea waves) and seiches (or lake waves), are not discussed hereafter. Readers can consult the available literature (e.g. Steinbrugge, 1982; Kanai, 1983, Okamoto, 1984; Bolt, 1999).

(i) Surface Rupture

Rupture of the ground surface may be induced by intense and long shaking as well as fault ruptures. These may generate deep cracks and large gaps (ranging in size from a few metres to several kilometres). Damage by fault rupture is more localized than the widespread damage caused by ground shaking. Nine kilometres of surface rupture along the Nojima fault on Awaji Island was observed in the 1995 earthquake in Japan (Figure 1.24). From left to right along the rupture shown in Figure 1.24, an earthquake-induced landslide covers a road, a fault scarp across a rice paddy and a right-lateral offset in a dirt road. The section of rice paddy to the right has been uplifted by more than 1 m; light damage was experienced by buildings even at very close distances to the fault.

The effects of major fault ruptures can be extreme on structures; buildings can be ripped apart. Cracks and gaps in the ground may also cause serious damage to transportation systems (highways, railways, ports and airports) and underground networks (water, wastewater and gas pipes, electric and telephone cables). Earthquake-induced ground shaking may cause cracking of the ground surface in soft, saturated soil (defined as 'lurching' or 'lurch cracking'). Movements of soil or rock masses at right angles to cliffs and steep slopes occur. Structures founded either in part or whole on such masses may experience significant lateral and vertical deformations.

(ii) Settlement and Uplift

Fault ruptures may cause large vertical movements of the ground. These movements in turn cause severe damage to the foundations of buildings, bridge footings and to underground networks. The collapse of

Figure 1.24 Fault rupture observed on northern Awaji Island during the 1995 Kobe (Japan) earthquake: aerial view with the fault rupture that cuts across the middle of the picture (*left*) and close-up showing both vertical and horizontal offset of the Nojima fault (*right*) (*courtesy of* Geo-Engineering Earthquake Reconnaissance, University of Southern California, Los Angeles)

Figure 1.25 Effects of ground settlements and uplift during the 1999 Kocaeli (Turkey) earthquake: flooding (*left*) and artificial waterfalls (*right*)

several approach structures and abutments of bridges was observed in the San Fernando (1971), Loma Prieta (1989), Northridge (1994) and Kobe (1995) earthquakes. Settlement, tilting and sinking of buildings have been observed in the aftermath of several earthquakes worldwide. Differential ground settlements may cause structural distress. Granular soils are compacted by the ground shaking induced by earthquakes, leading to subsidence. This type of ground movement affects dry, partially saturated and saturated soils with high permeability. Subsidence of 6–7 m was observed during the New Madrid earthquakes (1811–1812) in the Mississippi Valley in the USA. Subsidence of areas close to sea, lakes and river banks may cause flooding of ports, streets and buildings. In some cases, artificial waterfalls may also be generated by settlements and uplifts as shown in Figure 1.25, from the Kocaeli, Turkey, earthquake of 1999.

(iii) Liquefaction

Excessive build-up of pore water pressure during earthquakes may lead to the loss of stiffness and strength of soils. The excessive pore water pressure causes ejection of the soil through holes in the ground, thus creating sand boils. Figure 1.26 shows two examples of liquefaction during the 1998 Adana–Ceyhan (Turkey) and the 2001 Bhuj (India) earthquakes. The ejection of soil causes loss of

Figure 1.26 Sand boils due to the 1998 Adana–Ceyhan (Turkey) earthquake (*left*) and the 2001 Bhuj (India) earthquake (*right*)

Figure 1.27 Collapses due to soil liquefaction: settlement and tilting of buildings in the 1964 Niigata (Japan) earthquake (*left*) and soil boils and cracks at pier foundations of Nishinomiya-ko bridge in the 1995 Kobe (Japan) earthquake (*right*) (*courtesy of* National Information Service for Earthquake Engineering, University of California, Berkeley)

support of foundations and thus structures tilt or sink into the ground. Massive liquefaction-induced damage has been observed in the two Niigata earthquakes of 1964 and 2004, as well as the recent Pisco-Chincha (Peru) earthquake of 2007, as discussed below.

Retaining walls may tilt or break from the fluid pressure of the liquefied zone. Heavy building structures may tilt due to the loss of bearing strength of the underlying soil. During the 1964 Niigata, Japan, earthquake ($M_S = 7.5$), four-storey apartment buildings tilted 60 degrees on liquefied soils as shown in Figure 1.27. Similarly, in the 1989 Loma Prieta earthquake, liquefaction of the soils and debris used to fill in a lagoon caused major subsidence, fracturing and horizontal sliding of the ground surface in the Marina district in San Francisco.

Soil liquefaction may cause the floating to ground surface of pile foundations with low axial loads and underground lightweight storage tanks. In Kobe, lateral spreading damaged the pile foundations of several buildings and bridges (Figure 1.27) because of horizontal movements. Quay walls and sea defences in the port of Kobe were also affected by soil liquefaction.

(iv) Landslides

Landslides include several types of ground failure and movement, such as rockfalls, deep failure of slopes and shallow debris flows. These failures are generated by the loss of shear strength in the soil. Landslides triggered by earthquakes sometimes cause more destruction than the earthquakes themselves. Immediate dangers from landslides are the destruction of buildings on or in the vicinity of the slopes with possible fatalities as rocks, mud and water slide downhill or downstream. Electrical, water, gas and sewage lines may be broken by landslides. The size of the area affected by earthquake-induced landslides depends on the magnitude of the earthquake, its focal depth, the topography and geologic conditions near the causative fault, and the amplitude, frequency content and duration of ground shaking. During the 1964 Alaska earthquake, shock-induced landslides devastated the Turnagain Heights residential development and many downtown areas in Anchorage. One of the most spectacular landslides observed, involving about 9.6 million cubic metres of soil, took place in the Anchorage area. The scale of such landslides on natural slopes can be large enough to devastate entire villages or towns, such as the Huascaran Avalanche triggered by the Peru earthquake (1970, $M_w = 7.8$). Most of the more than 1,000 landslides and rockfalls occurred in the epicentral zone in the Santa Cruz Mountains during the 1989 Loma Prieta earthquake. One slide, on State Highway 17, disrupted traffic for about 1 month. In the 1994 Northridge earthquake, landslides that occurred in Santa Monica, along the Pacific Coast Highway, caused damage to several family houses built on the cliffs overlooking the ocean. This is shown in Figure 1.28. Relatively few landslides were triggered by the Hyogo-ken Nanbu earthquake in Japan. This is partly due to the fact that the earthquake occurred during the dry season. Landslides are often triggered by rainfall pressure generated inside fractured ground.

In the Kashmir earthquake of 8 October 2005, land-sliding and critical slope stability was a multi-scale problem that ranged from limited sloughing of a superficial nature to a scale that encompassed entire mountain sides (Durrani *et al.*, 2005). The land-sliding problem in the mountains of Azad Jammu and Kashmir and North West Frontier Province, Pakistan, has similarities to land-sliding that occurred in the mountains of Central Taiwan due to the 1999 Chi-Chi earthquake. Figure 1.28 shows a large-scale landslide in the Neela Dandi Mountain to the north of Muzaffarabad. The satellite image shows that the landslide blocked the Jhelum River.

Problem 1.5

The 17 August 1999 Kocaeli ($M_w = 7.4$) and 12 November 1999 Düzce ($M_w = 7.2$) earthquakes were the largest natural disasters of the twentieth century in Turkey after the 1939 Erzincan earthquake. These earthquakes caused severe damage and collapse especially of building structures. Figure 1.29 shows typical damage observed in the city of Izmit (Kocaeli earthquake). The collapse of a multi-storey reinforced concrete building in Izmit (Düzce earthquake) is also provided in the figure. Correlate the surveyed failure with the earthquake-induced ground effects illustrated in Section 1.4.2.

1.4.3 Human and Financial Losses

During the twentieth century, over 1,200 destructive earthquakes occurred worldwide and caused damage estimated at more than $10 billion (Coburn and Spence, 2002). If these costs are averaged over the century, annual losses are about $10 billion. Monetary losses from earthquakes are increasing rapidly. Between 1990 and 1999, annual loss rates were estimated at $20 billion, twice the average

Figure 1.28 Effects of a large landslide in Santa Monica in the 1994 Northridge (California) earthquake (*top*) (*courtesy of* National Information Service for Earthquake Engineering, University of California, Berkeley) and satellite view of extensive land-sliding during the 2005 Kashmir (Pakistan) earthquake in the Neela Dandi Mountain (*bottom*) (*courtesy of* Mid-America Earthquake Center, University of Illinois at Urbana-Champaign)

twentieth century annual losses. The Federal Emergency Management Agency released a study (FEMA 366, 2001) estimating annualized earthquake losses to the national building stocks in the USA at $4.4 billion, with California, Oregon and Washington accounting for $3.3 billion of the total estimated amount. An update of the above landmark study was released in 2006 (www.fema.gov) to include in the estimation of the annualized losses three additional features of earthquake risk analysis, i.e. casualties, debris and shelter. In the latter study, it is estimated that the annualized earthquake losses to the national building stock are $5.3 billion and about 65% is concentrated in the State of California. The largest earthquake in modern times in the USA was the 1964 Alaska Earthquake, measuring 8.4 on the Richter scale. The earthquake caused $311 million in damage and 115 fatalities. In a historical context, the largest recorded earthquakes in the contiguous USA are the New Madrid earthquakes of

Figure 1.29 Damage observed during the 17 August 1999 Kocaeli (*left*) and the 12 November 1999 Düzce (*right*) earthquakes in Turkey

1811 and 1812. In the USA, 39 out of 50 states (nearly 80%) are at risk from damaging earthquakes. The Central and Eastern States in the USA now recognize earthquakes as a major threat. In particular, the eight central States of Illinois, Arkansas, Indiana, Tennessee, Kentucky, Mississippi, Alabama and Missouri have dedicated considerable resources to work with FEMA and other earthquake engineering organizations to assess the possible impact of earthquakes and to mitigate as well as plan for response and recovery from their effects.

With regard to loss of life, on average 10,000 people per year were killed by earthquakes between 1900 and 1999 (Bolt, 1999). In 2001, three major earthquakes in Bhuj (India, $M_S = 7.9$), El Salvador ($M_S = 7.6$) and Arequipa (Peru, $M_S = 8.4$) caused more than 26,000 casualties. The Bam (Iran, $M_S = 6.6$) and Sumatra (Indian Ocean, $M_w = 9.3$) earthquakes, which occurred in 2003 and 2004, both on 26 December, caused more than 26,000 and 280,000 deaths, respectively. The Kashmir earthquake of 8 October 2005 caused over 85,000 deaths. The human death toll due to earthquakes between 1906 and 2005 is given in Figure 1.30 (www.usgs.gov). Over this 108-year period, deaths due to earthquakes totalled about 1.8 million. China accounted for more than 30% of all fatalities.

Figure 1.31 compares the human death toll due to earthquakes with that caused by other natural hazards (www.usgs.gov). It is observed from the figure that earthquakes rank second after floods; earthquakes account for about 3.6 million fatalities. If the death toll caused by tsunamis were added to that caused by earthquakes, the total figure would amount to around 4.5 million.

Monetary losses due to collapsed buildings and lifeline damage are substantial. Furthermore, the economic impact of earthquakes is increasing due to the expansion of urban development and the higher cost of construction. For example, the 1994 Northridge earthquake, which is said to be the most costly natural disaster in the history of the USA, caused \$30 billion in damage and \$800 billion replacement value on taxable property (Goltz, 1994). In this event 25,000 dwellings were declared uninhabitable, while buildings with severe and moderate damage numbered 7,000 and 22,000, respectively. Unexpected brittle fractures were detected in more than 100 steel-framed buildings as illustrated in Appendix B. Damage to the transportation system was estimated at \$1.8 billion and property loss at \$6.0 billion. In the above-mentioned earthquake, the most severe damage occurred to non-retrofitted structures, designed in compliance with seismic regulations issued in the 1970s.

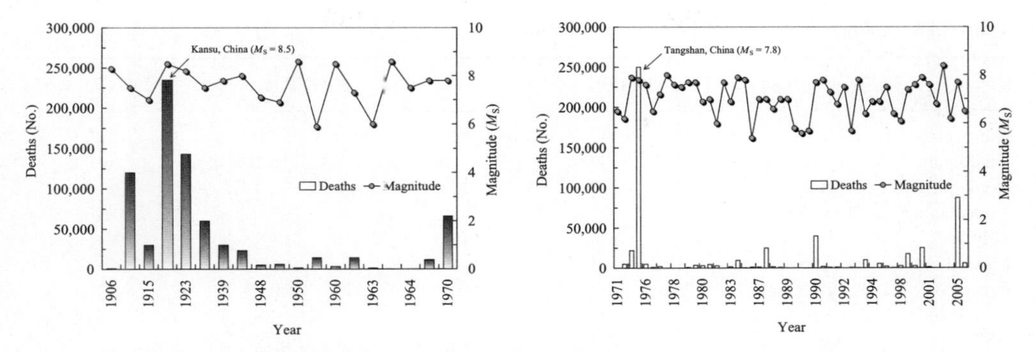

Figure 1.30 Human death toll due to earthquakes: 1906–1970 (*left*) and 1971–2005 (*right*)

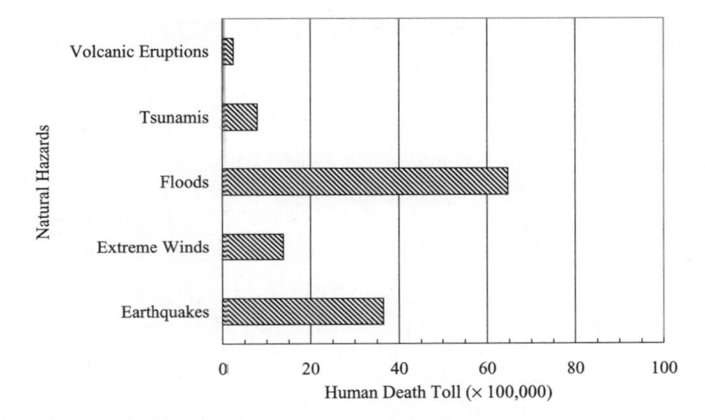

Figure 1.31 Human death toll caused by major natural hazards

Several reconnaissance reports have concluded that building collapses caused 75% of earthquake fatalities during the last century. Other major causes of death were fires and gas explosions, tsunamis, rockfalls and landslides. In the Loma Prieta earthquake, 42 out 63 deaths (about 63%) were attributed to bridge failures. However, in the 1995 Kobe earthquake in Japan, 73% of the deaths were caused by collapsed houses. The likelihood of the collapse of multi-storey RC structures in developing countries, where the quality of construction remains relatively substandard, is high.

Earthquake damage resulting in the collapse of monuments, historical places of worship and stately buildings represents an irreplaceable loss in terms of cultural heritage, while their restoration costs exceed by far the gross national product (GNP) of many affected nations. The expense of reconstructing the world famous vault of the Basilica at Assisi (Italy) with its early Renaissance frescoes caused serious repercussions for the national economy after 1997. Even more problematic are the implications for important heritage sites in seismically active developing countries. The earthquakes of Gujarat (India), Bam (Iran), Arequipa (Peru) and Yogyakarta (Indonesia), have caused major damage to invaluable historical sites that may or may not be restored over a number of years and at an extremely high cost.

One of the most severe consequences of earthquakes is the cost of recovery and reconstruction. It is instructive to note, however, that the absolute financial loss is less critical to an economy than the loss

Table 1.13 Earthquake financial losses (*after* Coburn and Spence, 2002).

Country	Earthquake	Year	Loss ($ bn)	GNP ($ bn)	Loss (% GNP)
Nicaragua	Managua	1972	2.0	5.0	40.0
Guatemala	Guatemala City	1976	1.1	6.1	18.0
Romania	Bucharest	1977	0.8	26.7	3.0
Yugoslavia	Montenegro	1979	2.2	22.0	10.0
Italy	Campania	1980	45.0	661.8	6.8
Mexico	Mexico City	1985	5.0	166.7	3.0
Greece	Kalamata	1986	0.8	40.0	2.0
El Salvador	San Salvador	1986	1.5	4.8	31.0
USSR	Armenia	1988	17.0	566.7	3.0
Iran	Manjil	1990	7.2	100.0	7.2

Key: GNP = gross national product.

as a percentage of the GNP. For example, in some 6 to 8 seconds, Nicaragua lost 40% of its GNP due to the 1972 Managua earthquake (Table 1.13), while the 800% higher bill ($17 billion versus $2 billion) from the Yerivan, Armenia earthquake constituted only 3% of the USSR's GNP (Elnashai, 2002).

The 'business interruption' element of earthquake impact has emerged lately as a major concern to industry and hence to communities. This is the effect of largely non-structural building damage (e.g. suspended light fixtures, interior partitions and exterior cladding), which affects businesses adversely, in turn leading to financial disruption and hardship (Miranda and Aslani, 2003). In several countries, such as the Mediterranean regions and Central America, where tourism is a vital industry, major economic losses have resulted from damage to hotels and negative publicity due to earthquakes. Another aspect of the economic impact is the 'loss of market share', which results from interruption to production in industrial facilities and difficulties in reclaiming the share of the market that the affected business previously held.

The consequences of direct financial losses, business interruption, and loss of market share on communities and industry have led major multinationals to create risk management departments in an attempt not only to reduce their exposure, but also to minimize insurance premiums. Global seismic risk management is therefore one of the highest growth areas in industry.

References

Abrahamson, N.A. (1991). Spatial coherency of ground motion from the SMART-1 array. *Geotechnical News*, 9(1), 31–34.

AIJ (1995). *Performance of Steel Buildings during the 1995 Hyogoken-Nanbu Earthquake*. Architectural Institute of Japan, Tokyo, Japan.

Ambraseys, N.N. (1959). A note on the response of an elastic overburden of varying rigidity to an arbitrary ground motion. *Bulletin of the Seismological Society of America*, 49(3), 211–220.

Ambraseys, N.N. (1985). Intensity-attenuation and magnitude–intensity relationships for Northwest European earthquakes. *Earthquake Engineering and Structural Dynamics*, 13(6), 733–778.

Ambraseys, N.N. (1989). *Long-term Seismic Hazard in the Eastern Mediterranean Region. Geohazards, Natural and Man-Made*. Chapman and Hall, New York, NY, USA.

Ambraseys, N.N. (2006). Comparison of frequency of occurrence of earthquakes with slip rates from long-term seismicity data: The cases of Gulf of Corinth, Sea of Marmara and Dead Sea Fault Zone. *Geophysical Journal International*, 165(2), 516–526.

Ambraseys, N.N. and Finkel, C. (1986). Corpus of Isoseismal Maps of Eastern Mediterranean Region. Engineering Seismology and Earthquake Engineering, Research Report No. ESEE 86/8, Imperial College, London, UK.

Ambraseys, N.N. and Melville, C.P. (1982). *A History of Persian Earthquakes*. Cambridge University Press, Cambridge, UK.

Ambrose, J. and Vergun, D. (1999). *Design for Earthquakes*. John Wiley & Sons, New York, NY, USA.

Andreason, K. and Rose, J.D. (1994). Northridge, California earthquake: Structural performance of buildings in San Fernando Valley, California (January 17, 1994). American Plywood Association, APA Report No. T94-5, Tacoma, Washington, DC, USA.

Applied Technology Council (ATC) (1998). Design, Retrofit and Performance of Nonstructural components. Proceedings of the ATC Seminar, Report No. 2-1, Redwood City, CA.

Baker, W.A., Brown, D.H. and Tissel, J.R. (1989). Loma Prieta earthquake, San Francisco Bay Area, October 17, 1989. American Plywood Association, APA Report No. T89-28, Tacoma, Washington, DC, USA.

Benioff, H. (1955). Mechanism and strain characteristics of the White Wolf fault as indicated by the aftershock sequence. In *Earthquake in Kern County, California During 1955*, Oakeshott, G.B. Editor, California Division of Mines, Bulletin No. 171CA, USA.

Ben-Menachem, A. (1961). Radiation patterns of seismic surface waves from finite moving sources. *Bulletin of the Seismological Society of America*, 51(2), 401–435.

Bertero, V.V. (2000). *Introduction to Earthquake Engineering*. (http://nisee.berkeley.edu/ebooks/).

Bolt, B.A. (1999). *Earthquakes*. 4th Edition, W.H. Freeman and Company, New York, NY, USA.

Bonilla, M.G., Mark, R.K. and Lienkaemper, J.J. (1984). Statistical relations among earthquake magnitude, surface rupture length, and surface fault displacement. *Bulletin of the Seismological Society of America*, 76(6), 2379–2411.

Broderick, B.M., Elnashai, A.S., Ambraseys, N.N., Barr, J.M., Goodfellow, R.G. and Higazy, E.M. (1994). The Northridge (California) Earthquake of 17 January 1994: Observations, Strong Motion and Correlative Response Analysis. Engineering Seismology and Earthquake Engineering, Research Report No. ESEE 94/4, Imperial College, London, UK.

Bukowski, W. and Scawthorn, C. (1995). Kobe Reconnaissance Report. *Fire. Earthquake Spectra, Supplement A*, 11(S1), 33–40. Earthquake Engineering Research Institute.

Clough, R.W. and Penzien, J. (1993). *Dynamics of Structures*. 2nd Edition, McGraw-Hill, New York, NY, USA.

Coburn, A. and Spence, R. (2002). *Earthquake Protection*. 2nd Edition, John Wiley & Sons, Chichester, UK.

Der Kiureghian, A. (1996). A coherency model for spatially varying ground motions. *Earthquake Engineering and Structural Dynamics*, 25(1), 99–111.

Dewey, J.F. (1972). Plate Tectonics. *Scientific American*, 226(5), 56–68.

Durrani, A.J., Elnashai, A.S., Hashah, Y.M.A., Kim, S.J. and Masud, A. (2005). The Kashmir earthquake of October 8, 2005. A quick look report. MAE Center Report No. 05-04, University of Illinois at Urbana-Champaign, IL, USA.

Elnashai, A.S. (2002). A very brief history of earthquake engineering with emphasis on developments in and from the British Isles. *Chaos Solitons & Fractals*, 13(5), 967–972.

Elnashai, A.S., Bommer, J.J., Baron, I., Salama, A.I. and Lee, D. (1995). Selected Engineering Seismology and Structural Engineering Studies of the Hyogo-ken Nanbu (Kobe, Japan) Earthquake of 17 January 1995. Engineering Seismology and Earthquake Engineering, Report No. ESEE/95-2, Imperial College, London, UK.

Elnashai, A.S., Bommer, J.J. and Martinez-Pereira, A. (1998). Engineering implications of strong motion records from recent earthquakes. Proceedings of the 11th European Conference on Earthquake Engineering, Paris, France.

Federal Emergency Management Agency (FEMA) (2001). HAZUS 1999 Estimated Annualized Earthquake Losses for United States. Report No. FEMA 366, Washington, DC, USA.

Finn, W.D.L. (1991). Geotechnical engineering aspects of microzonation. Proceedings of the 4th International Conference on Seismic Zonation, Stanford, CA, USA, pp. 199–259.

Goltz, J.D. (1994). The Northridge, California Earthquake of January 17, 1994: General Reconnaissance Report. National Centre for Earthquake Engineering Research, Report No. NCEER-94-0005, Buffalo, NY, USA.

Gutenberg, B. and Richter, C.F. (1936). Magnitude and energy of earthquakes. *Science*, 83(2147), 183–185.

Gutenberg, B. and Richter, C.F. (1956). Earthquake magnitude. Intensity, energy and acceleration. *Bulletin of the Seismological Society of America*, 46(1), 105–145.

Haricharan, R.S. and Vanmarcke, E.H. (1986). Stochastic variation of earthquake ground motion in space and time. *Journal of Engineering Mechanics, ASCE*, 112(2), 154–174.

Housner, G.W. (1965). Intensity of earthquake ground shaking near the causative fault. Proceedings of the 3rd World Conference on Earthquake Engineering, Wellington, New Zealand, Vol. 1, pp. 95–115.

Housner, G.W. (1973). Important features of earthquake ground motions. Proceedings of the 5th World Conference on Earthquake Engineering, Rome, Italy, Vol. 1, pp. CLIX–CLXVIII.

Kanai, K. (1983). *Engineering Seismology*. University of Tokyo Press, Tokyo, Japan.

Kanamori, H. (1977). The energy release in great earthquakes. *Journal of Geophysical Research*, 82(20), 2981–2987.

Kasahara, K. (1981). *Earthquake Mechanics*, Cambridge University Press, Cambridge, UK.

Kramer, S.L. (1996). *Geotechnical Earthquake Engineering*. Prentice Hall, Upper Saddle River, NJ, USA.

Krinitzsky, E. (1974). Fault assessment in earthquake engineering. Army Engineering Waterways Experiment Station, Miscellaneous Paper S-73-1, Vicksburg, MS, USA.

Lee, W.H.K., Kanamori, H., Jennings, P.C. and Kisslinger, C. (2003). *International Handbook of Earthquake and Engineering Seismology*. Academic Press, San Diego, CA, USA.

Luco, J.E. and Wong, H.L. (1986). Response of a rigid foundation to a spatially random ground motion. *Earthquake Engineering and Structural Dynamics*, 14 (6), 891–908.

Mallet, R. (1862). *Great Neapolitan Earthquake of 1857. The First Principles of Observational Seismology*. Vol. 1, Chapman and Hall, London, UK.

Manolis, G.D. and Koliopoulos, P.K. (2001). *Stochastic Structural Dynamics in Earthquake Engineering. Advances in Earthquake Engineering Series*. WIT Press, Southampton, UK.

Mark, R.K. and Bonilla, M.G. (1977). Regression analysis of earthquake magnitude and surface fault length using the 1970 data of Bonilla and Buchanan. U.S. Geological Survey, USGS Open File Report 77–164. Menlo Park, CA, USA.

Miranda, E. and Aslani, H. (2003). Building-specific loss estimation methodology. PEER Report No. 2003–03, Pacific Earthquake Engineering Research Center, University of California at Berkeley, Berkeley, CA, USA.

Nuttli, O.W. (1983). Average seismic source-parameter relations for mid-plate earthquakes. *Bulletin of the Seismological Society of America*, 73(2), 519–535.

Okamoto, S. (1984). *Introduction to Earthquake Engineering*. 2nd Edition, University of Tokyo Press, Tokyo, Japan.

Oliveira, C.S., Hao, H. and Penzien, J. (1991). Ground modelling for multiple input structural analysis. *Structural Safety*, 10(1–3), 79–93.

Osteraas, J. and Krawinkler, H. (1990). The Mexico earthquake of September 19, 1985 – Behavior of steel buildings. *Earthquake Spectra*, 5(1), 51–88.

Plafker, G. and Galloway, J.P. (1989). Lessons learned from the Loma Prieta California, Earthquake of October 17, 1989. USGS Circular 1045.

Purcaru, G. and Berckhemer, H. (1978). A magnitude scale for very long earthquakes. *Tectonophysics*, 49, 189–198.

Reiter, L. (1990). *Earthquake Hazard Analysis: Issues and Insights*. Columbia University Press, New York, NY, USA.

Richter, C.F. (1935). An instrumental earthquake scale. *Bulletin of the Seismological Society of America*, 25(1), 1–32.

Richter, C.F. (1958). *Elementary Seismology*. W.H. Freeman and Company, San Francisco, CA, USA.

Sano, T. and Pugliese, A. (1999). Parametric study on topographic effects in seismic soil amplification. Proceedings of 'Advances in Earthquake Engineering', Earthquake Resistant Engineering Structures II, Vol. 4, pp. 321–330.

Scawthorn, C., Eidinger, J.M. and Schiff, A.J. (2005). Fire Following Engineering. American Society of Civil Engineers, Technical Council on Lifeline Earthquake Engineering, Monograph No. 26, ASCE/NFPA, Reston, VA, USA.

Scholz, C.H. (1990). *The Mechanics of Earthquakes and Faulting*. Cambridge University Press, Cambridge, UK.

Scholz, C.H., Aviles, C. and Wesnousky, S. (1986). Scaling differences between large intraplate and interpolate earthquakes. *Bulletin of the Seismological Society of America*, 76(1), 65–70.

Seed, H.B., Idriss, I.M. and Kiefer, F.S. (1969). Characteristics of rock motions during earthquakes. *Journal of the Soil Mechanics and Foundations Division, ASCE*, 95(SM5), 1199–1218.

Silva, W.J. (1988). Soil response to earthquake ground motion. Electric Power Research Institute, Report No. EPRI-NP-5747, Palo Alto, CA, USA.

Singh, J.P. (1985). Earthquake ground motions: Implications for designing structures and reconciling structural damage. *Earthquake Spectra*, 1(2), 239–270.

Slemmons, D.B. (1977). State-of-the-art for assessing earthquake hazards in the United States: Report 6, faults and earthquake magnitude. Miscellaneous Paper S-173-1, US Army Corps of Engineers, Waterways Experiment Station, Vicksburg, MS, USA.

Somerville, P., McLaren, J.P., Mrinal, K.S. and Helmberg, D.V. (1991). The influence of site conditions on the spatial incoherence of ground motions. *Structural Safety*, 10(1–3), 1–13.

Somerville, P.G., Smith, N.F., Graves, R.W. and Abrahamson N.A. (1997). Modification of empirical strong motion attenuation relations to include the amplitude and duration effects of rupture directivity. *Seismological Research Letters*, 68(1), 199–222.

Sponheuer, W. (1960). *Methoden zur Herdtiefenbestimung in der Madkroseismik*. Akademic Verlag, East Berlin, Germany.

Steinbrugge, K.V. (1982). *Earthquakes, Volcanoes and Tsunamis: An Anatomy of Hazards*. Skandia America Group, New York, NY, USA.

Stewart, J.P., Chiou, S.J., Bray, J.O., Graves, R.W., Somerville, P.G. and Abrahamson, N.A. (2001). Ground motion evaluation procedures for performance-based design. Pacific Earthquake Engineering Center, Report No. PEER 2001/9, University of California, Berkeley, CA, USA.

Street, R.L. and Turcotte, F.T. (1977). A study of northeastern North America spectral moments, magnitudes and intensities. *Bulletin of the Seismological Society of America*, 67(3), 599–614.

Taylor, C. and VanMarcke, E. (2002). Acceptable risk processes: Lifelines and natural hazards. Technical Council on Lifeline Earthquake Engineering, ASCE, Monograph No. 21, Reston, VA, USA.

Taylor, C., Mittler, E. and LeVal, L. (1998). Overcoming Barriers: Lifeline Seismic Improvement Programs. Technical Council on Lifeline Earthquake Engineering, ASCE, Monograph No. 13, Reston, VA, USA.

Tocher, D. (1958). Earthquake energy and ground breakage. *Bulletin of the Seismological Society of America*, 48(1), 147–153.

Udias, A. (1999). *Principles of Seismology*. Cambridge University Press, Cambridge, UK.

Wells, D. L. and Coppersmith, K.J. (1994). New empirical relationships among magnitude, rupture length, rupture width, rupture area, and surface displacement. *Bulletin of the Seismological Society of America*, 84(4), 974–1002.

Wyss, M. (1979). Estimating maximum expectable magnitude of earthquakes from fault dimensions. *Geology*, 7(7), 336–340.

2

Response of Structures

2.1 General

The objectives of this chapter are to address and unify definitions of the fundamental response parameters considered to be most influential in structural earthquake engineering, and to highlight the factors influencing these fundamental response parameters. The parameters postulated in this book to be the basic building blocks of understanding and controlling earthquake response of structures are Stiffness, Strength and Ductility. As presented in the following chapter, stiffness is the most pertinent parameter in responding to the requirements of serviceability under the small frequent earthquake. In an analogous manner, strength is utilized to control the level of inelasticity under the medium-sized infrequent earthquake, hence it maps onto the damage control limit state. Finally, collapse prevention under the large rare earthquake is most affected by ductility, thus completing the hazard-limit state-response parameter triads discussed in Section 1.4 of Chapter 1. The material in this chapter is presented in a strictly hierarchical framework of material, section, member, connection and system characteristics most influential in affecting stiffness, strength and ductility. The chapter concludes with a treatment of the two important quantities of overstrength and damping, which are consequential to the three fundamental parameters discussed above. This chapter therefore articulates the general guidelines of Chapter 1 into operational quantities, and prepares for a thorough understanding of Chapters 3 and 4 on earthquake strong-motion and structural analysis tools, respectively.

2.2 Conceptual Framework

2.2.1 Definitions

In order to establish a common nomenclature and in recognition of the plethora of conflicting definitions in the literature, generic and rigorous definitions of the main terms used in this book are given herein. Focus is placed on the three response characteristics used hereafter as the most important parameters that describe the behaviour of structures and their foundations when subjected to earthquakes. These are stiffness, strength (or capacity) and ductility. Prior to defining the three quantities, it is instructive to reiterate the definition of two more fundamental quantities, namely 'action' and 'deformation'. The former is used in this book to indicate stress resultants of all types, while the latter is used to indicate strain resultants. The three quantities of stiffness, strength and ductility are treated in detail in subse-

Fundamentals of Earthquake Engineering Amr S. Elnashai and Luigi Di Sarno

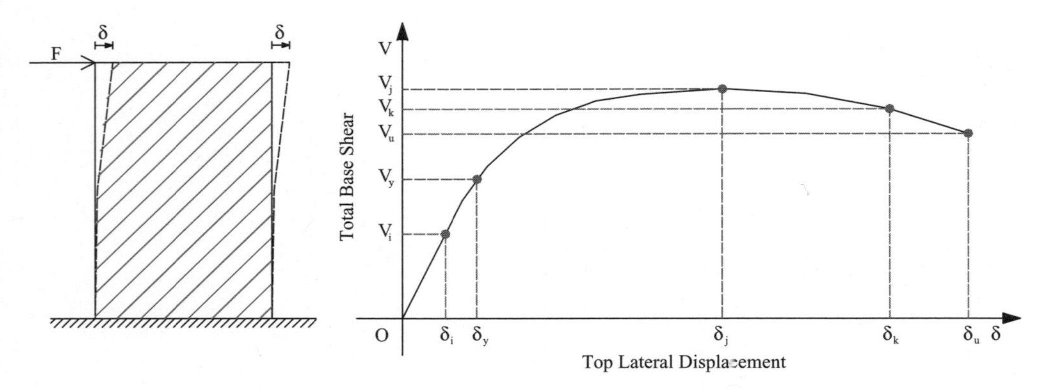

Figure 2.1 Typical response curve for structural systems subjected to horizontal loads

quent sections of this chapter, and are succinctly defined hereafter in order to permit a rational discussion of the conceptual framework for the whole text.

Stiffness is the ability of a component or an assembly of components to resist deformations when subjected to actions, as shown in Figure 2.1. It is expressed as the ratio between action and deformation at a given level of either of the two quantities and the corresponding value of the other. Therefore, stiffness is not a constant value. In Figure 2.1, K_i is the stiffness at a required deformation δ_i and corresponds to force resistance V_i. If increments or first derivatives of actions and deformations are used, the ensuing stiffness is the tangent value. If total actions and deformations are used, the ensuing stiffness is the secant value.

Strength is the capacity of a component or an assembly of components for load resistance at a given response station. It is also not a constant value, as shown in Figure 2.1. In this book, the term 'capacity' is preferred to the term 'strength' to represent both action resistance and the ability to endure deformation, or deformation capacity. In the figure, V_j and V_k are the force capacities corresponding to δ_j and δ_k, respectively. V_y, referred to as the yield strength, corresponds to the yield displacement δ_y, which is required for ductility calculations.

Ductility is the ability of a component or an assembly of components to deform beyond the elastic limit, as shown in Figure 2.1, and is expressed as the ratio between a maximum value of a deformation quantity and the same quantity at the yield limit state. In the figure, the displacement ductility μ is the ratio between the maximum or ultimate displacement δ_u and the yield displacement δ_y.

Demand is the action or deformation imposed on a component or an assembly of components when subjected to earthquake ground motion. This demand is not constant. It continuously varies as the structural characteristics vary during inelastic response. It also varies with the characteristics of the input motion.

Supply is the action or deformation capacity of a component or an assembly of components when subjected to earthquake ground motion. Therefore, the supply represents the response of the structure to the demand. It may continuously vary as the structural characteristics change during inelastic response. It also varies with the characteristics of the input motion. For inelastic systems, demand and supply are coupled.

2.2.2 Strength-versus Ductility-Based Response

Traditional force-based seismic design has relied on force capacity to resist the earthquake effects expressed as a set of horizontal actions defined as a proportion of the weight of the structure. In the

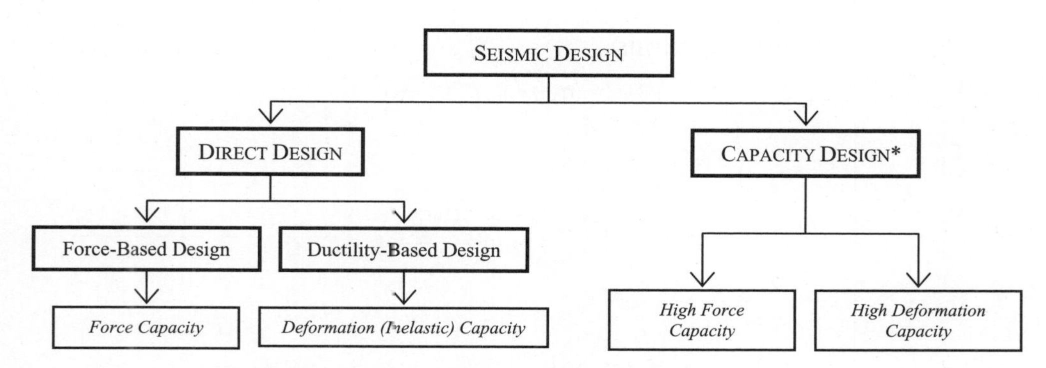

Figure 2.2 Different approaches to seismic design
Key: * = For capacity design high force and high deformation coexist)

past 20 to 30 years, there has been a tendency to substitute ductility (or inelastic deformation capacity) for strength (or force capacity). The latter approach was developed in recognition of the great uncertainty associated with estimating seismic demand. A ductility-designed structure is significantly less sensitive to unexpected increase in the force demand imposed on it than its strength-designed counterpart. In general, ductility-based structures are lighter and use less material, but more workmanship. A different way of dimensioning earthquake-resistant structures is 'capacity design' (Figure 2.2). Capacity design employs a mixture of members with high load capacity and members with high inelastic deformation capacity to optimize the response of the structural system. This is achieved by identifying a failure mechanism, the members and regions responsible for its development, and providing these members and regions with adequate ductility. In parallel, the rest of the structure is protected by providing it with adequate strength to ensure nearly elastic behaviour. The opposite of 'capacity design' is 'direct design', which is the dimensioning of individual components to resist the locally evaluated actions with no due consideration to the action-redistribution effects in the system as a whole. Direct design can be either ductility-based or strength-based. Capacity design, on the other hand, is based on both strength and ductility of components.

The difference between direct and capacity design is depicted in Figure 2.3. The maximum effects of both horizontal and vertical loads are computed through structural analysis.

In the direct approach, all design actions are the 'applied' quantities, calculated from the combination of static and seismic loads. In capacity design, one set of actions represents the ultimate capacity of the members, regions or mechanisms that are responsible for energy absorption, while the rest of the design actions are calculated to maintain equilibrium. In the figure, M_{c12}, M_{c11} and M_b are the moments on the two columns and the beam, respectively, evaluated from the applied actions. M_{bmax} is the maximum capacity of the beam, the member responsible for energy absorption in the weak-beam strong-column design approach, taking into account various sources of overstrength (e.g. unintentional increase in material properties, rounding-off of member or reinforcement dimensions, post-yield hardening, etc.). Design actions M'_{c12} and M'_{c11} are the product of M_{c12} and M_{c11}, and the ratio M_{bmax}/M_b. They are there for 'applied' actions.

2.2.3 Member-versus System-Level Consideration

Only in recent years has the earthquake engineering community taken the overall system response fully into account. Conventional seismic design recommends the dimensioning of members to resist the actions emanating from structural analysis where the dead and live loads are applied alongside a factored value of equivalent horizontal earthquake actions, as shown for example in Figure 2.3. The

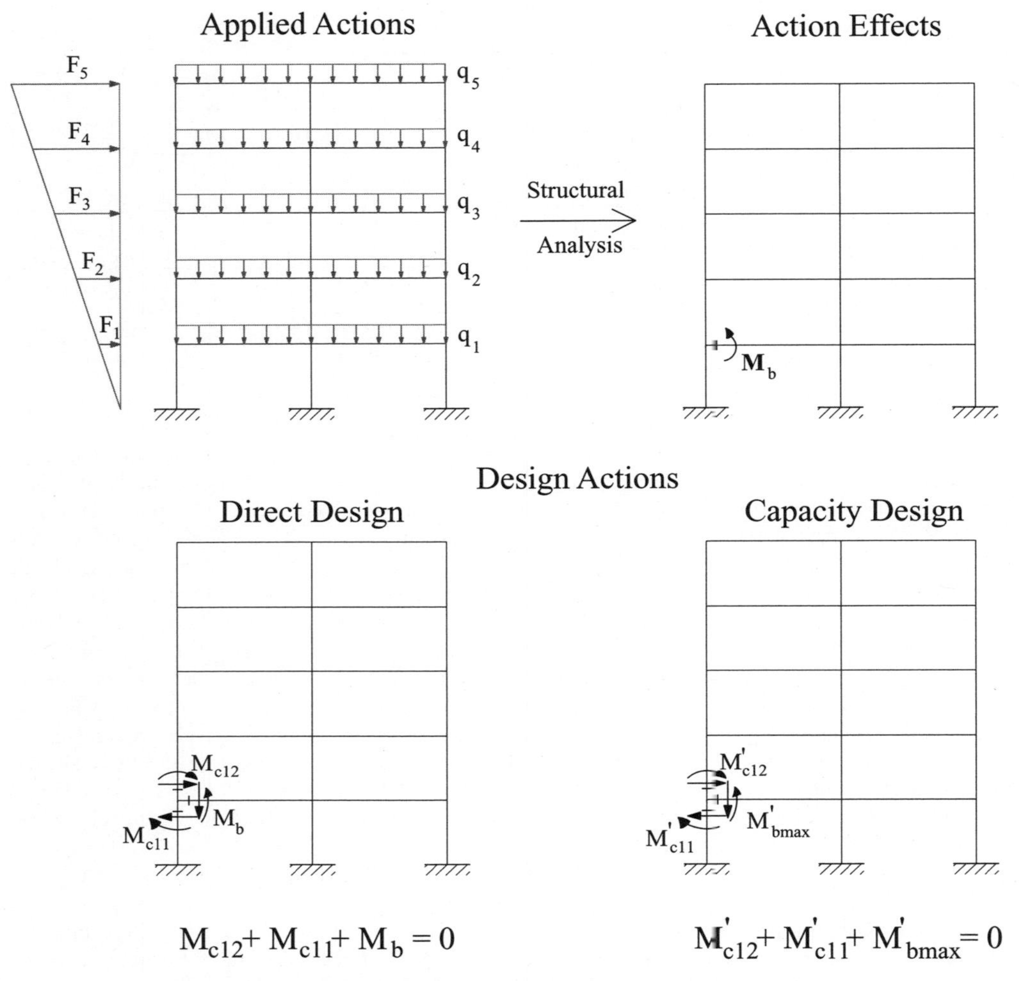

Figure 2.3 Comparison between direct and capacity design

interaction between member and system in structural earthquake engineering is complex, but its understanding is essential for effective seismic design. The hierarchical relationship between local and global structural response is provided in Figure 2.4; it is applicable to the three fundamental quantities of stiffness, strength and ductility.

Quantitative expressions linking local action–deformation characteristics to global response quantities can only be derived under idealized conditions and with a number of simplifying assumptions that limit their scope of application. A qualitative appreciation of the local–global interaction, and application to specific cases, are central to controlled seismic performance. The conceptual framework of system response may also be used to assess seismic demand and supply. Yield of one member or more does not necessarily feature in the system action–deformation response, and hence is not necessarily a system limit state.

The chain system proposed by Paulay and Priestley (1992) to describe the rationale behind the capacity design philosophy is effective as a basis for explanation, rather than application, of the concept. An

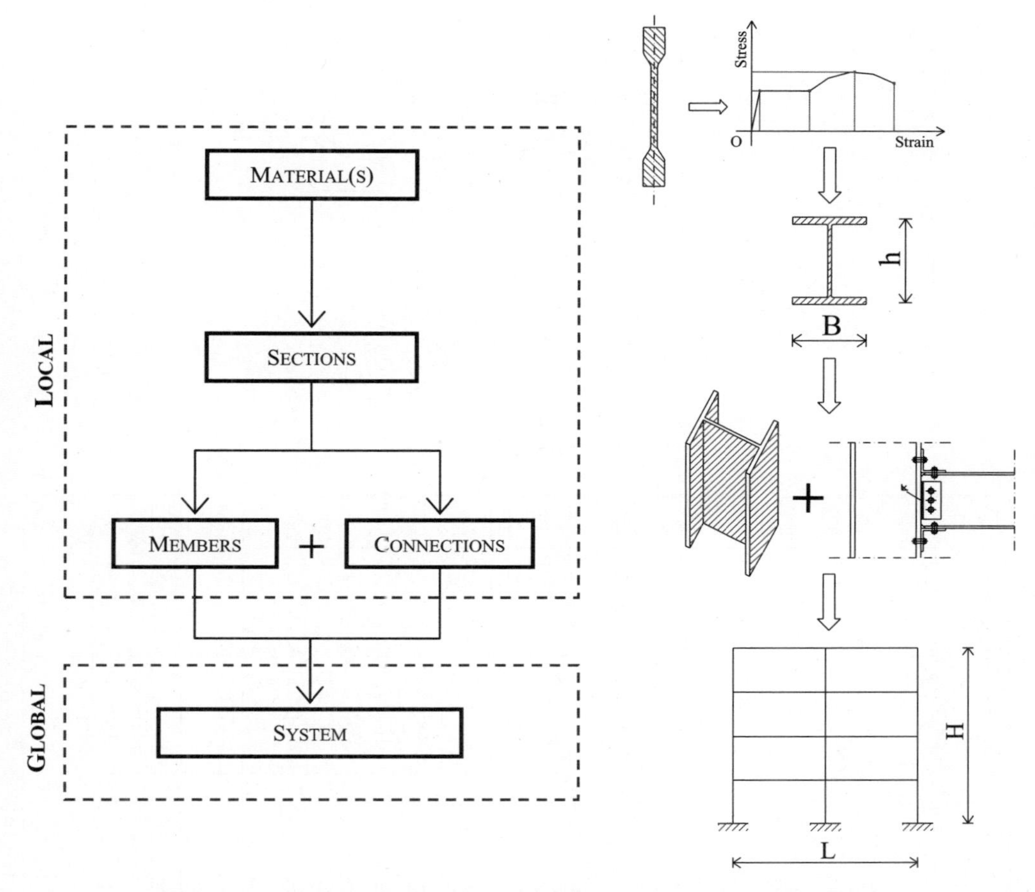

Figure 2.4 Hierarchical relationship between local and global structural response

in-series chain is inadequate to characterize the complex response of structures under earthquake loads. Networks, e.g. road networks, provide a basis for conceptual and pictorial description of the seismic behaviour of structures (Figure 2.5) and also prove that barriers between sub-disciplines are artificial. The transportation network combines parallel and series failure modes that may take place in structural systems. It also provides a visual description of other important aspects of seismic design, such as load path (direction of traffic flow), capacity (maximum flow capacity of a link) and plastic redistribution (likelihood of alternative routes in the case of traffic congestion).

In the network analogy, the demand on the network (earthquake actions) may be expressed by the origin-destination pairs imposed on the network. The capacity of roads to carry traffic symbolizes the capacity of structural members.

2.2.4 Nature of Seismic Effects

Unlike most other types of dynamic actions, earthquake effects are not imposed on the structure but generated by it. Therefore, two structures founded on the same soil a few metres apart may have to

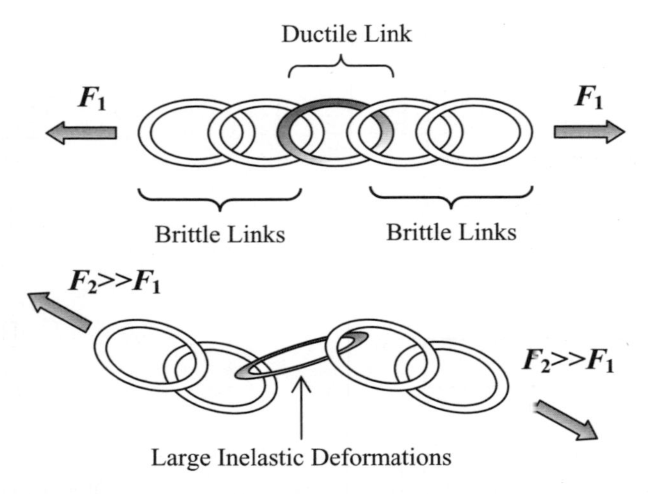

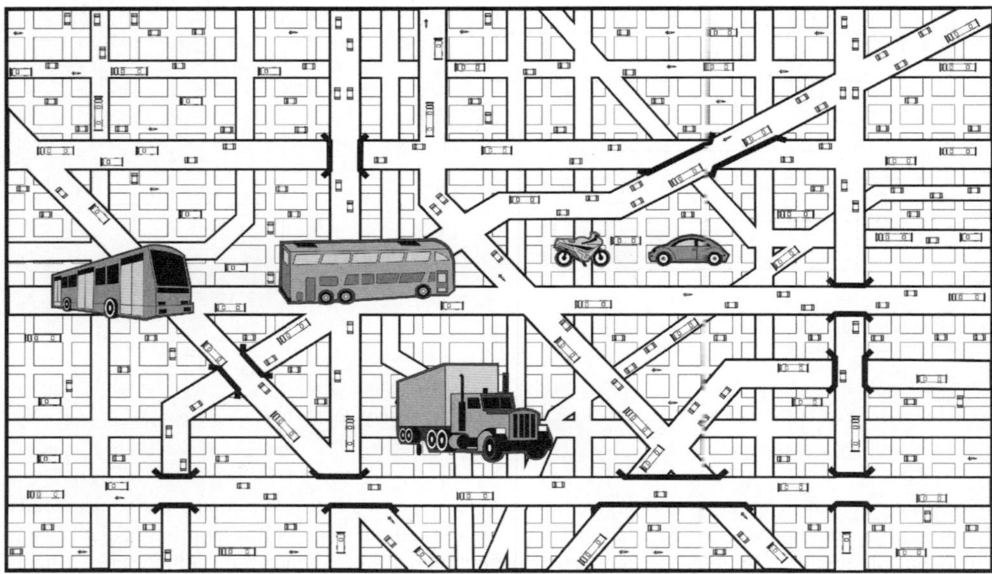

Figure 2.5 Capacity design analogy: chain (*top*) versus network system (*bottom*)

accommodate vastly different action and deformation demands, depending on their own mass, stiffness, strength and ductility. It is argued herein that the fundamental quantities are not period and damping, since period is a function of mass and stiffness (as well as strength in the inelastic range) and the main source of damping in earthquake engineering is energy absorption by inelastic deformation. Setting the 'mass' term aside, earthquake response is affected in a complex manner by stiffness, strength (or capacity) and ductility. In a simple version of this complex problem, stiffness dictates vibration periods, hence amplification. Changes in stiffness cause detuning of the structure and the input motion hence also affects amplification. Figure 2.6 provides an example of two single-degree-of-freedom (SDOF) systems, one stiffer than the other, but the dynamic amplification is such that the less stiff structure (taller pier) displaces less than the stiffer structure.

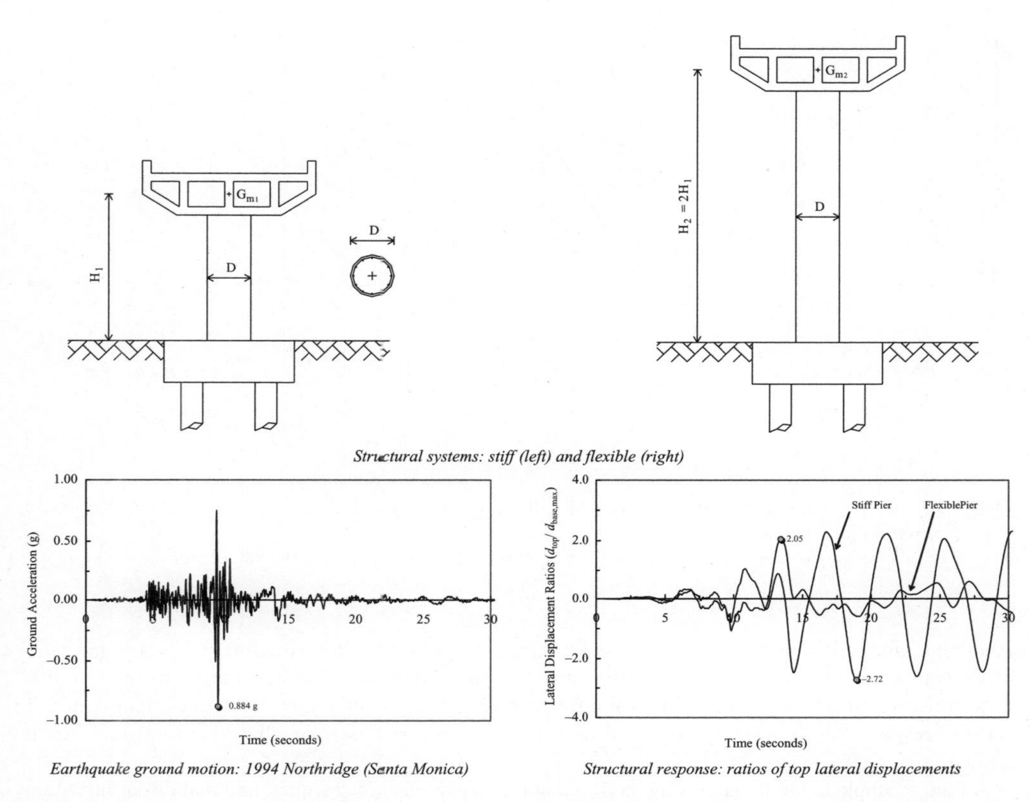

Figure 2.6 Seismic response of structural systems with different dynamic characteristics: stiff and flexible piers

Strength limits describe the region where the structure is able to sustain irreversible damage hence absorbs and dissipates the seismic action. Strength limits therefore lead to the next phase, which is inelastic ductile response. Ductility is an energy sink, therefore it could be considered equivalent damping. Its effect on structural response is, similar to damping, elongation of the response period and reduction in the vibration amplitude.

2.2.5 Fundamental Response Quantities

Stiffness and strength are not always related. For a single structural member, or a structure that employs only a single-mode structural system (e.g. frames only, trusses only or walls only), they are proportional. It is, however, instructive to explore cases where they are not proportional or their constant of proportionality can be changed. The motivation for so doing is to gain a deeper insight into the components of response, providing engineers with a set of tools that enable fine-tuning of both their understanding of seismic response and their ensuing design. A simple example of the decoupling of stiffness and strength is the concept of 'selective intervention' for seismic retrofitting (Elnashai, 1992) as shown pictorially in Figure 2.7. A structure that was designed using direct (strength or even ductility) design where it is required to transform it to a capacity-designed structure, the only requirement is to change the strength distribution without necessarily changing the stiffness distribution. Other scenarios are given in the latter reference and shown experimentally to be totally realistic (Elnashai and Pinho, 1998).

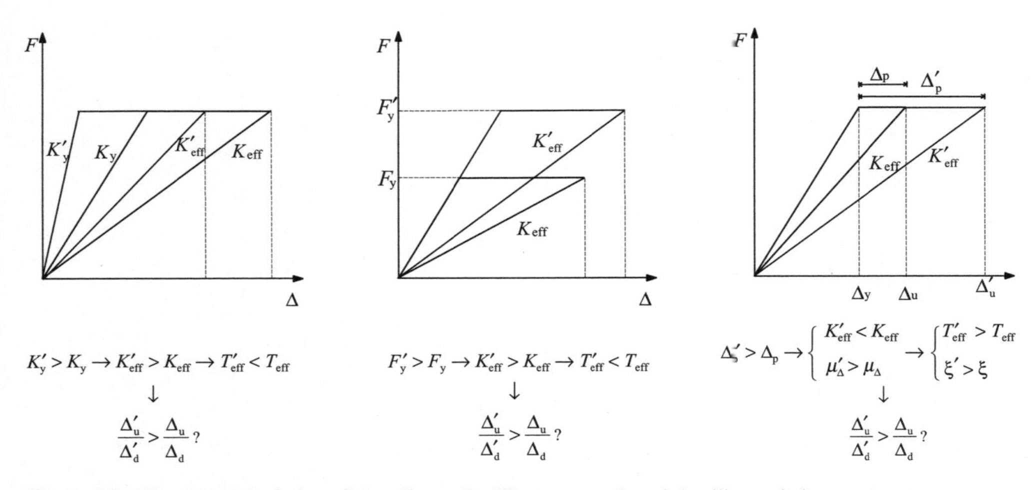

Figure 2.7 Conceptual depiction of the effects of stiffness, strength and ductility variations on system response in seismic retrofitting of structures
Key: K_{eff} is the secant stiffness at maximum deformation; the superscript indicates the value after intervention.

Another example is a mixed-mode frame-wall reinforced concrete (RC) structure. Changing the ratio of walls in a floor immediately changes the stiffness and strength in a disproportionate manner. In this book, the three fundamental quantities of stiffness, strength and ductility are used to explain issues of seismic response of structural systems within a framework that is somewhat distinct from current trends.

A final example that emphasizes the notion that the strength is not constant, and that different failure modes may be obtained from identical structures being subjected to different demands, is presented hereafter. In Figure 2.8, two RC walls are subjected to different loading regimes; one is subjected to monotonic or low-cycle cyclic loading while the other is subjected to severe cyclic loading.

The monotonically loaded wall would fail in flexure if its flexural capacity is reached before the web shear capacity. On the other hand, heavy cyclic loading of the other wall causes the opening of a horizontal crack that then precipitates a sliding shear failure mode. This example also emphasizes that the link between stiffness and strength may be, under some conditions, broken.

2.2.6 Social and Economic Limit States

Herein, in the context of establishing a common vocabulary through the articulation of a conceptual framework, generic limit states are discussed with regard to the effect of earthquakes on vulnerable communities. When subjected to small earthquakes, a society seeks the least disruption from damage. This may be considered an 'uninterrupted use' limit state, and is clearly most correlated with structures having adequate stiffness to resist undergoing large deformations. When subjected to medium earthquakes, a society would tolerate disruption to its endeavours, but would seek to minimize repair costs. This may be viewed as a 'controlled economic loss' limit state and is most related to the structure having adequate strength so that the damage is limited. Finally, when subjected to large earthquakes, a society would accept interruption, high economic loss, but would seek to minimize loss of life. This is a 'life safety' limit state and is most affected by the ductility of the structure that enables it to deform well into the inelastic range, without significant loss of resistance to gravity actions. The fundamental response characteristics of stiffness, strength and ductility are therefore clearly related to the most

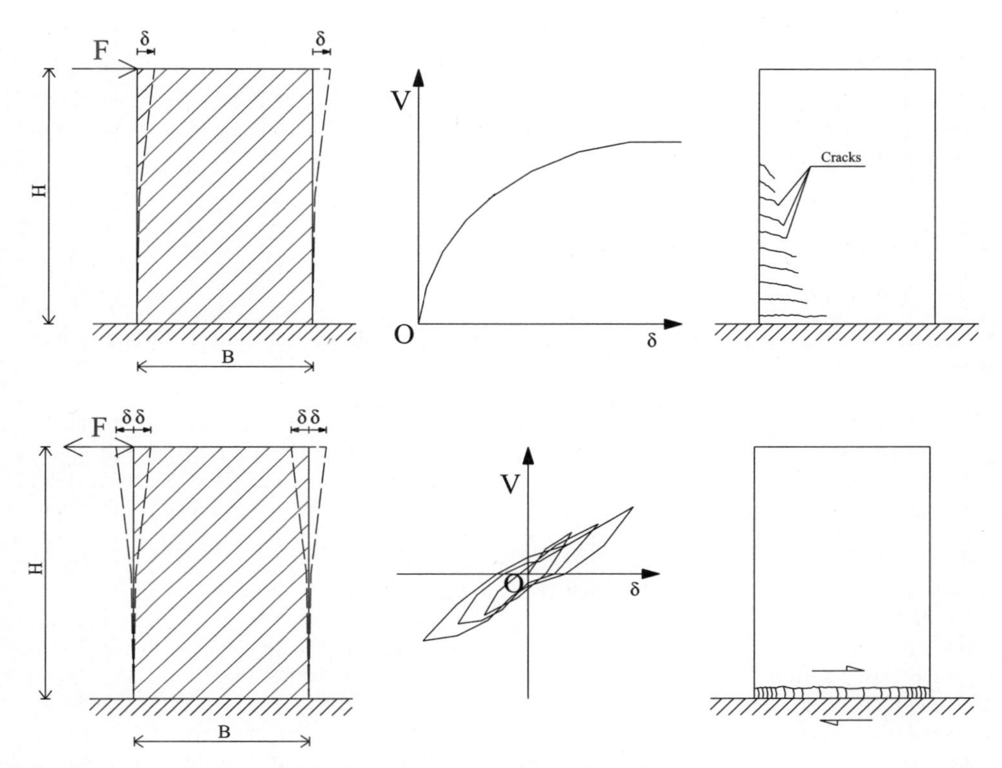

Figure 2.8 Structural RC walls under monotonic or low-cycle cyclic loading (*top*) and severe cyclic loading (*bottom*)

Table 2.1 Relationship between earthquakes, structural characteristics and limit states.

Return period (years)	Earthquake magnitude	Structural characteristics	Engineering limit state	Socio-economic limit state
~75–200	~4.5–5.5	Stiffness	Insignificant damage	Continued operation
~400–500	~5.5–6.5	Strength	Repairable damage	Limited economic loss
~2,000–2,500	~6.5–7.5	Ductility	Collapse prevention	Life loss prevention

important response limit states of continued use, damage control and life safety (Table 2.1), which are in turn related to the socio-economic considerations governing the reaction of communities.

The specific values given in Table 2.1 may be disputed on an engineering basis. For example, the earthquake magnitude associated with a return period is heavily dependent on the seismo-tectonic environment under consideration. Also, more than three limit states appear in many publications on earthquake engineering (e.g. Bertero, 2000; Bozorgnia and Bertero, 2004, among others). However, the above proposed framework is robust and utilizes fundamental structural response characteristics that are generically linked to response or performance targets and map onto social and economic requirements. Intermediate limit states are useful in specific cases, and are largely associated with the special requirements of stakeholders.

> **Problem 2.1**
>
> What are the differences between 'direct' and 'capacity' design? In a multi-storey reinforced concrete frame that is to be capacity-designed, state the sequence of dimensioning of each of the components of the frame, from the foundations to the roof.

2.3 Structural Response Characteristics

2.3.1 *Stiffness*

Stiffness defines the relationship between actions and deformations of a structure and its components. Whereas member stiffness is a function of section properties, length and boundary conditions, system stiffness is primarily a function of the lateral resisting mechanisms utilized, e.g. moment-resisting frames, braced frames, walls or dual systems, as illustrated in Appendix A. Relationships between geometry, mechanical properties, actions and deformations can be established from principles of mechanics. Their complexity depends on the construction material used. Cracking of concrete, yielding of reinforcement bars and other sources of inelasticity in RC structures pose problems in defining a fixed value of stiffness. For RC and masonry structures, the stiffness can be taken as the secant to the yield point or to any other selected point on the response curve. Slippage at connections, local buckling and yielding in steel structures are the counterparts to the above discussion on RC structures.

Figure 2.9 shows a plot of the structural response of a system subjected to lateral loads; the response curve is represented by base shear V versus top horizontal displacement δ. In the figure, the initial slope K_0 is the elastic stiffness of the structure, while the secant stiffness is the slope K_s of the line corresponding to a given level of load. The initial stiffness K_0 is higher than the secant stiffness K_s for conventional materials of construction. In the case of rubber and other special materials, used for example in devices for structural vibration control, the stiffness may increase as loads increase. For the latter, values of V-δ pairs are generally utilized to define the secant stiffness. Variations in stiffness in the inelastic range are often expressed by the tangent stiffness K_t, which is the slope of the tangent to the response curve in Figure 2.9 for a given V-δ pair. A decrease in the values of K_t indicates that softening of the structure is taking place. In analysis of inelastic structures, use is often made of secant stiffness to avoid dealing with negative tangent stiffness beyond the peak action resistance. Since inelastic response problems are solved by iterations, the solution will normally converge by using the secant stiffness even before reaching the point peak action resistance, but the rate of convergence will be lower than in the case of using tangent stiffness.

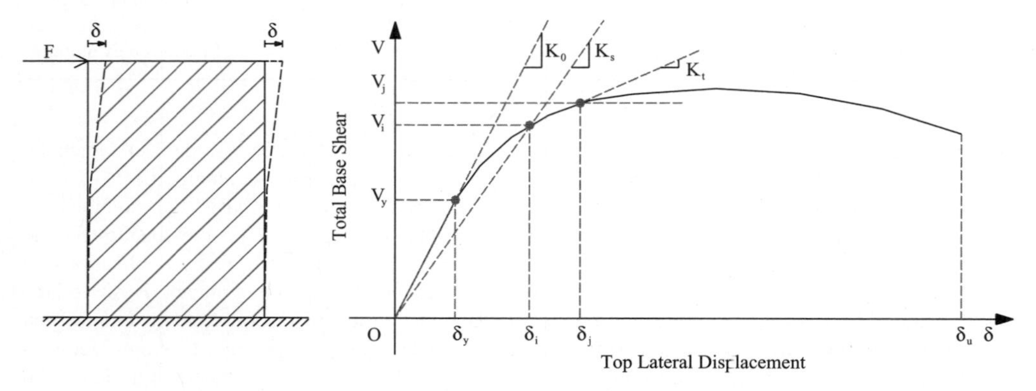

Figure 2.9 Definition of initial and secant structural stiffness

Several types of stiffness may be defined, depending on the nature of applied loads. Structures designed for vertical (gravity) loads generally possess sufficient vertical stiffness. Earthquakes generate inertial forces due to vibration of masses. Horizontal components of these inertial forces are often dominant; hence lateral (or horizontal) stiffness is of primary importance for structural earthquake engineers. The definition of the lateral stiffness, especially the secant value K_s, depends significantly on the region of interest in the response domain, i.e. the behaviour limit state of interest. The stiffness of a system is associated primarily with satisfaction of the functionality (or serviceability) of the structure under dynamic loads. High deformability (and hence low stiffness) drastically reduces the structural functionality.

In seismic design, adequate lateral stiffness is an essential requirement to control deformations, prevent instability (local and global), prevent damage of non-structural components and ensure human comfort during minor-to-moderate earthquakes. Human response to earthquakes is generally different from the discomfort induced by other environmental actions, e.g. strong winds (Mileti and Nigg, 1984; Durkin, 1985; Taranath, 1998). The reason is twofold. Firstly, earthquakes are less frequent than windstorms and have shorter duration; few seconds versus some minutes. Secondly, earthquakes may have serious psychological effects, such as trauma, on people.

Lateral stiffness is influenced by properties of construction material, section type, members, connections and systems, which are linked hierarchically as shown in Figure 2.4. Further discussion is given below.

2.3.1.1 Factors Influencing Stiffness

(i) Material Properties

Material properties that influence the structural stiffness are the elastic Young's modulus E and the elastic shear modulus G. In the inelastic range, the lateral stiffness depends still on the moduli E and G, not on initial, but rather tangent values. The material stiffness is often evaluated through the ratio of the elastic modulus E to the weight γ. Values of E/γ are 20–30 $\times$ 10^4 m for masonry and 200–300 $\times$ 10^4 m for metals as also outlined in Table 2.3. The specific elasticity E/γ of concrete is about 100–150 $\times$ 10^4 m. Construction materials with low values of E/γ lead to stiff structures, e.g. masonry buildings are stiffer than steel.

(ii) Section Properties

Section properties that affect the structural stiffness are the cross-sectional area A, the flexural moment of inertia I and the torsional moment of inertia J. Section area and flexural inertia primarily influence the axial, bending and shear stiffness of the system, for metal structures area (A) and moment of inertia (I and J) do not change with types and levels of applied loads. Conversely, for masonry and RC, the above properties are a function of the loading and boundary conditions. For example, the flexural moment of inertia I of RC rectangular members about the strong axis can be defined as shown in Figure 2.10; similarly, for the definition of the area A of RC cross sections. For elements in tension, it is generally assumed that only the steel reinforcement bars are effective because of the low tensile strength of concrete.

The stiffness of the section is significantly affected by modifications of its geometry. Figure 2.11 shows the variation of area A and flexural moment of inertia about the strong axis I obtained by increasing the size of beam and column members. In the figure, the subscript 1 refers to the original section, while the subscript 2 is for the new section (original and added component). The dimensionless results plotted in Figure 2.11 demonstrate that the increase in the inertia I is higher than the area A. The results emphasize that by jacketing members, the previous balance between axial, torsional and flexural stiffness, and strength, is disturbed, hence a full reassessment of the original design is warranted.

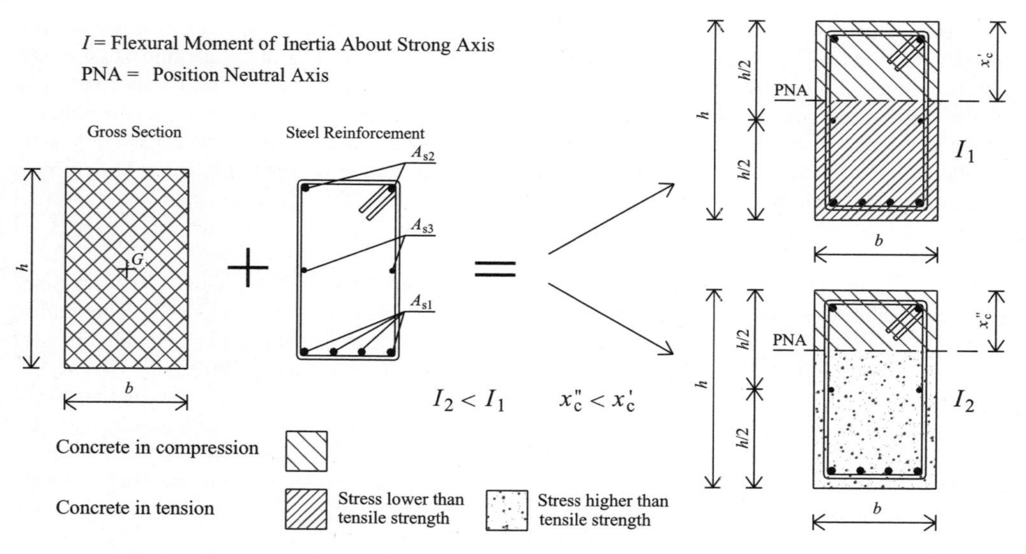

Figure 2.10 Definition of flexural moment of inertia I for reinforced concrete members

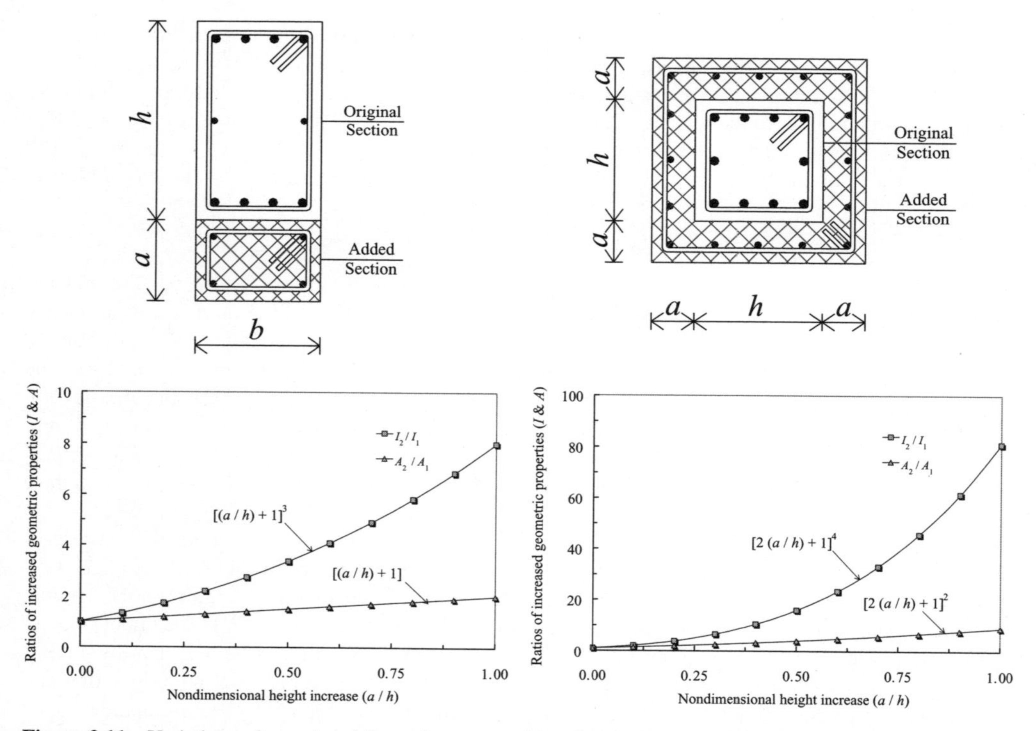

Figure 2.11 Variations of area A and flexural moment of inertia I for beam (*left*) and column (*right*) elements

The orientation of cross sections influences remarkably the lateral stiffness of a structural system. For several sections, such as rectangular, I- and T-shape, moment of inertia about principal axes, i.e. I_x and I_y, may be very different. Structural members with I- or T-sections are stiffer if loaded in the direction of higher inertia, referred to as the strong axis. Large variations between the lateral stiffness about orthogonal directions should be avoided in seismic design. Sections with ratios of I_x/I_y close to unity should be used due to the uncertainty inherent in the direction of earthquake ground motion. In bridge piers, for example, circular or square columns ($I_x/I_y = 1$) are preferable to rectangular sections with section aspect ratios larger than 3–4.

(iii) Member Properties

The lateral stiffness also depends on the type of structural members utilized to withstand earthquake loads. Structural walls are much stiffer in their strong axis than columns. Geometrical properties of structural components, such as section dimensions, height and aspect ratio, influence significantly their horizontal shear and flexural stiffness values. Flexural deformations are normally higher than shear deformations for relatively slender structural components. Flexural deformation dominance occurs if the aspect ratio h/B of rectangular sections of columns is less than 3–4 and the slenderness ratios H/B and H/h are greater than 4–5 in the case of walls. The relationship between horizontal displacement δ and applied load F for cantilevered walls is as follows (Figure 2.12):

$$\delta = \left(\frac{H^3}{3\,E\,I} + \frac{\chi\,H}{G\,A} \right) F \tag{2.1}$$

where E, I and H denote the Young's modulus, the moment of the inertia of the section about the axis of flexure under consideration and the wall length, respectively. Symbols A and G are the area of the section and the shear modulus, respectively. The factor χ is the section 'shear shape factor', which, for rectangular sections, is equal to 1.2.

By setting the flexural (k_f) and shear (k_s) stiffness of the wall as below:

$$k_f = \frac{3\,E\,I}{H^3} \tag{2.2.1}$$

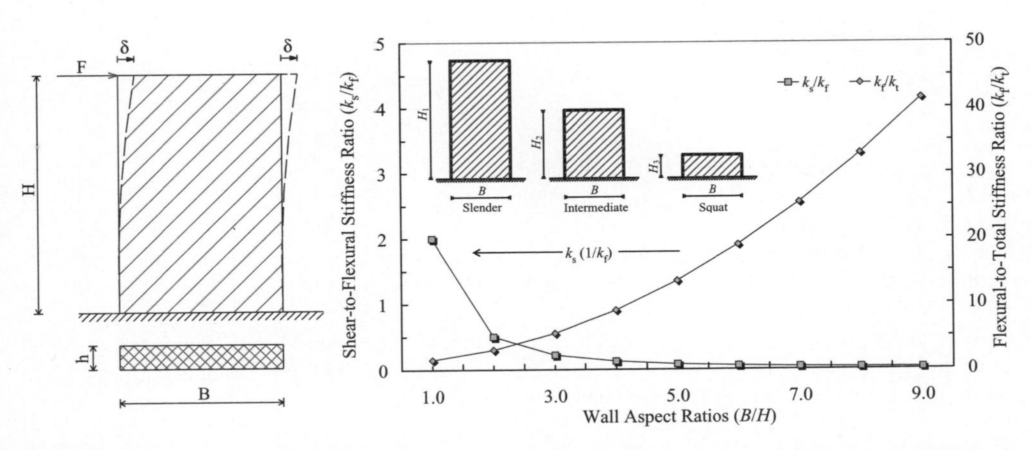

Figure 2.12 Structural wall under horizontal loads: wall layout (*left*) and variations of relative shear and flexural stiffness (*right*)

$$k_s = \frac{G\,A}{\chi\,H} \tag{2.2.2}$$

it follows that equation (2.1) can be rewritten as:

$$F = \frac{k_f\,k_s}{k_f + k_s}\delta \tag{2.3.1}$$

and the total lateral stiffness k_t of the wall is given by:

$$k_t = \frac{k_f\,k_s}{k_f + k_s} = \frac{k_f}{1 + \dfrac{k_f}{k_s}} \tag{2.3.2}$$

Equations (2.2.1) and (2.2.2) show that the lateral stiffness depends on material properties (E and G) in addition to section shape (A and I) and member geometry (H). Flexural (k_f) and shear (k_s) stiffness values are cubic and linear functions of the wall height H, respectively. Consequently, for a given horizontal force F, the displacement δ becomes 1/8 of its original value if the wall height is halved. In turn, for a given δ, the load carried is eight times higher. Equation (2.3.2) demonstrates that if the ratio k_f/k_s is much lower than 1.0, i.e. the shear stiffness k_s is much higher than the flexural stiffness k_f, the relationship between horizontal displacement δ and lateral force F is given by

$$\delta = \frac{H^3}{3\,E\,I}F \tag{2.4}$$

which may be derived from equations (2.1) and (2.3.1). The stiffness ratio k_f/k_s is expressed as a function of the geometric properties as follows:

$$\frac{k_f}{k_s} \approx \frac{1}{2}\left(\frac{B}{H}\right)^2 \tag{2.5.1}$$

or

$$\frac{k_f}{k_s} \approx \frac{1}{2}\left(\frac{h}{H}\right)^2 \tag{2.5.2}$$

depending on whether the strong or weak axis flexural moment of inertia of the wall is utilized (Figure 2.12). The above equations show the influence of the wall slenderness on its lateral stiffness. The higher the ratio H/h and H/B, the lower is the ratio k_f/k_s. Thus, for slender walls, the lateral displacements are mainly due to the flexural deformability. As a consequence of the above discussion, when horizontal earthquake forces are distributed among structural members, their effective flexural and shear stiffness should be considered.

Structural stiffness is also influenced by the type of connection between adjacent members or between structural components and the ground. The general relationship for the lateral bending stiffness k_f^* of the wall in Figure 2.12 can be expressed as follows:

$$k_f^* = \alpha\frac{E\,I}{H^3} \tag{2.6}$$

where the coefficient α depends on the boundary conditions of the structural member. Boundary conditions are analytical relationships that express the properties of connections between members, and between members and the ground. Common values of α in equation (2.6) are 3 (for members with edges fixed-free and fixed-pinned) and 12 (for members with edges fixed-fixed). The bending stiffness k_f^* increases as α increases.

(iv) Connection Properties

Connection behaviour influences significantly the lateral deformation of structural systems. For example, in multi-storey steel frames, 20–30% of the relative horizontal displacement between adjacent floors is caused by the deformability of the panel zone of beam-to-column connections (e.g. Krawinkler and Mohasseb, 1987; Elnashai and Dowling, 1991). Pinned connections are inadequate for unbraced frames, while rigid and semi-rigid connections can be used for both braced and unbraced frames. Laboratory tests on a two-storey steel frame with semi-rigid and fully rigid connections have demonstrated that a reduction of the connection stiffness by 50% and 60% leads to a reduction in the frame stiffness by 20% and 30%, respectively (Elnashai *et al.*, 1998). Numerical analyses of simplified models have shown that the lateral stiffness $K_{\text{semi-rigid}}$ of semi-rigid steel frames can be expressed as a function of the lateral stiffness K_{rigid} of rigid frames through the following:

$$\frac{K_{\text{semi-rigid}}}{K_{\text{rigid}}} = \frac{m(1+\zeta)+6}{m(1+\zeta)} \tag{2.7.1}$$

where m and ζ are dimensionless parameters given by:

$$m = \frac{(K_\varphi)_{\text{con}}}{(EI/L)_{\text{b}}} \tag{2.7.2}$$

and

$$\zeta = \frac{(EI/L)_{\text{b}}}{(EI/H)_{\text{c}}} \tag{2.7.3}$$

where K_φ is the connection rotational stiffness; I, L and H are the flexural moment of inertia, the beam span and column height, respectively; and E is Young's modulus of the material. It is generally assumed that connections with $m < 5$ are pinned, while rigid connections have $m > 18$. Semi-rigid connections are characterized by values of m ranging between 5 and 18 (Figure 2.13).

The stiffness of beam-to-column connections influences also the natural period of vibration of framed structures. Based on shaking table tests for a single-storey steel frame with flexible (double web angle), semi-rigid (top and seat angle with double web angle) and rigid (welded top and bottom flange with double web angle) connections, Nader and Astaneh (1992) suggested simplified relationships to compute the fundamental period T. These are as follows:

$$T = 0.085\, H^{(0.85-m/180)} \quad 5 < m < 18 \quad \text{(semi-rigid)} \tag{2.8.1}$$

$$T = 0.085\, H^{3/4} \quad m \geq 18 \quad \text{(rigid)} \tag{2.8.2}$$

where the dimensionless parameter m is expressed as in equation (2.7.2). In equations (2.8.1) and (2.8.2), H is the frame height, in metres. Figure 2.13 shows the influence of the stiffness of beam-to-column connection on the lateral stiffness and period elongation of frames.

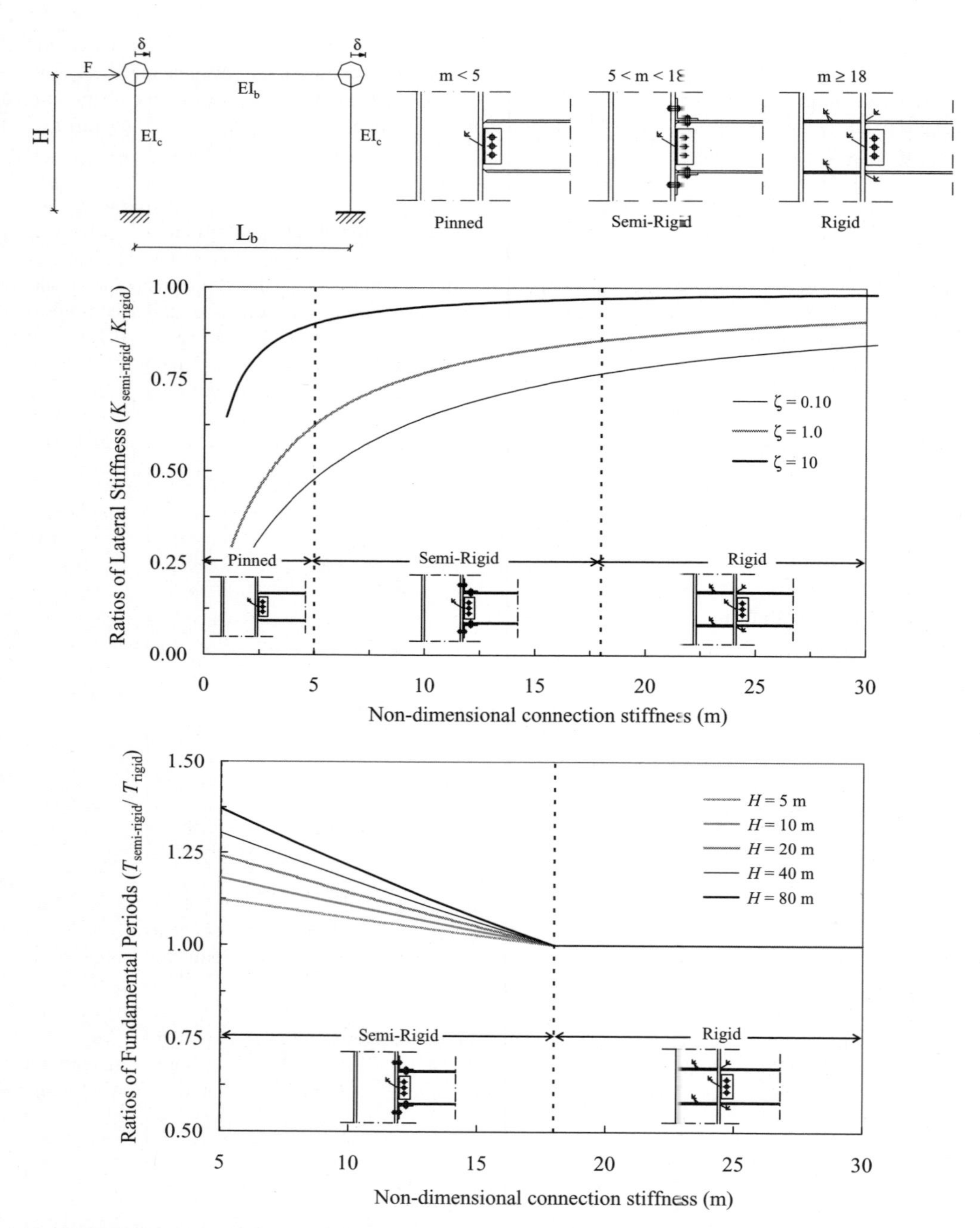

Figure 2.13 Influence of the connection flexibility on lateral stiffness of framed systems: frame layout (*top*), variations of lateral stiffness (*middle*) and fundamental period of vibration (*bottom*)

(v) System Properties

The lateral stiffness of a structure depends on the type of system utilized to withstand horizontal earthquake loads, the distribution of the member stiffness and the type of horizontal diaphragms connecting vertical members. For example, moment-resisting frames (MRFs) are generally more flexible than braced frames. The latter class includes concentrically (CBFs) and eccentrically (EBFs) braced frames. Structural walls are stiffer than all types of frames. Frames with rigid connections exhibit higher stiffness than those with semi-rigid connections (*see* also Figure 2.13). A detailed description of horizontal and vertical structural systems for earthquake resistance is provided in Appendix A. It suffices here to state that uniform distribution of stiffness in plan and elevation is necessary to prevent localization of high seismic demand. Soil-structure interaction should also be accounted for in the evaluation of the global system stiffness. This type of interaction reduces the stiffness of the superstructure and may alter the distribution of seismic actions and deformations under earthquake ground motions (e.g. Mylonakis and Gazetas, 2000, among others).

2.3.1.2 Effects on Action and Deformation Distributions

Inertial forces caused by earthquake motion are distributed among lateral resisting systems in the elastic range as a function of their relative stiffness and mass. The higher the stiffness, the higher the load attracted for a given target deformation. Stiffer elements and structural systems will reach their capacity earlier than their flexible counterparts. Significant reductions of the initial (elastic) stiffness may occur in construction materials, structural members and connections, when they are subjected to increasing loads. Repeated and reversed loading also reduces effective stiffness; an observation termed 'stiffness degradation'. Effects of stiffness on the distribution of actions and deformations are discussed below.

The lateral deformability of structural systems is measured through the horizontal drift. In buildings, storey drifts Δ are the absolute displacements of any floor relative to the base, while inter-storey drifts δ define the relative lateral displacements between two consecutive floors (Figure 2.14). The inter-storey drifts are generally expressed as ratios δ/h of displacement δ to storey height h. Drifts of the roof Δ normalized by the total height H of the building (roof drifts, Δ/H) are also used to quantify the lateral stiffness of structural systems. The roof drift ratio Δ/H may be considered δ/h averaged along the height and hence is not suitable for quantifying variations of stiffness in the earthquake-resisting system. In

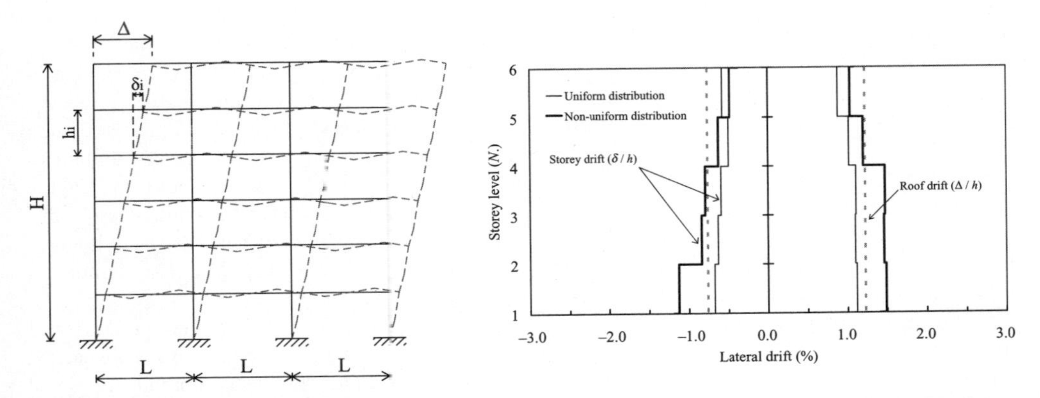

Figure 2.14 Lateral drifts of multi-storey buildings under earthquake loads: definition of inter-storey and roof drift (*left*) and their relationship for uniform and non-uniform lateral stiffness distribution along the frame height (*right*)

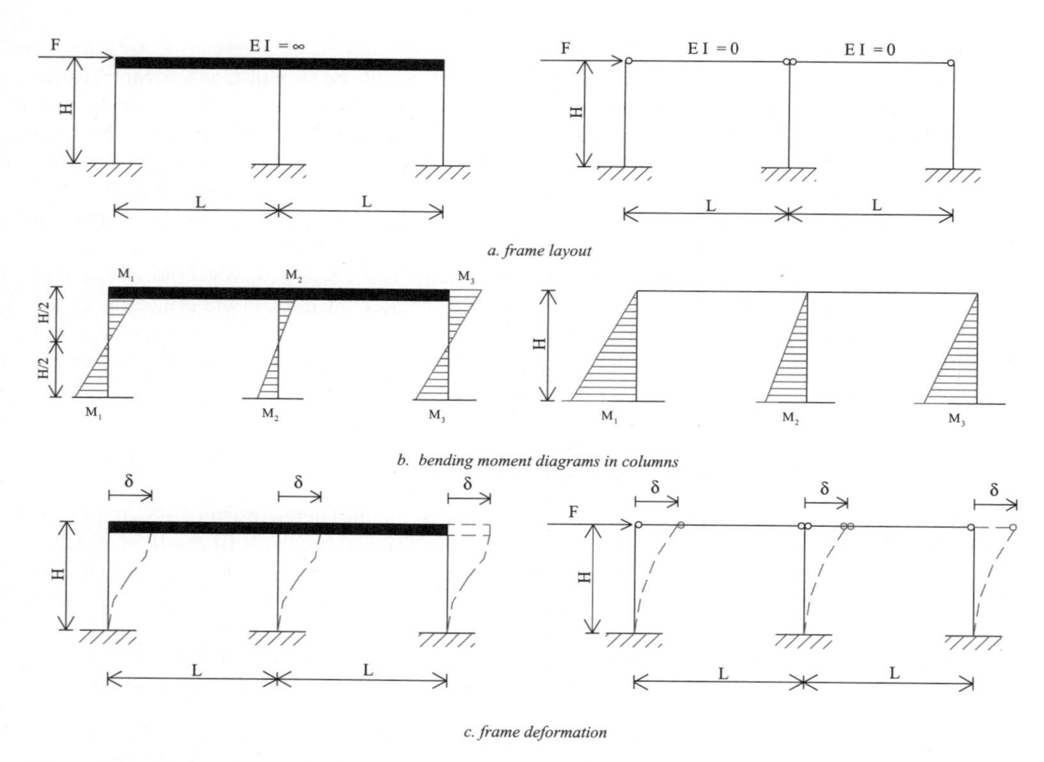

Figure 2.15 Effects of relative stiffness of beams and columns on the distribution of actions and deformations in single-storey frames

structures with evenly distributed mass and lateral stiffness, either δ/h or Δ/H may be employed because they are equivalent.

Inter-storey drifts are caused by flexural, shear and axial deformations of structural elements, e.g. beams, columns, walls and connections. Axial deformations due to shortening or elongation of members are generally negligible; flexural and shear deformations are the primary cause of non-structural damage, as illustrated in Section 2.3.1.3. The overall lateral deformation is affected by the structural system utilized. For example, in MRFs, axial deformations of both beams and columns are not significant. Conversely, axial deformations influence the lateral response of braced frames.

In addition to the importance of absolute stiffness, the relative stiffness of members within a structural system is of significance especially in seismic assessment, because it influences the distribution of actions and deformations. For example, beams with very low flexural stiffness, e.g. flat beams (Figure 2.15), do not restrain the rotation of the columns connected to them. On the other hand, deep beams provide effective restrain for columns in framed structures. If the flexural stiffness of beams is much higher than that of columns, the structure exhibits shear-frame response as displayed, for example, in Figure 2.15 for a multi-span single-storey frame loaded by horizontal force F.

The results of comparative analyses to investigate the behaviour of multi-storey frames with different relative stiffness of beams and columns are shown in Figures 2.16 to 2.18. The comparisons are carried out for structures subjected to vertical (Figure 2.16), horizontal (Figure 2.17) and combined vertical and horizontal (Figure 2.18) loads.

The frames shown in Figure 2.16 employ strong column-weak beams (SCWBs) and weak column-strong beams (WCSBs), respectively. Under gravity loads, these systems undergo negligible lateral

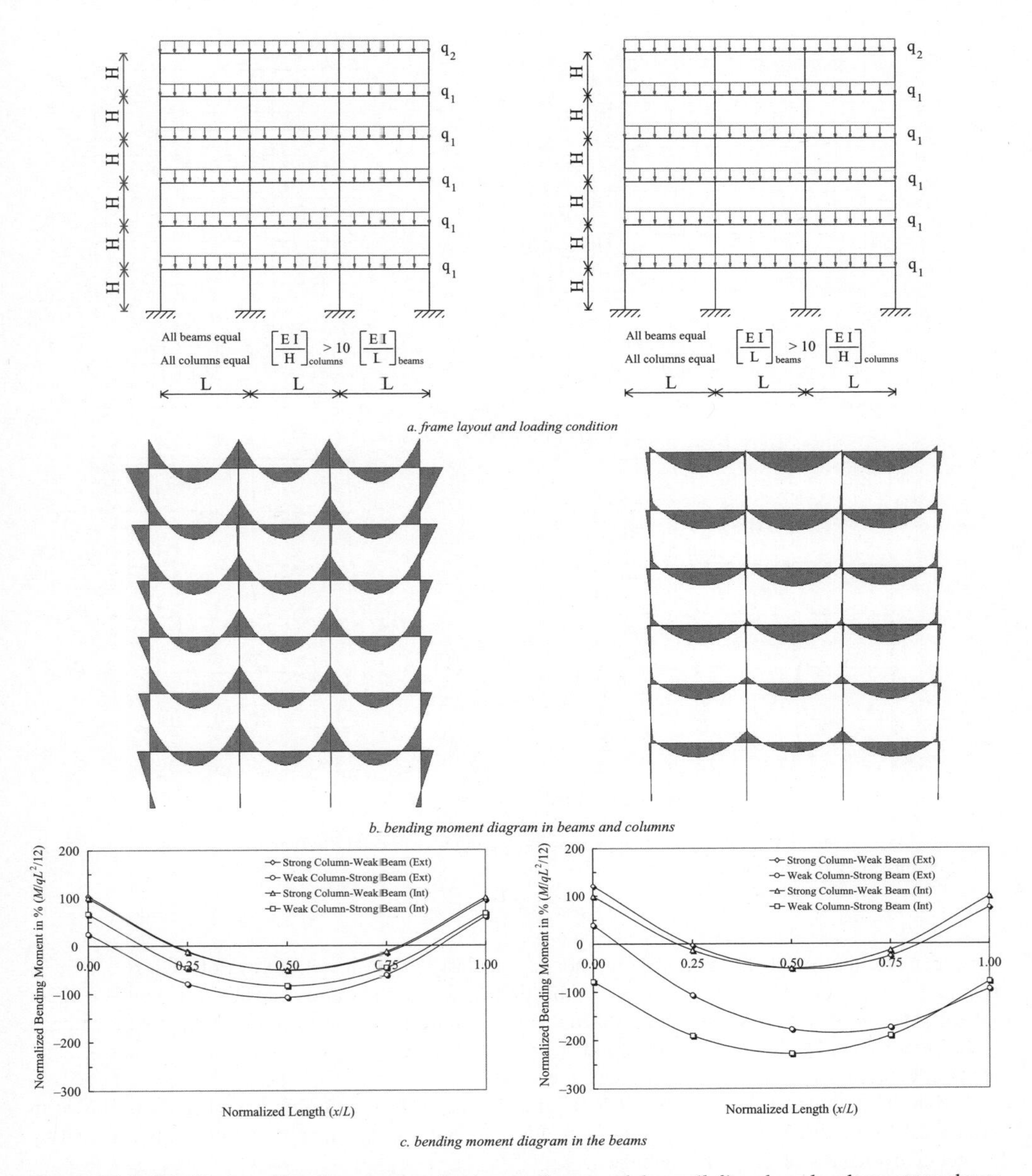

Figure 2.16 Distribution of bending moments in strong column-weak beam (*left*) and weak column-strong beam (*right*) multi-storey frames under gravity loads: beams at first (*bottom left*) and sixth (*bottom right*) floor

displacements, because they are symmetric structures with symmetric loads. The relative stiffness of beams and columns affects significantly the distribution of bending moments especially in the beams as shown in Figure 2.16; the values of the moments are normalized with respect to $qL^2/12$, where q is the uniformly distributed load at each level and L the beam span. In SCWB frames, the large bending stiffness of the columns reduces considerably the rotations at the ends of the beams. Consequently, the

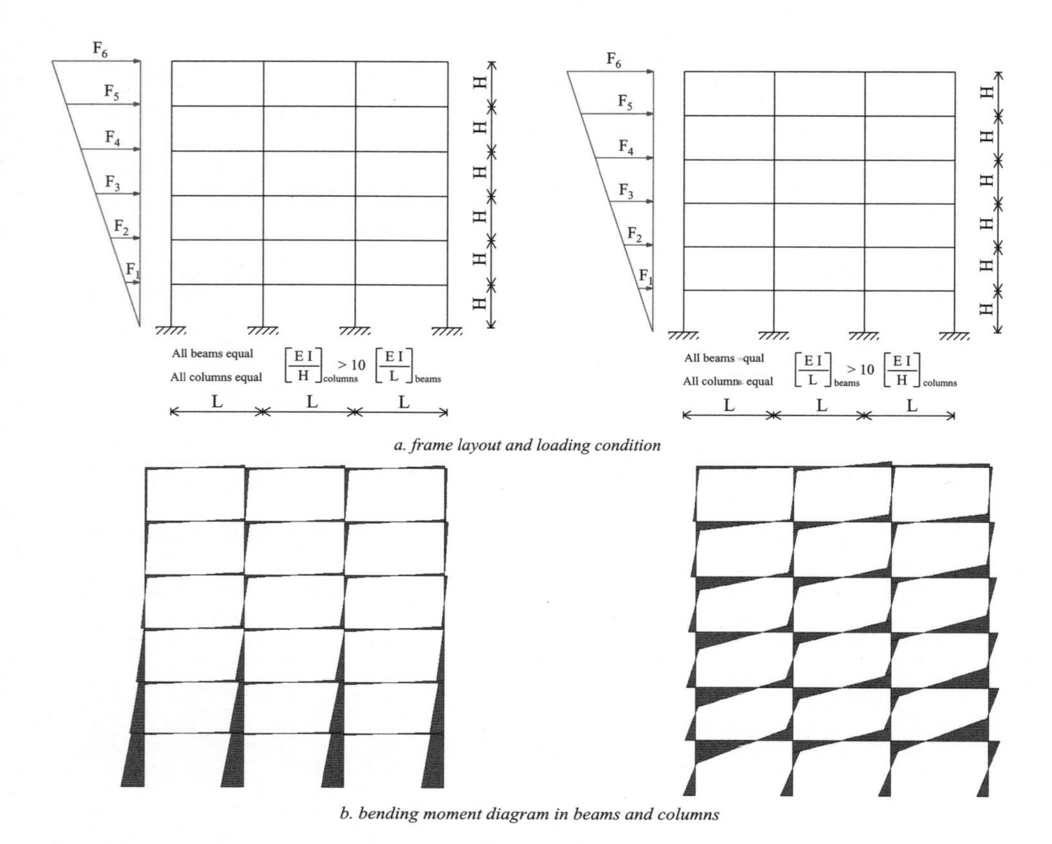

Figure 2.17 Distribution of bending moments in strong column-weak beam (*left*) and weak column-strong beam (*right*) multi-storey frames under horizontal loads

latter behave like members fixed at both ends, particularly in the lower storeys of high-rise frames. Bending moments at the beam-to-column connections are $qL^2/12$; at mid-spans the moment is $qL^2/24$. On the other hand, when WCSBs are utilized, beam response is similar to simply supported members. Under gravity loads, bending moments and shear forces on columns are often small. Values of axial loads in columns depend on tributary areas at each floor.

Action distribution in frames with SCWBs and WCSBs subjected to horizontal loads is shown in Figure 2.17. The distribution of bending moments, especially in columns, is significantly affected by the relative stiffness of the frame members.

In frames with SCWBs, the relatively low flexural stiffness of beams causes a shift upwards of the point of contra-flexure in columns as shown in Figure 2.17. This is typically observed at lower storeys. High values of bending moments can be expected in the columns at ground level. By increasing the flexural stiffness of beams, buildings behave like shear frames, as shown in Figure 2.15. The points of contra-flexure in the columns of shear frames are located at mid-height for both exterior and interior columns.

The distribution of moments caused by the combined effects of vertical and horizontal loads is shown in Figure 2.18 for SCWBs and WCSBs frames. By comparing such distributions with those provided in Figures 2.16 and 2.17, it is noted that, especially for systems with WCSBs, bending moments may

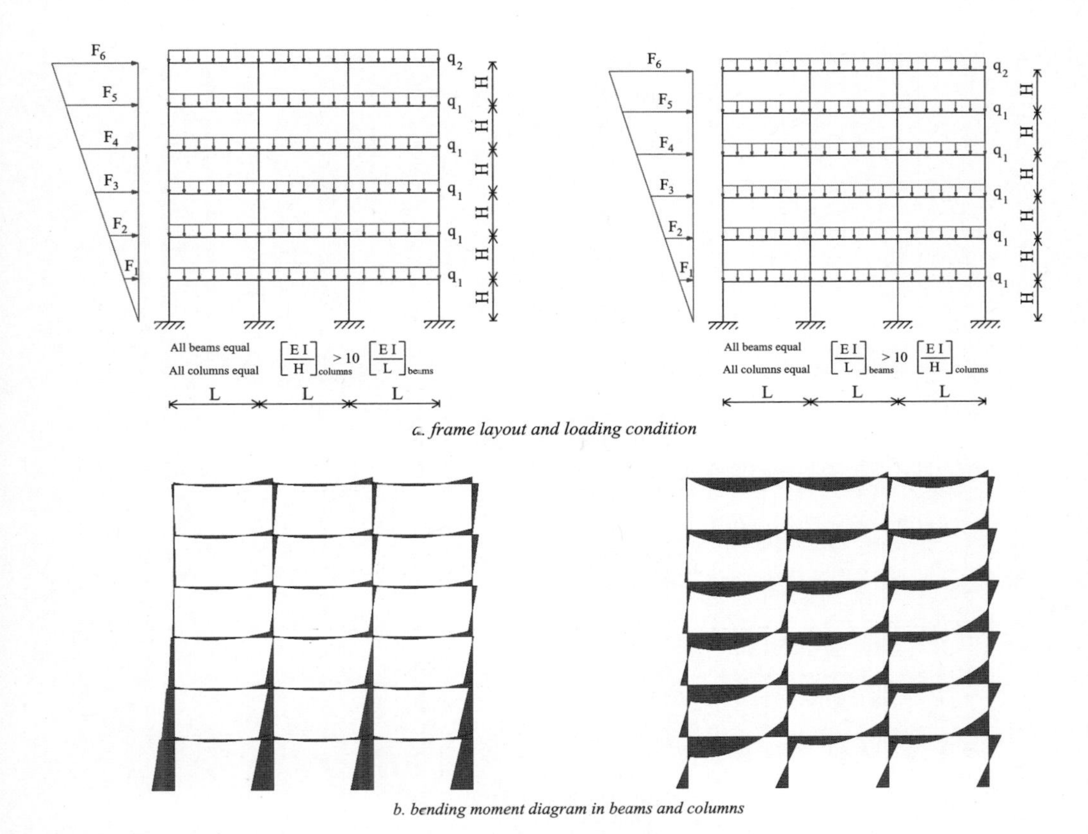

Figure 2.18 Distribution of bending moments in strong column-weak beam (*left*) and weak column-strong beam (*right*) multi-storey frames under gravity and horizontal loads

vary significantly at beam ends. At beam-to-column connections, the flexural moments due to seismic loads can become larger than negative moment values generated by gravity loads.

The distribution of the lateral deformations of the SCWB and WCSB frames is provided in Figure 2.19. The values are expressed in non-dimensional form. Drifts of the WCSB frame are generally higher than those of the SCWB frame, especially at higher storeys.

Irregularities, such as sharp variations of stiffness, may generate concentrations of displacement demands. Figure 2.20 displays the response of regular and irregular MRFs. The former employ beams and columns with the same sections at all storeys, while the latter have an abrupt change in the column sections of the second floor. The significant variation of column stiffness along the height causes a 'soft storey' in the irregular frame; large drifts are observed at the second floor as shown in Figure 2.20.

The above examples highlight the effect of stiffness distribution on the distribution of actions and deformations in framed systems subjected to lateral forces. When frames are used in combination with structural walls, the latter attract the majority of horizontal earthquake-induced forces at lower and intermediate stories. Vertical loads in frame-wall systems are distributed according to tributary areas. Horizontal earthquake accelerations induce inertial forces in structural systems, which are applied in the centre of mass of the structure (C_M). Restoring forces are generated by the reaction of the structure. These are applied in the centre of rigidity (C_R) of the lateral resisting systems. Centres of mass and

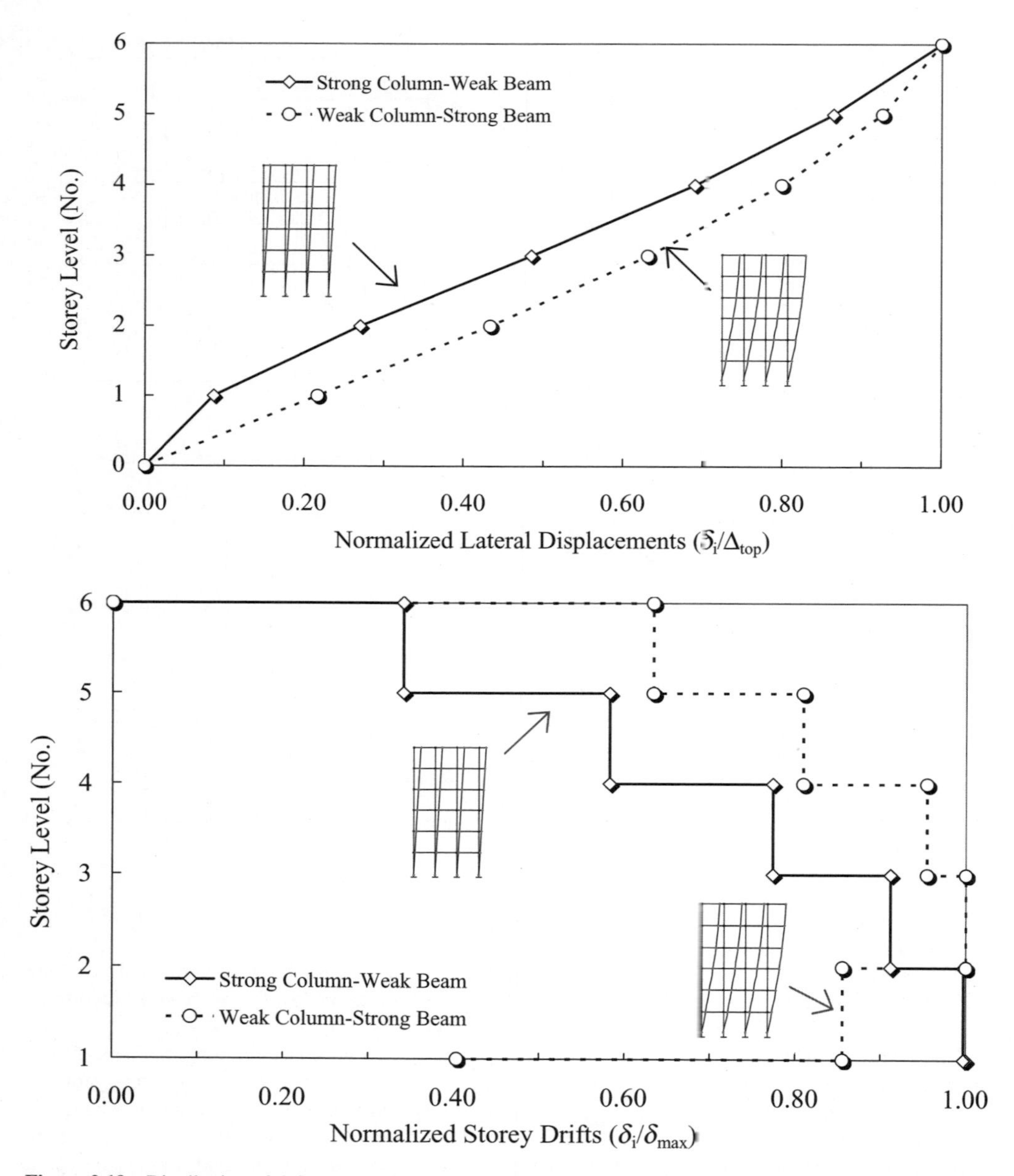

Figure 2.19 Distribution of deformations along the height in strong column-weak beam and weak column-strong beam multi-storey frames in Figure 2.17: storey (*top*) and inter-storey (*bottom*) drifts

rigidity may or may not coincide. If there is an offset (eccentricity, e) between C_M and C_R, torsional effects are generated.

Earthquakes impose dynamic loads with various amplitudes, which can cause deformations in structures well beyond their elastic threshold in alternate directions. Load reversals may also cause stiffness degradation and elongation of the period of vibration in the inelastic range. Extensive analytical work

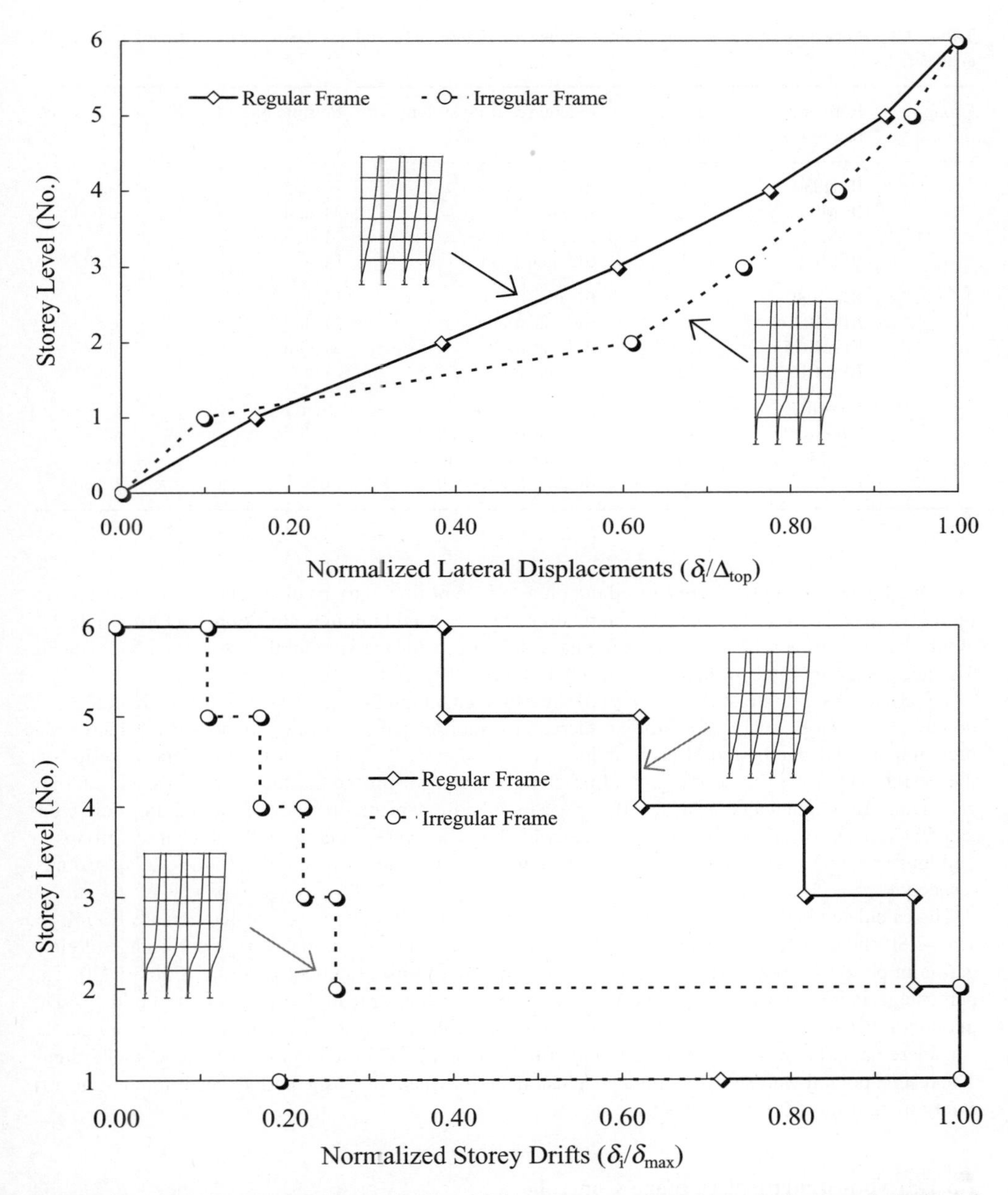

Figure 2.20 Comparison between the distribution of deformations in regular and irregular frames under horizontal loads: storey (*top*) and inter-storey (*bottom*) drifts

on the seismic response of RC buildings reported by Mwafy and Elnashai (2001) demonstrated that the spread of inelasticity may lead to significant decrease of the structural stiffness that in turn causes the fundamental period of vibration to elongate considerably. The distribution of the inertia forces along the building undergoes continuous changes as a result of stiffness and period variation. Static and

Table 2.2 Properties of the sample frames shown in Figure 2.21 and levels of seismic hazard used in the analyses.

Group (no.)	Reference name	Storeys (no.)	Lateral resisting system	Seismic hazard level	Natural period (s)
1	IF-H030	8	Irregular frame	High	0.674
	IF-M030	8	Irregular frame	Medium	0.654
	IF-M015	8	Irregular frame	Medium	0.719
	IF-L015	8	Irregular frame	Low	0.723
2	RF-H030	12	Regular frame	High	0.857
	RF-M030	12	Regular frame	Medium	0.893
	RF-M015	12	Regular frame	Medium	0.920
	RF-L015	12	Regular frame	Low	0.913
3	FW-H030	8	Regular frame-wall	High	0.538
	FW-M030	8	Regular frame-wall	Medium	0.533
	FW-M015	8	Regular frame-wall	Medium	0.592
	FW-L015	8	Regular frame-wall	Low	0.588

dynamic inelastic analyses were carried out on a sample of three groups of regular and irregular structural systems (Figure 2.21). The characteristics of the analysed buildings are summarized in Table 2.2, along with the different levels of seismic hazard assumed for the structural performance assessment. The design accelerations utilized for the frames were 0.10 g and 0.30 g.

To provide insight into the response of the investigated buildings, extensive analyses in the frequency domain, i.e. Fourier analyses, of the acceleration response at the top were conducted to identify the predominant inelastic period of each sample building. Figure 2.22 illustrates the calculated periods at the design and twice the design ground acceleration, along with the elastic period for each building calculated from eigenvalue analyses. It is clear that the fundamental periods of the buildings are elongated as a result of the spread of cracks and yielding. The average elastic periods for three groups of building are 0.69 (irregular frames), 0.90 (regular frames) and 0.56 (regular frame-walls) seconds, respectively.

The calculated inelastic periods at the design and twice the design ground acceleration are 1.30–1.46, 1.65–1.80 and 0.81–1.00 seconds, respectively. Figure 2.22 shows that the average percentage of elongation in period is 100% (irregular frames), 90% (regular frames) and 60% (regular frame-wall). The percentage increase is related to the overall stiffness of the structural system of the building. The maximum calculated elongation is recorded in the most flexible system where the first storey can be considered a soft storey, whereas the minimum elongation is observed in the stiff frame-wall system. The results point towards an important conclusion; employment of elastic periods leads to non-uniform safety margin for different structural systems.

2.3.1.3 Non-Structural Damage Control

Non-structural damage caused by earthquakes can be attributed to excessive lateral drifts of structural systems. In multi-storey buildings, correlations between large inter-storey drifts and non-structural damage are evident from analytical studies, laboratory tests and field observations. Structures may possess sufficient strength to withstand earthquake loads but have insufficient lateral stiffness to limit non-structural damage. Strength limit states do not provide adequate drift control especially for steel structures and for medium- to high-rise buildings with MRFs or narrow shear walls. Lateral deflections

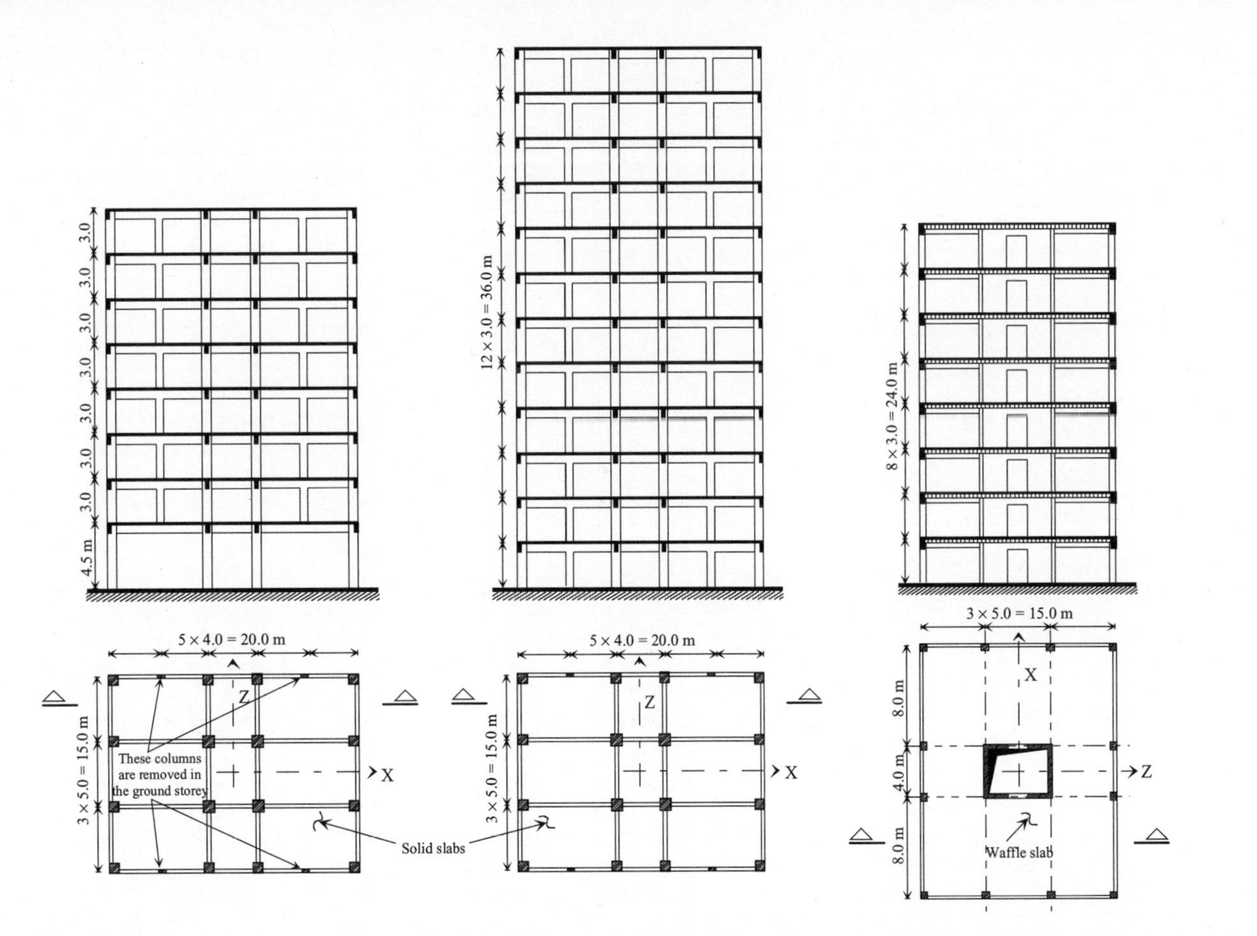

Figure 2.21 Plane and cross-sectional elevation of the buildings: 8-storey irregular frame buildings (*left*), 12-storey regular frame (*middle*) and 8-storey regular frame-wall (*right*) buildings

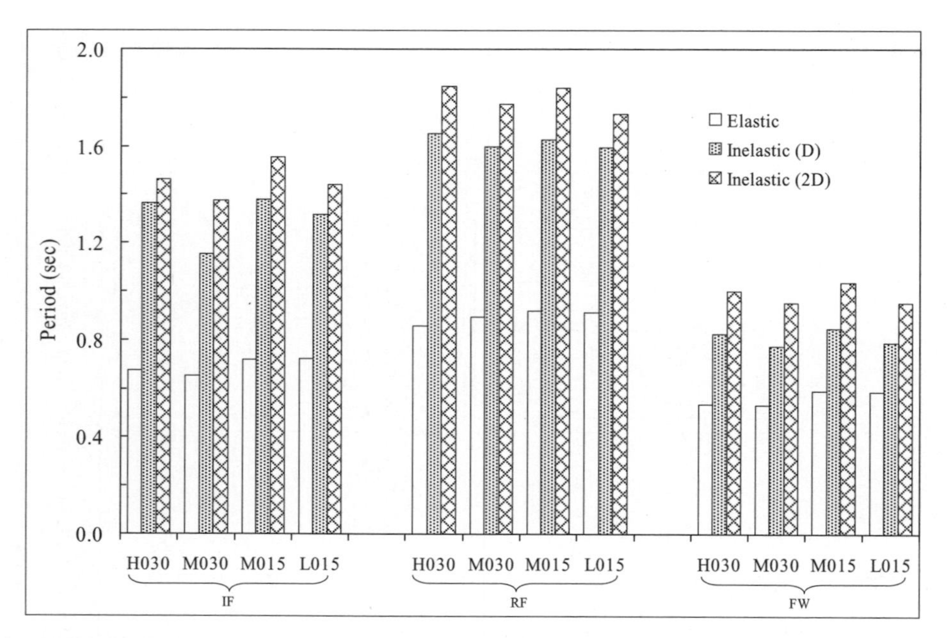

Figure 2.22 Elastic and inelastic (at the design, D, and twice the design, 2D, ground acceleration) predominant response periods of the buildings – average for the eight seismic actions
Key: IF = irregular frames; RF = regular frames; FW = frame-wall structures

may be used to control deformations up to and beyond the strength limit. Structural systems tend to behave linearly under low-magnitude earthquakes; at this stage, values of drifts vary between 0.5% and 1.0%. Analytical work by Ascheim (2002) has shown that the yield horizontal displacement of steel MRFs may vary between 1.0% and 1.2% of the height of the structure, while RC frames often yield at about half of the above values. Modern seismic design codes also include stringent drift limits to ensure adequate lateral stiffness of the structure and hence reduce the extent of non-structural damage. Sharp variations of stiffness in plan and elevation can cause damage concentrations and should be avoided.

Infill panels and brick walls influence the response of frames with low lateral stiffness, e.g. multi-storey steel frames. The more flexible the basic structural system is, the more significant the effects of non-structural components are (Moghaddam, 1990). Masonry and precast concrete infills are frequently used as interior partitions and exterior walls in steel, composite and RC structures. Their interaction with the bounding frame should be accounted for in the assessment of the seismic performance. Studies have demonstrated that the seismic behaviour of infilled structures may be superior to that of bare systems (Shing and Mehrabi, 2002); enhanced lateral stiffness and strength are readily achieved. While their capacity for gravity loads may be low, infills often act as shear walls and affect the seismic structural response in the following respects:

(i) *Stiffening of the structure:* the fundamental period of vibration of the bare system is shortened. Consequently, the dynamic amplification characteristics vary. The importance of this effect depends on the characteristics of the ground motion, which are discussed in detail in Section 3.4;
(ii) *Load path:* infills alter the lateral stiffness distribution of the structure and hence change the load flow illustrated in Section 2.3.2.2. Unexpected stress concentrations may also arise from the interaction of wall panels and bounding frames;

(iii) *Failure Mechanism:* shear failures can be generated by the presence of infills, especially in multi-storey frames and where incomplete panel infilling is used. In addition, 'pounding' and brittle failure of the walls can undermine the seismic performance of the structure.

Extensive experimental and numerical simulations on steel frames infilled with brick walls carried out by Moghaddam *et al.* (1988) showed that infilled frames have an increase in stiffness of 15 to 40 times over that of the bare steel frames. As a result of their contribution to the lateral stiffness of structures, infills undergo cracks and damage during earthquakes. Non-structural damage can be controlled by imposing stringent drift limits. Consequently, the difference in the relative stiffness between frames and infills is reduced. On the other hand, non-structural components in RC concrete structures often crack prematurely when subjected to alternating seismic loads. Experimental and analytical studies have demonstrated that infills continue to govern the overall response of the structure even after cracking during earthquakes (Klinger and Bertero, 1976; Bertero and Brokken, 1983; Fardis and Calvi, 1995; Fardis and Panagiotakos, 1997; Kappos *et al.*, 1998). Cracking due to low tensile strength of the masonry diminishes drastically the initial elastic stiffness of masonry panels; these are generally slender and possess low out-of-plane stiffness (Abrams and Angel, 1993). The presence of masonry infills may affect the response positively or negatively, depending on the bare frame period and its relationship with the dominant period of input motion.

Problem 2.2

An eight-storey reinforced concrete building is to be constructed to replace an existing condominium block that has collapsed during a major earthquake. Two options are available for the building lateral resisting system. These are provided in Figure 2.23 along with the lateral capacity of the sample structures obtained from inelastic pushover analysis. Calculate the elastic lateral stiffness and the secant lateral stiffness at ultimate limit states for both multi-storey structures. If the property owner decides to employ brittle partitions, which structural system is preferable and why? It may be assumed that both structures behave linearly up to the yield limit state.

2.3.2 Strength

Strength defines the capacity of a member or an assembly of members to resist actions. This capacity is related to a limit state expressed by the stakeholder. It is therefore not a single number and varies as a function of the use of the structure. For example, if the interested party decides that the limit of use of a structural member corresponds to a target sectional strain, then the strength of the member is defined as its load resistance at the attainment of the target strain. This may be higher or lower than the peak of the load–displacement curve, which is the conventional definition of strength (Figure 2.24). Target strains may assume different values depending on the use of structural systems. For instance, strains utilized in multi-storey frames for power plants may be lower than those employed in residential or commercial buildings. Target strains can be correlated to the risk of failure, which in turn depends on the use of the structure.

Strength is usually defined as a function of the type of applied action. Axial, bending and shear resistances are utilized to quantify the capacity of structures and their components in earthquake structural engineering. In the response curve shown in Figure 2.24, the shear capacity V of the system is defined with respect to either the shear at yield V_y or at maximum strength V_{max}. Alternatively, the shear capacity can be expressed at any intermediate point between V_y and V_{max}, e.g. V_i in Figure 2.24. Similarly, axial (N) and bending (M) resistances are evaluated through axial load-axial displacement

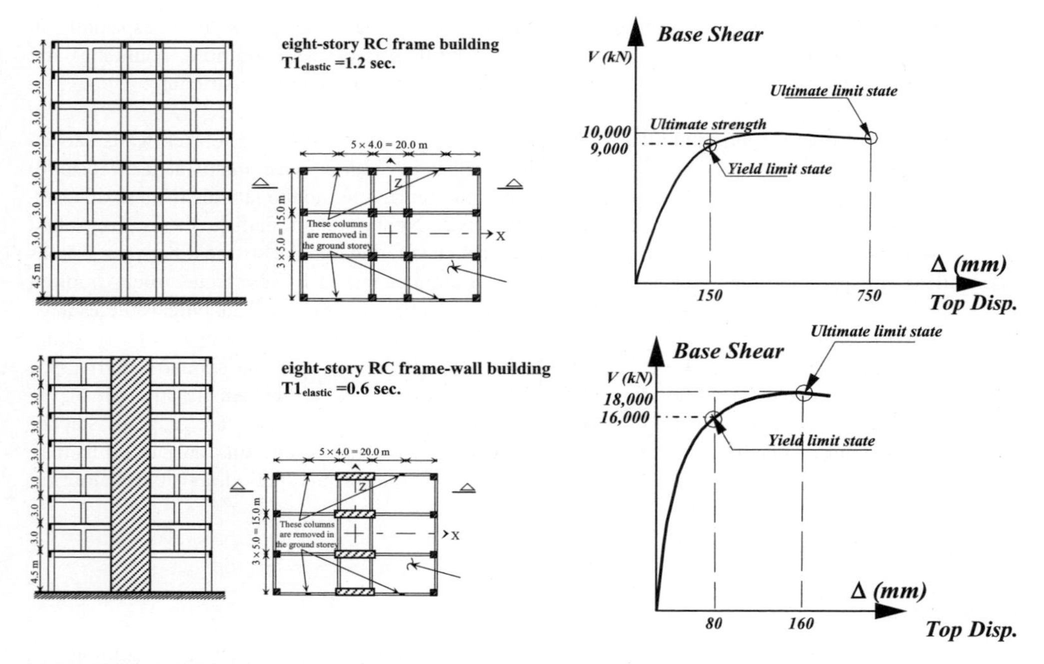

Figure 2.23 Reinforced concrete moment-resisting frame (*top*) and dual (moment-resisting frame and structural wall) system (*bottom*): layout (*left*) and capacity curve (*right*)

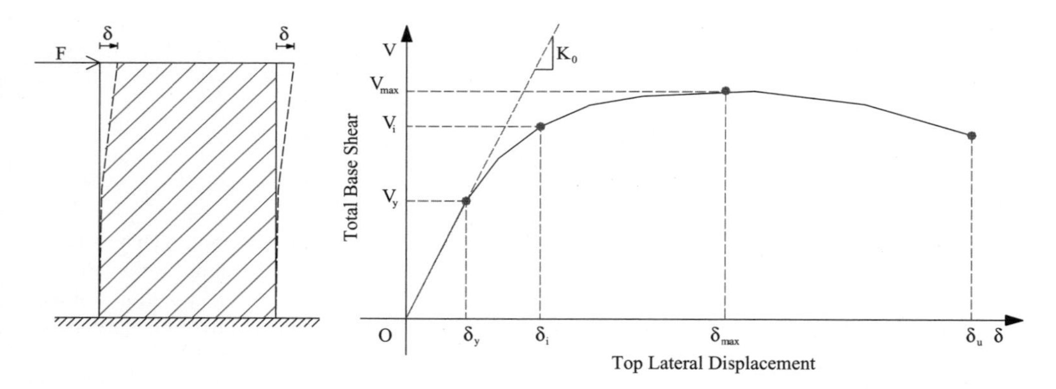

Figure 2.24 Definition of strength

and moment–rotation relationships, which can be represented by curves similar to that shown in Figure 2.24.

The definition of strength parameters is often more straightforward than that of stiffness. Relationships between geometry, mechanical properties and strength can be derived from principles of engineering mechanics. These relationships depend on the type of construction materials employed. Uncertainties in the evaluation of structural capacities are attributed to (i) the randomness in material properties, especially strength parameters, (ii) geometric properties, e.g. section and member size, and (iii) construction quality. Thorough control is imposed for materials, members and connections which are manufactured in part or totally in fabrication yards. Therefore, the uncertainty in the evaluation of the capacity of prefabricated structures is often lower than that of systems built *in situ*. In general, the randomness of physical properties (mechanical and geometric) is small in metal structures (coefficients of variation, or COVs, are typically between 4% and 6%) but can be significant for RC and masonry (COVs greater than 10–15%). Moreover, the construction quality of metal structures is frequently higher than that for RC and masonry. Reliable estimation of section, member and structure capacities require low values of COVs for both geometrical and mechanical properties.

Attainment of shear, axial and flexural capacities in gravity and earthquake-resistant systems can cause damage in structural components. Damage is related to the safety of the system but it does not necessarily lead to structural collapse. Collapse prevention is the behaviour limit state controlled by ductility, as illustrated in Section 2.3.3.3.

Horizontal seismic loads usually exceed wind loads, especially for low- and medium-rise structures in areas of medium-to-high seismic hazard. Earthquakes produce lateral forces proportional to the weight of the structure and its fixed contents; the resultant of seismic force is known as 'base shear'. Adequate shear, axial and flexural capacity is required to withstand storey and base shear forces. Bending effects, such as uplift and rocking, may be caused by horizontal forces due to masses located throughout the height of the structure; these effects are also referred to as 'overturning moment'. Combined horizontal and vertical loads in the event of ground motions increase the stress level in members and connections. If total stresses exceed the capacity, failure of structural components occurs; this corresponds to the structural damage limit state. Local damage, however, does not impair the integrity of the structure as a whole. Correlations between strength and structural damage are presented in Section 2.3.2.3.

As with lateral stiffness, the strength of a structure depends significantly on the properties of materials, sections, elements, connections and systems (reference is made to the hierarchical link shown in Figure 2.4) as discussed below.

2.3.2.1 Factors Influencing Strength

(i) Material Properties

The efficient use of material strength may be quantified by the 'specific strength', i.e. the strength-to-weight ratio σ/γ. Values of σ/γ for materials commonly used in structural earthquake engineering are provided in Table 2.3. Specific elasticity E/γ was also included for purposes of comparison. Fibre composites, wood and metals possess the highest values of specific strength; this renders them suitable for earthquake structural engineering applications. In the case of wood, the drawback is that member sizes required to achieve high levels of strength may be very large.

Construction materials may be isotropic, orthotropic or anisotropic, depending on the distribution of properties along the three principle axes. Some materials, such as structural steel and unreinforced concrete may be treated as isotropic. Laminated materials are usually orthotropic. Examples of anisotropic materials include masonry, wood and fibre-reinforced composites. Strength of materials is influenced by strain hardening and softening as well as strain rate effects (e.g. Paulay and Priestley, 1992; Bruneau *et al.*, 1998; Matos and Dodds, 2002, among others).

A loss of both strength and stiffness takes place in concrete as the strain increases; this is referred to as strain softening or strength and stiffness degradation. Strain softening can be reduced in RC systems by providing transverse confinement of concrete by either hoops or spirals. Circular hoops are more efficient than those with rectangular shapes because they uniformly confine the core concrete. The loss of bond between concrete and steel in RC structures under large alternating loads reduces strength and stiffness. Conversely, structural steel exhibits higher strength at large deformations, generally at strains ε greater than 10–15 ε_y, with ε_y being the strain at yield; this is known

Table 2.3 Specific stiffness and strength of some materials used in seismic design.

	Density (kN/m^3)	Young's modulus (GPa)	Strength (MPa)	Specific elasticity ($\times 10^4$ m)	Specific strength ($\times 10^2$ m)
Concrete					
Low strength	18–20	16–24	20–40	89–120	11–20
Normal strength	23–24	22–40	20–55	92–167	8–22
High strength	24–40	24–50	70–1,000	100–125	29–250
Masonry					
Concrete	19–22	7–10	5–15	37–45	3–20
Brick clay	16–19	0.8–3.0	0.5–4	5–16	0.3–2
Fibre composites					
Aramidic	14–16	62–83	2,500–3,000	443–519	1,786–1,875
Carbon	18–20	160–270	1,400–6,800	889–1,350	778–3,400
Glass	24–26	70–80	3,500–4,100	292–308	1,458–1,577
Wood	1.1–13.3	0.2–0.5^p	28–70^p	4–18^p	53–255^p
		7–12°	2–10°	90–636°	8–18°
Metals					
Mild	79	205	200–500	259	25–63
Stainless	80	193	180–480	241	23–60
Aluminium	27	65–73	200–360	240–270	74–133
Other alloys	40–90	185	800–1,000	205–462	111–200

Key: o = orthogonal to fibres; p = parallel to fibres; w = work hardened.

as strain hardening. Beyond the peak stress, around 100–150 ε_y, many types of steel start strain softening.

Under dynamic loads, the material strength increases with the increase in strain rate. During earthquakes, strain rates in steel structures vary between 10^{-2}/second and 10^{-} /second (Roeder, 2002). Steel structures susceptible to buckling are also affected by strain rate, even if their dynamic response is in the slow rate region (Izzuddin and Elnashai, 1993). Strain rates can be as high as 5×10^{-2}/second in RC structures with short periods (Paulay and Priestley, 1992). For the latter value of strain rate, the compression strength of concrete is increased by about 30% compared to the quasi-static strain rates. For the same values of strain rates, the increase in the yield strength of steel is about 20%. The stiffness of concrete is also a function of the strain rate; the increase of the stiffness is lower than that in compressive strength. Nevertheless, strain rates may be favourable for the stiffness degradation of concrete. In RC beam elements, where the seismic response is controlled by steel reinforcement bars, strain rates have minor effects (Fu *et al.*, 1991). Consequently, for RC structures, strain rates are likely to influence the response of members in which the behaviour is dominated by concrete failure. Both strain hardening and strain rate effects influence the overstrength of structural systems under earthquake loading. The overstrength is the structural characteristic that quantifies the difference between the required and actual strength of a material, a component, an assembly of components or a system, as further discussed in Section 2.3.4.

(ii) Section Properties

Lateral strength of structural systems is influenced by section properties. The area A of cross sections affects both axial and shear capacity, while flexural (I) and torsional (J) moments of inertia influence flexural and torsional capacity, respectively. For RC sections, the strength increases with the amount of steel longitudinal reinforcement. A, I and J for RC and masonry members vary with the type and value of applied loads, as illustrated in Section 2.3.1.1. The tensile strength f_t of these materials is much lower than their compression strength f_c. The tensile strength f_t is often less than 10–15% of the value of f_c. Consequently, when subjected to stress reversals due to earthquake ground motions, the response of RC and masonry members in compression is notably different from the behaviour in tension. To achieve cost-effective designs, the shape of cross sections should be selected as a function of the applied action and lateral resisting system. For example, rectangular sections are cost-effective for beams and columns to resist axial and shear loads, while wide-flange sections can be used for flexure with or without axial loads.

Section capacities depend on the interaction between different types of applied actions, e.g. axial load N, bending moment M and shear force V. For example, the flexural capacity of steel sections subjected to uniaxial bending M and axial compression loads N is lower than the capacity of sections in simple bending. On the other hand, for RC sections, the moments capacity M increases with compressive N up to the balanced failure (N_b, M_b), i.e. simultaneous crushing of concrete and yielding of steel bars. For $N > N_b$, the flexural moments M decrease as the compression axial loads N increase.

In structures designed for gravity loads only, bending moments in the columns are generally small. Under earthquake loads, horizontal loads induce high moments and shears in columns. Due to the randomly oriented direction of earthquake ground motions, columns in three-dimensional earthquake-resistant structures are subjected to reversing biaxial bending and tension-compression variations of axial loads, i.e. flexure about two orthogonal directions. The biaxial moment capacity of RC members is generally less than that under uniaxial bending. Interactions between axial loads N and bending moments M, especially if biaxial, reduce the capacity of column sections.

Shear-axial and shear-flexure interactions affect the seismic response of beams and columns in framed systems. The effects of these interactions considerably reduce the capacity of RC sections. Failure in RC beams is often caused by the interaction between flexure and shear actions. Similarly, in columns the response is influenced by both shear-axial and shear-flexure interactions. Large variations of axial loads may take place in columns of buildings and piers of bridges under earthquake ground motions. Reductions in axial load, or even tensile actions, erode the shear capacity of RC members (Lee and Elnashai, 2002). For steel structures the interaction between shear, axial and bending capacity is usually insignificant (e.g. Kasai and Popov, 1986, among many others).

The strength of steel cross sections may also be reduced by local buckling. Full axial, bending and shear capacities cannot be reached if local buckling occurs (reference is made to Section 2.3.3). Adequate width-to-thickness (b/t) and diameter-to-wall thickness (d/t) ratios for the plates, forming webs and flanges of the cross section, should be used to prevent local buckling.

(iii) Member Properties

System strength is affected by the properties of structural components. Columns generally possess lower flexural and shear strengths than structural walls. Slender walls are frequently used to increase lateral stiffness and strength in medium- to high-rise frames. Such walls can withstand high overturning moments and base shears, provided that their connections with the foundation systems do not fail. The position of steel reinforcement bars in the cross section of structural walls significantly affects lateral strength and ductility as further discussed in Section 2.3.3. For slender walls, experimental and numerical simulations have shown that the concentration of bars at the edges, rather than their distribution evenly over the width, improves the seismic performance (e.g. Paulay and Priestley, 1992). To facilitate the insertion of several longitudinal bars at the edges, slender walls often employ flanged section as displayed in Figure 2.25. The two flanges are also effective in transmitting the bending moments between the walls and adjacent beams in multi-storey dual systems. A minimum amount of steel bars should be placed along the width of the wall cross section to prevent undesirable shear failures. Conversely, in squat walls, steel longitudinal and transverse reinforcements are often uniformly distributed both in plan and elevation (Figure 2.25), since such walls will not exhibit a flexural mode of response that requires edge steel yielding. Reinforcement bars arranged in a grid limit the cracking and provide adequate shear strength to the wall.

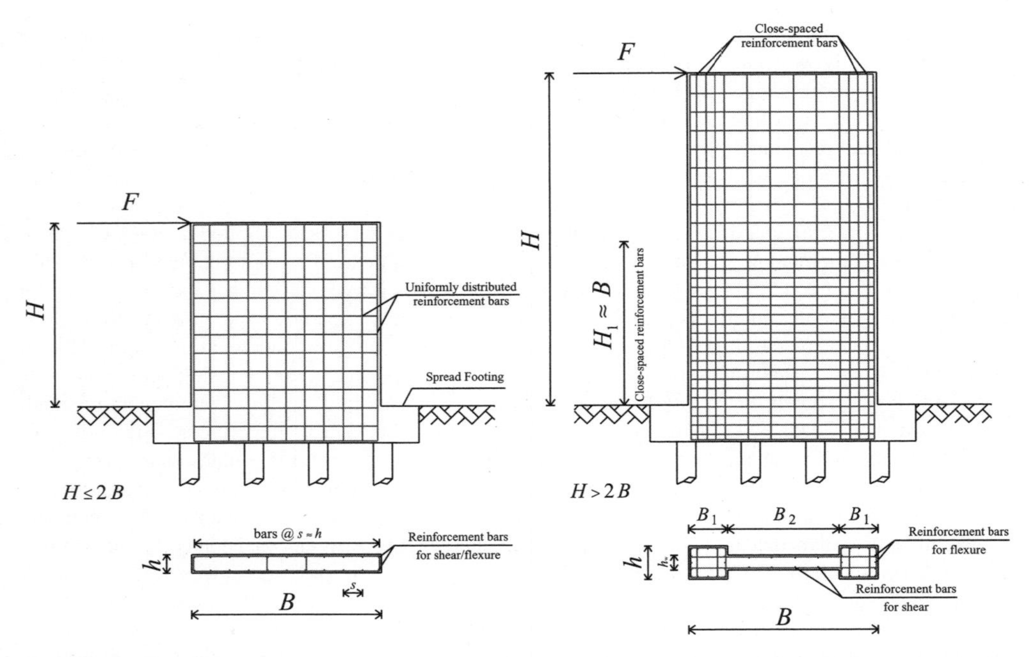

Figure 2.25 Distributions of steel reinforcement in structural walls to enhance lateral strength: squat (*left*) and slender (*right*) walls

Confinement of compressed concrete and prevention of steel bar buckling are also essential to reach the maximum member capacity of walls and columns. Seismic design of RC structures should ensure that member strengths are governed by flexure rather than shear or failure of bond and anchorage. This is allowed for energy dissipation in flexure as illustrated in Section 2.3.3.

(iv) Connection Properties

Under earthquake loads, high shear reversals are generated in beam-to-column and column base connections; both are critical components of framed systems. Stress concentrations in joints may be caused by their complex geometric layout and congestion, e.g. in RC structures several longitudinal steel bars from elements framing into them intersect. The loss of stiffness and strength of structural joints leads to the deterioration of stiffness and strength of the frame. Consequently, to achieve adequate global seismic performance, joint stresses and deformations should be limited to tolerable levels.

Connections between horizontal diaphragms and lateral force-resisting systems, e.g. frames and walls, considerably influence the global action and deformation capacity of the structure. For example, connections between RC flat slabs and columns should possess high shear capacity to prevent punching shear. Similarly, in steel and composite structures, the area in the neighbourhood of the beam-to-column connections is often strengthened with additional steel bars to prevent shear failure. In composite systems, the local and global strengths are also affected by the shear connectors between structural steel and concrete components. Stress concentrations at the connection of flat slabs and structural walls may lead to tearing of the slab, especially in flat-slab systems. Additional steel reinforcement can be placed in the slab to increase its shear strength. Weak connections between foundations and superstructures may cause sliding shear or overturning as indicated in Figure 2.26.

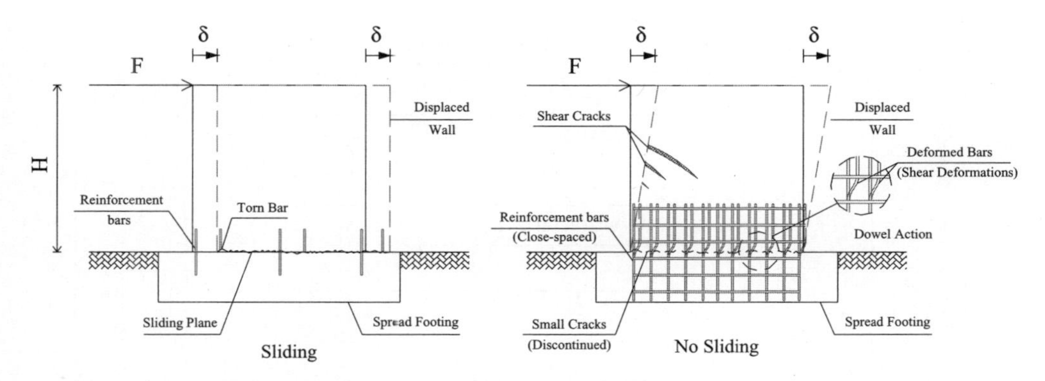

Figure 2.26 Squat wall with weak (*left*) and strong (*right*) foundation connections

Sliding is a common failure mode observed in masonry and wood buildings and is caused primarily by the low strength of fastenings. On the other hand, overturning of walls in masonry and wood systems is often caused by the inadequate resistance of connections between orthogonal structural walls. In RC structural walls, inadequate anchorage lengths of steel reinforcement bars in the foundation may endanger the flexural capacity. Similarly, insufficient bolt anchorages may undermine global lateral strength in both steel and composite structural walls.

(v) System Properties

The overall lateral earthquake resistance of a system is not the sum of the resistance of its components and the connections between them. Beams, columns, connections and infill panels, if any, interact in a complex manner. Their structural response is not amenable to simple parallel or series system representations, as emphasized in Section 2.2.3. Structures employing either unbraced (MRFs) or braced (CBFs or EBFs) frames as earthquake-resistant systems possess relatively high lateral strength, although their stiffness and ductility vary significantly as also discussed in Section 2.3.1.1 and in Appendix A. Cyclic loading may cause loss of resistance in structural components and the connections between them, especially shear and axial capacities, which, in turn, considerably lower the global strength of the system. Consistent distribution (near-constant ratio of supply-to-demand) of strength in plan and elevation are fundamental prerequisites to avoid concentration of high demand leading to concentrated damage. Interaction between structural and non-structural components, e.g. infill panels, may lead to localized damage in columns. Under lateral loads, the infills behave like struts and may generate high shear forces in the sections at the base and top of columns. In RC structure, adequate transverse reinforcement should be placed in such sections to withstand the additional shear demand imposed by the strut action. Infill panels may also contribute significantly to the storey horizontal strength in addition to the lateral stiffness and ductility. These effects on strength and stiffness are also illustrated in Sections 2.3.1 and 2.2.3, respectively. The enhancement of the storey shear capacity depends on the construction material and on the relative properties of frames and infills (e.g. Fardis and Panagiotakos, 1997; Al Chaar *et al.*, 2002, among many others). For example, in RC weak frame-strong infills, masonry panels contribute significantly to the lateral strength of the system. This contribution may be up to three times or more the corresponding bare frame strength (Mehrabi *et al.*, 1996). Conversely, in strong frames-weak infills, the increase of capacity is relatively lower than that in weak frame-strong infills, e.g. about 40–60%. In steel frames, the enhancement of lateral global strength due to the presence of infills is higher than in RC frames (Moghaddam, 1990). The increase in resistance due to infill panels varies as a function of their slenderness. The lower the slenderness, the higher is the contribution to the lateral resistance (Saneinejad and Hobbs, 1995).

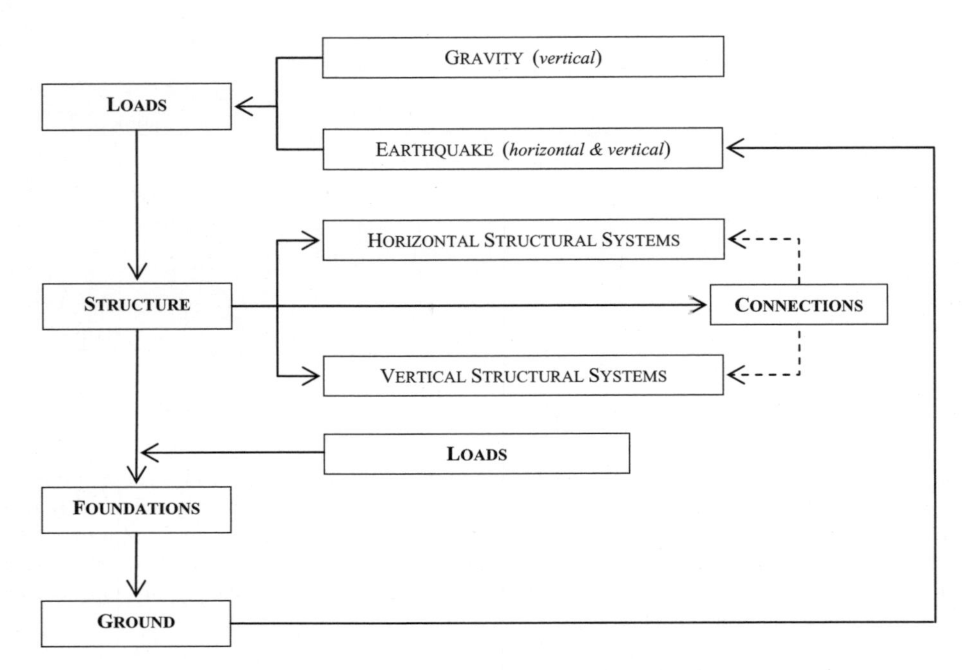

Figure 2.27 Path for vertical and horizontal loads

2.3.2.2 Effects on Load Path

Earthquake-resistant structures should be provided with lateral and vertical seismic force-resisting systems capable of transmitting inertial forces from the location of masses throughout the structure to the foundations. Structures designed for gravity loads have very limited capacity to withstand horizontal loads. Inadequate lateral resisting systems and connections interrupt the load path. Continuity and regular transitions are essential requirements to achieve adequate load paths as shown in Figure 2.27.

In framed structures, gravity and inertial loads generated at each storey are transmitted first to the beams by floor diaphragms (or slabs), then to columns and foundations (Figure 2.28). Mechanical and geometrical properties of beam-to-column and column-to-base connections may alter the load path. Continuity between structural components is vital for the safe transfer of the seismic forces to the ground. Failure of buildings during earthquakes is often due to the inability of their parts to work together in resisting lateral forces as illustrated in Appendix B. Structural damage may occur at any point in the system if lack of sufficient resistance exists at that location. Partial failure does not necessarily cause collapse of a structure. The link between structural components and connections is more complex than the in-series system shown in Figure 2.5, thus confirming the validity of the analogy between structure and road network discussed in Section 2.2.3. When it comes to structural damage, earthquakes are likely to find the 'weakest link' in any complex system and cause damage to the most vulnerable element.

Load paths depend on the structural system utilized to resist vertical and horizontal loads. Structural systems for gravity-load and lateral earthquake-resistance transfer applied actions to the ground through their components, but stresses induced in both systems are different. For example, gravity loads acting on cantilevered bridge piers result in axial actions only as shown in Figure 2.29. Stresses are uniformly distributed in the pier section. Concentrations of stresses are localized at the intersection of the pier and the spread footing used as foundations. In this critical section, vertical load P and self-weight of

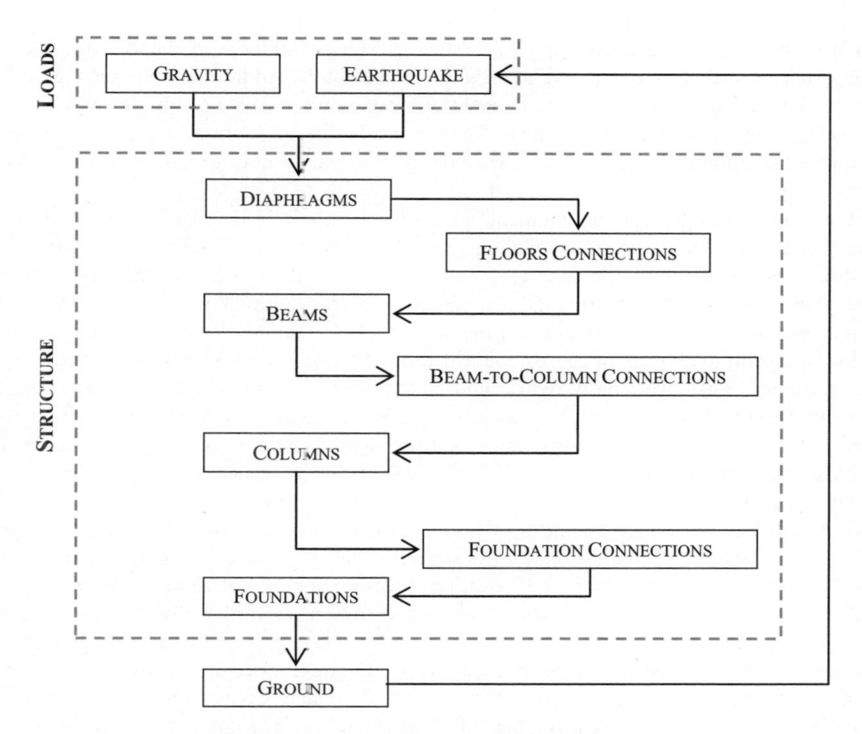

Figure 2.28 Load path in building structures

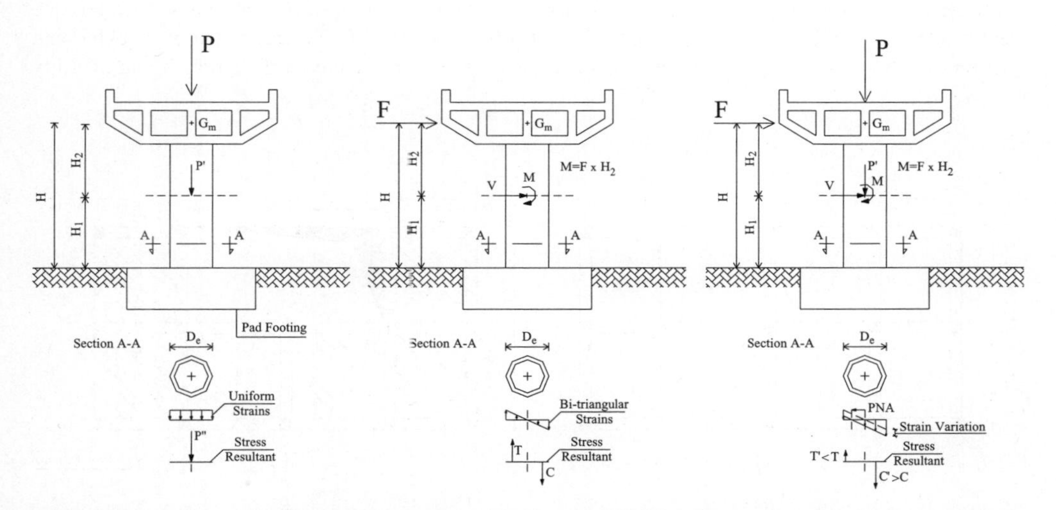

Figure 2.29 Bridge pier subjected to vertical loads (*left*), horizontal loads (*middle*) and combined vertical and horizontal loads (*right*)

the pier W produce the maximum (compressive) stresses. Horizontal forces F due to earthquakes induce bending moments, which vary linearly along the pier height H. Additional compressive and tensile stresses can be caused by the vertical component of ground motion. As a result, the stress distribution in the critical section becomes trapezoidal. Shear stresses due to F are also present. This example demonstrates that combined vertical and horizontal loads in the event of ground motions may increase the stress level within members and connections.

Combined axial load and bending moment in the section at base of the pier in Figure 2.29 may lead to failure because of concrete crushing. Yielding and buckling of the reinforcement bars may also occur. Due to low tensile strength of concrete, cracks appear at low values of horizontal forces. Insufficient shear resistance may also cause structural damage in the pier. Additionally, strength of the soil at pier foundation may be inadequate to withstand the additional vertical and lateral pressures due to earthquake loads. Cyclic loading may cause severe deterioration of the resistance of the RC pier. Shear strength and bond resistance rapidly reduce under large load reversals. Weak anchorage between the superstructure and its foundation may produce horizontal sliding, as discussed above. For very slender piers, e.g. $H/D > 10$ or $H/B > 10$ for circular and rectangular piers, respectively, the eccentricity of vertical loads caused by earthquake-induced horizontal loads may generate second-order P-Δ effects. The latter in turn increase the bending moment in the critical section.

Gravity and earthquake loads should flow in a continuous and smooth path through the horizontal and vertical elements of structures and be transferred to the supporting ground. Discontinuities are, however, frequently present in plan and elevation. Sidestepping and offsetting are common vertical discontinuities which lead to unfavourable stress concentrations as further illustrated in Appendix A. In plan, openings in diaphragms may considerably weaken slab capacities. This reduction of resistance depends on the location, size and shape of the openings. Figure 2.30 depicts an example of stress concentrations caused by a large opening for stairwells in a floor slab. Conversely, small openings do not jeopardize the load transfer at floor level; the diaphragm behaves like a continuous beam under uniform seismic forces. High stress concentrations may also exist at the connection between structural walls and slabs, as well as between columns and flat slabs. Adequate shear capacity should be provided at these connections to prevent localized failures.

Asymmetry in plan and elevation, generated by off-centre structural cores significantly alters the load transfer from the superstructure to the foundations (Paulay, 1998, 2001). Systems with asymmetries may lead to undesirable torsional effects and stress concentrations. Eccentricities between the centre of mass and the centre of resistance can also be generated by the occurrence of cracking and yielding in RC or masonry components. Eccentricities of all types increase the strength and ductility

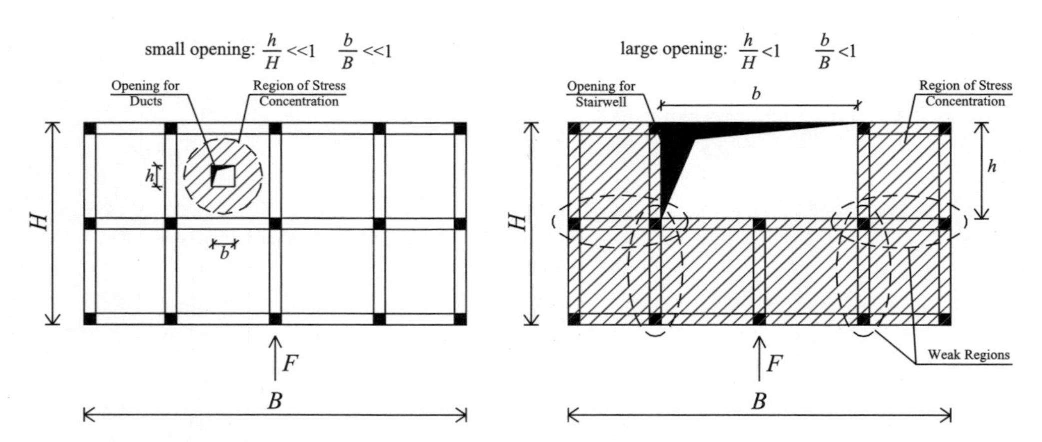

Figure 2.30 Stress concentrations caused by small (*left*) and large (*right*) openings in horizontal diaphragms. *Note.* Small openings do no endanger the in-plane strength of diaphragms

demand on perimeter columns in framed structures and on walls in dual systems as further discussed in Appendix A.

Load paths may also be significantly affected by masonry and concrete infills in framed structures as also discussed in Section 2.3.2.1. Under horizontal loads, as lateral deformations increase, bays of frames and infills deform in flexural and shear modes, respectively. Lateral displacement cycles generate alternating tension and compression diagonals in wall panels. Infilled frames behave like unidirectional braced systems. Relative displacements between frame and infills cause their separation along the tension diagonal while struts are formed on the compression diagonal. The above separation occurs at 50% to 70% of the shear capacity of the infill for RC frames (Paulay and Priestley, 1992), and at much lower values of lateral forces for steel frames (e.g. Bruneau *et al.*, 1998). The premature localized failure of masonry may lead to severe damage concentrations and even to collapse of the structure, since it may increase irregularity leading to concentration of demand. Stress concentrations may be generated at the intersection of infills with beams and columns. Strut action increases shear forces at both top and bottoms of columns. Failure of infills may cause the failure of the reinforced concrete frame or even the collapse of the entire structure. For example, premature failure of the infills in the first floor will cause soft storey and the structural collapse as observed after the 1999 Kocaeli (Turkey) earthquake (Elnashai, 1999).

Masonry and concrete infills are generally distributed non-uniformly in plan and elevation. Irregular layouts of infills may generate considerable torsional effects and lead to high stress concentrations, especially in columns.

2.3.2.3 Structural Damage Control

Strength is generally associated with the control of structural damage. Strength failure may be caused by the accumulation of stresses beyond the capacity of materials, members and connections in the structure. The occurrence of damage in structural components can also be associated with the onset of target values of strains (materials), curvatures (sections), rotations (elements and connections), inter-storey drifts (sub-system) and global drift (systems). Damage control can be achieved at both local and global levels. Target values of strains, curvatures, rotations and drifts are utilized to limit local and global damage. It is recommended that limit states at all levels are defined and continuously assessed since no one single quantity is suited to controlling all levels of damage.

In general, damage increases as the load and deformation resistance is lowered. In seismic design it is, however, cost-effective to allow for the occurrence of a limited amount of controlled structural damage. For example in RC structures, repairable damage includes spalling of concrete cover and formation of flexural cracks; fracture and buckling of reinforcement bars are not readily repairable. Damaged structures may or may not collapse depending on the inelastic deformation capacity (ductility), which is discussed in Section 2.3.3.

Under earthquake loads, stress concentrations may occur in critical regions of structures, e.g. sections with maximum bending and shear forces, high axial compression or net tensile forces. These concentrations are also typically observed where abrupt changes in the structural layout are present, as discussed in the previous section. Large reversing actions may lead to stiffness and strength deteriorations. Stiffness and strength degradation accelerate the occurrence of failure. In seismic areas, it is desirable that shear resistance should be significantly higher than flexural capacity, a target that can be achieved by applying capacity design. Shear strength and stiffness deteriorate much faster than their flexural counterparts. Shear effects often become dominant under large amplitude cyclic loads and failure occurs. This is the case for structural walls which may experience shear failure.

Limiting damage in beam-to-column and foundation connections is essential to achieve adequate performance of the structural system. Excessive cracking and bond deterioration should be prevented especially in RC joints. Reductions in joint shear capacities are detrimental to the seismic performance of framed systems. Damage in beam-to-column joints significantly increases the inter-storey drifts and may endanger the stability of the structure as a whole. Stress concentrations and additional shear generated by the strut action of masonry and concrete panels should be accounted for in the design and assessment of beam-to-column joints of infilled frames. In the case of poorly designed infilled frames, e.g. very strong walls in a weak frame, failure can occur due to the premature failure of beam-to-column joints or columns (e.g. Schneider *et al.*, 1998). Damage in steel beam-to-column connections has been observed in the mid-nineties, after a period when steel frames were considered far superior to RC alternatives. Damage was often associated with a particular connection configuration referred to as hybrid connection, where moments are resisted by top and seat angles, while shear is resisted by shear tabs welded to the column and bolted to the beam (e.g. Di Sarno and Elnashai, 2002).

Sliding of structures resting on shallow foundation blocks and sliding of tiled roofs in low-rise wood and masonry constructions are due to shear effects. High shear resistance and adequate anchorage between structural components can prevent these failure modes.

In framed systems with no structural walls, strength and stiffness often increase proportionally, especially in RC and composite structures. In steel frames, strength enhancement is generally higher than the increase in lateral stiffness. Unfavourable failure of members and structures may be caused by high stiffness and inadequate strength. For example, short stiff columns attract high shear loads as shown in Section 2.3.1.1. If these members do not possess sufficient shear resistance, failure may occur. In wall systems, the link between strength and stiffness can be broken. Consequently, the former structural response parameter can be increased without a commensurate increase in the latter. In so doing, in seismic retrofitting, either traditional interventions or novel techniques, e.g. wrapping with fibre-reinforced plastic (FRP) or post-tensioning with smart materials, e.g. FRP tendons or shape memory alloys wires, may be utilized.

Overturning moments caused by horizontal seismic loads tend to tip over the superstructure with or without its foundations. This mechanism is referred to as 'uplift'. Deep foundations are often more effective in resisting overturning moments than shallow footings because of friction activated along the lateral surface of embedded piles. Overturning can also generate net tension and excessive compression in columns. Axial actions induced by seismic horizontal forces can exceed those due to gravity loads. By reducing the axial compression in RC members, the shear capacity is lowered. Fracture and buckling of reinforcement steel bars (RC structures) and structural steel components (steel and composite structures) can be attributed to high overturning moments. Reductions of shear capacity in RC are also caused by vertical components of earthquake ground motions as observed in several bridge piers during past earthquakes and illustrated in Appendix B.

Problem 2.3

Consider the single-storey dual system shown in Figure 2.31. To distribute the seismic force F_y among lateral resisting elements, i.e. frames and structural walls, the following equation is employed:

$$V_{yi} = \frac{k_{yi}}{\sum_{i=1}^{M} k_{yi}} F_y \pm \frac{k_{yi}\, d_{xi}}{\sum_{j=1}^{M} k_{yj}\, d_{xj}^2} M_t \tag{2.9}$$

where k_{yi} are the lateral stiffness of the moment-resisting systems along the y-direction. The distances of these systems from the centre of stiffness C_R are d_{xj}; M_t is the torsional moment. Derive the relationship in equation (2.9). Does the relationship hold for both elastic and inelastic systems?

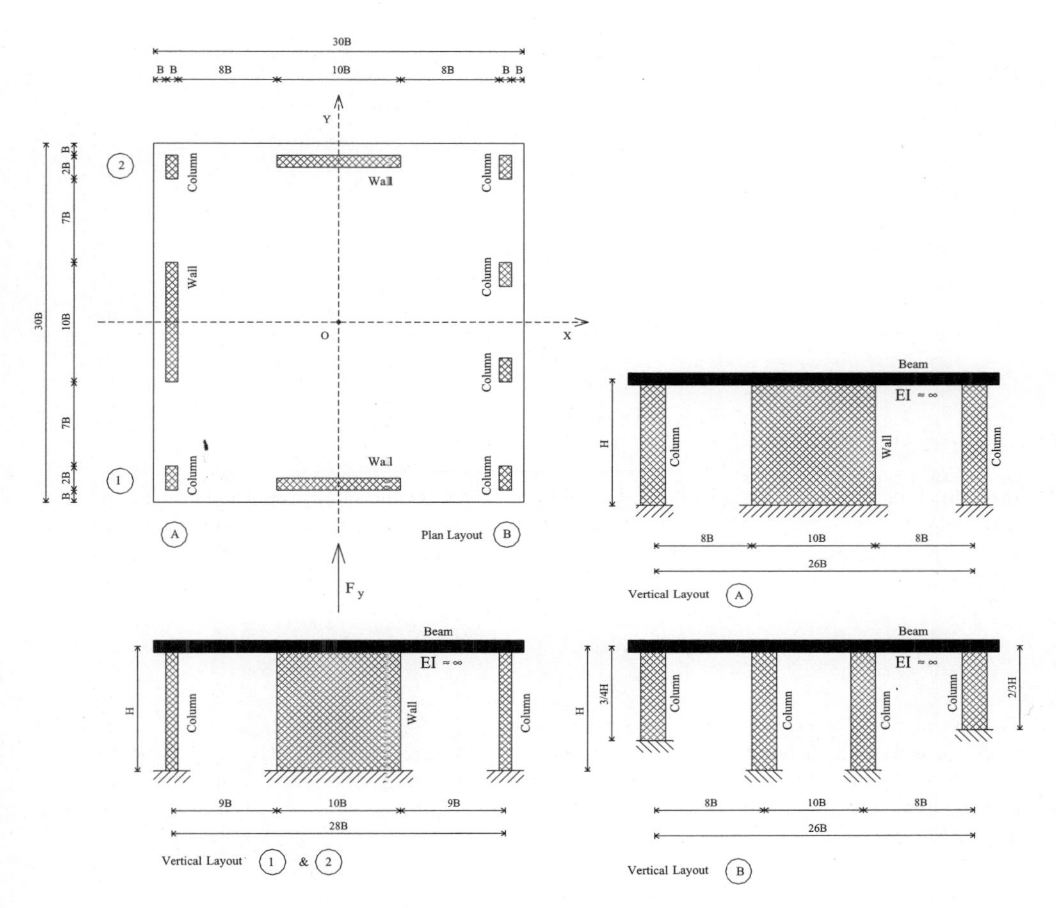

Figure 2.31 Single-storey lateral resisting system

2.3.3 Ductility

Ductility is defined as the ability of a material, component, connection or structure to undergo inelastic deformations with acceptable stiffness and strength reduction. Figure 2.32 compares the structural response of brittle and ductile systems. In the figure, curves A and B express force–displacement relationships for systems with the same stiffness and strength but distinct post-peak (inelastic) behaviour. Brittle systems fail after reaching their strength limit at very low inelastic deformations in a manner similar to curve A. The collapse of brittle systems occurs suddenly beyond the maximum resistance, denoted as V_{max}, because of lack of ductility. Conversely, curve B corresponds to large inelastic deformations, which are typical of ductile systems. Whereas the two response curves are identical up to the maximum resistance V_{max}, they should be treated differently under seismic loads. The ultimate deformations δ_u corresponding to load level V_u are higher in curve B with respect to curve A, i.e. $\delta_{u,B} >> \delta_{u,A}$.

Most structures are designed to behave inelastically under strong earthquakes for reasons of economy. The response amplitudes of earthquake-induced vibrations are dependent on the level of energy dissipation of structures, which is a function of their ability to absorb and dissipate energy by ductile deformations. For low energy dissipation, structural systems may develop stresses that correspond to relatively large lateral loads, e.g. accelerations of 0.5 g to 1.0 g were observed (Housner, 1956). Consequently,

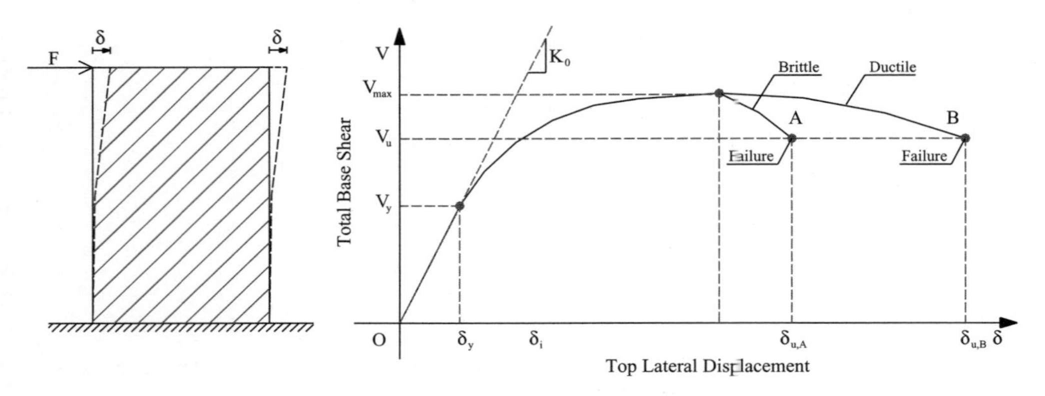

Figure 2.32 Definition of structural ductility

such structures should be designed to withstand lateral forces of the same proportion of their weight to remain in the elastic range. This is uneconomical in all practical applications with the exception of nuclear power plants, offshore platforms and water- and fluid-retaining structures, alongside other safety-critical structures.

The general analytical definition of displacement ductility is given below:

$$\mu = \frac{\Delta_u}{\Delta_y} \tag{2.10}$$

where Δ_u and Δ_y are displacements at ultimate and yield points, respectively. The displacements Δ may be replaced by curvatures, rotations or any deformational quantity. The ratio μ in equation (2.10) is referred to as 'ductility factor'. The following types of ductility are widely used to evaluate structural response:

(i) *Material ductility* (μ_ε) characterizes material plastic deformations.
(ii) *Section (curvature) ductility* (μ_χ) relates to plastic deformations of cross sections.
(iii) *Member (rotation) ductility* (μ_θ) quantifies plastic rotations that can take place in structural components such as beams and columns. This type of ductility is often also used for connections between structural members.
(iv) *Structural (displacement) ductility* (μ_δ) is a global measure of the inelastic performance of structural sub-assemblages or systems subjected to horizontal loads.

The conceptual relationship between local and global ductility is displayed in Figure 2.33, which reflects the hierarchical link between structural response levels illustrated in Figure 2.4 of Section 2.2.3.

The inelastic performance of structures may significantly vary with the displacement history (e.g. Akiyama, 1985; Wakabayashi, 1986; Usami *et al.*, 1992, among others). Consequently, under load reversals, the definition of ductility factor provided in equation (2.10) may not reflect the actual maximum deformations experienced by the structure because of the cyclic response under earthquake loads, residual plastic deformations, and cyclic stiffness and strength degradation. Alternatively, the following definitions may be adopted for the ductility factor:

(i) *Definition of ductility factor based on cyclic response:* the factor μ is related to the cyclic deformations as given below:

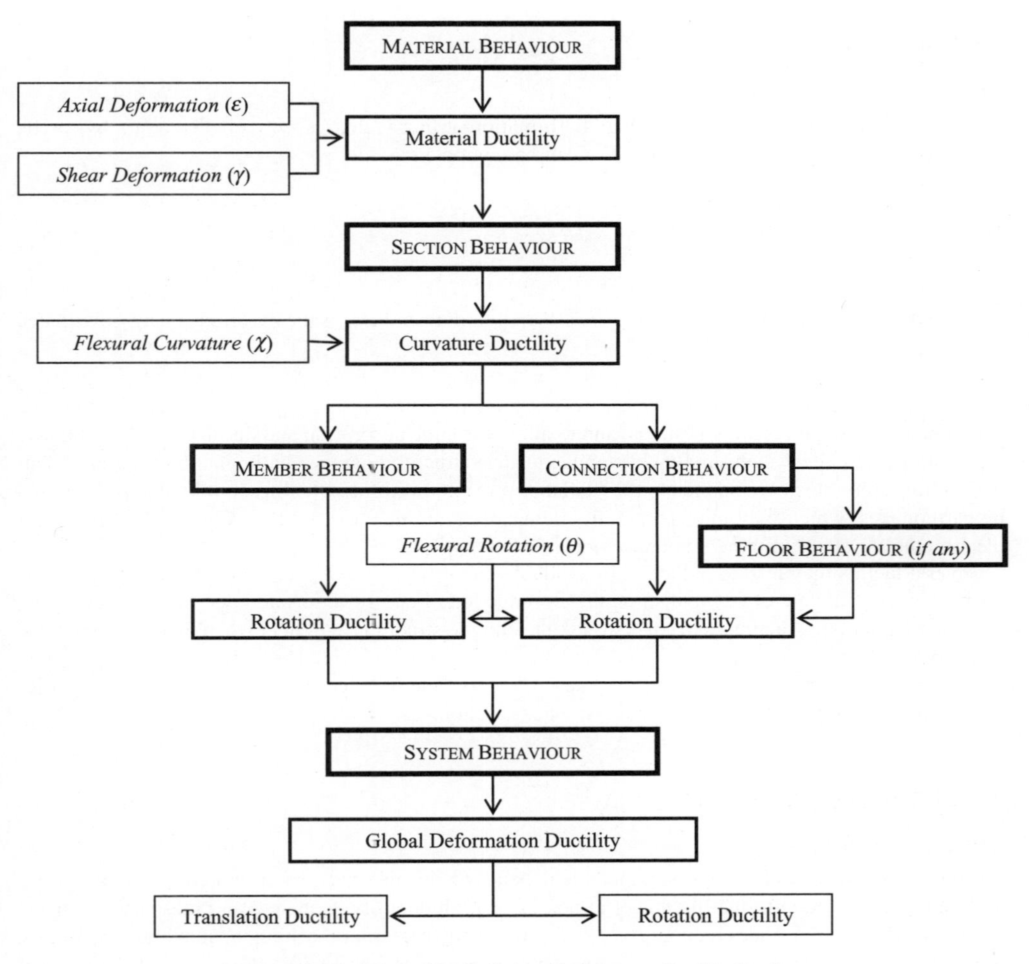

Figure 2.33 Hierarchical relationship between ductility levels

$$\mu = \frac{|\Delta_{max}^{+}| + |\Delta_{max}^{-}|}{|\Delta_{y}^{+}| + |\Delta_{y}^{-}|} \tag{2.11}$$

where Δ_{max}^{+} and Δ_{max}^{-} are the positive and negative ultimate deformations, respectively; Δ_{y}^{+} and Δ_{y}^{-} the corresponding deformation at the yield point.

(ii) *Definition of ductility factor based on total hysteretic energy:* the entire response history of the system is accounted for by the total hysteretic dissipated energy $E_{t,H}$ and the ductility factor can be expressed as below:

$$\mu = \frac{E_{t,H}}{E_{E}} \tag{2.12.1}$$

where E_{E} is the elastic energy, also referred to as 'strain energy', at yield and is given by:

$$E_E = \frac{1}{2} F_y \, \delta_y \tag{2.12.2}$$

where F_y and δ_y are the action and deformation at first yield, respectively. The total hysteretic energy dissipated before failure $E_{t,H}$ can be computed as follows:

$$E_{t,H} = \sum_{i=1}^{N} E_{i,H} \tag{2.12.3}$$

where the summation is over all cycles N up to failure and $E_{i,H}$ is the hysteretic energy dissipated in the ith cycle.

In seismic design, high available ductility is essential to ensure plastic redistribution of actions among components of lateral resisting systems, and to allow for large absorption and dissipation of earthquake input energy. Ductile systems may withstand extensive structural damage without collapsing or endangering life safety; this corresponds to the 'collapse prevention' limit state. Structural collapse is caused by earthquakes, which may impose ductility demand μ_d that may exceed the available ductility μ_a of the structural system. Imminent collapse occurs when $\mu_d > \mu_a$.

Several factors may lead to reduction of available ductility μ_a. These include primarily (i) strain rate effects causing an increase in yield strength, (ii) reduction of energy absorption due to plastic deformations under alternating actions, (iii) overstrength leading to structures not yield when they were intended to yield and (iv) tendency of some materials to exhibit brittle fracture. These factors may affect both local and global ductility. The effects of material, section, member, connection and systems properties on the structural ductility are discussed in the next section.

2.3.3.1 Factors Influencing Ductility

(i) Material Properties

The ductility of structural systems significantly depends on the material response. Inelastic deformations at the global level require that the material possesses high ductility. Concrete and masonry are brittle materials. They exhibit sharp reductions of strength and stiffness after reaching the maximum resistance in compression. Both materials possess low tensile resistance, which is followed by abrupt loss of strength and stiffness. The material ductility μ_ε can be expressed as the ratio of the ultimate strain ε_u and the strain at yield ε_y, i.e. $\mu_\varepsilon = \varepsilon_u/\varepsilon_y$. Consequently, the ductility μ_ε of concrete and masonry in tension is equal to unity, while μ_ε is about 1.5–2.0 in compression. For concrete, the higher the grade, the lower is the inelastic deformation capacity. Metals and wood exhibit much higher values of μ_ε. Mild steel has average values of material ductility of 15–20 if ultimate strains ε_u are limited to the incipient strain hardening ε_{sh}, i.e. $\varepsilon_u = \varepsilon_{sh}$ as shown in Figure 2.34. Values of μ_ε in excess of 60–80 can be obtained by using the deformation at ultimate strength. Similarly, metal alloys, such as aluminium and stainless steels, exhibit values of material ductility as high as 70–80. These alloys do not possess clear yield points and a conventional 'proof stress', i.e. stress corresponding to 0.2% residual strain, is utilized to define the elastic threshold (e.g. Di Sarno and Elnashai, 2003, among others).

Steel reinforcement can be utilized in plain concrete and masonry to enhance their ductility. Steel-confined concrete exhibits inelastic deformations 5–15 times higher than plain concrete (CEB, 1996). Strain at maximum compressive strength is about 0.3–0.4% for almost all grades of concrete. Unconfined concrete exhibits very limited ductility μ_ε in compression. Confinement limits the post-peak strength reduction, thus increasing the residual resistance. Ductility of concrete is significantly enhanced due to confinement provided by transverse steel reinforcement. Experimental simulations have also

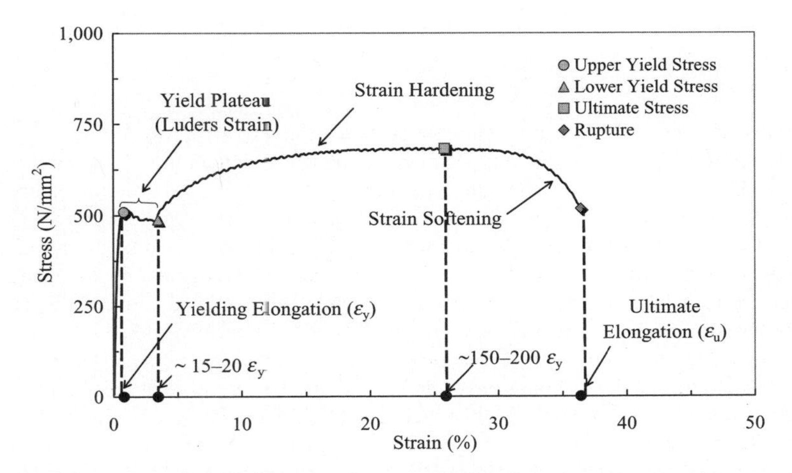

Figure 2.34 Inelastic material response for mild steel under monotonic loads

demonstrated that circular spirals confine concrete more effectively than rectangular or square hoops (Park and Paulay, 1975). Circular confinement bars provide uniform confining pressure around the circumference because of their shape. Confined concrete is subjected to multi-axial stress states, which is beneficial for both strength and ductility.

Earthquakes cause alternating loads, thus action–deformation relationships generate hysteretic loops. The cyclic inelastic response of materials should be used to evaluate the ductility μ_ε. In so doing, equations (2.11) and (2.12) can be utilized. Several factors influence the inelastic cyclic response of materials; the most common include stiffness and strength degradation. The latter reduces the energy dissipation capacity of the material. The amount of energy absorbed at a given deformation level corresponds to the total area under the action–deformation curves. This hysteretic energy absorption is often replaced by equivalent damping in analytical formulations. The dissipated energy in a cycle of deformation is referred to as 'hysteretic energy' or 'hysteretic damping' and is further discussed in Section 2.3.5. Ductility is directly related to energy dissipation; high ductility is required to dissipate large amounts of seismic energy. For conventional construction materials, high energy absorption is associated with high levels of damage, since energy can only be absorbed and dissipated by irreversible deformations. For novel and smart materials, such as shape memory alloys or viscous fluids, large amount of seismic energy can be dissipated with limited structural damage (e.g. Di Sarno and Elnashai, 2003, among others). Strain softening and strain hardening should be accounted for in the evaluation of inelastic response for masonry, RC and steel, as appropriate. Strain softening, which typically affects the post-peak response of plain concrete and masonry, involves loss of strength with increasing strain. Strain softening may be reduced by providing adequate transverse confinement of the material. Spreading of plasticity within members of both steel and RC structures is controlled by the strain hardening of steel as shown in the following sections. The higher the ratio between the ultimate and yield strength, f_u and f_y, respectively, the higher the spreading of inelasticity. Mild steel possesses values of f_u/f_y ratios, which typically range between 1.10 and 1.30. Other metals, such as aluminium alloys and stainless steels, exhibit much higher values of f_u/f_y, e.g. values of about 2.0–2.2 (Di Sarno *et al.*, 2003). Spreading of inelasticity is a convenient means to reduce concentrated inelastic demands, thus preventing brittle failures.

(ii) Section Properties

The ductile response of cross sections of structural members subjected to bending moments is generally measured by the curvature ductility μ_χ, which is defined as follows:

$$\mu_\chi = \frac{\chi_u}{\chi_y} \tag{2.13}$$

where χ_u and χ_y are the ultimate and yield curvatures, respectively.

In RC structures, the curvature ductility significantly depends on the ultimate concrete compressive strain ε_{cu}, the compressive concrete strength f_c, the yield strength of the steel reinforcement bars f_y, the stress ratio f_u/f_y of reinforcement steel, the ratio of compression-to-tension steel A_s'/A_s and the level of axial load $\nu = N/A_c f_c$. By increasing the ultimate concrete strains ε_{cu}, e.g. through transverse confinement, the curvature ductility is enhanced; thus confined concrete behaves in a ductile manner. The use of high-strength steels increases the yield curvature χ_y, while values of χ_u do not change. The net effect is that these types of steels reduce the curvature ductility μ_χ. Conversely, increases in the stress ratio f_u/f_y of reinforcement steel, increase the curvature ductility. Adding reinforcement steel bars in compression is beneficial to the ductile response of RC cross sections. The presence of axial compression loads increases the depth of the neutral axis, both at yield and at ultimate limit states. The yield curvature χ_y increases while the ultimate curvature χ_u decreases. Consequently, the ductility μ_χ is lowered. An increase in the normalized axial loads ν from 10% to 30% of squash load leads to a reduction in curvature ductility to one-third. Dimensionless axial loads ν in columns of RC framed structures should not exceed values of 0.15–0.20 to achieve adequate curvature ductility. Transverse confinement is an effective countermeasure to the reduction in ductility caused by axial loads. The effects of axial loads and confinement on the ratio χ_u/χ_y for RC cross sections are shown in Figure 2.35.

The variation of μ_χ with the aforementioned design parameters, for practical values of RC cross-section dimensions and steel reinforcement layouts, is summarized in Table 2.4.

Curvature ductility in RC members can also be affected by the presence of shear forces. Transverse confinement, which is used to confine plain concrete, increases the shear strength of structural components. Consequently, flexural inelastic response is not fully developed prior to shear distress. In steel structures, shear-flexure interaction does not generally affect the section ductility. On the other hand, the presence of axial loads considerably reduces the curvature ductility μ_χ in both steel and composite cross sections. As a result, dimensionless axial loads ν should not exceed 0.15–0.20 as for RC structures.

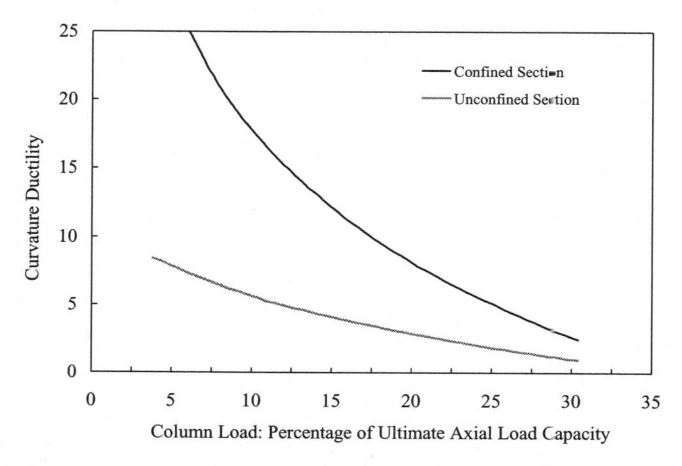

Figure 2.35 Variations of curvature ductility as a function of the level of axial loads and transverse confinement (*adapted from* Blume *et al.*, 1961)

Table 2.4 Variation of curvature ductility in reinforced concrete members as a function of different design parameters.

Parameters	Curvature ductility	
	Increment	Decrement
Ultimate concrete compressive strain (ε_{cu})	↑	↓
Compressive concrete strength (f_c)	↑	↓
Reinforcement steel yield strength (f_y)	↓	↑
Overstrength of steel reinforcement (f_u/f_y)	↑	↓
Compression to tension ratio of steel (A_s'/A_s)	↑	↓
Level of axial load ($\nu = N/A_c f_c$)	↓	↑

Key: ↑ = increase; ↓ = decrease.

To achieve high curvature ductility, it is essential to limit the depth of neutral axis in cross sections of plastic hinges of beam elements. For RC members, extensive experimental and numerical analyses have shown that the position of the neutral axis, expressed in dimensionless form, should not exceed 0.25 (CEB, 1996). This value ensures that curvature ductility as high as 10–15 can be reached. For steel structures, the curvature ductility may be undermined by the occurrence of local buckling. It is thus of importance to utilize sections with low slenderness. Generally, adequate width-to-thickness ratios are employed to ensure that the section behaviour is governed by plastic capacity rather than by local buckling.

(iii) Member Properties

An adequate metric for ductile behaviour of structural members is the rotation ductility factor μ_θ given by:

$$\mu_\theta = \frac{\theta_u}{\theta_y} \tag{2.14}$$

where θ_u and θ_y are the ultimate and yield rotations, respectively. These rotations are directly estimated from the ultimate and yield curvatures χ_u and χ_y, respectively, defined in the previous section. The rotations θ_u and θ_y are indeed computed by integrating the curvature distributions χ_u and χ_y along the member length.

Inelasticity is concentrated in flexural plastic hinges at the ends of beams and columns. It is often assumed that curvatures within plastic hinges are constant, thus allowing plastic rotations θ_p to be expressed as follows:

$$\theta_p = \chi_p L_p \tag{2.15}$$

where χ_p is the plastic curvature and L_p the length of the plastic hinge. Figure 2.36 depicts a typical bending moment distribution in a RC cantilever column. It is assumed that the structural member is moderately reinforced such that the moment–curvature relationship can be assumed elasto-plastic. The theoretical distribution of curvature is indicated by the broken lines in the figure. The abrupt change at the base of the component from χ_y to χ_u is not practically possible because strains in concrete cannot vary so rapidly. Likely distributions of yield and ultimate curvatures are given by the jagged thick line. These distributions lead to curvatures smaller than those predicted theoretically at points away from the fixed end. The underestimation is caused by the tension stiffening effect of concrete in the cross sections between flexural cracks. At the base, theoretical predictions provide values lower than those

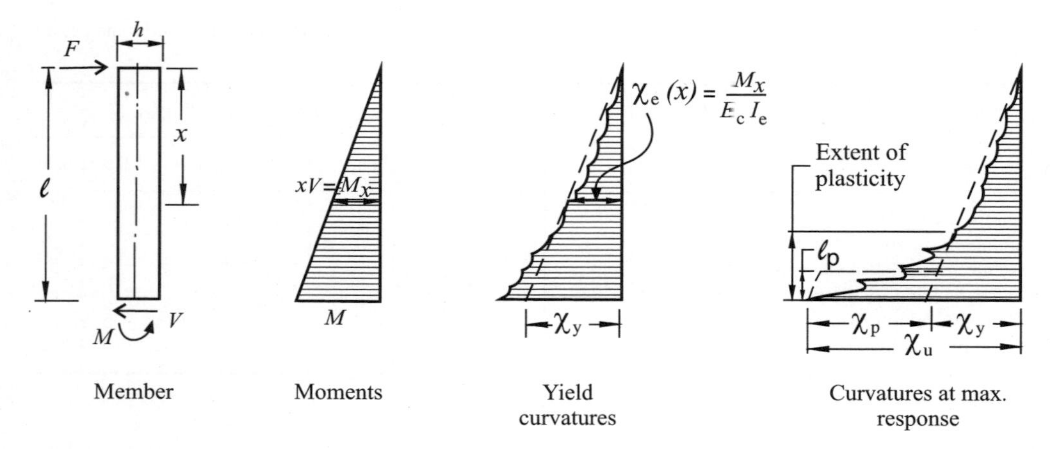

Figure 2.36 Bending moment diagram, curvature distribution and plastic hinge length in a cantilever column

estimated from the likely curvature distributions. The ductility of a frame member depends on the spreading of inelasticity, which takes place in the region corresponding to the plastic hinge of length L_p in Figure 2.36. Longitudinal steel bars elongate beyond the base of the cantilever member given in Figure 2.36 because of the finite bond stress. This elongation causes additional rotation and deflection in the member; this response is referred to as yield penetration. Additionally, interactions between flexural- and shear-induced cracks increase the spreading of plasticity in the critical region.

Plastic hinges should be located in beams rather than in columns since the columns are responsible for the gravity load resistance, hence the stability of the structure against collapse. Shear capacity of both beams and columns should always be higher than flexural strength to avoid brittle shear failure.

To ensure adequate rotational ductility (e.g. $\mu_\theta \geq 10$–15) in flexural plastic hinges, it is necessary to carefully detail critical regions (plastic hinges). For example, in RC members, it is essential to provide closely spaced stirrups, which confine effectively the concrete and use sufficient lap splices and anchorage lengths. For steel and composite members, cross sections employing plates with low width-to-thickness ratios in plastic hinge regions are necessary in order to avoid local buckling.

(iv) Connection Properties

The behaviour of connections (e.g. beam-to-column in MRFs, brace-to-column and brace-to-beam in either CBFs or EBFs, and those between superstructure and foundation systems) affects significantly the global ductile response of structures. In RC frames, the ductile behaviour of joints is a function of several design parameters, which include, among others: (i) joint dimension, (ii) amount of steel reinforcement, (iii) bond resistance, (iv) level of column axial loads, and (v) presence of slab and transverse beams framing into the connection. All other parameters being equal, by increasing joint dimensions lower shear stresses are generated. The advantage of increasing column depths is twofold. Joint shear stresses are considerably reduced and bond demands on longitudinal steel reinforcement of beam bars passing through the joint are minimized. Both effects prevent brittle failure modes in RC beam-to-column joints, i.e. loss of bond resistance along the joint boundary, inability to resist high stresses caused by perimeter bond actions and inability to sustain diagonal compression strut in the joint core. Brittle failure due to low shear capacity can be prevented by adequately confining the joint by hoops. In so doing, shear strength and bond resistance are enhanced. Occurrence of bond failure endangers the ductile behaviour of frames and should be prevented when designing RC joints. Similarly, the presence of slabs may erode the ductility of beam-to-column connections because of the additional shear demand caused by the raised beam moment. Effects of column axial loads reduce the total lateral drift at yield, which is beneficial for the ductile response (Kurose, 1987). Nevertheless, as the vertical stresses increase in the joint, the contribution of the diagonal compression strut to the shear resistance is lowered. The net effect is that by increasing the column axial loads, the joint ductility is impaired. Transverse beams enhance joint shear resistance and provide concrete confinement, which in turn improves the ductility.

The global ductility of steel and composite structures also depends on the response characteristics of the connections, especially beam-to-column and column-to-foundation connections. The ductility of beam-to-column connections is controlled by yield mechanisms and failure modes. For welded-flange shear-tab connections, yield is by flexural yielding of beam and shear yielding of the panel zone. Possible failure modes include fracture at welds or at weld access holes, lateral torsional buckling, local web and flange buckling, excessive plastic deformation of panel zone, beam, column web and flanges (Roeder, 2002). Multiple yield mechanisms may contribute to plastic rotations if their resistances are all lower than the strength of the critical failure mode of the connection. Multiple yield mechanisms rather than a single yield mechanism are generally desirable to achieve adequate seismic performance. By sizing members and connections, it can be ensured that the most desirable yield mechanism occurs first. On the other hand, failure modes cause fracture, tearing or deterioration of connection capacity. Similar to yield mechanisms, all connections have a number of likely failure modes. The critical failure mode is that with the lowest resistance of all possible modes for the given connection. The ductility and the inelastic performance of a connection are controlled by the proximity of the critical failure mode resistance to the controlling yield mechanism resistance. Connections with a controlling yield mechanism resistance significantly lower than the critical failure mode resistance develop considerable inelastic deformations and therefore exhibit high plastic rotations capacity.

Finally, the ductile response of the lateral resisting systems is dependent on the response characteristics of connections between foundation and superstructure. For example, for cantilever structural walls, the seismic detailing of the base connection is of great importance to achieve adequate roof lateral displacement ductility. For RC and composite walls, sufficient anchorage lengths of steel reinforcement bars and closely spaced transverse confinement of concrete in the lower part of the cantilever structure are essential requirements to account for in the design.

(v) System Properties

The most convenient parameter to quantify the global ductility of structural systems under earthquake loads is the displacement or translation ductility μ_δ, which is defined as given in equation (2.10). Displacement ductility factors μ_δ should be expressed as storey drift ductility rather than roof lateral displacements, as discussed in the preceding sections. Storey translational ductility is a measure of the ductility distribution along the height in multi-storey frames and can be utilized to detect localized inelastic demands in irregular structures. For example, two frames may possess the same values of roof translational ductility μ_δ although the distribution of the storey drift ductility is different along the height.

The evaluation of the deformation quantities δ_u and δ_y from action–deformation relationships, similar to those provided in Figure 2.32, is not always straightforward. Yield points are often ill-defined because of non-linearities and formation of plastic hinges in beams, columns and joints. Response curves of RC structures frequently do not present well-defined yield points because of cracking of concrete and sequential yielding of reinforcement bars, as also discussed in Section 2.3.2. Various definitions for yield deformations (Figure 2.37) have been proposed as summarized below (Park, 1988):

(a) Deformation corresponding to first yield;
(b) Deformation corresponding to the yield point of an equivalent elasto-plastic system with the same elastic stiffness and ultimate load as the real system;
(c) Deformation corresponding to the yield point of an equivalent elasto-plastic system with the same energy absorption as the real system;

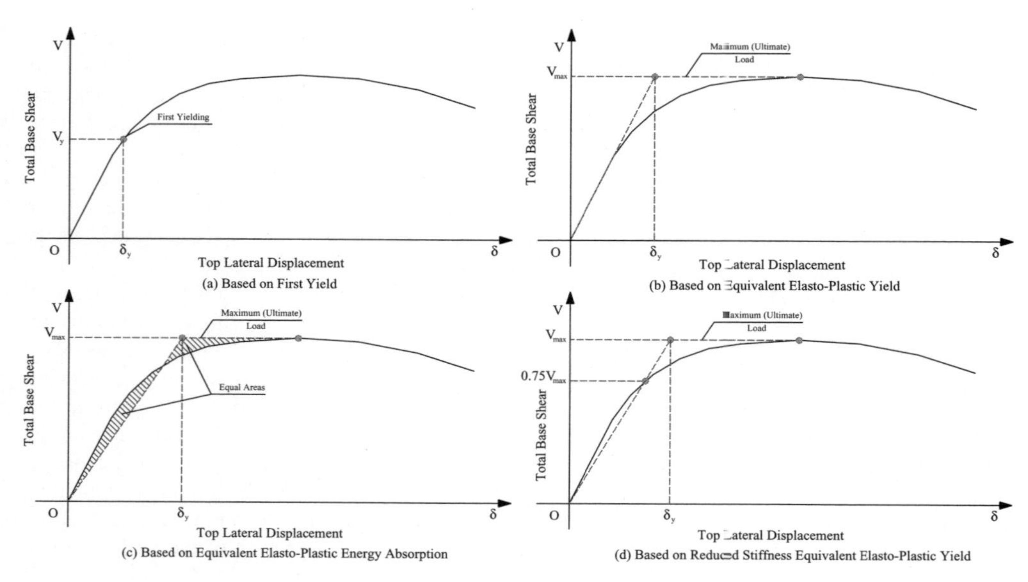

Figure 2.37 Definitions of yield deformations

(d) Deformation corresponding to the yield point of an equivalent elasto-plastic system with reduced stiffness computed as the secant stiffness at 75% of the ultimate lateral load of the real system.

The use of secant stiffness accounts for the reduction of structural stiffness due to cracking at the elastic limit; the latter is the most realistic definition for yield deformation in RC structures. Definitions given above are derived primarily for RC structures. Equivalence may be drawn to other materials to render them applicable to all materials of construction. Similarly, definitions for ultimate deformations are as follows (Park, 1988):

(a) Deformation corresponding to a limiting value of strain;
(b) Deformation corresponding to the apex of the load–displacement relationship;
(c) Deformation corresponding to the post-peak displacement when the load-carrying capacity has undergone a small reduction (often taken as 10–15%);
(d) Deformation corresponding to fracture or buckling.

Ductile structures usually have post-peak load strength and their load–deformation curves do not exhibit abrupt reduction in resistance, especially for MRFs. Definition of ultimate deformations given in Figure 2.38a,b may underestimate the actual structural response. Hence the most realistic definitions are those given in Figures 2.38c,d, because they account for the post-peak deformation capacity.

The global ductility of structural systems significantly depends on available local ductility. Large inelastic deformations and large amounts of energy dissipation require high values of local ductility (e.g. Elnashai, 1994, among others). Adequate inelastic behaviour of structures under severe earthquakes can only be achieved if curvature ductility factors μ_χ are much higher than displacement ductility factors μ_δ. It is therefore necessary to design RC, steel and composite structures with seismic details, which ensure that μ_χ-values are three to four times higher than μ_δ. Relationships between curvature and displacement ductility can be derived on the basis of structural mechanics. For example, for the cantilever bridge pier shown in Figure 2.39 and subjected to horizontal seismic force at the upper-deck level, the relationship between μ_δ and μ_χ can be expressed as follows:

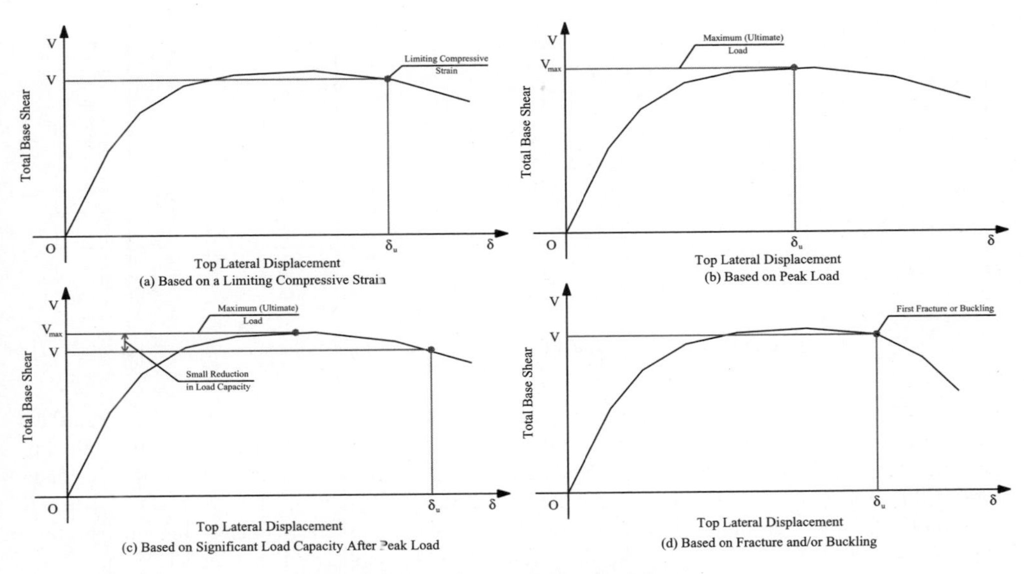

Figure 2.38 Definitions of ultimate deformations

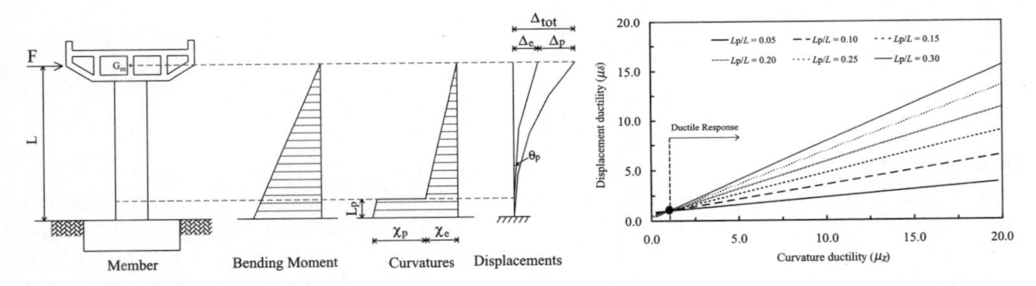

Figure 2.39 Relationship between local and global ductility for cantilever systems: free body and deflection diagrams (*left*) and variation of displacement ductility as a function of geometric layout (*right*)

$$\mu_\delta = 1 + 3\frac{L_p}{L}\left[(\mu_\chi - 1)\left(1 - 0.5\frac{L_p}{L}\right)\right] \tag{2.16}$$

where L_p and L are the plastic hinge length and the cantilever height, respectively. Thus, for ordinary cross sections of columns and piers, to obtain global ductility factors $\mu_\delta = 4$–5 the required μ_χ-values range between 12 and 16. The relationship in equation (2.16) accounts for total horizontal deflections δ_u generated solely by flexural deformations and assumes fixed base for the cantilever; the ultimate lateral displacement δ_u is given by:

$$\delta_u = \delta_y + \delta_p \tag{2.17}$$

where δ_y and δ_p are the yield and plastic displacements, respectively.

Shear deformations of the member, connection flexibility (*see* for example Section 2.3.1.1) and soil-structure interaction increase the yield displacement δ_y. Conversely, plastic displacements δ_p remain

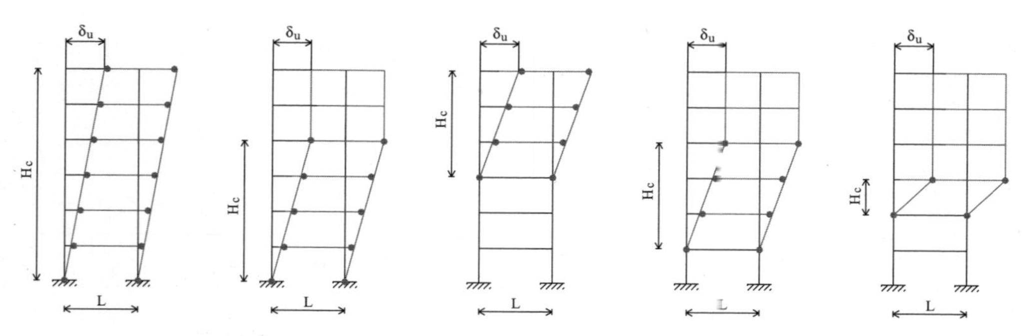

Figure 2.40 Typical plastic mechanisms for framed systems

unchanged because they are caused only by plastic rotations of the cantilever member. As a result, the displacement ductility factor μ_δ is reduced. To increase μ_δ, higher values of curvature ductility are thus required. Relationships similar to that in equation (2.16) can be derived for different boundary conditions of the structural system and combined effects of horizontal and vertical actions.

In multi-storey framed buildings, plastic lateral displacements δ_p are frequently higher than those estimated for simple cantilever systems as that shown in Figure 2.39. The displacements δ_p include the contribution from different sources of deformations, such as flexural and shear flexibilities in both beams and columns, joint flexibility, horizontal and rotational flexibility of the foundation system.

Inelastic lateral displacements of ductile frames are often larger at lower storeys, where P-Δ effects are also significant. Inelastic storey drifts are correlated to plastic hinge rotations θ_p; similarly, plastic roof drifts δ_p are related to θ_p through the following:

$$\delta_p = \delta_u - \delta_y = \theta_p H_c \tag{2.18}$$

where H_c is the sum of the inter-storey height of stories involved in the collapse mechanism as shown in Figure 2.40. Global mechanisms with plastic hinges at column base and within beams are preferred due to the higher energy dissipation capacity. Consequently, to ensure adequate energy dissipation and prevent dynamic instability of the system as a whole, plastic hinges at the base should possess high rotational ductility. Members with large slenderness ratios should be avoided and the level of axial loads should not exceed 25–30% of the plastic resistance in the columns. High axial compressive actions endanger the inelastic deformation capacity of structural members. Furthermore, variations of axial loads in columns due to overturning moments and vertical vibration modes increase the likelihood of local and global instability.

Global ductility of structures is also correlated to the capacity of lateral resisting systems. Relationships between strength and ductility are addressed in Section 2.3.6. It suffices to state here that in general for a given earthquake ground motion and predominant period of vibration, the global ductility increases as the yield level of the structural system decreases.

2.3.3.2 Effects on Action Redistribution

Inelastic response of structures subjected to earthquakes is primarily controlled by local and global ductility. Ductile systems may sustain inelastic deformations in the post-peak response domain as demonstrated by the action–deformation curve given in Figure 2.32. Failure of ductile structures does not correspond to the maximum resistance or formation of first plastic hinge in structural components. Ductility allows redundant structures, e.g. multi-storey MRFs, to dissipate energy and continue to resist seismic actions, while successive plastic hinges are formed. Due to the reduced stiffness in the dissipa-

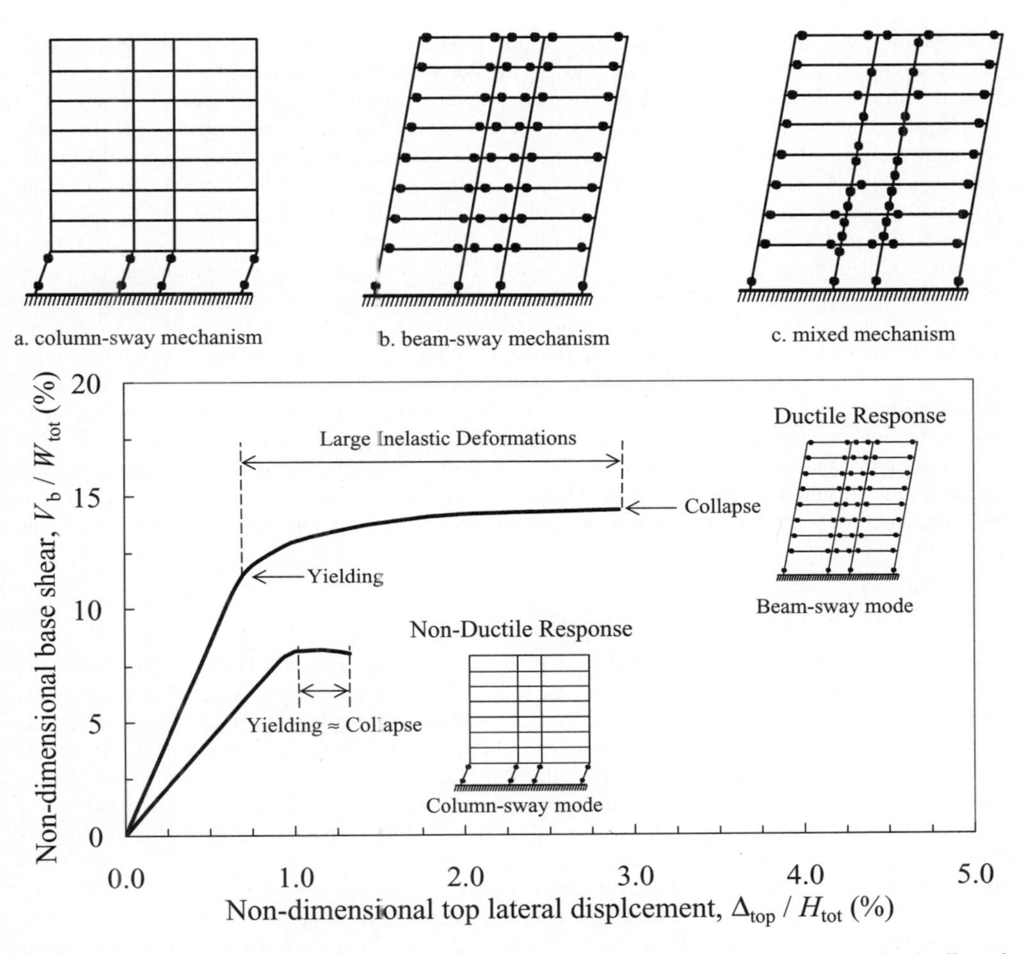

Figure 2.41 Energy-dissipating mechanisms for multi-storey frames (*top*) and response curves for ductile and non-ductile behaviour (*bottom*)

tive parts, and the relatively higher stiffness in the non-dissipative parts, actions migrate from the former to the latter, thus prolonging the life of the structure. This is referred to action redistribution.

It is highly desirable in seismic design to control the location of dissipative zones and the type of post-elastic response in these zones, i.e. it is essential to exercise 'failure mode control' which is also discussed in Section 2.3.3.3. To achieve ductile response and high energy absorption before structural collapse occurs, it is necessary to allow the formation of flexural plastic hinges in beams. Usual energy-dissipating mechanisms for multi-storey frames subjected to horizontal loads are provided in Figure 2.41.

Column-sway and beam-sway modes correspond to WCSB and SCWB design approaches, respectively. In the former, plastic hinges are located at column ends, while in the latter, flexural hinges occur in the beam. In a beam-sway mode, plastic hinges are required at column bases to generate global mechanisms. Figure 2.41 shows also the response curves for frames with column-sway and beam-sway mechanisms. The amount of seismic energy dissipated in beam-sway mechanisms is higher than that in column-sway. Frames with beam-sway modes are characterized by a ductile behaviour, while frames with column-sway modes exhibit non-ductile response. This demonstrates that buildings with SCWBs possess high action redistribution, while those with WCSBs are characterized by poor ductility and limited action redistribution. The reason for such structural performance is threefold. In frames with SCWB, the total number of plastic hinges is generally higher than in frames with WCSB. In weak-column structures, plastic deformations are often concentrated only in certain storeys along the height. For the same level of roof translation ductility, a relatively high storey ductility factor is required compared to the beam-sway mechanism. In both flexural and shear failure of columns, degradation is higher than in beam yield. Axial forces erode the ductility available in columns, while the levels of axial force in beams are negligible. Furthermore, systems with WCSB may experience severe damage in columns (Schneider *et al.*, 1993). Column failure leads to the collapse of the entire building.

The number of possible plastic mechanisms increases with the increase in the number of elements, and plastic hinges are likely to form at different locations in different earthquakes. Global frame response is often characterized by mixed-mode mechanisms, with hinges in beams and columns as shown in Figure 2.41. Mixed mechanisms are generally caused by material randomness, material strain hardening and overstrength due to the presence of slabs or other geometric characteristics.

To ensure that plastic hinges occur in beams rather than in columns, the latter are capacity-designed and hence they exhibit high strength. Beam-to-column connections can be designed to withstand actions higher than the capacity of the members framing into them, as required by the capacity design approach.

2.3.3.3 Structural Collapse Prevention

Prevention of structural collapse is a fundamental objective of seismic design. The definition of collapse may be expressed in terms of different response quantities, at local (e.g. strains, curvatures, rotations) and global (e.g. inter-storey and/or roof drifts) levels. Collapse implies that horizontal and vertical systems utilized to withstand effects of gravity and earthquake ground motions are incapable of carrying safely gravity loads. Generally, structural collapse occurs if vertical load-carrying elements fail in compression and if shear transfer is lost between horizontal and vertical elements, such as shear failure between flat slabs and columns. Collapse may also be caused by global instability. Individual storeys may exhibit excessive lateral displacements and second-order P-Δ effects significantly increase overturning moments, especially in columns at lower storeys.

Brittle structures, such as unreinforced masonry, fail when the maximum applied actions exceed the strength of the system. When failing in shear, masonry walls exhibit limited energy dissipation capacity, especially when subjected to high compression stresses (e.g. Tomazevic, 1999). In order to increase the lateral resistance and to improve the horizontal translational ductility, masonry walls can efficiently be reinforced with longitudinal and transverse steel bars. Reinforced masonry walls may exhibit adequate local and global ductility. The extent of inelastic excursions in reinforced masonry walls depends on the detailing adopted in the design. Global ductile response imposes high inelastic deformation demands at fibre level, as shown for example in equation (2.18).

Ductile steel structures, RC and composite systems do not collapse at the onset of the maximum strength. They sustain inelastic deformations and dissipate the input energy. They are safe as long as the required ductility capacity is available, i.e. $\mu_a > \mu_d$. Alternating actions may cause stiffness and strength deterioration, especially in RC members. The net effect is the erosion of the available ductility μ_a and hence the energy dissipation capacity is lowered. Experimental simulations have shown that collapse depends on the maximum displacement demand for well-detailed RC elements and structures without bond or shear failure (e.g. CEB, 1996). Similar response is observed for steel structures in which local buckling is inhibited.

Structural collapse prevention can be achieved through failure mode control. The latter is the basis for the capacity design of structures. In the capacity design approach, the designer dictates where the damage should occur in the system. The designer imposes a ductile failure mode on the structure as a whole. In so doing, the parts of the structure that yield in the selected failure mode are detailed for high energy absorption; these parts are termed 'dissipative components'. For MRFs designed in compliance with SCWB philosophy, beams are dissipative members. The remainder of the structure, e.g. columns and joints, is provided with the strength to ensure that no other yielding zones are likely to occur; these are 'non-dissipative components'. The design actions for the latter are derived from capacity design principles. Elements carrying vertical loads are designed with added strength. The capacity design factors that are used to define the design actions for the non-dissipative components are referred to as protection factors or overdesign factors. The overdesign factors should not be confused with the overdesign factor in assessment, as opposed to design. Overdesign in assessment is the ratio between the intended and the actual strength of a structure or a component. Protection factors may also be employed in the design of structural components where significant shear effects, compressive/tensile forces or brittle failure is expected. In steel and composite structures, spreading of inelasticity in column panel zones is often allowed. However, there is no general agreement among earthquake engineers on this issue; research is still ongoing (e.g. Di Sarno and Elnashai, 2002).

Bertero and Bertero (1992) compared the ductility required for two different failure modes of multi-storey MRFs, i.e. SCWB and WCSB. Framed systems with WCSBs are characterized by high values of imposed ductility, especially for flexible structures, e.g. with fundamental period of vibration $T > 1.5$–2.0 seconds. Experimental and numerical investigations have demonstrated that WCSB designs are not desirable in seismic regions (e.g. Schneider *et al.*, 1993, among others).

Failure mode control is significantly affected by material randomness, presence of non-structural components and quality control. Variations of mechanical properties depend on the construction material utilized, as discussed in Section 2.3.2.1. Values of COVs for material properties are generally lower than 15–20% and are often negligible compared to the randomness of both seismic input and quality control (Kwon and Elnashai, 2006).

Infilled walls, claddings and internal partitions can play an important role in the seismic response of structural systems and may alter the hierarchy in the failure mode sequence, e.g. beam before connections and columns in MRFs or braces before beams, connections and columns in CBFs. While not normally considered in the design, non-structural elements interact with the structural system and influence its performance. To achieve an adequate control of the failure mode, non-structural components should be accounted for in the analysis of the dynamic behaviour and in the seismic detailing of the dissipative components. Infilled systems were discussed in detail in preceding sections.

Failure modes that should be avoided are those involving sudden failure (e.g. brittle or buckling modes) and those involving total collapse due to failure of vertical load-carrying members. Common brittle failure modes are summarized in Table 2.5 categorized according to the material of construction.

Table 2.5 Typical brittle failure modes as function of common materials of construction.

Material of construction	Brittle failure modes
Reinforced Concrete	Buckling of reinforcement bars Bond or anchorage failure Member shear failure
Masonry	Out-of-plane bending failure Global buckling of walls Sliding shear
Structural Steel	Fracture of welds and/or parent material Bolt shear or tension failure Member buckling Member tension failure Member shear failure

Structures with high ductility capacity dissipate large amount of energy, thus allowing the control of progressive collapse of the structure.

Problem 2.4

Compare the bending moment capacity of sections at the base of reinforced concrete columns under monotonic and earthquake loads shown in Figure 2.42. Assume that stirrups may be either close-spaced or with large spacing. Is the axial load beneficial for the shear capacity of column members? Illustrate the answer with sketches.

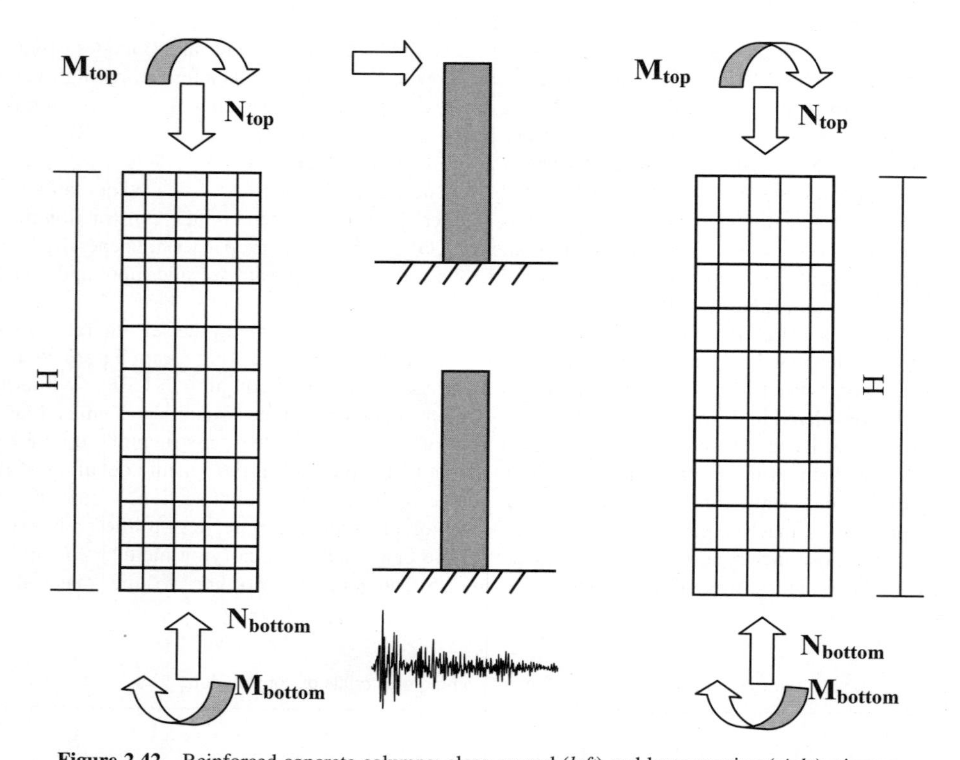

Figure 2.42 Reinforced concrete columns: close-spaced (*left*) and large spacing (*right*) stirrups

Problem 2.5

The structural response of the bridge pier shown in Figure 2.43 can be idealized as an elastic-perfectly plastic relationship. Assume that the yield bending moment ($M_y = V_y L$) and the elastic lateral stiffness ($k L$) of the pier are 480 kNm and 480 kN/m, respectively. Calculate the displacement ductility μ_δ of the pier corresponding to a top drift of 0.5 m. If the plastic hinge length L_p is equal to $0.1L$, compute the curvature ductility factor μ_χ for the cantilever pier.

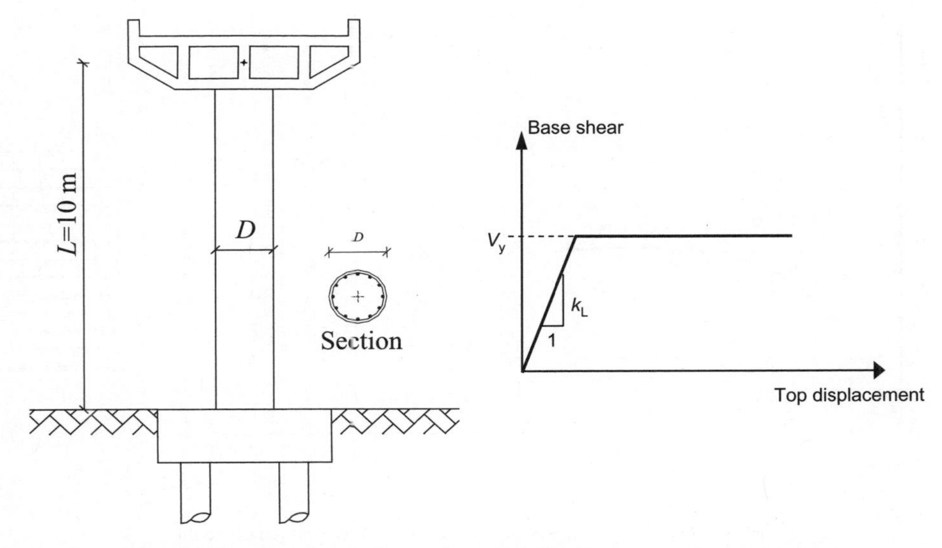

Figure 2.43 Bridge pier: layout (*left*) and idealized structural response (*right*)

2.3.4 Overstrength

Overstrength is a parameter used to quantify the difference between the required and the actual strength of a material, a component or a structural system. Structural overstrength is generally expressed by the 'overstrength factor' Ω_d defined as follows:

$$\Omega_d = \frac{V_y}{V_d} \tag{2.19}$$

where V_y and V_d are the actual and the design lateral strengths of the system, respectively. The Ω_d-factor is often termed 'observed overstrength' factor. The relationship between strength, overstrength and ductility is depicted in Figure 2.44.

For building structures, an additional measure relating the actual V_y to the elastic strength level V_e of lateral resisting systems has been suggested by Elnashai and Mwafy (2002) alongside the overstrength Ω_d in equation (2.19). The proposed measure Ω_i is given as:

$$\Omega_i = \frac{V_y}{V_e} \tag{2.20}$$

and is termed 'inherent overstrength' to distinguish it from the 'observed overstrength' Ω_d commonly used in the literature. The suggested measure of response Ω_i reflects the reserve strength and the anticipated behaviour of the structure under the design earthquake, as depicted in Figure 2.45. Clearly, in the case of $\Omega_i \geq 1.0$, the global response will be almost elastic under the design earthquake, reflecting the high overstrength of the structure. If $\Omega_i < 1.0$, the difference between the value of Ω_i and unity is an indication of the ratio of the forces that are imposed on the structure in the post-elastic range. Structures with $\Omega_i \geq 1.0$ should be treated with care since they may be amenable to redesign to achieve substantial economies without jeopardizing safety.

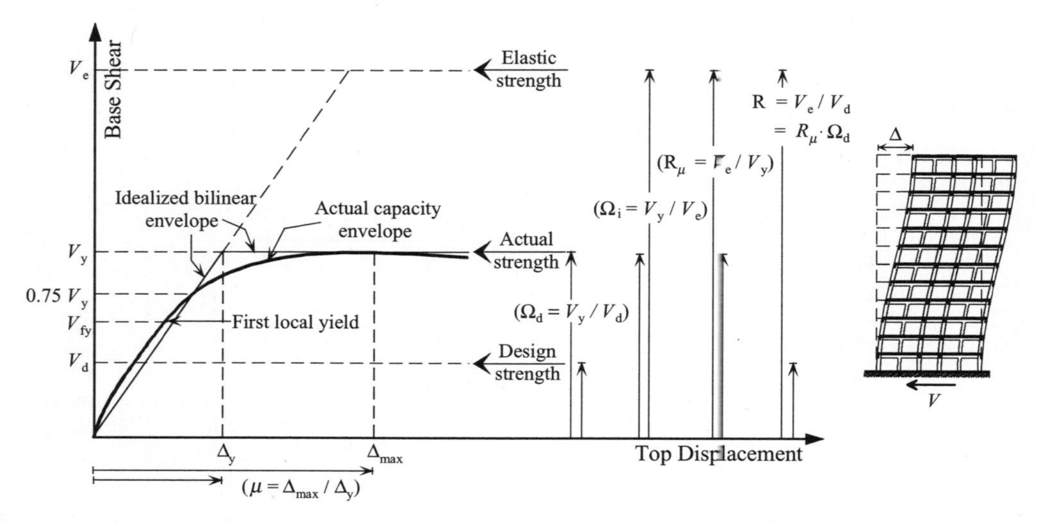

Figure 2.44 Relationship between strength, overstrength and ductility

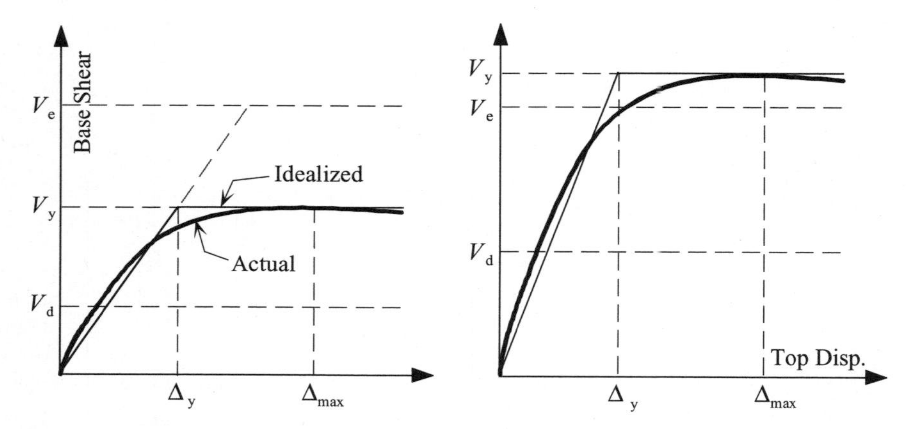

Figure 2.45 Different levels of inherent overstrength Ω_i: ductile response, $\Omega_i < 1.0$ (*left*), and elastic response under design earthquake $\Omega_i \geq 1.0$ (*right*)
Key: V_d = design base shear strength; V_e = elastic base shear strength; V_y = actual base shear strength; Δ = roof displacement

Experimental and numerical research on the performance of buildings during severe earthquakes have indicated that structural overstrength plays a very important role in protecting buildings from collapse (e.g. Whittaker *et al.*, 1990, 1999; Jain and Navin, 1995; Elnashai and Mwafy, 2002, among others). Similarly, high values of Ω_d-factors are generally essential for the survivability of bridge systems (Priestley *et al.*, 1996). Structural overstrength results from a number of factors (Uang, 1991; Mitchell and Paulter, 1994; Humar and Ragozar, 1996; Park, 1996). The most common sources of overstrength include:

(i) Difference between actual and design material strengths, including strain hardening;
(ii) Effect of confinement in RC, masonry and composite members;

(iii) Minimum reinforcement and member sizes exceeding design requirements;
(iv) Conservatism of the design procedures, e.g. utilizing the elastic period to obtain the design forces and ductility requirements;
(v) Effect of structural elements not considered in predicting the lateral load capacity (e.g. actual slab width contribution to beams, degree of interaction of shear connectors in composite systems);
(vi) Load factors and multiple load cases adopted in seismic design including accidental torsion;
(vii) Serviceability limit state provisions;
(viii) Structural redundancy;
(ix) Participation of non-structural elements in the earthquake response of structures.

The above factors show that a generally applicable and precise estimation of the overstrength is difficult to determine since many parameters contributing to it are uncertain. For example, the actual strength of materials, confinement effects, contribution of non-structural elements and the actual participation of some structural elements are factors leading to relatively high uncertainties (Humar and Ragozar, 1996). Randomness of mechanical properties of materials of construction leads to unexpected overstrength in the global structural response, which may undermine the failure mode control in capacity design (*see* Section 2.3.3). To control collapse mechanisms of ductile systems, randomness in the system capacity should be quantified and included in the design and assessment processes.

It is re-emphasized that overstrength has positive and negative consequences. Flexural overstrength in the beams of MRFs may cause storey collapse mechanisms due to failure in columns or brittle shear failure in beams. Non-structural elements also may cause shear failure in columns or soft storey failure (Park, 1996). Moreover, Ω_d-factors vary widely according to the period of the structure, the design intensity level, load cases other than seismic action, the structural system and the ductility level employed in the design. Moreover, structural redundancy is significantly influenced by element capacity ratios, types of mechanisms which may form, individual characteristics of building systems and materials, structure height, number of storeys, irregularity, torsional imbalance, diaphragm spans, number of lines of resistance and number of elements per line. Consequently, effects of redundancy under earthquake loads are not straightforward (Bertero and Bertero, 1999; Whittaker *et al.*, 1999; Wen and Song, 2003). This compounds the difficulties associated with evaluating structural overstrength factors accurately. Quantification of the actual overstrength can be employed to reduce the forces utilized in the seismic design of structures, hence leading to more economical studies, as described below.

Detailed analytical studies on the influence of overstrength factors on the seismic performance of multi-storey buildings were carried out by Elnashai and Mwafy (2002) and Di Sarno (2003), for RC and steel structures, respectively. In both studies, structural lateral capacity and overstrength was well assessed from inelastic analyses, such as pushover and dynamic response-history analyses (*see* Section 4.5). In particular, in the study by Elnashai and Mwafy (2002), the set of 12 RC buildings described in Figure 2.21 and Table 2.2 was investigated. Capacity envelopes of the RC structures obtained from response-history analyses, which are presented in Mwafy and Elnashai (2001, 2002), were utilized to evaluate overstrength factors. The envelopes were developed using regression analysis of the maximum roof drifts and base shears of eight seismic excitations for each building. Figure 2.46 shows the capacity envelopes for the three groups of buildings obtained from inelastic pushover analyses using a triangular lateral load distribution. The computed values show that for the sample structures, the strength at first indication of member yielding V_{fy} is notably higher than the design strength levels (refer also to Figure 2.44). The average V_{fy}/V_d ratio for irregular frames, regular frames and frame-wall systems is 1.33, 1.46 and 1.57, respectively. This ratio is relatively high, particularly for regular buildings.

The observed overstrength factors from inelastic static pushover and response-history analyses for the sample RC buildings are depicted in Figure 2.47. In the same figure, inherent overstrength factors computed from equation (2.20) are included. The estimated overstrength factors show that all sample

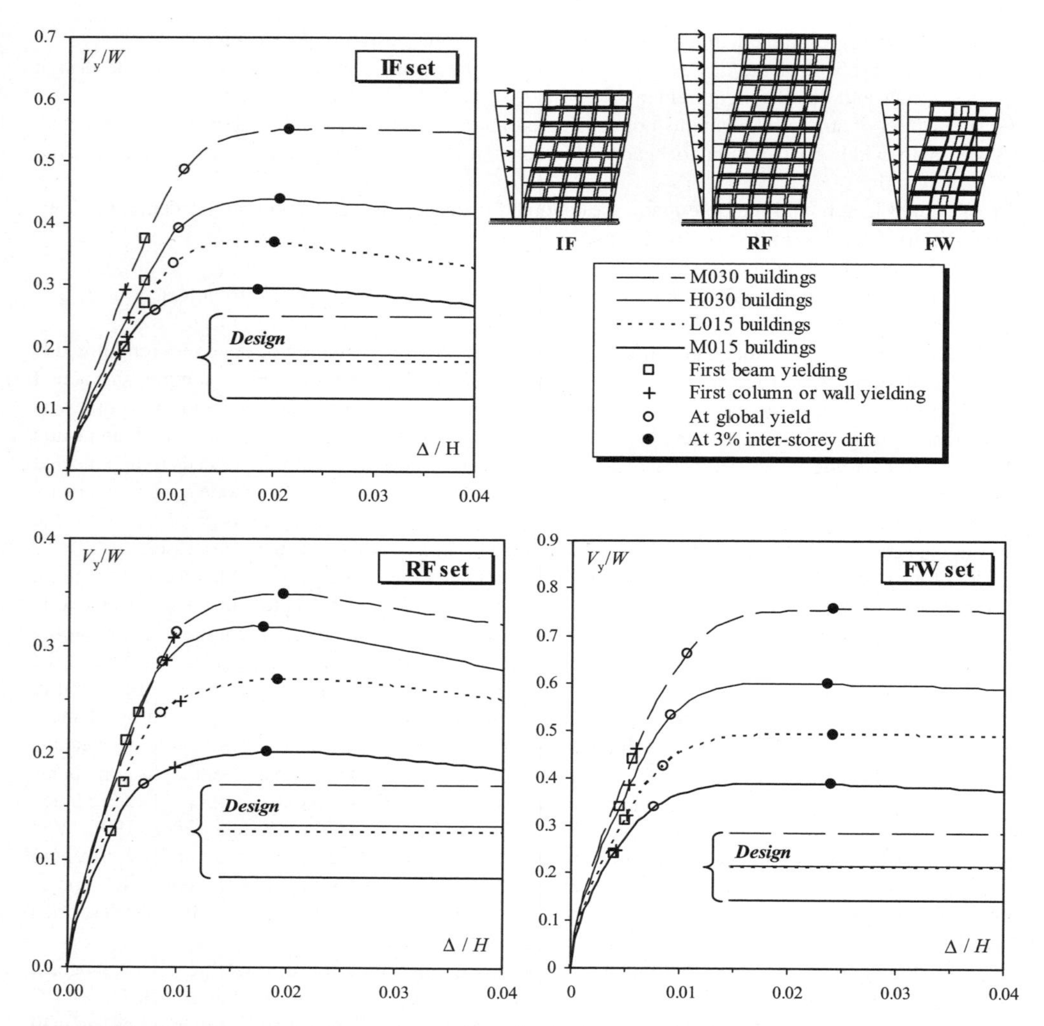

Figure 2.46 Comparisons between capacity envelopes for the 3 groups of buildings, obtained from inelastic static pushover analyses

Key: V_y = actual base shear strength; W = weight of the building; Δ = roof displacement; H = height of the building; IF = irregular frames; RF = regular frames; FW = frame-wall structures

structures exhibit values of Ω_d over 2.0. Frame-wall systems (group 3) have the highest level of overstrength, the results for irregular (group 1) and regular (group 2) frames being comparable. For the two buildings, designed to the same ductility level in each group, that designed to a lower seismic intensity exhibits higher overstrength, reflecting the higher contribution of gravity loads. Higher ductility level buildings display higher reserve strength.

Previous studies on RC structures have shown that low-rise buildings exhibit higher Ω_d-factors compared with medium-rise buildings (Mwafy, 2000). Therefore, a minimum overstrength of 2.0 can be used to characterize the seismic response of low- and medium-rise RC buildings. On the other hand, studies carried out on buildings designed to US seismic codes have indicated that the overstrength factor

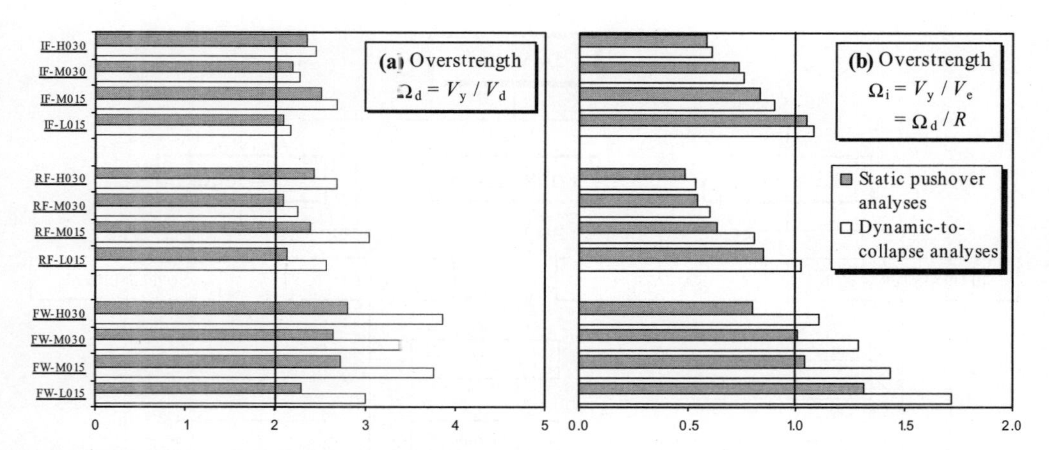

Figure 2.47 Observed (Ω_d) and inherent (Ω_i) overstrength factors for RC multi-storey buildings
Key: V_y = actual base shear strength; V_e = elastic base shear strength; IF = irregular frames; RF = regular frames; FW = frame-wall

Ω_d varied widely depending on the height of the building, the design seismic intensity and the structural systems (Whittaker *et al.*, 1990, 1999; Jain and Navin, 1995). The scatter of the overstrength factors computed for buildings in the USA is generally high. Overstrength values range between 1.8 and 6.5 for long- and short-period range structures.

Figure 2.47 shows that the values of Ω_i are quite high for the third group of buildings, namely the frame-wall structures. The strength levels of the four buildings of this group generally exceed the elastic strength V_e. The use of Ω_i-factors to quantify the structural overstrength highlights the over-conservatism of existing code provisions for structural wall systems, where minimum section sizes and reinforcements lead to an elastic response for this class of structure under the design intensity. The response of the buildings designed to a low ductility level in each group is likely to be elastic, which again reflects the conservatism of the code. For such a type of structure, no capacity design rules are applied, although some requirements to enhance the ductility are imposed. Figure 2.47 also demonstrates that, contrary to the conventional definition of overstrength Ω_d, the values of Ω_i display clearly the expected higher overstrength of the irregular-frames group of buildings compared with the regular-frames group. However, for the buildings designed for the same seismic intensity in each group, the higher-ductility-level buildings show lower values of Ω_i, reflecting the higher reliance on seismic design when the design ground motion is high. Values of inherent overstrength Ω_i are consistent with the results of the overstrength Ω_d in terms of the higher values for the buildings designed for lower seismic intensity.

Steel and composite frames are generally more flexible than RC frames. The overstrength for steel frames increases with the building height since the design is likely to be governed by stiffness, e.g. storey drift limitations (Uang, 1991). An extensive parametric study was carried out by Di Sarno (2003) on a set of nine perimeter MRFs designed according to the US seismic design practice for different earthquake hazard levels, namely low (Boston), intermediate (Seattle) and high (Los Angeles). Inelastic static pushovers and response-history analyses were employed to assess the seismic performance of the frames with different heights, i.e. low-rise (3 storeys), medium-rise (9 storeys) and high-rise (20 storeys). It was observed that for the sample steel frames, the computed overstrength factors are on average 2.80, thus leading to higher values than those estimated for RC buildings. The computed inherent overstrength factors also demonstrate the amount of inelasticity that occurs in steel low- and medium-rise frames. The highest values of observed and inherent overstrength factors, about 8 and 0.9, respectively, are observed for the set of MRFs located in low seismic hazard. The design of the latter frames is governed by stiffness requirements at serviceability due to wind rather than seismic loads.

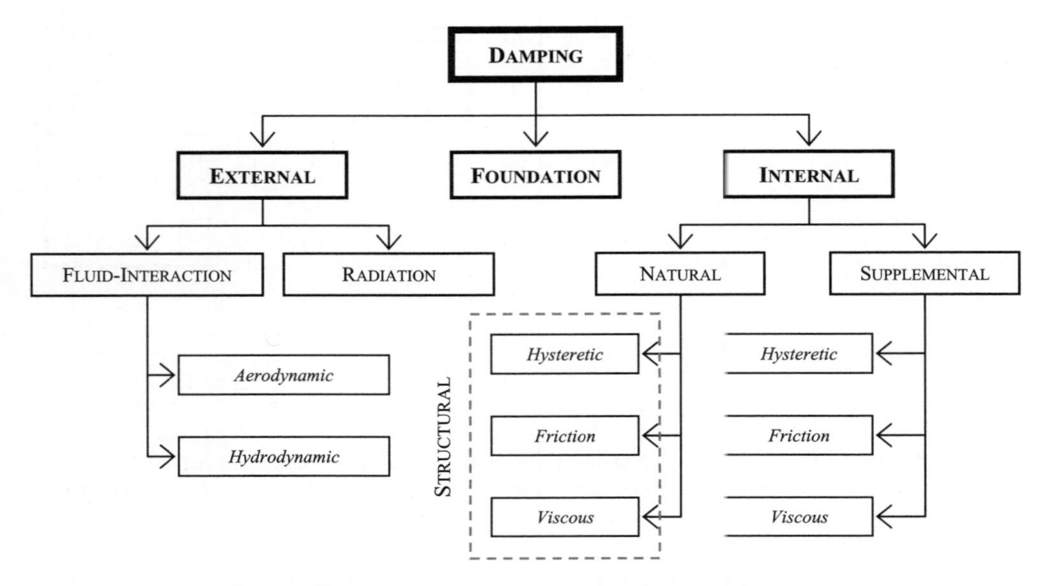

Figure 2.48 Sources of damping mechanisms in structural systems

2.3.5 Damping

Damping is utilized to characterize the ability of structures to dissipate energy during dynamic response. Unlike the mass and stiffness of a structure, damping does not relate to a unique physical process but rather to a number of possible processes. Damping values depend on several factors, among these are vibration amplitude, material of construction, fundamental periods of vibration, mode shapes and structural configurations (Bachmann *et al.*, 1995).

Seismic energy transmitted to structures can be dissipated through different damping mechanisms as shown in Figure 2.48. Primary sources of damping are, however, as follows:

(i) *Structural damping:* due to energy dissipation in materials of construction, structural components and their connections;
(ii) *Supplemental damping:* due to energy dissipation of devices added to structural systems to increase their damping;
(iii) *Foundation damping:* due to the transfer of energy from the vibrating structure to the soil, through the foundations;
(iv) *Radiation damping:* due to radiation of seismic waves away from foundations.

External damping may be aerodynamic and hydrodynamic caused by interaction between structure and surrounding air and water, respectively. The latter mechanisms are generally negligible compared to other types of damping in earthquake response of structures (Hart *et al.*, 1973; Hart, 1996). Inelastic deformations of the ground in the vicinity of foundations, caused by soil hysteresis, and seismic wave propagation or radiation result in two fundamentally different damping mechanisms associated with soils, namely foundation and radiation damping. Soil-structure interaction may significantly contribute towards the overall damping. This depends on several site and structural characteristics (e.g. Roesset *et al.*, 1973; Novak, 1974; Tsai, 1974). When the soil is infinitely rigid, then the foundation damping may be neglected. This section focuses on internal or structural damping. Supplemental damping can be added to structures to enhance their dissipation capacity and hence reduce actions and deformations.

Several types of energy dissipation devices can provide hysteretic, friction and viscous damping; these are cost-effective for seismic retrofitting of structures (e.g. Di Sarno and Elnashai, 2005). The latter devices are also being increasingly used for new structures because of their ability to considerably reduce storey displacements, accelerations and shears (Soong and Spencer, 2002).

Structural damping is a measure of energy dissipation in a vibrating system that results in bringing the structure back to a quiescent state. It is associated with absorption of seismic energy in structural components. It also accounts for material viscosity and friction at connections and supports. In structural components, the energy imparted by earthquakes is dissipated mainly through hysteretic damping characterized by action–deformation loops. Such loops express action–deformation relationships of materials, sections, members, connections or systems under alternating loads. For hysteretic damping, the dissipation varies with the level of displacement, but it is constant with the velocity. The amount and mechanisms of material hysteretic damping vary significantly depending on whether the material is brittle, such as concrete and masonry, or ductile, e.g. metals. For RC energy dissipation is due to opening and closing of cracks but the material remains held together by the steel. In masonry, there is also sliding along the cracks; hence the hysteretic damping of masonry is lower than that of RC. Whereas hysteretic damping is complex and cannot be expressed in simple forms, it is almost always represented in dynamic analysis as equivalent viscous damping, which is proportional to the velocity. This form of damping conveniently allocates a parameter to the velocity term in the dynamic equilibrium equations that matches the mass and stiffness terms associated with acceleration and displacement, respectively.

Friction or Coulomb damping results from interfacial mechanisms between members and connections of a structural system, and between structural and non-structural components such as infills and partitions. It is independent of velocity and displacement; its values significantly depend on the material and type of construction. For example, in steel structures, the contribution of friction damping in bolted connections is higher than welded connections. In infilled masonry walls, friction damping is generated when cracks open and close. In other materials, e.g. for concrete and masonry, this type of damping cannot be relied upon because of the degradation of stiffness and strength under cyclic load reversals.

Values of hysteretic damping ξ_m for common materials of construction are outlined in Table 2.6. These are expressed as ratios of the critical damping. It is observed that ξ_m increases with the amplitude of action or deformation. The values in Table 2.6 are, however, approximate estimates of damping for different construction materials.

For relatively small values of damping, e.g. less than 10–15%, hysteretic, viscous and friction damping can be conveniently expressed by 'equivalent viscous damping' c_{eq} as follows (Jennings, 1968):

Table 2.6 Hysteretic damping for different construction materials (*after* Bachmann *et al.*, 1995).

Material	Damping, ξ_m (%)
Reinforced concrete	
Small amplitudes (un-cracked)	0.7–1.0
Medium amplitudes (fully cracked)	1.0–4.0
High amplitudes (fully cracked) but no yielding of reinforcement	5.0–8.0
Pre-stressed concrete (un-cracked)	0.4–0.7
Partially stressed concrete (slightly cracked)	0.8–1.2
Composite	0.2–0.3
Steel	0.1–0.2

$$c_{\mathrm{eq}} = \xi_{\mathrm{eq}}\, c_{\mathrm{cr}} \tag{2.21}$$

where c_{cr} is the critical damping coefficient and ξ_{eq} the equivalent damping ratio defined as:

$$\xi_{\mathrm{eq}} = \xi_0 + \xi_{\mathrm{hyst}} \tag{2.22}$$

in which ξ_0 corresponds to the initial damping in the elastic range and ξ_{hyst} indicates the equivalent viscous damping ratio that represents the dissipation due to the inelastic hysteretic behaviour. Equations (2.21) and (2.22) are written for a substitute structure, i.e. a single-degree-of-freedom (SDOF) elastic system, the characteristics of which represent the inelastic system; substitute structures are further discussed in Section 4.4.

Procedures to estimate the viscous damping ξ_0 are based on the measurement of the amplitude decay from laboratory tests or on real buildings on site (Jeary, 1996; Fang *et al.*, 1999; Blandon and Priestley, 2005). The values of ξ_0 may vary in practice between 2% and 5%. On the other hand, the equivalent viscous damping ξ_{hyst} corresponding to the hysteretic response can be computed from the following (Jacobsen, 1930):

$$\xi_{\mathrm{eq}} = \frac{1}{4\pi}\frac{\omega_1}{\omega}\frac{E_{\mathrm{Diss}}}{E_{\mathrm{Sto}}} = \frac{1}{2\pi}\frac{A_{\mathrm{hyst}}}{F_0\, u_0} \tag{2.23}$$

where E_{Diss} is the energy loss per cycle ($E_{\mathrm{Diss}} = A_{\mathrm{hyst}}$) and E_{Sto} represents the elastic strain energy stored in an equivalent linear elastic system, or viscous damper, as shown pictorially in Figure 2.49. The terms ω_1 and ω are the natural frequency of the system and the frequency of the applied load, respectively. F_0 is the force corresponding to the deformation parameter u_0. Equation (2.23) shows that the coefficient ξ_{eq} depends on ω; hence, when values of ξ_{eq} are provided, the relative circular frequency of the applied load should also be specified.

The magnitude of equivalent damping c_{eq} and hence ξ_{eq} can be estimated from the hysteretic response represented by the action–deformation cyclic curves as shown in Figure 2.49. The dissipated energy is equal to the area enclosed inside an entire loop. However, to use this approach it is necessary to assume that both inelastic and equivalent elastic systems are subjected to harmonic excitations. Moreover, the energy dissipated by both systems in one cycle at peak response, i.e. shaded areas in Figure 2.49, is equal. Inelastic and equivalent elastic systems should also have the same initial period of vibration, which corresponds to the 'resonance' between the excitation and the SDOF structure. The above hypotheses ensure that the loops used to apply equation (2.23) are complete and that a closed-form

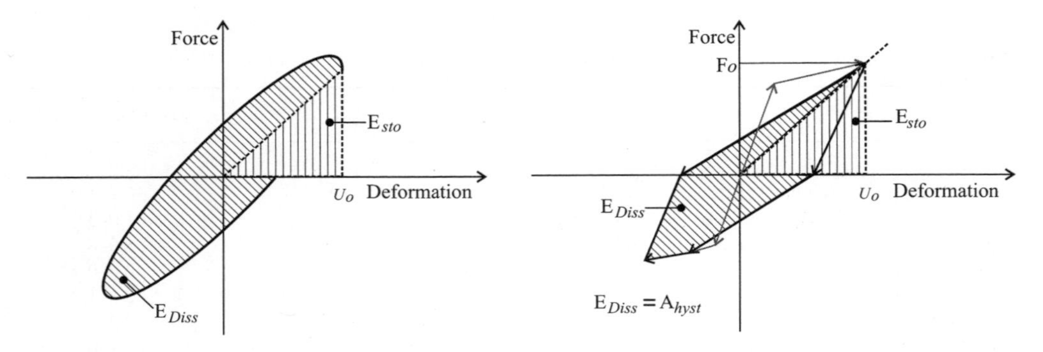

Figure 2.49 Dissipated and stored forces for viscous damping (*left*) and hysteretic cycles (*right*)

solution for the displacement can be obtained. The strain energy E_{Sto} stored in the system is given by:

$$E_{Sto} = \frac{1}{2} k\, u_0^2 \tag{2.24.1}$$

where k is the stiffness of the equivalent elastic system. The dissipated energy E_{Diss} in one cycle of oscillation, corresponding to the area inside the hysteretic loop in Figure 2.49, can be computed as follows:

$$E_{Diss} = 2\,\pi\,\xi \frac{\omega}{\omega_1} k\, u_0^2 \tag{2.24.2}$$

in which is ω and ω_1 are the circular frequencies as for equation (2.23). When the areas inside the loops in Figure 2.49 are made equal and equation (2.24.1) is substituted in equation (2.24.2), then equation (2.23) is obtained.

The definition of equivalent damping ξ_{eq} in equation (2.23) is based on a sinusoidal response of a SDOF system; this is also known as Jacobsen's approach. It is clear, however, that response of structures to earthquake ground motions cannot be adequately represented by steady-state harmonic response, and that an unknown error will be introduced in the estimation of displacements, based on the approximations made in Jacobsen's approach (e.g. Priestley and Grant, 2005). It was also determined that the latter approach is non-conservative for structures with high hysteretic energy absorption. There are several references that report equations for equivalent viscous damping factors (e.g. Fardis and Panagiotakos, 1996; Calvi, 1999; Miranda and Ruiz-Garcia, 2002; Priestley, 2003; Lin *et al.*, 2005). An extensive numerical study carried out by Borzi *et al.* (2001) demonstrated that relationships to compute the equivalent viscous damping ξ_{eq} for inelastic systems subjected to earthquake ground motions are significantly influenced by the hysteretic rule of the structure and the ductility level μ. Equation (2.23) was written for elastic-perfectly plastic (or bilinear) and hysteretic hardening-softening (or trilinear) model as follows:

$$\xi_{eq} = \alpha\left(1 - \frac{1}{\mu}\right) \tag{2.25}$$

where α is 0.64 for bilinear hysteretic model when all the cycles of the load reversals have the same amplitude up to the target ductility. Values of the parameter α are summarized in Table 2.7 for values of ductility μ varying between 2 and 6 Differences in hysteretic behaviour are represented by variations in α. Nil entries in Table 2.7 indicate that structures with highly degrading response ($K_3 = -20\%$ and $-30\%\ K_y$) could not have ductility capacity of 4 or more.

Alternatively, the viscous damping ratios ξ_{eq} for different materials of construction and lateral resisting systems may be computed as follows:

$$\xi_{eq} = \xi_0 + \alpha\left(1 - \frac{1}{\mu^{\beta}}\right) \tag{2.26}$$

in which ξ_0 is the initial damping as specified in equation (2.23), while α and β are two-model coefficients. Values of α and β are summarized in Table 2.8 for common earthquake-resistant structures and construction material. The values of ξ_{eq} in equations (2.25) and (2.26) are expressed in percentage.

Viscous damping ratios ξ_{eq} increase in proportion to the natural frequency of vibration in structural systems (Wakabayashi, 1986). Figure 2.50 provides values of structural viscous equivalent damping

Table 2.7 Percentage of equivalent damping ξ_{eq} and α-values in equation (2.25).

	$\mu = 2$		$\mu = 3$		$\mu = 4$		$\mu = 6$	
	ξ_{eq} (%)	α	ξ_{eq} (%)	α	ξ_{eq} (%)	α	ξ_{eq} (%)	α
EPP	8	0.16	13	0.19	16	0.21	19	0.23
$K_3 = 0$	11	0.22	15	0.23	18	0.23	20	0.24
$K_3 = 10\% K_y$	10	0.19	13	0.20	15	0.20	17	0.20
$K_3 = -20\% K_y$	13	0.27	20	0.31	28	0.37	—	—
$K_3 = -30\% K_y$	15	0.31	27	0.41	—	—	—	—

Key: EPP = elastic-perfectly plastic model. K_y and K_3 are the secant and post-yield stiffness of the primary curve of the hysteretic hardening-softening model; μ = ductility

Table 2.8 Values of α and β in equation (2.26).

Structural system	α	β
Concrete frame	$\frac{120}{\pi}$	0.5
Concrete columns and walls	$\frac{95}{\pi}$	0.5
Precast walls and frames	$\frac{25}{\pi}$	0.5
Steel members	$\frac{120}{\pi}$	1.0

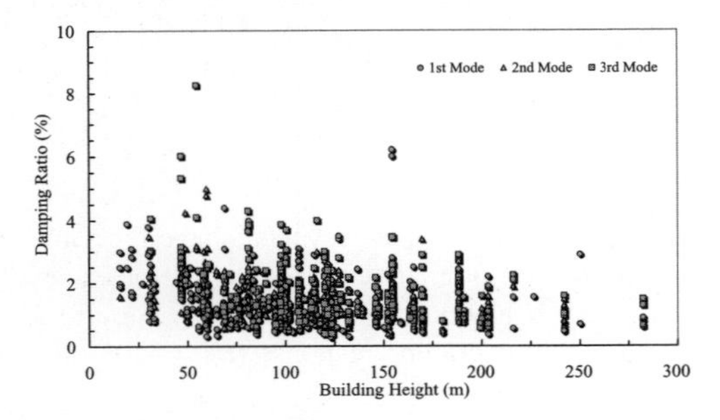

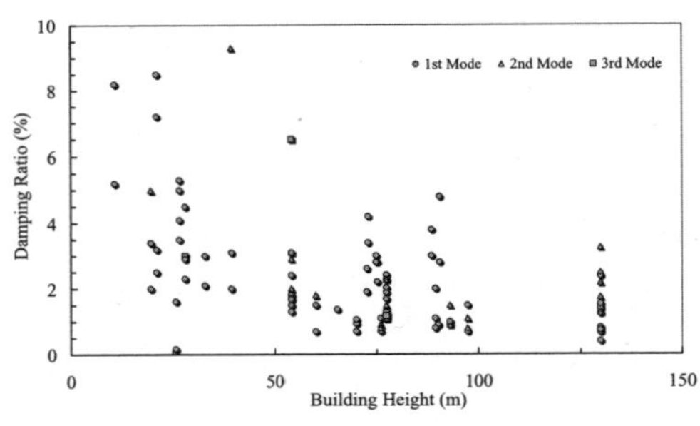

Figure 2.50 Structural damping ratios measured in existing buildings: steel (*left*) and reinforced concrete (*right*) structures (*adapted from* Suda *et al.*, 1996)

measured in existing buildings for the first three modes of vibration; data refer to steel and RC structures (Suda *et al.*, 1996). It is evident that an inelastic response trend exists and that the data exhibit large scatter. The scatter is, however, considerably lower in steel structures. Damping ratios in the first mode of RC buildings are higher than those in steel, e.g. 5–7% versus 2–4% for heights less then 40 m. The measurements provided in the figure also show that the higher the building, the lower the damping ratio for the first mode of vibration.

Extensive experimental and numerical studies carried out by Jeary (1986, 1996) showed that the mechanism of structural damping exhibits peculiar characteristics at low and high amplitudes. Data

Table 2.9 Structural damping for different structural systems (*after* ESDU, 1991).

Structural system (type)	Structural damping (%)		
	Minimum	Maximum	Mean
Buildings	0.5	5.0	2.75
Steel towers, unlined, welded construction	0.4	0.7	0.55
Steel tower, unlined, bolted construction	0.6	1.0	0.80
Steel tower, unlined welded, elevated on steel support structure	0.3	0.5	0.40
Concrete tower	0.5	1.2	0.85
Concrete tower with internal partitions	0.1	2.5	1.30
Steel bridges	0.3	1.0	0.65
Reinforced concrete bridges	0.5	2.0	1.25
Prestressed concrete bridges	0.3	1.0	0.65

recorded on different types of structure, such as buildings, chimneys and dams, showed that damping increases with amplitude but has a constant range or plateau at low and high amplitudes. The transition between these amplitudes is highly non-linear. These findings have been confirmed by several field measurements (e.g. Fang *et al.*, 1999; Satake *et al.*, 2003).

It is instructive to compare structural damping exhibited by buildings under earthquakes with other structural systems. Table 2.9 provides minimum, maximum and mean values of equivalent viscous structural damping for several forms of constructions employing different materials; the values summarized in the table are derived from ESDU (1991).

The values in Table 2.9 show large variability; the range is 0.1% to 2.5% for concrete towers and 0.5% to 5.0% for buildings. Higher values are associated with various sources of energy dissipation including high redundancy, overstrength and interaction between structural and non-structural components.

2.3.6 Relationship between Strength, Overstrength and Ductility: Force Reduction Factor 'Supply'

Design requirements for lateral loads, such as winds or earthquakes, are fundamentally different from those for vertical (dead and live) loads. While design for wind loads is a primary requirement, due to the frequency of the loading scenario, seismic design deals with events with lower probability of occurrence as discussed in Section 3.2. It may therefore be highly uneconomical to design structures to withstand earthquakes for the performance levels used for wind design. For example, building structures would typically be designed for lateral wind loads in the region of 1% to 3% of their weight. Earthquake loads may reach 30–40% of the weight of the structure, applied horizontally. If concepts of plastic design used for primary loads are employed for earthquake loads, extremely heavy and expensive structures will ensue. Therefore, seismic design, by necessity, uses concepts of controlled damage and collapse prevention as illustrated in Sections 2.3.2.3 and 2.3.3.3, respectively. Indeed, buildings are usually designed for 15–20% only of the elastic earthquake forces V_e. This is illustrated in Figure 2.44, where the elastic and inelastic responses are depicted, and the concept of equal energy is employed to reduce the design force from V_e to V_d (denoting elastic and design force levels, respectively). Therefore, damage is inevitable in seismic response and design. It is the type, location and extent of damage that is the target of the design and detailing process in earthquake engineering. The ratio between elastic V_e base and seismic design V_d shears is defined as 'force reduction factor' R:

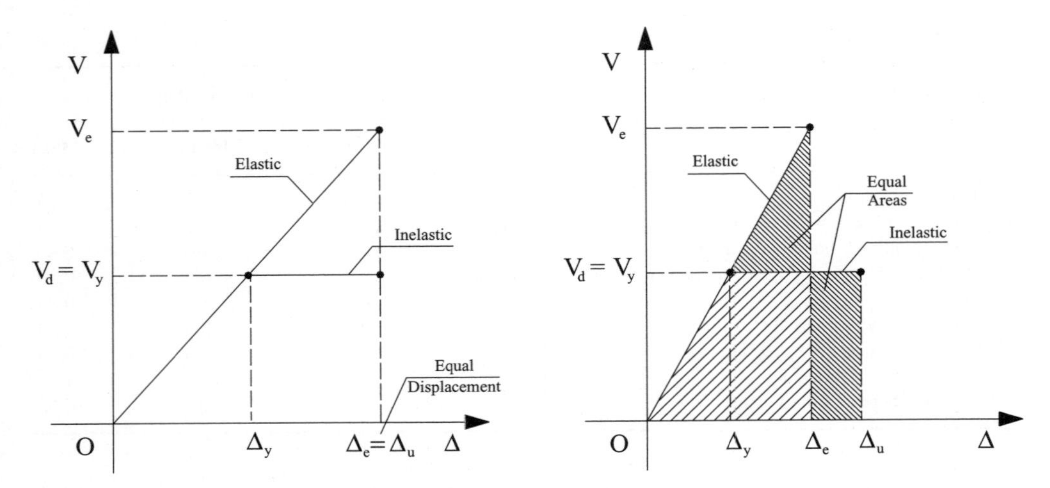

Figure 2.51 Base shear–lateral displacement relationships for inelastic single-degree-of-freedom systems with long (*left*) and intermediate (*right*) periods
Key: Δ_e = elastic displacement; Δ_y = yield displacement; Δ_u = ultimate displacement;
V_d = design base shear; V_e = elastic base shear; V_y = yield base shear; V_u = ultimate base shear

$$R = \frac{V_e}{V_d} \tag{2.27}$$

The values of R-factors computed from equation (2.27) correspond to force reduction factors 'supply'. They express the energy absorption and dissipation capacity of structural systems. Force reduction 'supply' factors, referred to as response modification factors R, q-factors or behaviour factors, are employed in all seismic codes worldwide for the design of ductile structures. R-factors 'demand' values are discussed in Section 3.4.4.

Force reduction factors 'supply' are related to the strength, overstrength, ductility and damping characteristics of structures. Relationships between R-factors and damping are not straightforward. Consequently, force reduction factors are often expressed as a function of the system resistance, overstrength Ω_d and translation ductility μ_δ factors as discussed below.

Series of analyses of SDOF systems with a linear elastic-perfectly plastic load–displacement response with varying levels of yield were undertaken and the results grouped for short (e.g. $T < 0.5$ seconds), intermediate (e.g. $0.5 < T < 1.0$ seconds) and long (e.g. $T \geq 1.0$ seconds) period structures (Newmark and Hall, 1969). Figure 2.51 shows the comparison between the response of these systems and elastic systems. For long-period structures, it is clear that the maximum displacement of the inelastic system is almost constant, regardless of the value of yield force (ignoring the very low levels of V_y, which are impractical). Therefore, a criterion based on 'equal displacements' may be used to link the two systems. This assumption leads to the following relationships between displacements at ultimate Δ_u and elastic Δ_e conditions:

$$\Delta_u = \Delta_e \tag{2.28.1}$$

which corresponds to the following ratios between actions (V_y and V_e) and deformations (Δ_y and Δ_e) at yield and elastic:

$$\frac{V_y}{\Delta_y} = \frac{V_e}{\Delta_e} \tag{2.28.2}$$

Therefore, it follows that:

$$V_y = \frac{V_e}{\Delta_u}\Delta_y = \frac{V_e}{\Delta_u/\Delta_y} \tag{2.28.3}$$

The inelastic (or design) base shear V_y of the new system is therefore given by:

$$V_y = \frac{V_e}{\mu} \tag{2.29}$$

and lateral displacement Δ_u can be computed using equation (2.28.1).

For the intermediate period systems, the displacement Δ_u increases with decreasing yield action V_y. Here, a criterion based on 'constant (or equal) energy' proves useful. By equating the energy absorbed by the elastic and inelastic systems, the following ensues (Figure 2.51):

$$\frac{1}{2}V_e\,\Delta_e = \frac{1}{2}V_y\,\Delta_y + V_y\,(\Delta_u - \Delta_y) \tag{2.30.1}$$

which leads to:

$$V_y^2 = \frac{V_e^2}{(2\Delta_u - \Delta_y)/\Delta_y} \tag{2.30.2}$$

The inelastic (or design) base shear V_y and lateral displacement Δ_u of the new system are therefore given by:

$$V_y = \frac{V_e}{\sqrt{2\mu - 1}} \tag{2.31.1}$$

$$\Delta_u = \frac{\mu}{\sqrt{2\mu - 1}}\Delta_e \tag{2.31.2}$$

For short-period structures, e.g. $T < 0.5$ seconds, there is no reduction in design forces, i.e. $V_y = V_e$, which corresponds to elastic design. The ductility required to reduce elastic base shears V_e is extremely high and seismic detailing is often impractical.

The above expressions, especially equations (2.30.1), (2.30.2), (2.31.1) and (2.31.2), point towards the following relationship:

$$R = \frac{V_e}{V_y}\Omega_d \tag{2.32}$$

where Ω_d is the observed overstrength factor defined in Section 2.3.4, while V_e and V_y are the elastic and the actual strength, respectively, as also displayed in Figure 2.51. In turn, the inherent overstrength Ω_i is related to the *R*-factor supply and Ω_d as given below:

Problem 2.6

Rank the components circled below (Figure 2.52) according to overstrength factors Ω_d to render the structure ductile (higher energy dissipation capacity):

- Beam, $\Omega_{d,bf}$;
- Column, $\Omega_{d,cf}$;
- Beam-column joint, $\Omega_{d,js}$.

$$R = \frac{\Omega_d}{\Omega_i} \tag{2.33}$$

The seismic performance of structural systems is satisfactory if the R-factors supply exceed the R-factor demands, discussed in Section 3.4.4.

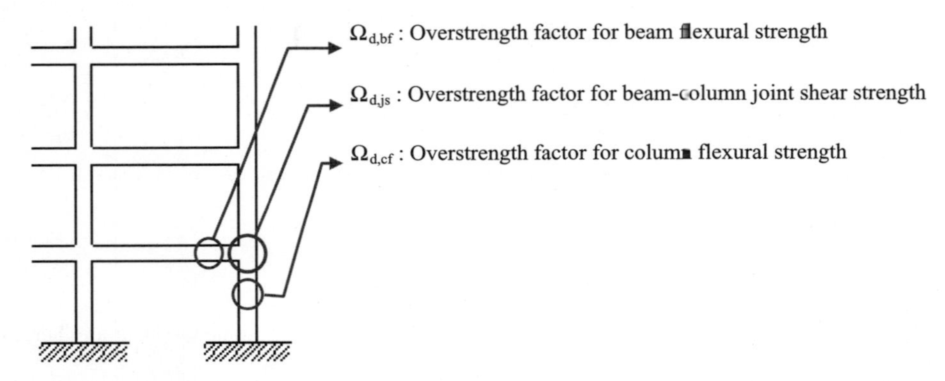

Figure 2.52 Overstrength factors employed for the design of multi-storey moment-resisting frames

Problem 2.7

The inelastic behaviour of two medium-rise steel moment resisting frames (MRFs) is assessed by the pushover curves provided in Figure 2.53. Response parameters of these frames are summarized in Table 2.10. Determine yield and ultimate deformations according to Section 2.3.3.1, as appropriate. Compare the computed values of Δ_u and Δ_y with those in Table 2.10. Determine observed Ω_d and inherent Ω_i overstrength factors for the frames. Compute also R-factors supply and translation ductility μ_δ. Comment on the results.

Table 2.10 Response parameters of assessed frames.

Frame (label)	Period (seconds)	V_d/W (%)	V_y/W (%)	V_u/W (%)	Δ_u/Δ_y (–)
MRF_1	2.53	4.04	10.18	14.28	4.10
MRF_2	3.63	1.51	7.53	8.02	1.52

Key: V_d = base design shear; V_y = base actual shear; V_u = base shear at collapse; W = seismic weight; Δ_y = roof drift at yield; Δ_u = roof drift at collapse.

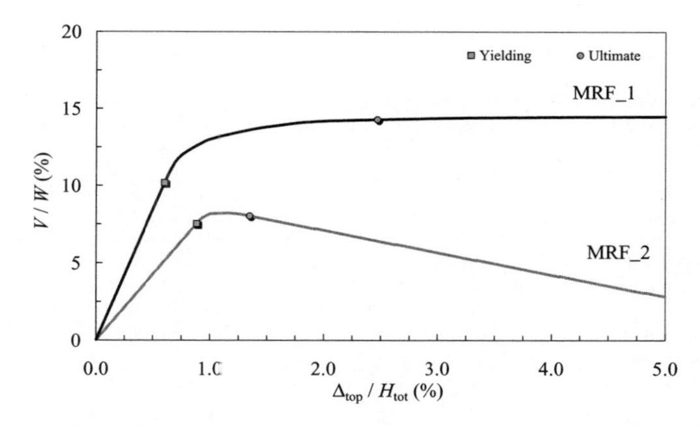

Figure 2.53 Response curve for steel moment-resisting frames: frame located in high (*MRF_1*) and low (*MRF_2*) seismicity areas
Key: V = base shear; W = seismic weight; Δ = roof drift; H_{tot} = total frame height; MRF = moment-resisting frame

References

Abrams, D.P. and Angel, R. (1993). Strength, behavior and repair of masonry infills. Proceedings of the Structures Congress '93, ASCE, Vol. 2, pp. 1439–1444.

Akiyama, H. (1985). *Earthquake-Resistant Limit-State Design for Buildings*. University of Tokyo Press, Tokyo, Japan.

Al Chaar, G., Issa, M. and Sweeney, S. (2002). Behavior of masonry-infilled nonductile reinforced concrete frames. *Journal of Structural Engineering, ASCE*, 128(8), 1055–1063.

Ascheim, M.A. (2002). Seismic design based on the yield displacement. *Earthquake Spectra*, 18(4), 581–600.

Bachmann, H., Ammann, W.J., Deischl, F., Eisenmann, J., Floegl, I., Hirsch, G.H., Klein, G.K., Lande, G.J., Mahrenholtz, O., Natke, H.G., Nussbaumer, H., Pretlove, A.J., Rainer, J.H., Saemann, E.U. and Steinbeisser, L. (1995). *Vibration Problems in Structures. Practical Guidelines*. Birkhauser Verlag, Basel, Switzerland.

Bertero, V.V. (2000). Performance-based seismic engineering: Conventional vs. innovative approaches. Proceedings of the 12th World Conference on Earthquake Engineering, Auckland, New Zealand, CD-ROM.

Bertero, R.D. and Bertero, V.V. (1992). Tall Reinforced Concrete Structures: Conceptual Earthquake Resistant Design Methodology. Earthquake Engineering Research Center, Report No. UCB/EERC/92-16, University of California, Berkeley, CA, USA.

Bertero, R.D. and Bertero, V.V. (1999). Redundancy in earthquake-resistant design. *Journal of Structural Engineering, ASCE*, 125(1), 81–88.

Bertero, V.V. and Brokken, S. (1983). Infills in seismic resistant building. *Journal of Structural Engineering, ASCE*, 190(6), 1337–1361.

Blandon, C.A. and Priestley, M.J.N. (2005). Equivalent viscous damping equations for direct displacement based design. *Journal of Earthquake Engineering*, 9(2), 257–278.

Blume, J.A., Newmark, N.M. and Corning, L.H. (1961). *Design of Multistory Reinforced Concrete Buildings for Earthquake Motions*. Portland Cement Association, IL, USA.

Borzi, B., Calvi, G.M., Elnashai, A.S., Faccioli, E. and Bommer, J.J. (2001). Inelastic spectra for displacement-based design. *Soil Dynamics and Earthquake Engineering*, 21(1) 47–61.

Bozorgnia, Y. and Bertero, V.V. (2004). *Earthquake Engineering. From Engineering Seismology to Performance-Based Engineering*. CRC Press, Boca Raton, FL, USA.

Bruneau, M., Uang, C.M. and Whittaker, A. (1998). *Ductile design of steel structures*. McGraw-Hill, New York, NY, USA.

Calvi, G.M. (1999). A displacement-based approach for vulnerability evaluation of classes of buildings. *Journal of Earthquake Engineering*, 3(3), 411–438.

Comite Euro-Internationale du Beton (CEB) (1996). *RC Frames Under Earthquake Loading. State of the Art Report*. Thomas Telford, London, UK.

Di Sarno, L. (2003). Performance-based assessment of steel perimeter frames. Proceedings of the International Conference 'Extreme Loading on Structures', Toronto, Canada, CD-ROM.

Di Sarno, L. and Elnashai, A.S. (2002). Seismic retrofitting of steel and composite building structures. Mid-America Earthquake Center Report, CD Release 02-01, University of Illinois at Urbana-Champaign, IL, USA.

Di Sarno, L. and Elnashai, A.S. (2003). Innovative strategies for seismic retrofitting of steel and composite structures. *Journal of Progress in Structural Engineering and Materials*, 7(3), 115–135.

Di Sarno, L. and Elnashai, A.S. (2005). Special metals for seismic retrofitting of steel buildings. *Journal of Progress in Structural Engineering and Materials*, 5(2), 60–76.

Di Sarno, L., Elnashai, A.S. and Nethercot, D.A. (2003). Seismic performance assessment of stainless steel frames. *Journal of Constructional Steel Research*, 59(10), 1289–1319.

Durkin, M.E. (1985). Behavior of building occupants in earthquakes. *Earthquake Spectra*, 1(2), 271–283.

Elnashai, A.S. (1992). Effect of member characteristics on the response of RC structures. Proceedings of the 10th World Conference on Earthquake Engineering, Madrid, Spain, pp. 3275–3280.

Elnashai, A.S. (1994). Local ductility in steel structures subjected to earthquake LoadIng. In *Behaviour of Steel Structures In Seismic Areas*, F.M. Mazzolani and V. Gioncu, Eds., STESSA '94, E&FN Spon, pp. 133–148.

Elnashai, A.S (1999). The Kocaeli (Turkey) Earthquake of 17 August 1999: Assessment of Spectra and Structural Response Analysis. Engineering Seismology and Earthquake Engineering, Report No. ESEE/99-3, Imperial College, London, UK.

Elnashai, A.S. and Dowling, P.J. (1991). Seismic design of steel structures. In *Steel Subjected to Dynamic Loading. Stability and Strength*, R. Narayanan and T.M. Roberts, Eds., Elsevier Applied Science, London, England.

Elnashai, A.S. and Mwafy, A.M. (2002). Overstrength and force reduction factors of multistorey reinforced-concrete buildings. *The Structural Design of Tall Buildings*, 11(5), 329–351.

Elnashai, A.S. and Pinho, R. (1998). Repair and retrofitting of RC walls using selective techniques. *Journal of Earthquake Engineering*, 2(4), 525–568.

Elnashai, A.S., Elghazouli, A.Y. and Danesh Ashtiani, F.A. (1998). Response of semi-rigid steel frames to cyclic and earthquake loads. *Journal of Structural Engineering, ASCE*, 124(8), 857–867.

Engineering Sciences Data Unit (ESDU) (1991). Structural Parameters Used in Response Calculations (Estimation of Numerical Values). Item No.91001, ESDU International, London, UK.

Fang, J.Q., Li, Q.S., Jeary, A.P. and Liu, D.K. (1999). Damping of tall buildings: Its evaluation and probabilistic characteristics. *The Structural Design of Tall Buildings*, 8(2), 145–153.

Fardis, M.N. and Calvi, G.M. (1995). Effects of infills on the global response of reinforced concrete frames. Proceedings of the 10th European Conference on Earthquake Engineering, A.A. Balkema, Rotterdam, Vol. 4, pp. 2893–2898.

Fardis, M.N. and Panagiotakos, T.B. (1996). Hysteretic damping of reinforced concrete elements. Proceedings of the 11th World Conference on Earthquake Engineering, Elsevier, Paper No. 464.

Fardis, M.N. and Panagiotakos, T.B. (1997). Seismic design and response of bare and masonry-infilled reinforced concrete buildings – Part II: Infilled structures. *Journal of Earthquake Engineering*, 1(3), 475–503.

Fu, H.C., Erki, M.A. and Seckin, M. (1991). Review of effects of loading rate on reinforced concrete. *Journal of Structural Engineering, ASCE*, 117(12), 3660–3679.

Hart, G.C. (1996). Random damping in buildings. *Journal of Wind Engineering and Industrial Aerodynamics*, 59(2–3), 233–246.

Hart, G.C., Lew, M. and Julio, R.D. (1973). High-rise building response: Damping and period non-linearities. Proceedings of the 5th World Conference on Earthquake Engineering, Rome, Italy, Vol. 2, pp. 1440–1444.

Housner, G.W. (1956). Limit design of structures to resist earthquakes. Proceedings of the 1st World Conference on Earthquake Engineering, Berkeley, CA, USA, Vol. 5, pp. 1–11.

Humar, J.L. and Ragozar, M.A. (1996). Concept of overstrength in seismic design. Proceedings of the 11th WCEE, Acapulco, Mexico, CD-ROM, Paper No. 639.

Izzuddin, B.A. and Elnashai, A.S. (1993). The influence of rate-sensitivity on the response of steel frames. Proceedings of Eurodyn '93, Trondheim, Norway, pp. 617–624.

Jacobsen, L.S. (1930). Steady forced vibration as influenced by damping. *Transactions of ASME*, 52(1), 169–181.

Jain, S.K. and Navin, N. (1995). Seismic overstrength in reinforced concrete frames. *Journal of Structural Engineering, ASCE*, 121(3), 580–585.

Jennings, P.C. (1968). Equivalent viscous damping for yielding structures. *Journal of Engineering Mechanics Division, ASCE*, 94(EM-1), 131–165.

Jeary, A.P. (1986). Damping in tall buildings. A mechanism and a predictor. *Earthquake Engineering and Structural Dynamics*, 14(5), 733–750.

Jeary, A.P. (1996). Description and measurement of non-linear damping in structures. *Journal of Wind Engineering and Industrial Aerodynamic*, 59(2–3), 103–114.

Kappos, A.J., Stylianidis, K.C. and Michailidis, C.N. (1998). Analytical models for brick masonry infilled R/C frames under lateral loading. *Journal of Structural Engineering, ASCE*, 2(1), 59–87.

Kasai, K. and Popov, E.P. (1986). A study on the seismically resistant eccentrically steel frames. Earthquake Engineering Research Center, Report No. UCB/EERC-86/01, University of California, Berkeley, CA, USA.

Klinger, R.E. and Bertero, V.V. (1976). Infilled frames in earthquake-resistant construction. Report No. EERC/76-32. Earthquake Engineering Research Center, University of California, Berkeley, CA, USA.

Krawinkler, H. and Mohasseb, S. (1987). Effects of panel zone deformations on seismic response. *Journal of Constructional Steel Research*, 8 (3), 233–250.

Kurose, Y. (1987). Recent studies on reinforced concrete deep-beam-column assemblages under cyclic loads. Phil M. Ferguson Structural Engineering Laboratory, Report No. 87-8, Civil Engineering, University of Texas, Austin, TX, USA.

Kwon, O.S. and Elnashai, A.S. (2006). The effect of material and ground motion uncertainty on the seismic vulnerability curves of RC structure. *Engineering Structures*, 28(2), 289–303.

Lee, D.H. and Elnashai, A.S. (2002). Inelastic seismic analysis of RC bridge piers including flexure-shear-axial interaction. *Structural Engineering and Mechanics*, Vol. 13(3), 241–260.

Lin, Y.Y., Miranda, E., Chnag, K.C. (2005). Evaluation of damping reduction factors for estimating elastic response of structures with high damping. *Earthquake Engineering and Structural Dynamics*, 34(11), 1427–1443.

Matos, C.G. and Dodds, R.H. (2002). Probabilistic modelling of weld fracture in steel frame connections. Part II: Seismic loadings. *Engineering Structures*, 24(6), 687–705.

Mehrabi, A.B., Shing, P.B., Schuller, M.P. and Noland, J.L. (1996). Experimental evaluation of masonry-infilled RC frames. *Journal of Structural Engineering, ASCE*, 11(3), 228–236.

Mileti, D.S. and Nigg, J.M. (1984). Earthquakes and human behavior. *Earthquake Spectra*, 1(1), 89–106.

Miranda, E. and Ruiz-Garcia, J. (2002). Evaluation of the approximate methods to estimate maximum inelastic displacement demands. *Earthquake Engineering and Structural Dynamics*, 31(3), 539–560.

Mitchell, D. and Paulter, P. (1994). Ductility and overstrength in seismic design of reinforced concrete structures. *Canadian Journal of Civil Engineering*, 21(6), 1049–1060.

Moghaddam, H.A. (1990). Seismic design of infilled frames. Proceedings of the 9th European Conference on Earthquake Engineering, Kucherenko Tsniisk of the USSR Gosstroy, Moscow, Vol. 3, pp. 3–8.

Moghaddam, H.A., Dowling, J.P. and Ambraseys, N.N. (1988). Shaking table study of brick masonry infilled frames subjected to seismic excitations. Proceedings of the 9th World Conference on Earthquake Engineering, Tokyo, Japan, Vol. VIII, Paper No. SI-13, pp. 913–918.

Mwafy, A.M. (2000). Seismic Performance of Code-Designed RC Buildings. PhD Thesis, Imperial College, London, UK.

Mwafy, A.M. and Elnashai, A.S. (2001). Static pushover versus collapse analysis of RC buildings. *Engineering Structures*, 23 (5), 407–424.

Mwafy, A.M. and Elnashai, A.S. (2002). Calibration of force reduction factors of RC buildings. *Journal of Earthquake Engineering*, 6(2), 239–273.

Mylonakis, G. and Gazetas, G. (2000). Seismic soil-structure interaction: Beneficial or detrimental. *Journal of Earthquake Engineering*, 4(3), 277–301.

Nader, M.N. and Astaneh, A.A. (1992). Seismic behaviour and design of semi-rigid steel frames. Report No. UCB/EERC-92/06, University of California, Berkeley, CA, USA.

Newmark, N.M. and Hall, W.J. (1969). Seismic design criteria for nuclear reactor facilities. Proceedings of the 4th World Conference on Earthquake Engineering, Santiago, Chile, B-4, pp. 37–50.

Novak, M. (1974). Effect of soil on structural response to wind and earthquake. *Earthquake Engineering and Structural Dynamics*, 3(1), 79–96.

Park, R. (1988). Ductility evaluation from laboratory and analytical testing. Proceedings of the 9th World Conference on Earthquake Engineering, Tokyo-Kyoto, Japan, Vol. VIII, pp. 605–616.

Park, R. (1996). Explicit incorporation of element and structure overstrength in the design process. Proceedings of the 11th WCEE, Acapulco, Mexico, CD-ROM, Paper No. 2130.

Park, R. and Paulay, T. (1975). *Reinforced Concrete Structures*. John Wiley & Sons, New York, NY, USA.

Paulay, T. (1998). Torsional mechanisms in ductile building systems. *Earthquake Engineering and Structural Dynamics*, 27(10), 1101–1121.

Paulay, T. (2001). Some design principles relevant to torsional phenomena in ductile buildings. *Journal of Earthquake Engineering*, 5(3), 273–308.

Paulay, T. and Priestley, M.N.J. (1992). *Seismic Design of Reinforced Concrete and Masonry Buildings*. John Wiley & Sons, New York, NY, USA.

Priestley, M.J.N. (2003). Myths and Fallacies in Earthquake Engineering Revisited. The Mallet Milne Lecture, IUSS Press, Italy.

Priestley, M.J.N. and Grant, D.N. (2005). Viscous damping in seismic design and analysis. *Journal of Earthquake Engineering*, 9(2), 229–255.

Priestley, M.J.N., Seible, F. and Calvi, G.M. (1996). *Seismic design and retrofit of bridges*. John Wiley & Sons, New York, NY, USA.

Roeder, C.W. (2002). Connection performance for seismic design of steel moment frames. *Journal of Structural Engineering, ASCE*, 128(4), 517–525.

Roesset, J., Whitman, R.V. and Dobry, R. (1973). Modal analysis for structures with foundation interaction. *Journal of Structural Division, ASCE*, 99(ST3), 399–416.

Saneinejad, A. and Hobbs, B. (1995). Inelastic design of infilled frames. *Journal of Structural Engineering, ASCE*, 121(4), 634–650.

Satake, N., Suda, K., Arakawa, T., Sasaki, A. and Tamura, Y. (2003). Damping evaluation using full scale data of buildings in Japan. *Journal of Structural Engineering, ASCE*, 129(4), 470–477.

Schneider, S.P., Roeder, C.W. and Carpenter, J.E. (1993). Seismic behavior of moment resisting steel frames: Experimental studies. *Journal of Structural Engineering, ASCE*, 119(ST6), 1885–1902.

Schneider, S.P., Zagers, B.R. and Abrams, D.P. (1998). Lateral strength of steel frames with masonry infills having large openings. *Journal of Structural Engineering, ASCE*, 124(8), 896–904.

Shing, P.B. and Mehrabi, A.B. (2002). Behaviour and analysis of masonry-infilled frames. *Progress in Structural Engineering and Materials*, 4(3), 320–331.

Soong, T.T. and Spencer, B.F. Jr. (2002). Supplemental energy dissipation: State-of-the-art and state-of-practice. *Engineering Structures*, 24(3), 243–259.

Suda, K., Satake, N., Ono, J. and Sasaki, A. (1996). Damping properties of buildings in Japan. *Journal of Wind Engineering and Industrial Aerodynamics*, 59(2–3), 383–392.

Taranath, B.S. (1998). *Steel, Concrete and Composite Design of Tall Buildings*. McGraw-Hill, New York, NY, USA.

Tomazevic, M. (1999). *Earthquake Resistant Design of Masonry Buildings. Series on Innovation in Structures and Construction.* Vol. 1, A.S. Elnashai and P.J. Dowling, Eds., Imperial College Press, London, UK.

Tsai, N.C. (1974). Modal damping for soil-structure interaction. *Journal of Engineering Mechanics Division, ASCE*, 100(EM2), 323–341.

Uang, C.M. (1991). Establishing R (or R_w) and C_d factors for building seismic provisions. *Journal of Structural Engineering, ASCE*, 117(1), 19–28.

Usami, T., Mizutani, S., Aoki, T. and Itoh, Y. (1992). Steel and concrete-filled steel compression members under cyclic loading. In *Stability and Ductility of Steel Structures Under Cyclic Loading*, Y. Fukumoto and G. Lee, Eds., CRC Press, Boca Raton, FL, USA, pp. 123–138.

Wakabayashi, M. (1986). *Design of Earthquake-Resistant Buildings*. McGraw-Hill, New York, NY, USA.

Wen, Y.K. and Song, S.H. (2003). Structural reliability/redundancy under earthquakes. *Journal of Structural Engineering, ASCE*, 129(1), 56–67.

Whittaker, A.S., Uang, C.M. and Bertero, V.V. (1990). An experimental study of the behavior of dual steel systems. Report No. UCB/EERC-88/14, University of California, Berkeley, CA, USA.

Whittaker, A., Hart, G. and Rojahn, C. (1999). Seismic response modification factors. *Journal of Structural Engineering, ASCE*, 125(4), 438–444.

3

Earthquake Input Motion

3.1 General

Earthquake response of structures and their foundations is an outcome of the complex interaction between the random input ground motion and the continuously changing dynamic characteristics of the system subjected to the ground motion. Therefore, to arrive at reliable assessment of assets, a complete understanding of both input motion and structural system, and their interaction, is required. Following the structurally oriented Chapter 2, in this present chapter, a simple but comprehensive outline of earthquake strong motion is given. Selection of return periods and probabilities of a certain ground-motion parameter being exceeded during the lifetime of an asset is discussed. Ground-motion models (or attenuation relationships) relating the intensity of ground shaking to the distance from the source are reviewed and their regional characteristics studied. Different commonly used forms of input motion representations, as outlined in Figure 3.1, and their ranges of applicability are discussed. Both time and frequency domain representations are addressed. The input characterizations presented are suitable for the whole range of applications, from simple code design to inelastic response history analysis. The material presented in this chapter provides the 'demand' side of the earthquake engineering design and assessment, while the next chapter provides, along with Chapter 2, the 'supply' or capacity side. Finally, the strong-motion characterization provided in this chapter maps onto the methods of structural analysis in Chapter 4.

3.2 Earthquake Occurrence and Return Period

It is of importance to estimate the frequency of occurrence of earthquakes that are likely to occur in an area that may influence the construction site during the lifetime of the intended facility. Account should be taken of the uncertainty in the demand imposed by the earthquake, as well as the uncertainty in the capacity of the constructed facility. Current seismic design approaches deal with uncertainties associated with structural demand and capacity by utilizing probabilistic analysis (e.g. Cornell *et al.*, 2002).

Earthquakes are usually modelled in probabilistic seismic hazard assessment as a Poisson process. The Poisson model is a continuous time, integer-value counting process with stationary independent increments. This means that the number of events occurring in an interval of time depends only on the length of the interval and does not change in time. Recent developments in hazard analysis employ time-dependent models that account for the occurrence of an earthquake in estimating the probability of occurrence of subsequent earthquakes (e.g. Lee *et al.*, 2003, among others). In the conventional approach described therein, the probability of an event occurring in the interval is independent of the

Fundamentals of Earthquake Engineering Amr S. Elnashai and Luigi Di Sarno

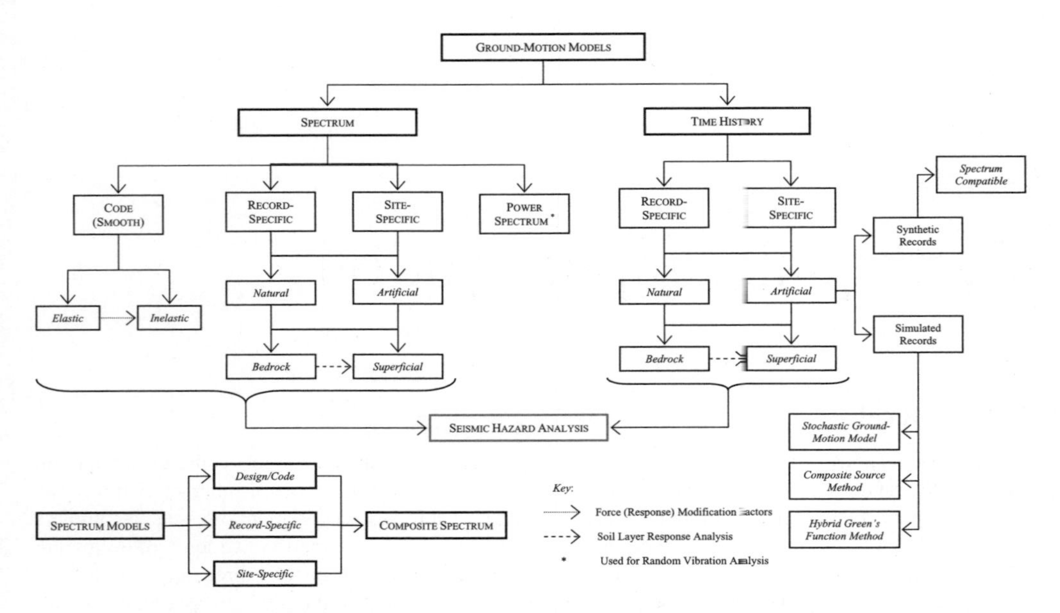

Figure 3.1 Earthquake ground-motion representations for seismic structural assessment

history and does not vary with the site. Thus, each earthquake occurs independently of any other seismic event. This model cannot include fore- and aftershocks. Notwithstanding, the Poisson model is simple because it is defined by a single parameter N, which expresses the mean rate of occurrence of an earthquake per unit of time.

The probability of earthquake occurrence modelled by the Poisson's distribution is as follows:

$$P[m > M, t_r] = \frac{(N\,t_r)^n\ e^{-N\,t_r}}{n!} \tag{3.1}$$

where P is the probability of having n earthquakes with magnitudes m greater than M over a reference time period t_r in a given area. The parameter N is the expected number of occurrences per unit time for that area, i.e. the cumulative number of earthquakes greater than M.

Recurrence relationships express the likelihood of earthquakes of a given size occurring in the given source during a specified period of time, for example one year. Therefore, the expected number of earthquakes N in equation (3.1) can be estimated by statistical recurrence formulae. Gutenberg and Richter (1954) developed the following frequency–magnitude relationship:

$$\ln N = a - b\,M \tag{3.2}$$

in which a and b are model constants that can be evaluated from seismological observational data through least-square fit. They describe the seismicity of the area and the relative frequency of earthquakes, respectively. The magnitude scale often adopted is the local or Richter magnitude, i.e. $M = M_L$. Usually an upper bound on magnitude is placed, based on the characteristics of the source and/or the maximum historical earthquake (e.g. Reiter, 1990).

The recurrence model, as given in equation (3.2), is the simplest mathematical formulation for describing earthquake recurrence. However, it has been found that it may lead to mismatches between predicted and observed values (e.g. Kramer, 1996). The predictive model in equation (3.2), which represents a Gumbel extreme type 1 distribution, matches with observed data for different tectonic zones in the range of intermediate magnitudes, e.g. $M_S \approx 6$ to 7.5, but it over-predicts the probability of occurrence at large magnitudes. In addition, for small magnitudes the agreement between predicted and measured values is rather poor because equation (3.2) envisages continuous slips. Consequently, more accurate models have been proposed (e.g. Coppersmith and Youngs, 1990, among others).

By combining equations (3.1) and (3.2), the probability of earthquake occurrence can be expressed in the form:

$$P[m > M, t_r] = 1 - e^{-e^{(a-b\,M)t_r}} \tag{3.3}$$

The return period T_R, defined as the averaged time between the occurrence of earthquakes with a magnitude m greater than M, can be estimated as follows:

$$T_R = \frac{1}{N} = -\frac{t_r}{\ln(1-P)} \tag{3.4.1}$$

or, using equation (3.3):

$$T_R = \frac{1}{N} = e^{(b\,M-a)} \tag{3.4.2}$$

thus, for example, earthquakes with a return period of 100 years equate to 60% of desired probability of being exceeded for a reference period of 100 years (Figure 3.2). Indeed, $P = P\,[m > M, t_r]$ in equation (3.4.1) expresses the desired probability of being exceeded during a reference period of time t_r.

Smaller magnitude events occur more often than larger magnitude events and generally are expected to produce less damage, as also discussed in Section 4.7. Longer return periods translate into a lower probability of earthquake occurrence, with higher potential of economical loss. The latter scenario is referred to as low probability-high consequence event. The relationship between the return period T_R, the lifetime of the structure and the desired probability of the estimate being exceeded $P\,[m > M, t_r]$ is plotted in Figure 3.2. Variations of the peak horizontal acceleration with the annual probability of being exceeded are also included for three percentiles, i.e. 15th, 50th and 85th.

Design earthquake loads are based on ground motion having a desired probability of being exceeded during the lifetime of the facility, as displayed in Tables 2.1 and 4.10. Seismic codes generally assume a lifetime of about 50 years for ordinary buildings and the probability of being exceeded equals 10%.

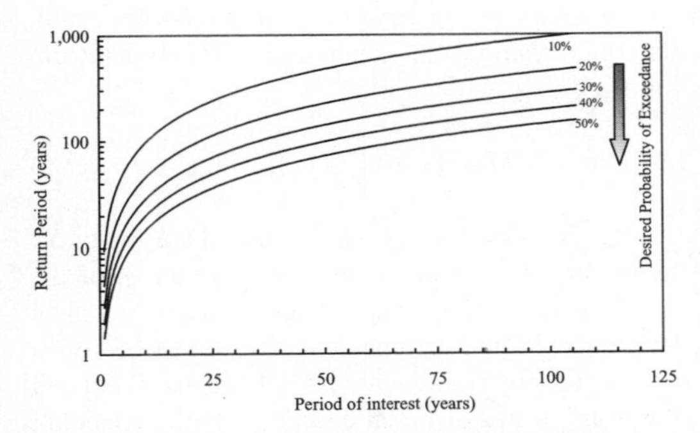

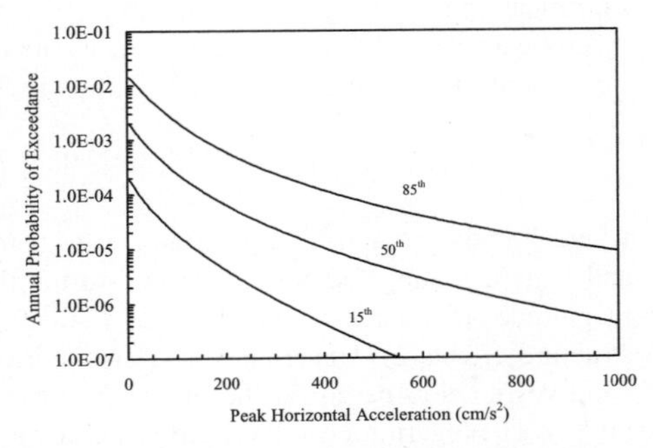

Figure 3.2 Relationship between return period, lifetime of the structure and desired probability of being exceeded (*left*) and hazard curves for peak horizontal ground accelerations (*right*)

Consequently, the return period is about 475 years, which can be computed from equation (3.4.1). For a facility lifetime larger than 80–90 years, the probability of the estimate being exceeded P can be assumed approximately equal to the period of interest divided by the return period. For example, for 100 years and 1,000 years as the return period, the probability P is about 10%. The current discussion ties in with Table 2.1.

Problem 3.1

A long-span suspension bridge is going to be built in an active seismic region in Japan. The structural earthquake engineer can choose the design ground-motion parameter with respect to three return periods: 475, 950 and 2,500 years. Which is the most suitable return period to select and why? What is the associated probability of the peak ground acceleration for the return period being exceeded?

3.3 Ground-Motion Models (Attenuation Relationships)

The 'attenuation' of earthquake ground motions is an important consideration in estimating ground-motion parameters for assessment and design purposes. Ground-motion models (or attenuation relationships) are analytical expressions describing ground-motion variation with magnitude, source distance and site condition, which account for the mechanisms of energy loss of seismic waves during their travel through a path as discussed in Sections 1.1 through 1.3. Attenuation relationships permit the estimation of both the ground motion at a site from a specified event and the uncertainty associated with the prediction. This estimation is a key step in probabilistic and deterministic seismic hazard analysis (Cornell, 1968). There are a number of ground-motion models that have been developed by various researchers. Relationships based on peak ground-motion parameters (peak ground acceleration, PGA; peak ground velocity, PGV; peak ground displacement, PGD) and spectral acceleration, velocity and displacement parameters (S_a, S_v and S_d), presented in Section 3.4.3, are generally employed in structural earthquake engineering.

The proliferation of strong-motion recording equipment over the past 50 years has provided large databanks of earthquake records. The most basic ground-motion models express PGA as a function of magnitude and epicentral distance. Several formulae include other parameters to allow for different site types (e.g. rock versus soft soil) and fault mechanisms. These are developed by fitting analytical expressions to either observational or synthetic data, depending on the availability of strong-motion records for the region under investigation. Ground-motion attenuation relationships are derived either empirically, utilizing natural earthquake records, or theoretically, employing seismological models to generate synthetic ground motions that account for source, path and site effects. These approaches may overlap. Empirical approaches generally match the data to a functional form derived from the theory; in turn, theoretical approaches often use empirical data to determine values of parameters. The functional form for ground-motion attenuation relationships is as follows:

$$\log(Y) = \log(b_1) + \log[f_1(M)] + \log[f_2(R)] + \log[f_3(M,R)] + \log[f_4(E_i)] + \log(\varepsilon) \tag{3.5}$$

where Y is the ground-motion parameter to be computed, for example PGA, PGV, PGD, S_a S_v or S_d, and b_1 is a scaling factor. The second-to-fourth terms on the right-hand side are functions f_i of the magnitude M, source-to-site distance R, and possible source, site and geologic and geotechnical structure effects E_i. Uncertainty and errors are represented by the parameter ε. Equation (3.5) is an additive function based on the model for ground-motion regression equations defined by Campbell (1985). The logarithm can be expressed either a natural 'ln' or in a different base, e.g. 'log', depending on the formulation. The above equation also accounts for the statistical log-normal distribution of the ground-motion parameter Y. Peak ground-motion parameters decrease as the epicentral distance

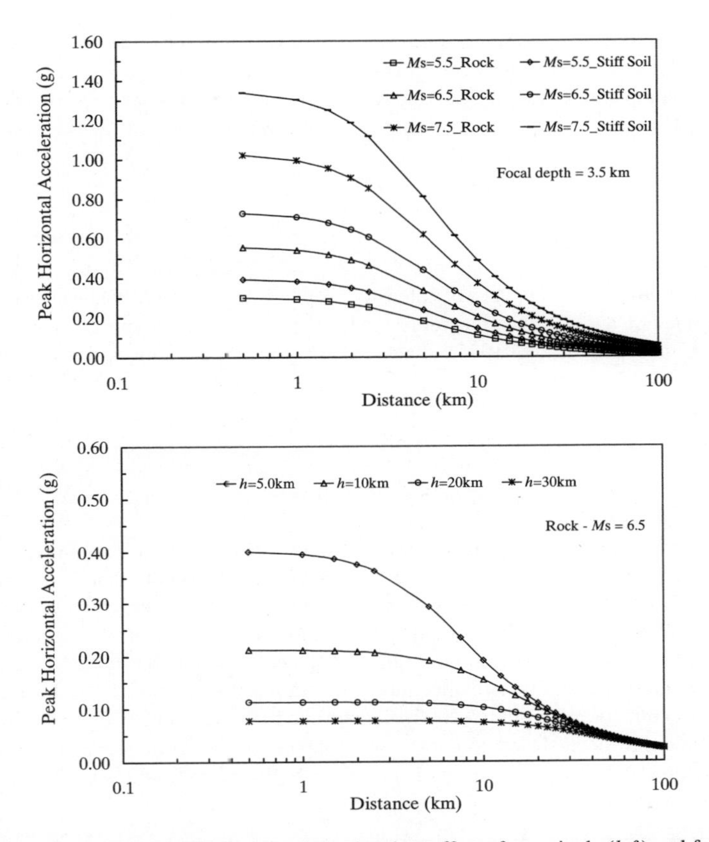

Figure 3.3 Attenuation of peak ground horizontal acceleration: effect of magnitude (*left*) and focal depth (*right*)

increases. The attenuation depends, however, on the magnitude; these variations may be expressed by equation (3.5). Figure 3.3 shows variations of peak ground horizontal acceleration with magnitude and the effect of focal depth.

A number of reviews of attenuation studies were made in the past (e.g. Trifunac and Brady, 1976; Idriss, 1978; Boore and Joyner, 1982; Campbell, 1985; Joyner and Boore, 1988, 1996; Ambraseys and Bommer, 1995). These provide useful summaries of the methods used, results obtained and problems associated with strong-motion attenuation relationships. A comprehensive worldwide summary of such relationships was compiled by Douglas (2001, 2002, 2004, 2006). The latter study discusses also data selection, processing and regression methods, alongside the forms of equations used. These equations are continuously reviewed and updated as more instrumental records become available and more refined mathematical models for ground motion are employed. However, it should be noted than in comparison to attenuation relationships based on magnitude, fewer studies attempt to relate ground-motion parameters to an intensity scale. Such studies were carried out in the late 1970s and early 1980s. Some of these formulations relating intensity to peak ground accelerations are discussed hereafter. These equations are clearly too simplistic and do not account reliably for parameters influencing earthquake damage potential. The subjective and discrete nature of intensity scales, presented in Section 1.2.1, does not allow an accurate description of structural damage. Moreover, in several cases, it is not straightforward to define a damage indicator on the basis of ground-motion parameters (Jennings, 1985).

Such intensity–ground motion relationships are nonetheless important so as to make use of historical data for which only observed intensity information exists.

Closed-form relationships between PGA and relevant intensity scales have been established in Japan and in the USA. These are given by Kanai (1983) as follows:

$$\text{PGA} = 0.25 \cdot 10^{0.50\, I_{\text{JMA}}} \tag{3.6.1}$$

$$\text{PGA} = 0.91 \cdot 10^{0.31\, I_{\text{MM}}} \tag{3.6.2}$$

in which I_{JMA} and I_{MM} are the values of intensity in the JMA and MM scales (*see* Section 1.2.1), respectively. In the above equations, the values of PGA are expressed in cm/s^2.

Similarly, Trifunac and Brady (1975) suggested the following relationships for horizontal peak ground acceleration and velocity:

$$\text{PGA} = 1.02 \cdot 10^{0.30\, I_{\text{MM}}} \tag{3.7.1}$$

$$\text{PGV} = 0.23 \cdot 10^{0.25\, I_{\text{MM}}} \tag{3.7.2}$$

where the values of PGA and PGV are in cm/s^2 and cm/s, respectively. Equations (3.7.1) and (3.7.2) are applicable for I_{MM} ranging between IV and X. It is instructive to note that for every unit increase in intensity, the PGA and PGV increase by more than 100% and 80%, respectively.

The use of the above relationships should not be indiscriminate but limited to those cases where historical observation data, based mainly on intensity values, are available.

A recent and significant addition to the library of strong-motion models is the New Generation Attenuation (NGA) due to Power *et al.* (2006). This ground-motion model is arguably the most robust model available and has been shown to apply in many parts of the world. Readers are referred to the source literature on the NGA ground-motion model (e.g. Boore and Atkinson, 2007; Campbell and Bozorgnia, 2007) where details of its background and use are given.

Problem 3.2

Modified Mercalli intensity I_{MM} of IX was assigned to an area of about 80 km long and 30 km wide during an earthquake that occurred in the Western United States. Compute the peak ground acceleration (PGA) from this earthquake. Compute the value of PGA by using both equations (3.6.2) and (3.7.1) and comment on the results. Estimate the intensity I_{JMA} according the earthquake in the Japanese Meteorological Agency (JMA) scale.

3.3.1 Features of Strong-Motion Data for Attenuation Relationships

The strong-motion data set (or catalogue) used for attenuation relationship derivation has to fulfil a number of requirements. First, all magnitudes should be uniformly recalculated using consistent approaches. Second, all distances have to be defined uniformly. It is necessary to use the distance from the closest point on the causative fault to the measuring site, not the epicentral distance. This is particularly important when considering large-magnitude earthquakes at short-to-medium distances.

Calculation of the above-mentioned distance is an involved task that requires deep knowledge of the local tectonic setting, especially when there is no surface manifestation of the fault. Moreover, the data set should be well populated and reasonably represent distributions in magnitude, distance and soil condition; otherwise the ensuing attenuation relationship will exhibit statistical bias. Strong-motion records in databanks may have errors due to instruments and digitization. Since the short- and long-period errors present in each record are unique for each type of instrument and digitization procedure, and because of the random nature of the errors, each accelerogram should ideally be corrected individually. Records from analogue instruments are particularly affected by long-period (or low frequency, e.g. less than 0.5 Hz) errors because of the digitization stage, which is not required for records from digital instruments. Low-frequency errors affect the peak ground velocity as well as the corresponding spectral values.

3.3.2 Attenuation Relationship for Europe

Comprehensive and systematic seismological studies in Europe aimed at defining ground-motion models for seismic hazard assessment and structural engineering applications were conducted by Ambraseys (1975). A great deal of research has been conducted since then at Imperial College, London, and attenuation relationships have been formulated for Europe and the Middle East (e.g. Ambraseys and Bommer, 1991, 1992; Ambraseys and Simpson, 1996; Ambraseys *et al.*, 1996). Other studies have focused on local areas with a relatively high occurrence of earthquakes. These areas include Greece (e.g. Skarlatoudis *et al.*, 2004), Italy (e.g. Rinaldis *et al.*, 1998), Turkey (e.g. Kalkan and Gulkan, 2004) and Romania (e.g. Stamatovska, 2002). Revised attenuation relationships for European countries and some regions in the Middle East have been formulated for both horizontal peak ground acceleration by Ambraseys *et al.* (2005a) and for vertical peak ground acceleration by Ambraseys *et al.* (2005b). The ground-motion model for the horizontal PGA is given by:

$$\log(\text{PGA})_h = 2.522 - 0.142\, M_w + (0.314\, M_w - 3.184)\log\sqrt{57.76 + d^2} + 0.137\, S_S + 0.050\, S_A - 0.084\, F_N + 0.062\, F_T - 0.044\, F_O \qquad (3.8.1)$$

with PGA expressed in m/sec^2 and d is the distance (in km) to the projection of the fault plane on the surface. The latter does not require a depth estimate, generally associated with large errors. The coefficients S_A and S_S are obtained from Table 3.1. These are given as a function of the soil type. Three types of soil conditions were considered: rock, stiff and soft. The term M_w in the equation above indicates the moment magnitude.

The coefficients F_N, F_O and F_T in equation (3.8.1) are related to the focal mechanism of the earthquakes, which were classified by using the method proposed by Frohlich and Apperson (1992). Such method does not require distinction between the main and the auxiliary planes. The values of the *F*-factors should be selected from Table 3.2.

The standard deviations σ for equation (3.8.1) depend on the earthquake magnitude M_w:

$$\text{(intra-plate)} \qquad \sigma_1 = 0.665 - 0.065\, M_w \qquad (3.8.2)$$

$$\text{(inter-plate)} \qquad \sigma_2 = 0.222 - 0.022\, M_w \qquad (3.8.3)$$

Table 3.1 Values of coefficients for equation (3.8.1).

Soil type	Shear wave velocity, v_S (in m/s)	S_A	S_B
Rock	$v_s > 750$	0	0
Stiff	$360 < v_s \leq 750$	1	0
Soft	$180 < v_s \leq 360$	0	1

Table 3.2 Values of coefficients for equation (3.8.1).

Focal mechanism	F_N	F_O	F_T
Normal	1	0	0
Odd	0	1	0
Strike-slip	0	0	0
Thrust	0	0	1

Table 3.3 Values of soil coefficient S in equation (3.9.1).

Soil type	Soil coefficient (S)
Rock	0.751
Hard soil	0.901
Medium soil	1.003
Soft soil	0.995

In deriving equation (3.8.1), the magnitudes of the suite of natural records were not converted from other scales because this increases the uncertainty in the magnitude estimates. In order to obtain a viable distribution of records at all magnitudes, records from earthquakes with $M_w < 5$ were not considered. This also excludes records from small earthquakes that are unlikely to be of engineering significance. The data set includes records with magnitude M_w ranging between 5.0 and 7.6, with distances $d <$ 100 km. Therefore, the possible bias due to non-triggering instruments and the effects of anelastic decay in different regions were reduced. Moreover, most ground motions were obtained from free-field stations although some were recorded from either basements or ground floors of relatively light structures that are unlikely to modify the motion from that of the free field.

3.3.3 Attenuation Relationship for Japan

Several studies have attempted to define analytical models for the ground-motion parameters in Japan (e.g. Iwasaki *et al.*, 1980; Kawashima *et al.*, 1986; Fukushima *et al.*, 1995; Kamiyama, 1995). Some have also concentrated on specific areas of the country, such as the Kanto region (e.g. Tong and Katayama, 1988). Takahashi *et al.* (2000) proposed the following attenuation relationship for Japan:

$$\log(\text{PGA}) = 0.446\, M_w - 0.00350\, d - \log\left(d + 0.012\cdot 10^{0.446\, M_w}\right) + 0.00665(h-20) + S \quad (3.9.1)$$

where PGA is given in cm/s^2. The terms d and h are focal distance and depth (in km), respectively. S is a coefficient depending on the soil type (rock, hard, medium and soft medium were considered in the regression analyses, as given in Table 3.3). However, in many circumstances, the site conditions of records used were either unknown or uncertain. In such cases, the mean site term S can be assumed equal to 0.941. M_w is the moment magnitude.

Model errors σ in equation (3.9.1) were computed as follows:

$$\sigma = \sqrt{\sigma_1^2 + \sigma_2^2} \quad (3.9.2)$$

where σ_1 and σ_2 are residuals for inter- and intra-plate earthquakes, respectively; values decrease with increasing magnitude. It may be assumed that the total scatter σ is equal to 0.24.

3.3.4 Attenuation Relationships for North America

Several attenuation relationships have been derived for North America since the early 1970s (e.g. Esteva and Villaverde, 1973; McGuire, 1978; Joyner and Boore, 1981, 1988; Boore *et al.*, 1997; Chapman, 1999, among others). Most of these are calibrated for western USA conditions, which exhibit high recurrence of earthquakes. Two relationships for PGA are presented hereafter. They can be employed for the central and eastern areas of the USA and in western North America, respectively.

(i) Central and Eastern United States

The Mid-America Earthquake Center developed a ground-motion model to predict horizontal PGA in the central and eastern United States (CEUS) region (Fernandez and Rix, 2006). The attenuation relationship is based on a stochastic method and employs three source models, i.e. Atkinson and Boore (1995), Frankel *et al.* (1996) and Silva *et al.* (2003). It was developed for soil sites in the Upper Mississippi Embayment in the New Madrid seismic zone, which is a low probability-high impact source of earthquakes. The region has been hit by three great earthquakes in 1811 and 1812 (e.g. Reiter, 1990).

The ground-motion model is defined by the following equations:

$$\ln(\text{PGA}) = c_1 + c_2\, M_w + c_3\,(M_w - 6)^2 + c_4 \ln(R_M) + c_5 \max\left[\ln\left(\frac{R}{70}\right), 0\right] + c_6\, R_M \qquad (3.10.1)$$

where the equivalent distance term R_M is given by:

$$R_M = R + c_7 \exp(c_8\, M_w) \qquad (3.10.2)$$

and the logarithmic standard deviation of PGA, termed $\sigma_{\ln(\text{PGA})}$, is considered to be magnitude-dependent. It is obtained from the following equation:

$$\sigma_{\ln(\text{PGA})} = c_9\, M_w + c_{10} \qquad (3.10.3)$$

In the above equations, R is the epicentral distance (in km), M_w is the moment magnitude and c_1 through c_{10} are the regression coefficients. The value of the peak ground acceleration PGA is expressed in g. In equations (3.10.1) and (3.10.2), the epicentral distance R is the distance from the observation point to the surface projection of the hypocentre. The c_i coefficients, which depend on the source model, the stress drop, the soil profile, dynamic soil properties and depth, can be found in Fernandez (2007). These coefficients are computed for epicentral distances uniformly distributed between 1 and 750 km for eight values of magnitudes M_w varying between 4 and 7.5. In equations (3.10.1) and (3.10.2), the equivalent distance term R_M accounts for the increase in travelling distance by the seismic waves due to the increase in fault rupture size. The exponential term in equation (3.10.2) accounts for the magnitude-dependence of the energy release. The effects of inelastic soil behaviour are incorporated in the above attenuation relationship. A number of ground-motion models to predict the horizontal components of PGAs in the CEUS for rock sites can be found in the literature (e.g. Atkinson and Boore, 1995, 2006; Frankel *et al.*, 1996; Silva *et al.*, 2003).

(ii) Western North America

Boore *et al.* (1997) and Boore (2005) formulated the following equation to predict peak ground accelerations in western North America:

$$\log(\text{PGA}) = b_1 + b_2(M_w - 6) + b_3(M_w - 6)^2 + b_4\, R + b_5 \log(R) + b_V(\log v_S - \log v_A) \qquad (3.11.1)$$

Table 3.4 Values of soil types used for deriving equations (3.12.1) and (3.12.2).

Soil type	Shear wave velocity, v_S (in m/s)
Class A	$v_s > 750$
Class B	$360 < v_s \leq 750$
Class C	$v_s \leq 360$

in which R is the focal distance given by:

$$R = \sqrt{d^2 + h^2} \tag{3.11.2}$$

where d and h are the epicentral distance and the focal depth, respectively; they are both expressed in kilometres. The value of the peak ground acceleration PGA is expressed in g.

Shallow earthquakes are those for which the fault rupture has a depth of 20 km or less. M_w is the moment magnitude. Coefficients b_1 through b_5 in equation (3.11.1) depend on the component of ground motion used. For randomly oriented horizontal components, the PGA is given by:

$$\log(\text{PGA}) = -0.105 + 0.229(M_w - 6) - 0.778\log(R) - 0.371(\log v_S - \log v_A) \tag{3.12.1}$$

and the focal depth h in equation (3.12.1) should be assumed equal to 5.57 km; the value of v_A is 1,400 m/sec. Thus, the resulting scatter σ is 0.160. On the other hand, for larger horizontal components, equation (3.11.1) should be modified as follows:

$$\log(\text{PGA}) = -0.038 + 0.216(M_w - 6) - 0.777\log(R) - 0.364(\log v_S - \log v_A) \tag{3.12.2}$$

in which the focal depth h should be 5.48 km, the value of v_A is 1,390 m/sec and the resulting scatter is 0.144.

Site conditions are accounted for in equations (3.12.1) and (3.12.2) by the average shear-wave velocity to a depth of 30 m (v_S, in m/sec). Three soil types were considered in the study; values for $v_{S,30}$ are summarized in Table 3.4.

It is worth mentioning that in the derivation of the above attenuation relationships, most of the earthquake ground motions were recorded at epicentral distances less than 80 km, thus extrapolations should be assessed carefully on the basis of engineering judgement.

3.3.5 Worldwide Attenuation Relationships

Attempts to provide ground-motion models applicable worldwide were initiated in the 1980s (Aptikaev and Kopnichev, 1980; Campbell, 1985) and continued during the 1990s (Campbell, 1993, 1997; Sarma and Srbulov, 1996, 1998). In some cases the attenuation relationships were derived for specific fault rupture mechanisms, such as subduction zones (Youngs *et al.*, 1988, 1997; Crouse, 1991) or extensional regimes (Spudich *et al.*, 1997, 1999). Formulae for intra-plate regions have also been proposed by Dahle *et al.* (1990). Comprehensive analytical studies based on large data sets of records for both horizontal and vertical components have been carried out by Bozorgnia *et al.* (2000), Campbell and Bozorgnia (2003) and Ambraseys and Douglas (2003); the latter is presented herein.

The form of the equation to predict horizontal peak ground accelerations PGA_h is as follows:

Table 3.5 Values of coefficients for equation (3.13).

Soil type	Shear wave velocity, v_S (in m/s)	S_A	S_S
Rock	$v_s > 750$	0	0
Stiff soil	$360 < v_s \leq 750$	1	0
Soft soil	$180 < v_s \leq 360$	0	1
Very soft soil	$v_s \leq 180$	0	1

$$\log(\text{PGA})_h = -0.659 + 0.202\, M_S - 0.0238\, d + 0.020\, S_A + 0.029\, S_S \qquad (3.13)$$

in which the epicentral distance d is in km; the scatter σ is 0.214. Coefficients S_A and S_S account for the effects of soil condition. Four soil categories were considered (rock, stiff soil, soft and very soft soil); they are classified on the basis of the average shear wave velocity to 30-m depth $v_{S,30}$. Values can be obtained from Table 3.5. Focal depths h are not greater then 20 km ($1 \leq h \leq 19$ km). The value of PGA in equation (3.13) is expressed in m/sec^2.

Equation (3.13) assumes decay associated with anelastic effects due to large strains. Consequently, in the above relationships, both terms log(d) and d [*see* equation (3.5), where d = R] are not utilized because their strong correlation does not permit a simple summation.

Problem 3.3

For an earthquake of magnitude M = 7.0 and depth 25 km, calculate the peak ground acceleration at a site 50 km from the epicentre using the attenuation relationships for Europe, Japan and western North America (randomly oriented horizontal components). The fault mechanism is normal and the seismic waves travel through a thick layer of rock (v_s = 780 m/sec). Compare the results with the prediction of the worldwide attenuation relationship. Plot the curve of the above attenuation relationships for M = 7 and comment on the plots.

3.4 Earthquake Spectra

3.4.1 Factors Influencing Response Spectra

The shape of earthquake (acceleration, velocity or displacement) spectra is influenced by a number of factors, which are similar to those affecting earthquake ground-motion characteristics, outlined below:

(i) Magnitude;
(ii) Source mechanism and characteristics;
(iii) Distance from the source of energy release;
(iv) Wave travel path;
(v) Rupture directivity;
(vi) Local geology and site conditions.

Some factors are more influential than others and therefore are selected for discussion hereafter. The three fundamental parameters influencing spectra are magnitude, distance and site conditions. Ideally, strong motions used to derive uniform hazard spectra should be uniformly distributed in the space of these three parameters. This is an onerous requirement that is often impossible to be complied with. Thus, compromises in design situations are almost always necessary. Moreover, parameterized equations for spectral ordinates require the knowledge of a set of magnitude–distance pair, which is not always available for an engineering project. Therefore, seismic codes recommend spectra that are dependent on peak ground parameters and the soil conditions only, as discussed in Section 3.4.5.

Similar to ground-motion models illustrated in Section 3.3, extensive statistical analyses of the spectra at different periods have been conducted. Spectral models are derived and are expressed by equations that are a function of magnitude, distance and soil conditions, leading to spectral ordinates at different periods. For example, spectra from different attenuation relationships are shown in Figure 3.4 for different magnitudes and constant distance as well as different distances and the same magnitude. In both cases, the spectra are normalized to the same PGA. It is clear that the effect of distance can be compensated for by scaling, whereas the actual shape changes for different magnitudes. However, this might not be the case of inelastic and degrading systems, where the shaking duration will have an effect, as discussed in Section 3.6. Hence, it is not possible to compensate for magnitude effects by scaling only since the spectral shape changes.

In Figure 3.5, the attenuation relationship of Ambraseys *et al.* (1996) was used to calculate acceleration spectra for a magnitude 5.5 earthquake at a distance of 10 km on three sites: rock, stiff and soft soil. It is demonstrated that the amplification characteristics are distinct. Moreover, the acceleration amplifications for soft soils extend over a larger period range than the amplifications for the other two soil categories. The longer the predominant period of vibration of the site, the greater is the period at which the response spectrum high amplification region occurs. The shape of the spectrum is also different, but not drastically so. This is not the case for other studies, especially from the USA (e.g. Douglas, 2001, 2006). On average, magnification in the short-period range in Europe, on stiff soil, is about 1.4 that on rock, with a value of about 1.7 for soft soil is observed. The corresponding values for the USA are 2 and 3 (Boore *et al.*, 1993). It is not clear why such large differences exist. The fact remains that site condition must be taken into account in deriving spectra since no process exists for scaling spectra to account for soil condition.

Filtering of acceleration traces from earthquakes may substantially affect the characteristics of the ensuing motion. This is the main motivation behind the increasing deployment of digital instruments. However, acceleration spectra are much more tolerant to filter corner frequencies than displacement spectra as displayed in Figure 3.6.

Therefore, it is reassuringly concluded that acceleration spectra may be derived with little effort dedicated to processing of the acceleration trace, with regard to filter frequency. In applications where the displacement applied at the base of the structure is of significance, such as in the case of non-synchronous motion or for deformation-based design, careful filtering is essential; otherwise, unrealistic net static displacements between support points may ensue.

3.4.2 Elastic and Inelastic Spectra

Strong-motion records are three-component (two horizontal components and a vertical component) time histories recorded by accelerometers in analogue or digital form. These records may be used to conduct response-history dynamic analyses and derive response spectra. The latter are described herein.

A response spectrum is a plot of the maxima of the acceleration, velocity and displacement response of single-degree-of-freedom (SDOF) systems with various natural periods when subjected to an earthquake ground motion. A family of curves is usually calculated for a given excitation, showing the effect of variation of the structural damping. For many practical structural applications, it is sufficient to employ the maximum (or 'spectral') values of the above response parameters rather than their values at each instant during the time history. Earthquake input may be defined by response spectra of various forms, i.e. elastic, inelastic, parameterized and smoothed as shown in Figure 3.1. Such forms are

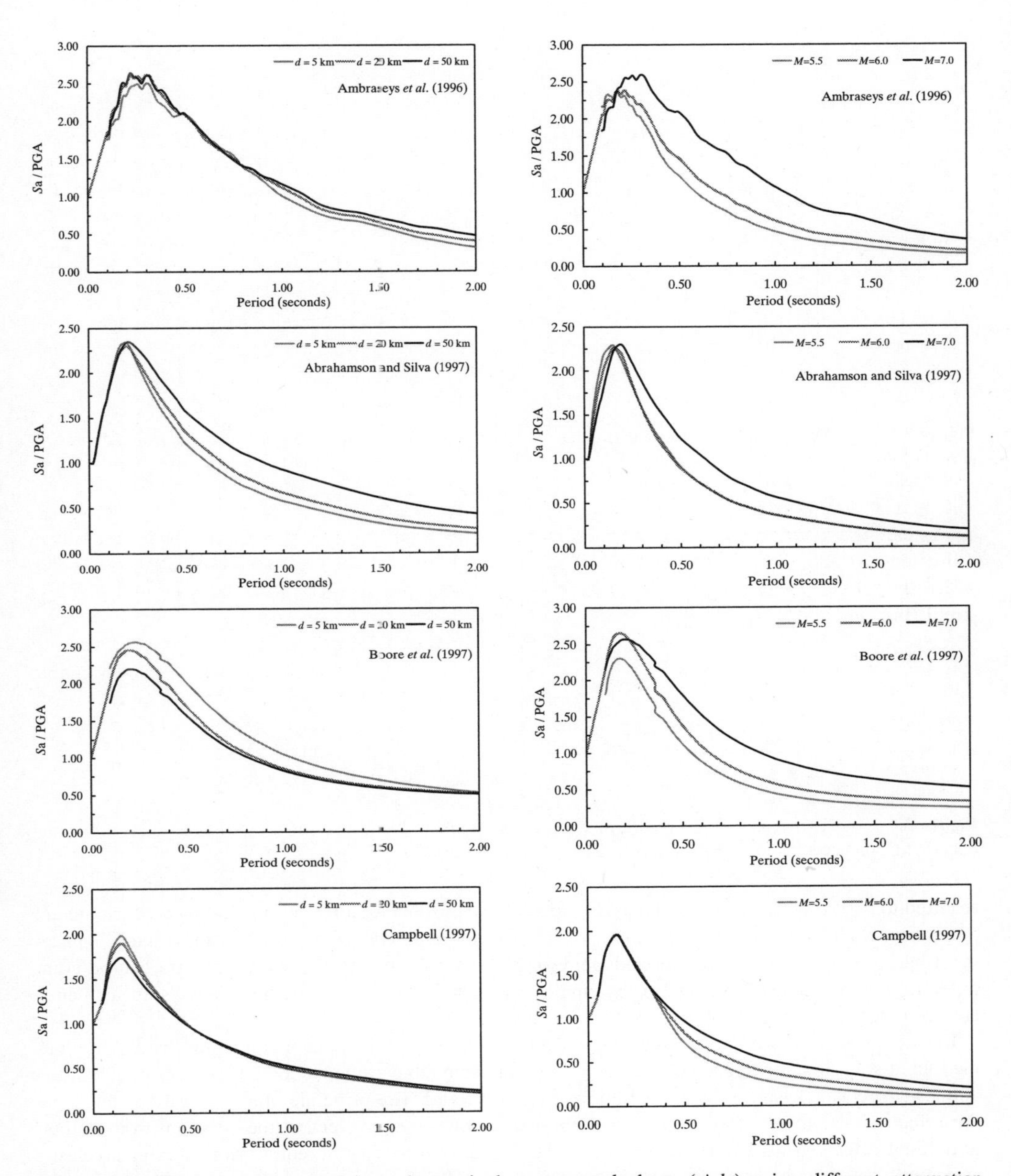

Figure 3.4 Effects of distance (*left*) and magnitude on spectral shape (*right*) using different attenuation relationships

Key: The spectra on the left are normalized for 5, 20 and 50 km, while the spectra on the right are normalized at 10 km for magnitude 5.5, 6.0 and 7.0

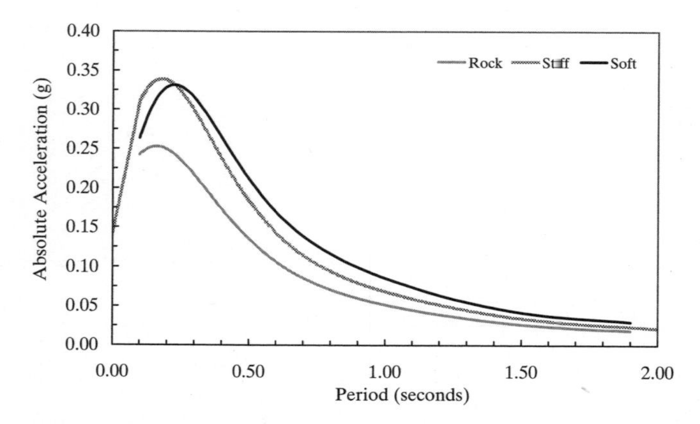

Figure 3.5 Spectra for a magnitude 5.5 earthquake at 10 km for different site conditions (*adapted from* Ambraseys *et al.*, 1996)

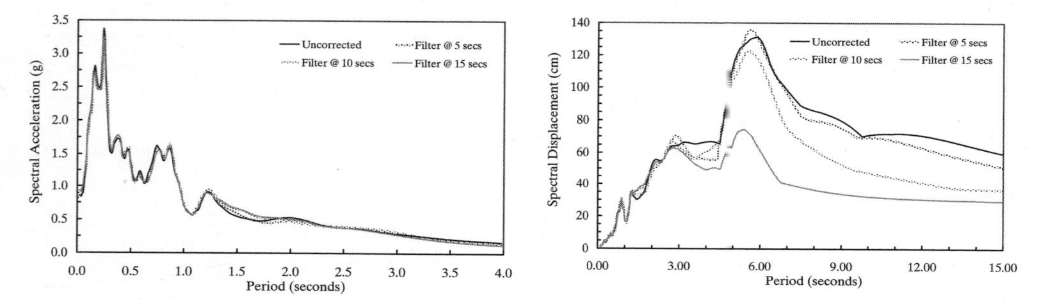

Figure 3.6 The 1978 Tabas (Iran) record filtered at 5, 10 and 15 seconds cut-off, as well as baseline correction only. Effect on acceleration (*left*) and displacement (*right*) spectra

required to perform modal spectral analysis and adaptive pushover with spectrum scaling as illustrated in Sections 4.5.1.1 and 4.5.2.2. They are also essential for capacity spectrum assessment and displacement-based design (e.g. Bozorgnia and Bertero, 2004, among others). Response spectra can be computed from earthquake accelerograms by employing one of several computer programs; some of which are presented in Section 3.8.1.

Elastic response spectra are derived analytically by evaluating the Duhamel integral, which provides the total displacement response of SDOF systems subjected to earthquake loading. Since superposition applies (for elastic system), the convolution integral is valid. The principle of superposition states that the effect of a number of simultaneously applied actions is equivalent to the superposition of their individual effects considered one at a time. The equation of dynamic equilibrium for linear elastic structural systems with mass m, stiffness k and damping c is as follows:

$$m\,\ddot{u} + c\,\dot{u} + k\,u = -m\,\ddot{u}_g \tag{3.14}$$

where the term $\ddot{u}_g$ is the ground acceleration. Thus, equation (3.14) expresses the equilibrium of inertial $m\ddot{u}$, damping c and elastic ku forces, and the earthquake loading $-m\ddot{u}_g$. It can be demonstrated by using principles of structural dynamics that the maximum value of the displacement S_d, defined as 'spectral displacement', is equal to (e.g. Chopra, 2002):

$$S_{\mathrm{d}} = [u(t)]_{\max} = \frac{\left[\int_0^t \ddot{u}_{\mathrm{g}}\, \mathrm{e}^{-\xi\,\omega\,(t-\tau)} \sin[\omega_{\mathrm{d}}(t-\tau)]\mathrm{d}\tau\right]_{\max}}{\omega} \tag{3.15}$$

in which τ is a time variable chosen arbitrarily within the duration of the strong-motion and ω the natural frequency of the undamped system. Moreover, ω_{d} is the damped circular frequency given as:

$$\omega_{\mathrm{d}} = \omega\sqrt{1-\xi^2} \tag{3.16}$$

while ξ the viscous damping of the oscillator expressed as a percentage of the critical value c_{crit}. Note that $c_{\mathrm{crit}} = 2m\omega$ and $\xi = c/c_{\mathrm{crit}}$. Ordinary structural systems exhibit viscous damping which ranges between 0.5% and about 10% as given in Tables 2.6 and 2.9. As a consequence, the values of undamped and damped frequencies in equation (3.16) are similar and hence ω can be used instead of ω_{d}. Displacement response spectra are essential for displacement-based design. Extensive analytical work has been conducted by Bommer and Elnashai (1999) and Tolis and Faccioli (1999) to derive parameterized displacement spectra. The latter spectra are discussed in Section 3.4.3.1.

On the other hand, the maximum velocity S_{v} can be approximated, assuming harmonic motion, by the product of the spectral displacement S_{d} and the fundamental frequency ω of the SDOF:

$$S_{\mathrm{v}} = \omega\, S_{\mathrm{d}} \tag{3.17}$$

which is defined as 'spectral pseudo-velocity' and corresponds to the integral at the numerator in equation (3.15). The prefix 'pseudo' shows that S_{v} is not the actual peak velocity, which would be obtained from differentiating the displacement expression. Nonetheless, for the practical range of damping in structural earthquake engineering mentioned earlier and for low-to-medium period systems, pseudo-velocity spectra are a close approximation of the true relative velocity spectra.

In a comprehensive study by Sadek *et al.* (2000) based on statistical analysis of 40 damped SDOFs subjected to 72 ground motions, it was shown that the above approximation holds for periods in the neighbourhood of 0.5 seconds as shown in Figure 3.7. However, differences are observed as the period and the damping ratios increase. Velocity spectra are of importance in seismic design because they are a measure of the energy transmitted into the oscillator (Housner, 1956).

Similarly, the 'spectral pseudo-acceleration' S_{a} is expressed as follows:

$$S_{\mathrm{a}} = \omega\, S_{\mathrm{v}} = \omega^2\, S_{\mathrm{d}} \tag{3.18}$$

Thus, the acceleration spectrum is derived by multiplying each ordinate of the velocity spectrum by the natural frequency ω of the SDOF. However, for structures with supplemental devices, e.g. with passive and/or active dampers or base isolation devices, the differences between maximum absolute acceleration and S_{a} increase as a function of the natural period T as shown in Figure 3.7. The true absolute acceleration spectra can be computed by differentiating twice the displacement expression from, for example, the Duhamel integral. It is instructive to note that the acceleration response spectra are related directly to the base shear used in seismic design and hence they are generally implemented in force-based codes of practice, e.g. equations (3.34) and (3.35.1) provided in Section 4.6.3. The relevant alternative for displacement-based design is the relative displacement spectrum.

The procedure to derive elastic spectra is schematically depicted in Figure 3.8. The computational scheme can be summarized as follows:

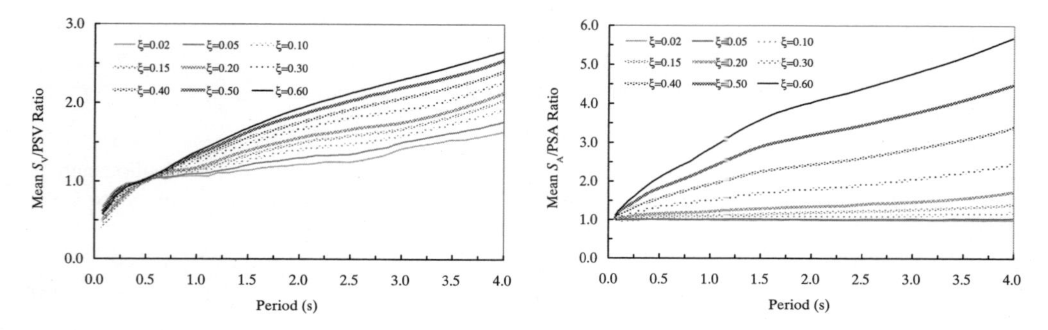

Figure 3.7 Mean ratio of maximum relative velocity to pseudo-velocity (*left*) and maximum absolute acceleration to pseudo-acceleration (*right*) as a function of the damping (*adapted from* Sadek *et al.*, 2000)

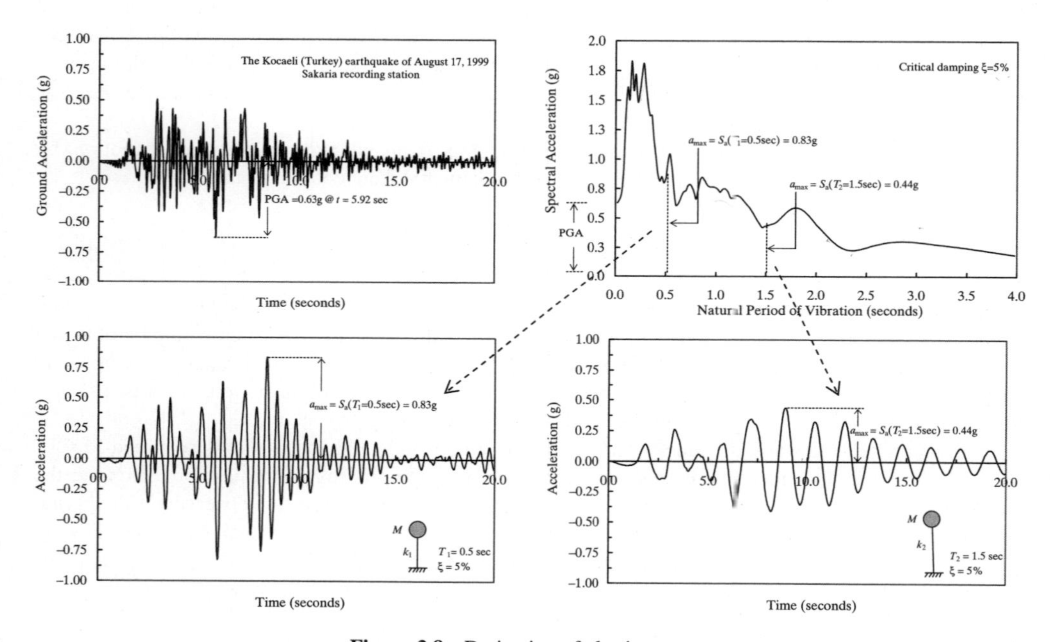

Figure 3.8 Derivation of elastic spectra

(i) Select the earthquake record from databanks, e.g. using those given in Section 3.7;

(ii) Select a T-ξ pair, i.e. the fundamental period of vibration and the damping ratio for the SDOF. Values of interest for structural earthquake engineering applications range between 0.01 and 5 seconds for T, for very rigid and very flexible structures, respectively, and 0% to 20% for ξ, for lightly and highly damped systems, respectively as illustrated in Section 2.3.5;

(iii) Select a numerical method to integrate the equation of motion as expressed, for example, in equation (3.14). Several reliable methods are available in the literature as also discussed in Section 4.5.1.2: their numerical stability and accuracy are reviewed in several textbooks (e.g. Hughes, 1987; Bathe, 1996);

(iv) Compute the response history for the given earthquake record. The peak value as the spectral displacement is S_d;

(v) Compute the pseudo-velocity S_v and pseudo-acceleration S_a by using equations (3.17) and (3.18), respectively. Alternatively, the true maximum relative velocity and absolute acceleration can be determined by means of numerical algorithms;

(vi) Select a new T-ξ pair and repeat steps (i) to (v);

(vii) Plot the maxima of response versus the fundamental period or frequency for various damping values. Structural earthquake engineers are generally more familiar with spectral response-period format.

Figure 3.9 shows the response spectra for the 1940 Imperial Valley (117 El Centro Station, closest distance to the fault rupture $d = 12\,\text{km}$) and the 1994 Northridge (24087 Arleta, Nordhoff Fire Station, closest distance to the fault $d = 3.9\,\text{km}$) earthquakes, which are representative of strong motions registered for stations far and close to the seismic source, respectively. The records are North-South horizontal components. Two common features can be observed for S_d and S_v. Spectral ordinates for all

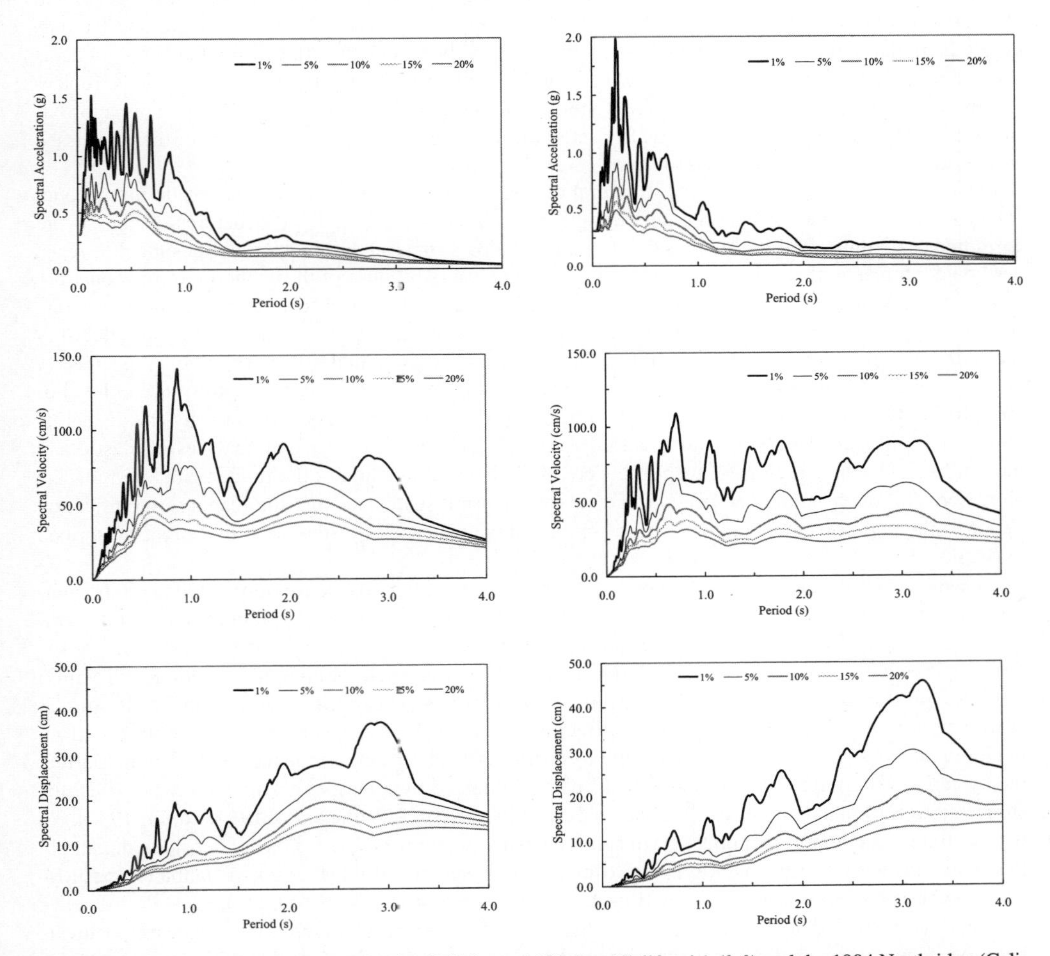

Figure 3.9 Elastic response spectra for the 1940 Imperial Valley (California) (*left*) and the 1994 Northridge (California) (*right*) earthquakes for various damping values (1%, 5%, 10%, 15% and 20%): acceleration (*top*), velocity (*middle*) and displacement (*bottom*)

damping levels increase with the period from zero to some maximum value and then descend to converge at the values of the peak ground displacement (PGD) and peak ground velocity (PGV), respectively, at long periods. The damping smoothes the local peaks in the response curves. The value of S_a is equal to the peak ground acceleration (PGA) at $T = 0$ second (i.e. for rigid structures) and for long periods (i.e. for very flexible structures), the response tends to zero asymptotically. These qualitative aspects can be generalized to all earthquake records. Differences in shape between long and short station-to-source distances response spectra are related to the frequency content of the input motion as mentioned in Section 1.3.1. The former are generally broadband signals while the latter are narrowband, pulse-like records. The short distance records often exhibit characteristics of the seismic source and are referred to as near-source strong-motion (Bolt, 1996).

The use of elastic spectra derives from dynamic analysis in the frequency domain approach. In the latter approach, the multi-degree-of-freedom (MDOF) system is considered a compendium of SDOFs with periods given by the period of vibration of individual modes of the MDOF system as discussed in detail in Section 4.5.1.1. Values of mass and stiffness are calculated from the mass and stiffness distribution of the structure and the relevant mode shape. Once the elastic spectrum is derived, modal forces on a MDOF system may be easily calculated. If the fundamental mode is dominant, as in most regular structures with periods of vibration up to about 1.0–1.5 seconds, replacing the modal mass by the total mass will yield an upper bound on the seismic force. This approximation is the basis of the simple equivalent lateral force procedure used in codes, as further discussed in Section 4.5.3.

Elastic spectra are useful tools for structural design and assessment. They, however, do not account for inelasticity, stiffness reduction and strength degradation experienced by structures during severe earthquakes as illustrated in Sections 2.3.1.2 and 2.3.2.3. Structural systems are not designed to resist earthquake forces in their elastic range, but for very few cases because of the economy of the construction. Concepts of energy absorption and plastic redistribution are used to reduce the elastic seismic forces by as much as ~80%. The inelastic behaviour of structures can be quantified by the ductility factor μ, defined in Section 2.3.3. High μ-values correspond to large inelastic deformations; for linearly elastic systems, the ductility factor is unity. Thus, inelastic spectra for a target ductility μ, i.e. level of inelasticity, were estimated simply by dividing the ordinates of the elastic spectra by the R-factors (Newmark and Hall, 1969), as illustrated in Section 3.4.4. Through extensive analyses of elastic and inelastic spectra, three regions of response were identified as a function of the fundamental period. The above breakthrough reference opened the door for intense research relating the response modification factor to the period of the structure and to significant characteristics of the input motion, as described in Section 3.4.4.

Inelastic spectra depend not only upon the characteristics of the ground motion, but also on the nonlinear cyclic characteristics of the structural system. This complicates the problem for structural earthquake engineers. The reduction of the elastic spectra by employing R-factors given in Section 3.4.4 is the simplest and popular approach to derive inelastic spectra. Such an approach is employed within codes of practice for seismic design to evaluated design base shears as illustrated in Section 4.5.3. However, this approach makes use of static concepts to scale the elastic spectrum, obtained from dynamic analysis. It is, as such, insensitive to characteristics of the earthquake motion, which affect the hysteretic damping. More accurate results can be obtained by inelastic dynamic analysis of SDOFs subjected to earthquake input (e.g. Elghadamsi and Mohraz, 1987; Vidic *et al.*, 1994; Fajfar, 1995). To demonstrate these important points, elastic and inelastic spectra for two records are considered, namely the 1994 Northridge earthquake (Sylmar Hospital station) and the 1995 Hyogo-ken Nanbu (Kobe JMA station). The plots are shown for a ductility factor μ of 2 and 4 in Figure 3.10.

The plots provided in Figure 3.10 demonstrate that lower accelerations and hence force levels have been generated in the inelastic systems, i.e. curves for ductility 2 and 4. The latter is due to the energy absorption by hysteresis. A comprehensive discussion of the R-factors is provided in Section 3.4.4.

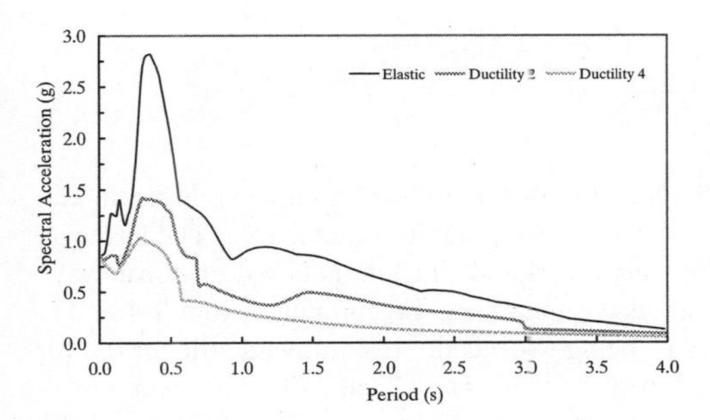

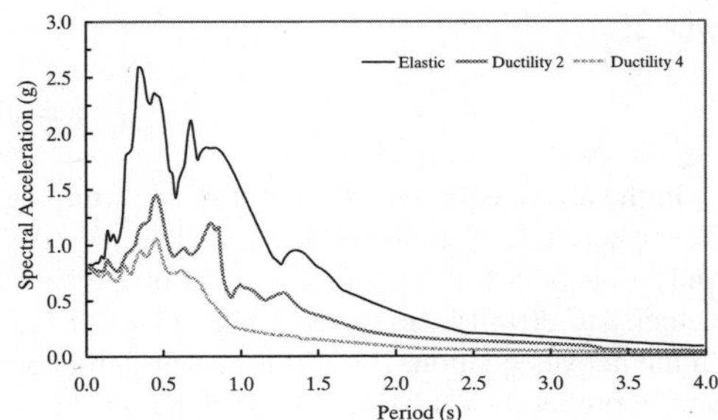

Figure 3.10 Elastic and constant ductility spectra for the 1994 Northridge (California) (*left*) and 1995 Kobe (Japan) (*right*) earthquakes

Problem 3.4

Plot the elastic and inelastic acceleration response spectra (ductility $\mu = 2$ and $\mu = 4$) for the 1994 Northridge (California) earthquake, Newhall Fire Station and the 1995 Hyogo-ken Nanbu (Japan) earthquake, Kobe University Station. Also, calculate the inelastic spectra for a record comprising the same records as above, applied twice in series in the same analysis. Comment on the difference in results if any between the two earthquake records and between the single and double application of the same record. For both elastic and inelastic spectra, assume a viscous damping of 1%.

3.4.3 Simplified Spectra

Deriving earthquake-specific spectra is often of limited use for analysis and assessment of structural systems, since earthquake characteristics vary even at the same site and when affected by the same source. Therefore, probabilistic spectra are generally derived to represent hazard scenarios for seismic design. There are different approaches to accomplish this and the reader is referred to the extensive literature on seismic hazard analysis (e.g. Reiter, 1990; Bozorgnia and Bertero, 2004, among others). It is often necessary to derive uniform hazard spectra; spectra with the probability of any ordinate being exceeded by the actual earthquake are the same regardless of period. Spectra derived from attenuation relationships and those derived from ground-motion parameters directly may be used with confidence in defining the force demand imposed on structures and are discussed below.

3.4.3.1 Spectra from Attenuation Relationships

The earliest frequency-dependent attenuation relationships for response spectral ordinates were published by Johnson (1973) and a large number of equations have since appeared in the technical literature. The majority of the available equations employ the spectral pseudo-velocity PSV as the predicted variable although there are also a number of attenuation relationships for acceleration response ordinates $S_a(T)$. For example, from the European scene, the attenuation relationships of Ambraseys *et al.* (1996) can be used to derive smooth acceleration spectra. These are based on the Imperial College Strong-Motion Databank (ICSMD) containing about 10,000 worldwide records, from which 416 high-quality records for Europe were selected. All magnitudes were uniformly recalculated. The relationship, derived for European earthquakes, is given by:

$$\log(S_a) = -1.48 + 0.265\, M_s - 0.922 \log(R) + 0.117\, S_A + 0.124\, S_S + 0.25\, P \qquad (3.19.1)$$

where:

$$R = \sqrt{d^2 + h_o^2} \tag{3.19.2}$$

In the above equation, d is the distance from the fault and h_o is 3.5; both values are expressed in km. The spectral acceleration S_a (T) in equation (3.19.1) is for 5% damping; its value is in g. The terms S_A and S_S account for the soil properties of the site; the classification of the site is based on shear-wave velocity to 30 m (v_S, in m/sec). The values of the factors S_A and S_S can be computed from Table 3.5. In the above equation, P is 1.0 for a 16% probability of being exceeded, i.e. 84th percentile, or 0.0 for a 50% probability of being exceeded, i.e. median. Recently, equations (3 8.1) and (3.28) by Ambraseys *et al.* (2005a,b) were proposed to estimate horizontal and vertical spectral accelerations for Europe and the Middle East, respectively. In the latter attenuation relationships, the values of S_a are expressed in m/sec^2.

Extensive analytical studies by Bommer and Elnashai (1999) and Tolis and Faccioli (1999) have concentrated specifically on displacement spectra. These studies were prompted by the advent of deformation-based seismic design. The significance of explicitly deriving displacement spectra is that the derivation of displacements from velocity or acceleration using simple harmonic motion conversions may be inaccurate, as demonstrated in Figure 3.11 where the error increases with period and for high values of spectral displacements on soft sites. Another limitation on the use of current attenuation relationships for spectral ordinates to provide the input for deformation-based design is the fact that the majority of the available equations only provide spectral ordinates for 5% damping. A notable exception to this is the equations presented by Boore *et al.* (1993, 1994), which predict spectral ordinates for damping ratios of 2, 5, 10 and 20%. However, these equations only predict spectral ordinates at response periods up to 2.0 seconds. Mohammadioun (1994) also reports regressions on ordinates of PSV for damping levels of 0, 2, 5, 10 and 20% up to periods of 5.0 seconds, but the coefficients for

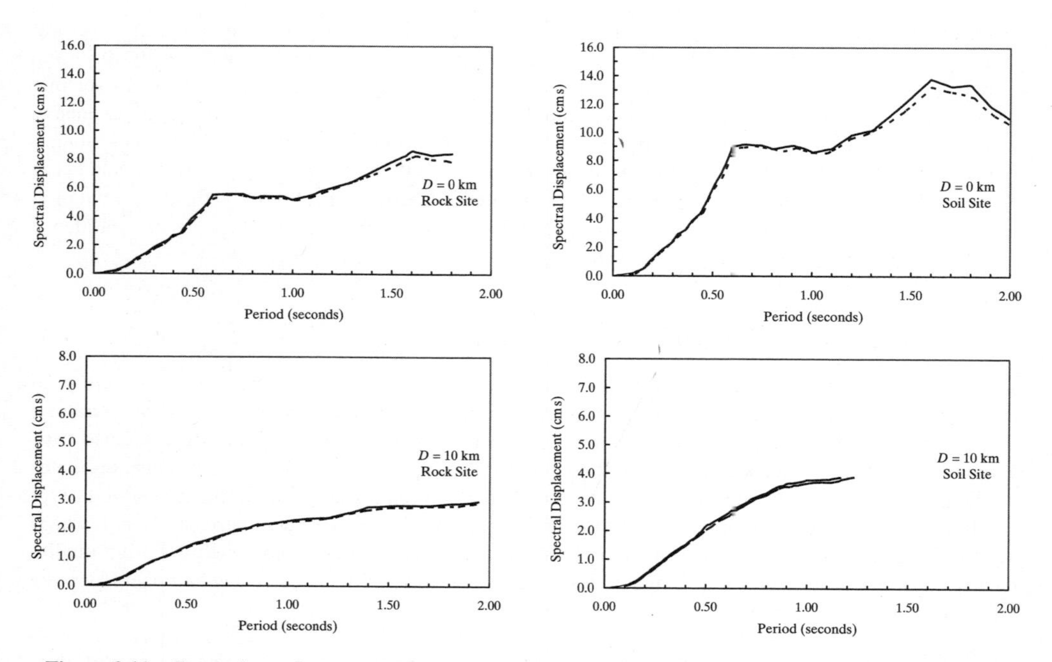

Figure 3.11 Comparison of spectra obtained from acceleration (*dotted*) or directly from displacement (*solid*)

the equations, which are a function of magnitude and distance only, are not presented. Since ductility-equivalent damping, employed in direct deformation-based design, may be up to 30% for fixed-base structures, displacement spectra specific to such applications are required (Borzi *et al.*, 2001).

In order to derive attenuation relationships for the prediction of response spectra to use in deformation-based design, it is necessary to compile a data set of high-quality accelerograms for which the associated source, path and site parameters are uniformly and accurately determined. It would be preferable to employ recordings from digital accelerographs (Tolis and Faccioli, 1999). However, the number of available digital accelerograms is relatively low and hence while these data may provide more accurate values for the spectral ordinates, it would be difficult to find correlations between these ordinates and the parameters characterizing the earthquake source, travel path and recording site. The data set presented by Ambraseys *et al.* (1996) consists of 422 triaxial accelerograms generated by 157 shallow earthquakes with surface magnitude M_s between 4.0 and 7.9 and is the basis for the attenuation relationships discussed herein. In view of the fact that the study reported in Bommer and Elnashai (1999) was concerned with the long-period response spectrum and that small-magnitude earthquakes do not produce significant long-period ground motion, it was decided to impose a higher magnitude limit on the data set. The removal of weak- and low-amplitude records from the data set, in order to obtain better signal-to-noise ratios, would not be acceptable since it would introduce a statistical bias. However, the removal of all the earthquakes with magnitude below the chosen lower limit of $M_s = 5.5$ does partially achieve this objective. Thus, the final data set consisted of 183 accelerograms from 43 shallow earthquakes. For three of the recording stations, each of which contributed only one record, the site classification is unknown. For the remaining 180 accelerograms, the distribution among the three site classifications, i.e. rock, stiff soil and soft soil, is 25:51:24, which compares favourably with the distribution of the original data set of Ambraseys *et al.* (1996) which is 26:54:20. Regression analyses were performed on the horizontal displacement spectral ordinates for damping ratios of 5, 10, 15, 20, 25 and 30% of critical damping. The regression model used for S_D ordinates (expressed in centimetres) was the same as that employed by Ambraseys *et al.* (1996) for acceleration spectral ordinates. At each period, the larger spectral ordinate from the two horizontal components of each accelerogram was used as the dependent variable. Each component record was only used for regressions up to a period of 0.1 second less than the long-period cut-off employed in processing that record. As a result, for periods greater than 1.8 seconds, there was a reduction in the number of data points available for each regression. At a response period of 3.0 seconds, the data set was reduced from 183 to 121 accelerograms. It was decided not to perform the regressions for periods longer than 3.0 seconds since the number of usable spectral ordinates becomes insufficient. Regression analysis was also performed on the larger values of peak ground displacement (in cm) from each record, using the same attenuation model as above. Although it is not possible to make direct comparisons because of the use of different definitions for the parameters, this regression predicts values of PGD very similar to those presented by Bolt (1999) for the near-field, but more rapid attenuation with distance was observed.

From inspection of a large number of displacement response spectra for the six specified damping levels of 5, 10, 15, 20, 25 and 30%, it was concluded that a general, idealized format would be as shown in Figure 3.12. The smoothed spectrum for each damping level comprises six straight-line segments and is defined by four control periods along with their corresponding amplitudes. The amplitude corresponding to T_E is the peak ground displacement. Only that part of the spectrum up to periods of 3.0 seconds is considered because longer periods would require use of hitherto unavailable digital recordings of a sufficiently large number. For displacement attenuation over longer periods, the reader is referred to Tolis and Faccioli (1999), where the 1995 Kobe strong motion was used to derive longer-period ordinates. The results of the work by Bommer and Elnashai (1999) are summarized in Tables 3.6 and 3.7.

Inspection of the predicted spectral ordinates shows that the shape of the spectra is strongly influenced by magnitude and site classification, but far less so by distance. It was observed that the decrease of

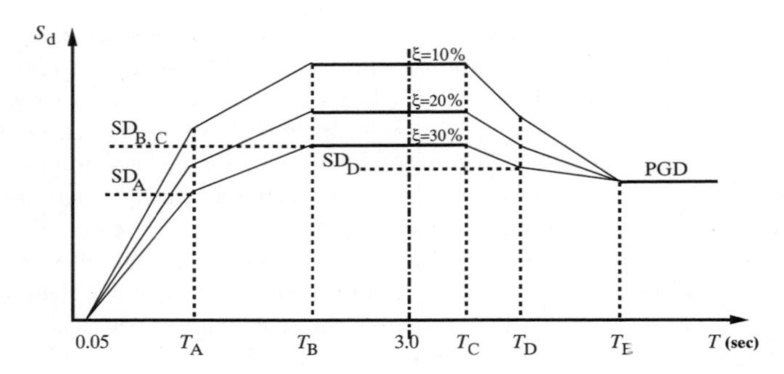

Figure 3.12 Idealized displacement spectrum shape

Table 3.6 Control periods T_A and T_B for design spectra as function of magnitude M_S and soil type.

M_s	T_A			T_B		
	Rock	Stiff	Soft	Rock	Stiff	Soft
5.5	0.75	0.75	0.75	2.00	2.30	2.70
6.0	0.85	0.85	0.85	2.15	2.30	2.80
6.5	1.00	1.00	1.00	2.30	2.50	2.90
7.0	1.40	1.40	1.40	2.50	2.70	3.00*
7.5	1.90	1.90	1.90	3.00	3.00*	3.00*

* Actual control period is probably slightly greater than 3.0 seconds.

Table 3.7 Control ordinates SD_A and SD_B for design spectra as function of magnitude M_S and soil type.

M_s	SD_A			SD_B		
	Rock	Stiff	Soft	Rock	Stiff	Soft
5.5	2.2	3.1	4.0	3.1	4.6	5.8
6.0	3.7	5.0	6.5	5.8	8.4	11.0
6.5	6.7	8.7	11.2	10.8	15.9	20.8
7.0	14.8	20.1	25.0	20.3	28.7	38.6
7.5	34.1	46.7	55.0	37.0	55.8	70.0

the spectral ordinates with distance is reasonably constant across the period range and similar for all three site categories. Therefore, simple reduction factors could be calculated. The 30% damped spectra for distances up to 50 km from the source can be obtained simply by multiplying the ordinates by the appropriate factor F_d taken from Table 3.8.

The next stage was to establish the amplification factors to be applied to the control ordinates in order to obtain the displacement spectra for damping levels from 5 to 30% of critical damping. These factors F_ξ are presented in Table 3.9. Therefore, using the values presented in the four tables (Table 3.6 through Table 3.9) and interpolating where necessary, design displacement spectra for rock, stiff soil and soft soil sites for magnitudes between 5.5 and 7.5 and distances up to 50 km can easily be

Table 3.8 Distance factors F_d for spectral ordinates as function of the source distance d.

Distance, d (km)	0	5	10	15	20	30	40	50
F_d	1.00	0.621	0.352	0.245	0.187	0.127	0.095	0.075

Table 3.9 Damping ratio factors F for spectral ordinates as function of the damping value ξ.

damping, ξ (%)	5	10	15	20	25	30
F_ξ	1.90	1.55	1.35	1.20	1.10	1.00

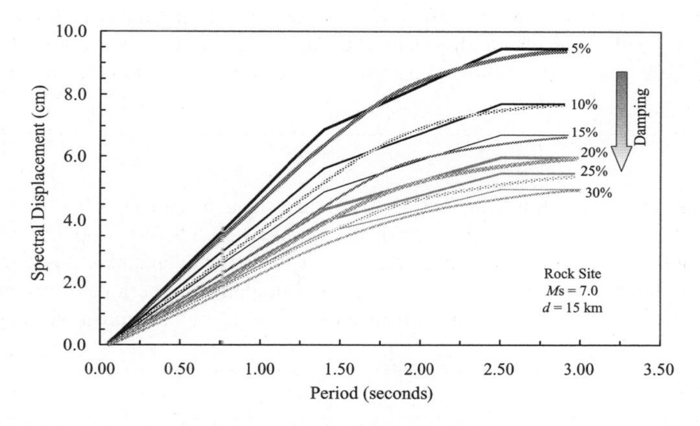

Figure 3.13 Comparison between attenuation relationships and idealized shape

constructed. Simpler, but less accurate formulations were developed by Bommer *et al.* (2000) for code applications.

Extensive comparisons between the parameterized and the actual spectra were undertaken (Bommer and Elnashai, 1999); a sample is shown in Figure 3.13. The simplified relationships are deemed sufficiently accurate and cover a wide range of source, site and structure characteristics.

The data set above was employed in a study by Borzi *et al.* (1998), alongside the same attenuation model. Two structural response models were employed in evaluating constant ductility spectra for all records. These were an elastic-perfectly plastic (EPP) and a hysteretic hardening-softening (HHS) model; the latter is similar though not identical to the former for $K_3 = 0$. It is defined by two force levels, yield and maximum, with a reduced stiffness between the two, as compared to the elastic stiffness. In the HHS model, K_y and K_3 are the secant and the post-yield stiffness values used to define the primary curve. Hysteretic models are characterized by the definition of a primary curve, unloading and reloading rules. The primary curve for a hysteretic force–deformation relationship is defined as the envelope curve under cyclic load reversals. For non-degrading models, the primary curve is taken as the response curve under monotonic load. Unloading and reloading branches of the HHS model have been established by a statistical analysis of experimental data (Saatcioglu and Ozcebe, 1989). Table 3.10 summarizes the median values of equivalent damping for various ductility levels for both EPP and HHS models employed in the study.

Table 3.10 Equivalent damping ratios for single-degree of freedom systems (*after* Borzi *et al.*, 1998).

	$\mu = 2$	$\mu = 3$	$\mu = 4$	$\mu = 6$
EPP	8%	13%	16%	19%
HHS, $K_3 = 0$	11%	15%	18%	20%
HHS, $K_3 = 10\%\ K_y$	10%	13%	15%	17%
HHS, $K_3 = -20\%\ K_y$	13%	20%	28%	—
HHS, $K_3 = -30\%\ K_y$	15%	27%	—	—

Key: EPP = elastic-perfectly plastic model; HHS = hysteretic hardening-softening model.
Ky and K_3 are the secant and post-yield stiffness of the primary curve of the HHS model.

The equivalent damping values given in Table 3.10 are recommended for use with the elastic response spectra given before, for periods of up to 3.0 seconds. Nil entries in Table 3.10 indicate that structures with highly degrading response ($K_3 = -20\%$ and $-30\%\ K_y$) would not have ductility capacity of 4 or more. If a more refined value of ductility-damping transformation is sought, the relationships given by Borzi *et al.* (1998) as a function of magnitude, distance, soil condition and period should be consulted.

Problem 3.5

Draw the displacement spectrum with ductility $\mu = 4$ on a site at a distance of 15 km from the source, on stiff soil and subjected to an earthquake of magnitude $M = 6.5$. What options are available to the designer if it is deemed necessary to decrease the displacement demand imposed on the structure below the value implied by the above spectrum, if the fundamental periods of vibration are about 1.7 and 3.0 seconds? For the inelastic displacement spectrum, assume a viscous damping of 0.5%.

3.4.3.2 Spectra from Ground-Motion Parameters

By plotting the response spectra of an ensemble of earthquake records, normalized by the relevant ground-motion parameter, Newmark and Hall (1969) derived statistical values of the amplification factors for acceleration, velocity and displacement. These amplification factors, expressed as ratios between peak ground parameter and peak response of the system, are provided in Table 3.11.

To establish the elastic response spectrum for the full range of periods, use is made of a four-way log paper. The resulting spectrum is referred to as a 'tripartite plot', since it includes the three spectral forms, and is shown in Figure 3.14. This is made possible by the simple relations between spectral acceleration, velocity and displacement given in equation (3.18). Indeed, the tripartite plot is based upon the following relationships:

$$\log S_{\mathrm{v}} = \log \omega + \log S_{\mathrm{d}} \tag{3.20.1}$$

$$\log S_{\mathrm{v}} = \log \omega - \log S_{\mathrm{a}} \tag{3.20.2}$$

Equations (3.20.1) and (3.20.2) indicate that the logarithm of the spectral velocity is linearly related to the logarithm of the natural frequency ω, provided the spectral displacement or acceleration remains constant. The slope is either +1 or −1 depending on whether the displacement or the acceleration is assumed as a constant. Thus, to draw the elastic spectrum for use, for example in spectral analysis illustrated in Section 4.5.1.1, the following steps are required:

Table 3.11 Spectral amplification factors for 84th percentile confidence (*after* Netmark and Hall, 1969).

Damping, ξ (%)	Amplification factors		
	Acceleration	Velocity	Displacement
0	6.4	4.0	2.5
1	5.2	3.2	2.0
2	4.3	2.8	1.8
5	2.6	1.9	1.4
7	1.9	1.5	1.2
10	1.5	1.3	1.1

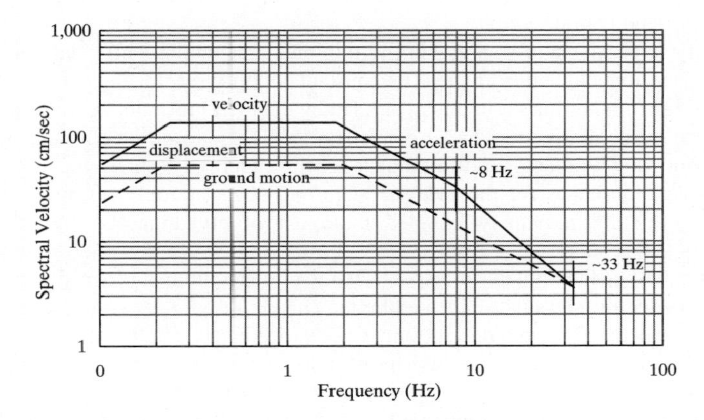

Figure 3.14 Tripartite plot, ground-motion parameters and elastic spectrum

(i) Draw straight lines passing through the values of the ground parameters, each measured on the appropriate axis. Generally a tripartite plot begins as a log-log plot of spectral velocity versus period. Spectral acceleration and spectral displacement axes are then superimposed on the plot at 45° angles.

(ii) For the required damping ξ, find the amplification factors from Table 3.11. Values equal to 2% for steel and 5% for reinforced concrete (RC) are recommended to perform elastic analyses; further values are provided in Section 2.3.5.

(iii) Calculate the response parameters as the product of each ground parameter and the relevant amplification factor.

(iV) Draw straight lines passing through the values calculated under (iii) above, each measured on the appropriate axis.

(v) At a frequency of 8 Hz, draw a straight line from the response spectrum to meet the ground-motion spectrum at a frequency of 33 Hz.

Step (v) is essential, since it represents the response of very stiff structures, where the response acceleration approaches the ground acceleration and the response relative displacement approaches zero. For long-period structures, i.e. flexible systems, the response displacement approaches the ground displacement. Other approaches for the development of an elastic spectrum from ground-motion parameters may be considered (e.g. Bozorgnia and Bertero, 2004, among others); some formulations are displayed in Figure 3.1. Techniques also exist for the construction of approximate inelastic spectra, using a ductility factor μ to scale the elastic spectrum.

3.4.4 Force Reduction Factors (Demand)

The reduction factor 'demand' is defined as the ratio between the elastic Sa_{elastic} and the inelastic $Sa_{\text{inelastic}}$ response spectral ordinates corresponding to a specific period T, i.e.:

$$\text{Force Reduction Factor} = \frac{Sa_{\text{elastic}}(T)}{Sa_{\text{inelastic}}(T)} \tag{3.21}$$

thus, it expresses the ratio of the elastic strength demand to the inelastic strength demand for a specified constant ductility μ and period.

The behaviour factor 'demand' represents the minimum reduction coefficient corresponding to a specific level of ductility obtained from inelastic constant ductility spectra and elastic spectra at a given period. The elastic spectral ordinate should be divided by the inelastic counterpart for a value of ductility expected for the structural system under consideration. The ratio of the elastic-to-inelastic spectra changes with period, ductility factor and earthquake record. Figure 3.15 shows the ratio between elastic and inelastic spectral ordinates for the 1999 Chi-Chi (TCU074 station) Taiwan earthquake for five specified values of ductility μ, namely 2, 3, 4, 5 and 6. It is observed that:

(i) A point on these R-factor plots corresponds to the force requirement for a system with the shown period if its ductility was that indicated by the curve chosen.
(ii) The R-factor demand is not constant but rather varies considerably with period.
(iii) At very short periods, the R-factor is almost unity, increasing with increasing period.
(iv) For low levels of ductility ($\mu = 2$ for example), the statically derived relationships of $R = 1$, $R = \mu$ and $R = \sqrt{2\mu - 1}$ hold quite well.
(v) The statically derived values are vastly distinct from the actual R-factors for higher ductility levels ($\mu = 4$ or larger).

It is important to note that the plots shown in Figure 3.15 represent the likely force reduction demand that will be expected from a structure. On the other hand, the force reduction supply is that obtained

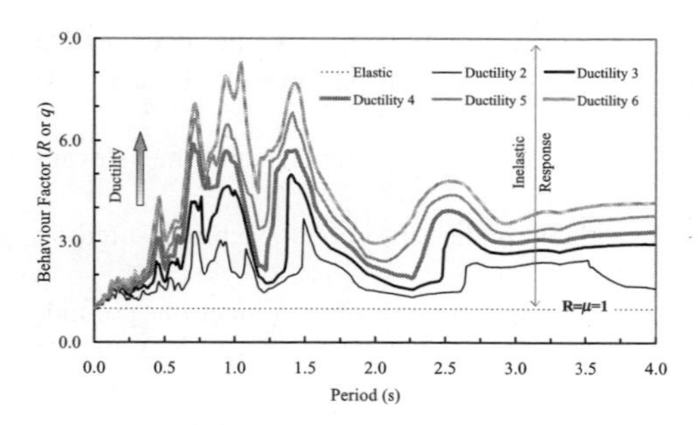

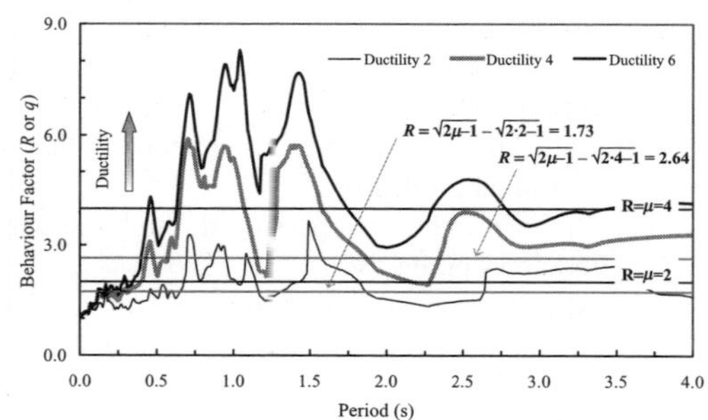

Figure 3.15 Ratio between elastic and inelastic spectra (behaviour factors, R) evaluated for the 1999 Chi-Chi (Taiwan) earthquake: variation with the level of ductility μ (*left*) and comparison with statically determined values (*right*)

from a detailed analysis of the structure. It follows that the supply should be equal or greater than the demand for a safe seismic response in the inelastic range. Thus, the above discussion points towards the necessity of studying the dynamic inelastic response of structures, in an attempt to quantify the behaviour factors (R or q) necessary for the derivation of design forces from elastic forces. Finally, inelastic spectra are generally derived by assuming elastic-plastic hysteretic models. However, SDOF systems with different force–deformation relationships can also be employed, e.g. bilinear with hardening and with stiffness and strength degradation, to more accurately represent the response of real structures. In general, elastic-plastic non-degrading SDOF systems exhibit higher energy absorption and dissipation than degrading systems. Therefore, estimates of force reduction factors based on the former are sometimes unconservative (i.e. they overestimate R, hence underestimate the design force). Use of R-factors based on elastic-plastic response should therefore be treated with caution, especially for high levels of inelasticity.

The relationship between displacement ductility and ductility-dependent behaviour factor has been the subject of considerable research. A few of the most frequently used relationships reported in the technical literature are discussed below.

(i) Newmark and Hall (1982)

The force reduction factor R_μ is defined as the ratio of the maximum elastic force to the yield force required for limiting the maximum inelastic response to a displacement ductility μ. In this early study R_μ was parameterized as a function of μ (Newmark and Hall, 1982), as also discussed in Section 2.3.6. It was observed that in the long-period range, elastic and ductile systems with the same initial stiffness reached almost the same displacement. As a consequence, the force reduction (or behaviour) factor can be considered equal to the displacement ductility. This is referred to as the 'equal displacement' region. For short-period structures, the ductility is higher than the behaviour factor and the 'equal energy' approach may be adopted to calculate force reduction. This approach is based on the observation that the energy associated with the force corresponding to the maximum displacement reached by elastic and inelastic systems is the same, as explained in Section 2.3.6 of Chapter 2. The proposed relationships for behaviour factor are:

$$R_\mu = 1 \qquad \text{when } T < 0.05\,\text{s} \tag{3.22.1}$$

$$R_\mu = \sqrt{2\mu - 1} \qquad \text{when } 0.12\,\text{s} < T < 0.5\,\text{s} \tag{3.22.2}$$

$$R_\mu = \mu \qquad \text{when } T > 1.0\,\text{s} \tag{3.22.3}$$

while a linear interpolation is suggested for intermediate periods. The above is the first and simplest formulation used in practice. It has endured over the years due to its intuitive nature and has been confirmed by other studies, such as Uang (1991) and Whittaker *et al.* (1999). Further refinement though is warranted since the force reduction factor may be a function of period within the regions defined by equations (3.22.1) to (3.22.3).

(ii) Krawinkler and Nassar (1992)

A relationship was developed for the force reduction factor derived from the statistical analysis of 15 western USA ground motions with magnitude between 5.7 and 7.7 (Krawinkler and Nassar, 1992). The records were obtained on alluvium and rock site, but the influence of site condition was not explicitly studied. The influence of behaviour parameters, such as yield level and hardening coefficient α, was taken into account. A 5% damping value was assumed. The equation derived is given as:

$$R_\mu = [c(\mu - 1) + 1]^{1/c} \tag{3.23.1}$$

where:

Table 3.12 Values of the constants in equation (3.23.2).

Hardening value	Model parameters	
α (%)	a	b
0	1.00	0.42
2	1.01	0.37
10	0.80	0.29

$$c(T, \alpha) = \frac{T^a}{1+T^a} + \frac{b}{T} \tag{3.23.2}$$

in which α is the strain-hardening parameter of the hysteretic model, and a and b are regression constants. Values of the constants in equation (23) were recommended for three values of hardening α as in Table 3.12.

(iii) Miranda and Bertero (1994)

The equation for the force reduction factor introduced by Miranda and Bertero (1994) was obtained from a study of 124 ground motions recorded on a wide range of soil conditions. The soil conditions were classified as rock, alluvium and very soft sites characterized by low shear wave velocity. A 5% of critical damping was assumed. The expressions for the period-dependent force reduction factors R_μ are given by:

$$R_\mu = \frac{\mu - 1}{\Phi} + 1 \tag{3.24.1}$$

where Φ is calculated from different equations for rock, alluvium and soft sites as shown below:

$$\Phi = 1 + \frac{1}{10\,T - \mu\,T} - \frac{1}{2\,T}\exp\left[-1.5(\ln T - 0.6)^2\right] \quad \text{for rock site} \tag{3.24.2}$$

$$\Phi = 1 + \frac{1}{12\,T - \mu\,T} - \frac{2}{5\,T}\exp\left[-2(\ln T - 0.2)^2\right] \quad \text{for alluvium site} \tag{3.24.3}$$

$$\Phi = 1 + \frac{T_1}{3\,T} - \frac{3\,T_1}{4\,T}\exp\left\{-3[\ln(T/T_1) - 0.25)]^2\right\} \quad \text{for soft site} \tag{3.24.4}$$

where T_1 is the predominant period of the ground motion. The latter corresponds to the period at which the relative velocity of a linear system with 5% damping is maximum within the entire period range.

(iv) Vidic et al. (1994)

The reduction coefficients R_μ calculated by Vidic *et al.* (1994) were approximated by a bilinear curve. In the short-period range, the reduction factor increases linearly with the period from 1.0 to a value that is almost equal to the ductility factor. In the remaining part of the period range, the reduction factor is constant. To calculate the reduction factor, a bilinear response model and a stiffness-degrading 'Q-model' were employed. A mass-proportional damping and an instantaneous stiffness-proportional damping were assumed. In this work, the standard records from California and Montenegro 1979 were chosen as being representative for 'standard' ground motion, i.e. severe ground motion at moderate

Table 3.13 Values of the constants in equations (3.25.1) to (3.25.3).

Model	Damping	c_1	c_2	c_R	c_T
Bilinear	Mass-proportional	1.35	0.75	0.95	0.20
Bilinear	Instantaneous stiffness proportional	1.10	0.75	0.95	0.20
Q-model	Mass-proportional	1.00	0.65	1.00	0.30
Q-model	Instantaneous stiffness proportional	0.75	0.65	1.00	0.30

epicentral distance, with a duration ranging between 10 and 30 seconds and predominant period between 0.3 and 0.8 seconds. The proposed formulation of reduction factor, based on the rather small sample size but augmented by a sensitivity analysis for special strong-motion features, is:

$$R_\mu = c_1 (\mu - 1)^{c_R} \frac{T}{T_0} + 1 \quad \text{when } T < T_0 \tag{3.25.1}$$

$$R_\mu = c_1 (\mu - 1)^{c_R} + 1 \quad \text{when } T \geq T_0 \tag{3.25.2}$$

where T_0 is the period dividing the period range into two portions. It is related to the predominant period of the ground motion T_1 by means of:

$$T_0 = c_2\, \mu^{c_T} T_1 \tag{3.25.3}$$

The coefficients c_1, c_2, c_R and c_T in the above equations depend on the hysteretic behaviour, either bilinear or with degrading stiffness, and damping, e.g. time dependent or independent. The values of the model parameters are outlined in Table 3.13.

The values of the constants in Table 3.13 are for 5% damping. Moreover, in both models, i.e. bilinear and with degrading stiffness, 10% hardening was assumed after yielding.

(v) Borzi and Elnashai (2000)

The formulations of R-factors discussed in preceding subsections were significant steps forward at the time they were derived. Areas of further possible improvement were identified by Borzi and Elnashai (2000) as:

(i) Improvement of data set by using a large number of records well distributed in terms of magnitude, distance and soil conditions from a wide range of seismo-tectonic environments;
(ii) Use of comprehensively represented hysteretic models exhibiting hardening and softening behaviour;
(iii) Using regression curves focusing on uniform distribution of target reliability across the period range and giving simple code-amenable expressions.

A large strong-motion data set was used to derive response modification factors (demand) taking into account the three points above. Two structural hysteretic models were utilized for the analytical study, namely an elastic-perfectly plastic (EPP) and a hysteretic hardening-softening (HHS) models, as illustrated in Section 3.4.3.1. Following the definition of response modification, or force reduction, regression analyses for the evaluation of the ratio between the elastic and inelastic acceleration spectra were undertaken. The influence of ductility and input motion parameters, especially magnitude, distance and soil conditions, on the force reduction factors was studied utilizing an EPP hysteretic model. It was

Table 3.14 Constants for trilinear behaviour factors spectra.

	$\mu = 2$				$\mu = 3$				$\mu = 4$				$\mu = 6$			
	T_1	T_2	q_1	q_2	T_1	T_2	q_1	q_2	T_1	T_2	q_1	q_2	T_1	T_2	q_1	q_2
EPP	0.20	0.79	2.06	2.20	0.21	0.78	2.89	3.31	0.22	0.87	3.59	4.34	0.25	0.99	4.81	6.13
$K_3 = 0$	0.20	0.56	2.20	2.51	0.25	1.67	3.10	4.09	0.27	1.55	3.76	5.45	0.29	1.26	4.78	7.79
$K_3 = 10\%\ K_y$	0.21	0.54	2.04	2.33	0.27	1.80	2.78	3.62	0.29	1.64	3.25	4.56	0.33	1.54	3.93	6.10
$K_3 = -20\%\ K_y$	0.26	0.26	2.43	2.43	0.24	1.76	2.83	3.93	0.25	1.69	3.25	5.12	—	—	—	—
$K_3 = -30\%\ K_y$	0.26	0.26	2.42	2.42	0.24	1.85	2.76	3.81	—	—	—	—	—	—	—	—

Key: EPP = elastic-perfectly plastic model; "μ = ductility
K_y and K_3 are the secant and post-yield stiffness of the primary curve of the hysteretic hardening-softening model

observed that the influence of input motion parameters on elastic and inelastic acceleration spectra is similar and significant. However, the effect cancels out for their ratio. Ductility is the most significant parameter, influencing the response modification factor. Consequently, analyses to define period-dependent behaviour factor functions for all the ductility levels and all structural models were undertaken. The average values and the standard deviations were calculated considering various combinations of input motion parameters. The period-dependent *R*-factor functions calculated were further approximated with a trilinear spectral shape. The *R*-factor is equal to 1.0 at zero period and increases linearly up to a period T_1, which is defined as the period at which the force reduction factor reaches the value q_1. A second linear branch is assumed between T_1 and T_2. The value of the reduction coefficient corresponding to T_2 is denoted herein q_2. For periods longer than T_2, the behaviour factor maintains a constant value equal to q_2:

$$q = (q_1 - 1)\frac{T}{T_1} + 1 \qquad \text{when } T \le T_1 \tag{3.26.1}$$

$$q = q_1 + (q_2 - q_1)\frac{T - T_1}{T_2 - T_1} \qquad \text{when } T_1 < T \le T_2 \tag{3.26.2}$$

$$q = q_2 \qquad \text{when } T > T_2 \tag{3.26.3}$$

The values q_1, q_2, T_1 and T_2 that define approximate spectra for all ductility levels and hysteretic parameters are summarized in Table 3.14, as they are obtained by a piece-wise linear regression, for the sample EPP and HHS models.

To demonstrate the reasonable fit of the trilinear representation to the regression force reduction factors spectra, the standard deviation σ of the ratio γ between the approximate and the original spectral values was studied. The standard deviation was calculated for all branches of the approximate spectra and across the whole period range. These values are provided in Table 3.15.

It is observed that the dispersion of γ is close to the global standard deviation. This has an important consequence, from a practical point of view, as the *R*-factor spectra proposed herein correspond to constant seismic design reliability over the whole period range, a feature not previously achieved. Finally, the coordinates of the points that allow the definition of the approximate spectra were expressed as a function of ductility and given as:

$$T_1 = b_{T1} \tag{3.27.1}$$

$$T_2 = a_{T2}\,\mu + b_{T2} \tag{3.27.2}$$

Table 3.15 Standard deviations of γ (ratio of approximate to accurate q- or R-factors).

	$\mu = 2$				$\mu = 3$				$\mu = 4$				$\mu = 6$			
	σ_1	σ_2	σ_3	σ	σ_1	σ_2	σ_3	σ	σ_1	σ_2	σ_3	σ	σ_1	σ_2	σ_3	σ
EPP	2.1	1.9	2.0	2.0	3.4	2.1	2.8	2.8	4.7	2.5	3.5	3.6	7.0	2.8	4.2	5.0
$K_3 = 0$	3.1	1.6	2.7	2.6	4.5	3.6	3.3	3.9	6.2	3.8	3.9	4.8	8.5	3.0	6.2	6.4
$K_3 = 10\% K_y$	3.8	1.6	2.8	2.9	6.4	3.4	3.5	4.7	9.0	3.4	3.4	6.1	2.8	1.7	2.5	2.6
$K_3 = -20\% K_y$	3.6	4.7	—	4.4	2.6	3.2	3.7	3.2	2.9	3.1	4.3	3.4	—	—	—	—
$K_3 = -30\% K_y$	3.7	4.9	—	4.6	2.9	3.1	4.1	3.3	—	—	—	—	—	—	—	—

Key: EPP = elastic-perfectly plastic model; μ = ductility
K_y and K_3 are the secant and post-yield stiffness of the primary curve of the hysteretic hardening-softening model

Table 3.16 Values of the constants in equations (3.27.1) to (3.27.4).

	b_{T1}	a_{T2}	b_{T2}	a_{q1}	b_{q1}	a_{q2}	b_{q2}
EPP	0.25	0.153	0.60	0.69	0.90	1.01	0.24
$K_3 = 0$				0.55	1.37	1.33	0
$K_3 = 10\% K_y$				0.32	1.69	0.96	0.51
$K_3 = -20\% K_y$				0.38	1.67	1.24	0
$K_3 = -30\% K_y$				0.29	1.83	1.21	0

Key: EPP = elastic-perfectly plastic model
K_y and K_3 are the secant and post-yield stiffness of the primary curve of the hysteretic hardening-softening model

$$q_1 = a_{q1}\,\mu + b_{q1} \tag{3.27.3}$$

$$q_2 = a_{q2}\,\mu + b_{q2} \tag{3.27.4}$$

where b_{T1}, a_{T2}, b_{T2}, a_{q1}, b_{q1}, a_{q2} and b_{q2} are constants. It was observed that the control periods of the approximate spectral shape do not depend on the hysteretic behaviour, hence, a single set of calibration constants may be used. On the other hand, it was seen that different values of a_{q1}, b_{q1}, a_{q2} and b_{q2} correspond to the different hysteretic behaviour patterns, thus necessitating a more complex formulation, as given in Table 3.16.

The effect of this simplification on the correlation between parameterized and actual response modification factors is small. The above formulation is derived using a much wider data set with consistent distributions in the magnitude, distance and site condition spaces, than the data used in previous studies. The data set included approximately 400 records. Moreover, the idealization proposed above leads to uniform hazard or reliability force reduction factors. Therefore, they are consistent with 'uniform hazard response spectra' commonly used in seismic design codes and discussed in Section 3.4.5.

(vi) Comparison Between Response Modification Factor Models

Comparisons between R-factors computed by utilizing the formulations illustrated in Sections 'i' through 'v' are provided in Figure 3.16 for values of target ductility in the range of 2 to 6.

It is observed that variations are significant mainly for low-period structural systems, e.g. with periods less than ~0.5 second. For long-period systems, e.g. periods greater than ~1.0 second, all formulations tend towards a constant value. The relationship by Miranda and Bertero (1994) seems to provide results

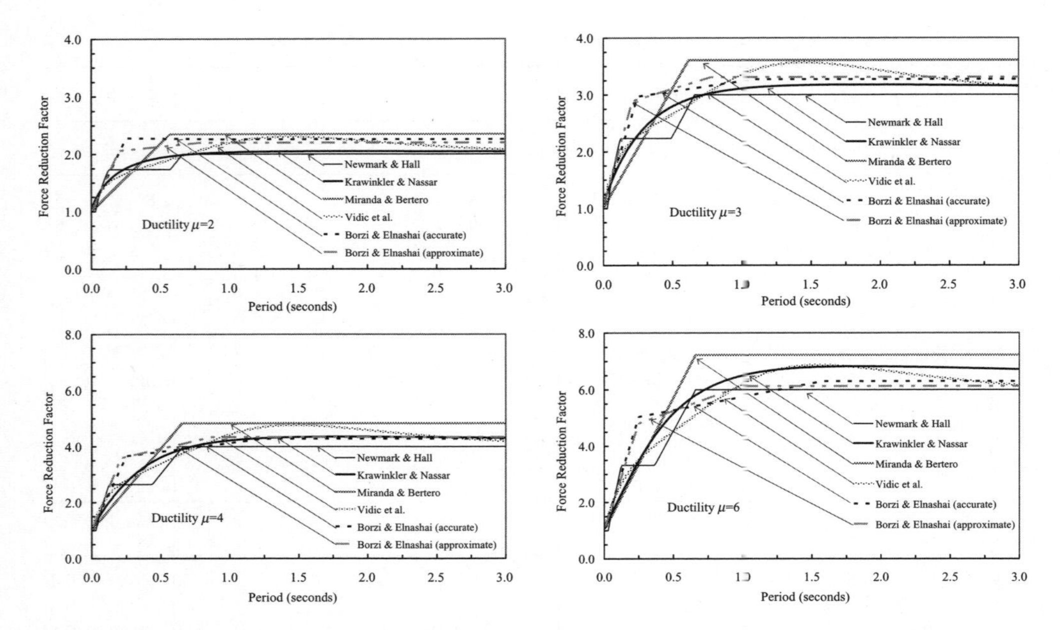

Figure 3.16 Comparison of force reduction factors (demand) derived by using different formulations for elastic-perfectly plastic systems on rock site: values of ductility μ are 2, 3, 4 and 6 (*from top-left to bottom-right*)
Key: For the formulation by Borzi and Elnashai (2000), both the piece-wise linear regression force reduction factors spectra (accurate) and the trilinear approximation (approximate) are plotted

on the unconservative side of other models, while the Newmark and Hall (1982) yields what seems to be a safe lower bound. The other models give intermediate results, especially in the intermediate and long-period ranges.

Problem 3.6

Calculate the elastic (5% damping) and inelastic acceleration spectral ordinates (ductility $\mu = 2$ and $\mu = 4$) for the 1979 Imperial Valley (California), Highways 98-115 Station and the 1995 Hyogo-ken Nanbu (Japan) Port Island Station. Plot all elastic and inelastic (ductility $\mu = 2$ and $\mu = 4$), and plot the force reduction factors. Compare the force reduction factor for a structure with 1.0-second vibration period from the plots above with the values predicted by the equations given in Section 3.4.4. For inelastic spectra, assume a viscous damping of 0.5%.

3.4.5 Design Spectra

Response elastic and inelastic spectra for a specific earthquake record are of importance primarily for structural assessment. They can be used to obtain the response of a structure to only ground motions with similar characteristics, such as magnitude, source mechanism, soil conditions and epicentral distance. Spectra for measured ground motions show irregular shapes. Thus, for design applications, response spectra obtained from records with similar characteristics are averaged and smoothed. Smoothing is necessary because of the difficulties encountered in determining the exact frequencies and mode shapes of the structure during severe earthquakes, when the dynamic response is likely to be highly inelastic.

It is important to note the fundamental difference between elastic and inelastic spectra on the one hand and design spectra on the other (Figure 3.1). Elastic and inelastic spectra, presented in Section 3.4.2, are 'computed' quantities that are mathematically based and reproducible by parties other than those who derived them. Design spectra, on the other hand, include features that are decided upon by code committees or other interested parties, and are therefore not necessarily reproducible by others. For example, a design spectrum may include features that prevent unconservative estimates of design actions, or protect against the adverse effects of errors in calculating periods of vibration. Therefore, design spectra, expressed in terms of acceleration response versus period, should not be used to derive displacement spectra, since they carry features that may violate basic theoretical principles.

Since the peak ground acceleration, velocity and displacement from various earthquake records generally differ, the computed response cannot be averaged on an absolute basis. Therefore, various procedures are used to normalize response spectra before averaging is carried out. The general procedure for generating statistically based averaged spectra is summarized as follows:

(i) Select a set of ground motions on the basis of their magnitude, distance and site conditions.
(ii) Generate response spectra in terms of acceleration, velocity and displacement, as appropriate for the seismic structural design.
(iii) Average the response spectra derived in Step (ii). Curves are generally fit to match computed mean spectra.
(iv) Evaluate the design response spectrum with desired probability of being exceeded on the basis of the relationships derived in the previous steps.

Site-specific design spectra can also be generated by employing ground-motion models (attenuation relationships) of response spectral ordinates as discussed in Section 3.3, or by advanced numerical modelling of the energy release and travel path associated with the ground motion, as illustrated in Section 3.5.3. Site-specific design spectra can also be provided as uniform hazard spectra as discussed above where the probability of each ordinate being exceeded is uniform. The curves, evaluated statistically, correspond to all magnitude–distance pairs contributing to the distribution of the spectral values, for all periods and damping levels considered.

From a structural engineering viewpoint, a design or smoothed spectrum is a description of seismic design forces or displacements for a structure having a certain fundamental period of vibration and structural damping. The first earthquake design spectrum was developed by Housner (1959). Thereafter, Newmark and Hall (1969) recommended simplified linear forms to represent earthquake design spectra. Design spectra can be either elastic or inelastic. The latter are employed to evaluate design forces and displacements for structural systems responding inelastically under earthquake loading. Inelastic design spectra can be obtained either directly (e.g. Mahin and Bertero, 1981; Vidic *et al.*, 1994; Fajfar, 1995) or by scaling elastic spectra through force reduction factors presented in Section 3.4.4. Scaled elastic spectra are provided in seismic design codes of practice. Such spectra are generally average acceleration response spectra that have been smoothed using control periods, which are either 2 or 3 depending on the code. The basic curves employ 5% damping; however, simplified expressions exist to obtain spectra for different damping values, e.g. equation (4.39) of Chapter 4. Moreover, the standard design response spectra are based on fixed spectral shapes, which vary as a function of the soil site conditions, e.g. rock, stiff and soft soils. Earthquake magnitude and source distance are also used, e.g. in the USA, to characterize the design spectra.

Design spectra are provided in the codes as normalized spectral curves; thus, the design spectra for a given site are computed by multiplying the spectral shapes by zone factors obtained from contour maps. Effective peak accelerations are sometimes used to scale the normalized spectra. Indeed, the fixed spectral shape is usually presented as normalized to 1.0-g ground acceleration, which is the response acceleration at zero period. Spectra may be presented in several formats, such as spectral ordinates (acceleration, velocity and displacements) versus period, tripartite plots and spectral acceleration versus spectral displacement. The latter form is termed 'composite spectrum' and is employed in the capacity spectrum assessment method (e.g. Fajfar, 1998; Freeman, 1998; Chopra and Goel, 1999, among many others). Alternatively, spectral values can be plotted as a function of frequency.

3.4.6 Vertical Component of Ground Motion

The vertical component of earthquake ground motion has generally been ignored in structural earthquake engineering. This is gradually changing due to the increase in near-source records obtained recently, coupled with field observations confirming the possible destructive effects of high vertical vibrations (Papazoglou and Elnashai, 1996).

The occurrence of vertical component of ground motion is mainly associated with the arrival of vertically propagating compressive P-waves, while secondary, shear S-waves are the main cause of horizontal components as discussed in Section 1.1.3. The wavelength of P-waves is shorter than that of S-waves, which means that vertical ground motion is associated with higher frequencies than its horizontal motion counterpart. Near the source of an earthquake, ground motion is characterized mainly by source parameters and rupture dynamics. The P-wave spectrum has a higher corner frequency than that of the S-wave. P and S corner frequencies gradually shift to lower frequencies as waves propagate away from the source due to the differentially stronger attenuation of higher frequencies. Consequently, the vertical motion amplitudes will attenuate at a faster rate. The behaviour of these two components of ground motion is often characterized by the vertical-to-horizontal v/h peak ground acceleration ratio.

Normally, the vertical component of ground motion has lower energy content than the horizontal component over the frequency range of interest. However, it tends to have all its energy concentrated in a narrow, high frequency band, which can prove damaging to engineering structures with vertical periods within this range. It has been observed that v/h ratios frequently are greater than 1.0 near the source of an earthquake (Abrahamson and Litehiser, 1989; Bozorgnia *et al.*, 1994; Ambraseys and Douglas, 2000, 2003; Bozorgnia and Bertero, 2004). Table 3.17 provides examples of earthquake ground motions with high vertical components: it may be observed that the v/h ratios can be greater than 2.0 (e.g. the 1976 Gazli earthquake in the former USSR).

The 1994 Northridge earthquake in California produced v/h ratios as high as 1.70 and the Hyogo-ken Nanbu (Japan) earthquake of 1995 exhibited peak v/h ratios of up to 2.00. Characteristics of some records possessing a strong vertical component measured during the Northridge and Kobe earthquakes are given in Tables 3.18 and 3.19, respectively. Records are ranked as a function of epicentral distances d for each station.

Moreover, there is strong evidence that the vertical component (assuming it is due to P-waves) is not strongly influenced by non-linear site effects in the way that horizontal S-waves are, which would provide a reasonable explanation for the following observation. During the 1995 Hyogo-ken Nanbu (Kobe, Japan) earthquake, liquefaction at the vertical array at Port Island caused an abrupt reduction

Table 3.17 Sample of earthquakes with high vertical component.

Earthquake	Country	Date	Peak horizontal acceleration (g)	Peak vertical acceleration (g)	Vertical-to-horizontal ratio
Gazli	ex-USSR	17 May 1976	0.622	1.353	2.17
Coyote Lake	USA	6 August 1979	0.256	0.420	1.64
Loma Prieta	USA	17 October 1989	0.424	0.514	1.21

Table 3.18 Northridge records possessing a strong vertical component (*after* Broderick *et al.*, 1994).

Station	*d* (km)	HPGA (g)	VPGA (g)	VPGA/HPGA
Tarzana, Cedar Hill Nursery	5	1.82	1.18	0.65
Arleta, Nordhoff Avenue Fire Station	10	0.35	0.59	1.69
Sylmar, County Hospital	16	0.91	0.60	0.66
Newhall, LA County Fire Station	20	0.63	0.62	0.98

Key: d = source distance; HPGA = horizontal peak ground acceleration; VPGA = vertical peak ground acceleration.

Table 3.19 Kobe records possessing a strong vertical component (*after* Elnashai *et al.*, 1995).

Station	*d* (km)	HPGA (g)	VPGA (g)	VPGA/HPGA
JMA Station	18	0.84	0.34	0.41
Port Island Array	20	0.35	0.57	1.63
Kobe University	25	0.31	0.43	1.39

Key: d = source distance; HPGA = horizontal peak ground acceleration; VPGA = vertical peak ground acceleration.

in the horizontal ground shaking, but the vertical motion continued to be amplified through the liquefied layer. The vertical-to-horizontal peak ground acceleration ratio was 1.63 (Table 3.19).

Vertically propagating dilatational waves are amplified in a manner identical to that of vertically propagating shear waves. Consequently, the vertical component of motion can be linearly amplified from bedrock to the surface up to very high levels, leading to the widely observed high v/h ratios near the source. For example, Table 3.20 shows the results of the study by Ambraseys and Simpson (1995) that involved worldwide records generated at source distances $d \leq 15\,\mathrm{km}$ by relatively large inter-plate earthquakes with magnitude $M_S \geq 6.0$ and significant vertical accelerations ($a_v \geq 0.10\,\mathrm{g}$). The identified data set consists of 104 records in the magnitude range $6.0 \leq M_S \leq 7.6$. A simple linear regression analysis, which is fully justifiable in the small distance range considered, was carried out. The results in Table 3.20 show the 84.1% confidence limit.

The computed values clearly show that the assumption that the vertical peak is 2/3 of the horizontal component suggested by Newmark *et al.* (1973) can be a serious underestimate, especially for short distances from the source, e.g. less than 15 to 20 km.

3.4.7 Vertical Motion Spectra

The commonly used approach of taking the vertical spectrum as 2/3 of the horizontal, without a change in frequency content, has been superseded (Elnashai and Papazoglou, 1997; Collier and Elnashai, 2001). The effect of vertical motion is currently subject to re-evaluation and independent vertical spectra have been proposed for implementation within codes of practice, for example in Europe. Below, two alternatives are given for obtaining a more realistic vertical spectrum than seismic codes have hitherto employed with the exception of recent European seismic standards (Eurocode 8, 2004).

Ambraseys *et al.* (2005b) proposed vertical ground-motion parameters attenuation relationships of the same form as those given in Section 3.4.3.1. The relationship for 5% damping spectral acceleration S_a can be expressed as follows:

Table 3.20 Attenuation of vertical-to-horizontal ratio for 84.1% confidence limit (*after* Ambraseys and Simposon, 1995).

	All (104 records) $\sigma = 0.48$		Thrust (53 records) $\sigma = 0.46$		Strike-slip (43 records) $\sigma = 0.53$		Europe (23 records) $\sigma = 0.36$	
M_S	$d = 0.0$	$d = 15.0$	$d = 0.0$	$d = 15.0$	$d = 0.0$	$d = 15.0$	$d = 0.0$	$d = 15.0$
6.0	1.28	1.15	1.43	0.96	1.32	1.32	0.77	0.75
6.5	1.37	1.24	1.54	1.07	1.40	1.40	0.98	0.97
7.0	1.46	1.32	1.64	1.17	1.48	1.48	1.20	1.18
7.5	1.54	1.41	1.75	1.28	1.56	1.56	1.41	1.39

Key: d = source distance (in kilometres); σ = standard deviation from regression analysis.

Table 3.21 Relationship between vertical and horizontal peak ground acceleration.

Magnitude, M_s	Distance (in km)			
	0	15	30	≥100
5.5	0.72	0.58	0.48	0.39
6.0	0.80	0.67	0.52	0.42
6.5	0.89	0.76	0.56	0.45
7.0	0.98	0.84	0.61	0.49
7.5	1.06	0.93	0.66	0.53

$$\log(S_a)_v = 0.835 - 0.083\,M_w + (0.206\,M_w - 2.489)\log\sqrt{31.36 + d^2} + 0.078\,S_S + 0.046\,S_A - 0.126\,F_N + 0.005\,F_T - 0.082\,F_O \quad (3.28)$$

where S_a is expressed in m/sec^2 and d is the distance (in kilometres) to the projection of fault plane on surface. All the other quantities in equation (3.28) are as in equation (3.8.1). The values of standard deviation σ for equation (3.28) are $\sigma_1 = 0.262$ and $\sigma_2 = 0.100$, for intra- and inter-plate earthquakes, respectively. Note that equation (3.28) can also be utilized to estimate vertical peak ground accelerations; the values of PGA are in m/sec^2.

Alternative spectra were derived by Elnashai and Papazoglou (1997), specifically for vertical motion, taking narrower (in general, higher) magnitude and distance ranges, so that the results are not biased towards the over-represented distant events. In its simplest form, the proposal in the above reference starts from the horizontal peak ground acceleration, obtained from relationships such as that given in equation (3.8.1), then evaluates the vertical peak ground acceleration from Table 3.21 (obtained by combining the results of Ambraseys and Simpson, 1996, with those of Abrahamson and Litehiser, 1989).

A spectral model may thereafter be used to derive the vertical elastic spectrum, where the corner periods are not a function of soil condition but are fixed to 0.05 and 0.15 seconds. Moreover, the amplification, regardless of soil condition, is 4.2 for 2% damping and 3.[illegible]5 for 5% damping. Interpolation may be used between these two values. The proposed relationship is provided graphically in Figure 3.17 as vertical-to-horizontal ratios.

Inelastic vertical spectra are difficult to deal with, because they involve identifying sources of energy dissipation and redistribution potential, as well as the difference between vertical motion upwards and

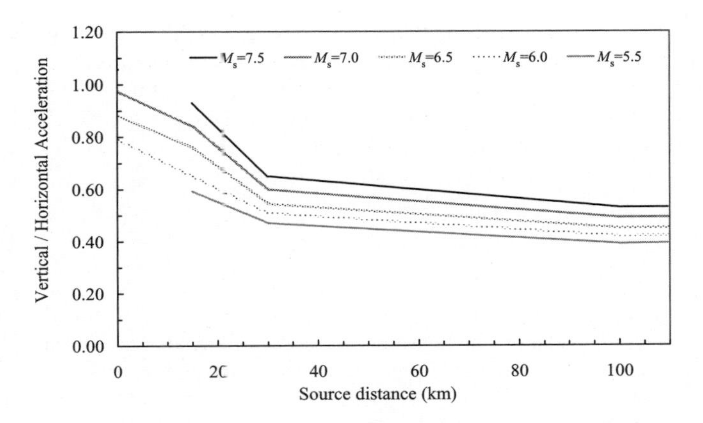

Figure 3.17 Vertical-to-horizontal peak ground acceleration ratios

downwards. A conservative assumption is that the response modification factor in the vertical direction is unity. Further details concerning the vertical spectra and combination rules for horizontal and vertical earthquake effects may be found in Collier and Elnashai (2001).

> **Problem 3.7**
>
> Evaluate the vertical component of the ground motion for an earthquake with $M_w = 6.5$ at a site located 15 km from the projection of fault plane on surface; the fault mechanism is normal. Assume that site conditions can be either soil or rock. Use the attenuation relationship given in equation (3.28). Evaluate the ratio of the horizontal and vertical peak ground accelerations by utilizing the curves provided in Figure 3.17. Comment on the results.

3.5 Earthquake Records

There are three approaches to obtaining earthquakes or earthquake-like ground-motion records (time histories) for purposes of assessment by advanced analyses in structural engineering. Natural records of earthquakes have increased exponentially in the past decade or so leading to the availability of high-quality strong-motion data from several sources (Bommer and Acevedo, 2004). Another approach is to generate a random signal that fits, with a certain degree of approximation, a target spectrum. Finally, use of mathematical source models to generate time series that look like earthquake strong motion, is increasing in popularity since the ensuing records resemble natural records more than signals generated to fit a target spectrum. These three options, i.e. natural and artificial records, and those based on mathematical formulations, are further described below. The relevant selection criteria are also reviewed. In so doing, verifiable and robust input motions for inelastic time-history analysis currently required by codes for many types of structure can be provided.

3.5.1 Natural Records

When using natural earthquake records, most codes recommend the use of a minimum of three to seven different accelerometer recordings that exhibit reasonable amplification in the period range of the structure, scaled appropriately as discussed in Chapter 4. Otherwise, artificially generated records can be used, provided that the distribution of frequencies associated with high energy is relevant to the fundamental period of the structure. This can be ensured by generating a record that conforms to an approved spectral shape. From a structural engineer's viewpoint, it is instructive to note that the features of strong motions that affect structural response are many and their inter-relationship is complex. It is thus of importance to highlight the regional differences in strong-motion data and the criteria for the selection of natural records. These two aspects are discussed hereafter.

3.5.1.1 Regional Differences

As a consequence of the proliferation of strong-motion databanks (e.g. Section 3.7), region-specific ground-motion models (attenuation relationships) have been derived. These studies have generated interest in regional differences in the characteristics of earthquake strong motion. A comprehensive worldwide review can be found in Douglas (2001, 2006). In order to undertake valid comparisons of strong-motion characteristics from different regions, records obtained under identical circumstances, i.e. magnitude, depth, fault mechanism, travel path and site characteristics, are needed from each region. This is clearly either very unlikely to be achieved or outright impossible.

The results of some studies suggest that regional differences in terms of strong-motion characteristics in seismically active areas are quite small. The attenuation relationships for peak ground acceleration, derived for western North America by Joyner and Boore (1981) and for Europe and adjacent areas by Ambraseys *et al.* (1996), predict rather similar results. For a given magnitude and distance pair, the difference between the two predictions is usually less than the standard deviation in the employed attenuation relationships. A study by Spudich *et al.* (1997) has examined strong-motion attenuation in seismically active zones of tectonic extension. The study concluded that, in general, peak ground acceleration and acceleration response spectral ordinates are lower in such regions than in other tectonically active areas. This supports the conclusion of McGarr (1984) who observed that PGAs in extensional regimes to be about 2/3 of the values encountered in compressional regimes. However, it is intuitively noted that regional differences in elastic response spectra may not carry over to inelastic spectra, where the influence of hysteretic energy absorption and the continuous change in response periods could possibly overtake regional differences on strong-motion records.

In the absence of a strong-motion databank for a specific region, it is necessary to select accelerograms from other regions that have produced significant strong-motion recordings. An early decision is needed regarding whether inter-plate or intra-plate earthquake records are sought. It follows that a definition of intra-plate and inter-plate earthquakes is needed. Dahle *et al.* (1990) derived attenuation equations for use in intra-plate regions by performing regressions on a data set from earthquakes that they classified as intra-plate. This data set includes records from regions such as the eastern USA, Australia and Germany, but also regions such as Greece, Italy and Yugoslavia. Dahle *et al.* (1990) classify areas as intra-plate on the basis of remoteness from active tectonic plate boundaries. However, some of these aforementioned regions are areas of appreciable tectonic deformation. Parts of mainland Japan, including the region of the 1995 Kobe earthquake, are also referred to as intra-plate in some studies on the same basis (e.g. Wesnousky *et al.*, 1984), although, the latter is also an area of active tectonic deformation. Johnston *et al.* (1994) distinguish intra-plate regions that are not being actively deformed and refer to these as stable continental regions. The latter definition is more convincing.

The differences between strong-motion characteristics in intra-plate and inter-plate regions are usually attributed to source and path effects. In terms of the path effects, anelastic (engineering seismology term for inelastic) attenuation is generally assumed to be greater in the more fragmented inter-plate regions. It has been shown that earthquake ground motions attenuate less rapidly in the eastern USA than in California (Atkinson and Boore, 1997). Abrahamson and Litehiser (1989) include a term in their attenuation equation for peak acceleration, which implies that anelastic attenuation is only of importance in inter-plate regions. As a result, for identical earthquakes at 100-km distance, peak horizontal acceleration would be 17% lower in an inter-plate region than in an intra-plate region.

It is important to note that conclusions drawn from attenuation relationships are influenced by the distribution of the data sets on which the regression has been performed, as mentioned in Section 3.3.1. For example, the magnitude-distance space occupied by the stable continental region accelerograms presented by Free (1996) and the western North American data set of Boore *et al.* (1993) are mutually exclusive. Furthermore, there may be other coupled factors involved. Atkinson and Boore (1997) have reported different site effects at rock locations in California and eastern USA. These considerations notwithstanding, Free (1996) observed that ground motions from intra-plate earthquakes have response spectral amplitudes that are appreciably higher than those from inter-plate earthquakes for high frequencies (greater than 10 Hz); although for lower frequencies the amplitudes are similar. However, there is very significant scatter in the attenuation relationships derived by Free (1996) and also important differences from one stable continental region to another. The above discussion points towards possible significance of regional characteristics of strong motion, but there are no universally accepted rules. The earthquake structural engineer should exercise caution and study carefully the seismo-tectonic environment pertinent to the project or study at hand in order to select records that truly represent the likely scenarios.

3.5.1.2 Selection Criteria

The ideal procedure for the selection of strong motions for use in analysis is to obtain records generated in conditions that are identical to those of the seismic design scenario. Bolt (1978) showed that if all the characteristics of the design earthquake could be matched to those of a previous earthquake, the probability of the characteristics of the record matching would be unity. The design earthquake, however, is usually defined in terms of only a few parameters; hence, it is difficult to guarantee that the selected records would closely match all of the characteristics of the design earthquake at the source, throughout the path and on to the site surface. Furthermore, even if the design earthquake scenario was defined in all aspects, it is unlikely that a record could be found in available databanks (*see*, for example, Section 3.7), which simultaneously matches all of the characteristics. However, Bolt (1978) also demonstrated that as the number of characteristics of a previous earthquake that match those of the design earthquake increases, the probability of the records matching increases rapidly. Therefore, to select records with a reasonable probability of bracketing the structural response, it is necessary to identify the most important parameters that characterize the conditions under which an earthquake record is produced and match as many of these as possible to the design earthquake scenario, taking as a primary measure the effect on inelastic structural response. It is emphasized that records giving seemingly consistent response parameters in the elastic response range, i.e. with low coefficients of variation, may yield much higher variations due to structural period shifts.

The parameters that characterize the conditions under which strong-motion records are generated can be grouped into three sets, representing the earthquake source, the path from the source to the recording site and the nature of the site. The important parameters in the above sets are as follows:

(i) *Source*: magnitude, rupture mechanism, directivity and focal depth;
(ii) *Path*: distance and azimuth;
(iii) *Site*: surface geology and topography.

The above list is not exhaustive, but it does include the parameters that have been established as having a notable influence on ground-motion characteristics. These parameters influence different characteristics of the recorded motion in different ways and to different degrees. Hence, the most appropriate selection parameters depend on which characteristics of the selected motion are considered most important from a structural response viewpoint.

The selection process is also a function of the objective of assembling a strong-motion record suite. For example, if the records are required for a specific site subjected to a well-defined hazard, normally characterized by magnitude, distance and site condition, the selection process would be distinct from the case where an engineering firm requires a number of records to be used in routine analysis of a wide variety of structures of yet-to-be-defined characteristics. Assuming that the two examples represented above cover many application examples for structural earthquake engineers, simple procedures can be defined for each as discussed hereafter.

(i) Matching a Design Scenario

Matching a design scenario often corresponds to reconciling magnitude–distance–soil condition triads. There are, however, uncertainties in magnitude and distance calculations. Therefore, it is reasonable to select records within ranges from the design event to increase the possibility of finding a viable record suite. Guidelines are lacking but searching within a range of ±0.3 magnitude unit and ±20–40 km distance is reasonable. Databanks of well-constrained records are examined, e.g. those presented in Section 3.7, provided the origin of the magnitude and distance calculation, as well as the site condition categorization, are known and accepted and a number of records within this range are selected. Matching the magnitude is more important than matching the distance. This is supported by the discussion on elastic spectra of Section 3.4.2. For inelastic cases, the increase in distance would lead to an increase in duration, which will in turn affect the response. Thus, the records obtained are unlikely to have the design PGA. Scaling, using a recommended procedure as illustrated in Section 3.5.4, should then be applied to arrive at a set of records that will consistently test the structures intended for the site. This is one of the simplest procedures for selecting records to fulfil a predetermined design scenario.

(ii) A Suite of Records for Design Office Applications

Whereas this is not a technically robust requirement, it is often requested by the non-specialist practice. For example, it is suitable for design office use, where limited knowledge of earthquake engineering and engineering seismology exists. Also, it is a useful approach for investigating seismic response in non-specific region applications (e.g. Broderick and Elnashai, 1996, among others).

Records are selected on the basis of the peak ground acceleration (PGA) to the peak ground velocity (PGV), i.e. the ratio PGA/PGV (Zhu *et al.*, 1988). The rationale behind this is that near-source shallow earthquakes or records measured on rock, will exhibit high acceleration peaks of short duration, leading to low-velocity cycles. These records will give high values of PGA/PGV. Deep or distant earthquakes or records measured on soft ground will have lower acceleration values, but individual cycles are of longer duration, leading to high-velocity waves. These will yield low PGA/PGV ratios. Intermediate scenarios in both senses will yield intermediate values of PGA/PGV.

With regard to structural response, high values of PGA/PGV records will be more critical for stiffer structures, while more flexible structures will be strongly shaken by low PGA/PGV records. Therefore, selection of records based on PGA/PGV, with a reasonable number in each of the regions low, medium and high, will ensure that the ensemble is capable of imposing high demands on structures in a wide range of periods and will implicitly include many of the engineering seismology features related to source characteristics, travel path and site conditions, as mentioned above. The approximate ranges of PGA/PGV ratios determining the low, medium and high ranges are as follows:

$$\text{low} \quad PGA/PGV < 0.8 \tag{3.29.1}$$

$$\text{medium} \quad 0.8 \le PGA/PGV \le 1.2 \tag{3.29.2}$$

$$\text{high} \quad 1.2 < PGA/PGV \tag{3.29.3}$$

where the acceleration PGA is expressed in g and the velocity PGV in m/s. There are many other approaches to earthquake record selection in the published literature (e.g. NRC, 2001; Naeim *et al.*, 2004; Iervolino and Cornell, 2005, among others). However, they are more complex than those discussed above, and are not necessarily more representative of future scenarios, due mainly to the high uncertainty in the characteristics of future earthquakes.

3.5.2 Artificial Records

Artificial acceleration records are an option for generating signals that satisfy engineering criteria unrelated to the physics of earthquake stress wave generation and propagation. Accelerograms can be mathematically simulated through random vibration theory. Both stationary and non-stationary random processes have been suggested (Shinozouka and Deodatis, 1988; Deodatis and Shinozouka, 1989). Indeed, strong motions include transitional phases at initial and final stages, respectively, moving from being stationary to the maximum shaking and vice versa. These transitional stages are non-stationary signals. Small earthquakes can also be described with such processes. By contrast, the middle portion of large earthquakes, i.e. the nearly uniform part of the vibration, can be modelled by means of stationary processes, such as white noise (Bycroft, 1960) or Gaussian white noise (Boore, 1983).

The most widely used approach is to develop a white noise signal with a response spectrum that matches a target response spectrum with a predefined accuracy, for example with 3% to 5% margin of error. The target spectrum is normally either a uniform hazard spectrum or a code spectrum. An example of such acceleration signals is shown in Figure 3.18, where a response spectrum-compatible accelerogram is displayed. The level of accuracy of the match is a function of the number of iterations carried out during the generation process.

Three elements are necessary to generate synthetic accelerograms: (i) power spectral density, (ii) random phase angle generator and (iii) an envelope function. Indeed, the simulated motion can be calculated as the sum of several harmonic excitations. Thus, the consistency of the artificial motion is assessed through an iterative algorithm, which examines the frequency content. This check can be carried out either with the response spectrum of the signal or its power spectral density. A detailed description of the procedures for generating artificial records is given in Clough and Penzien (1993), for example. Several computer programs that generate such records have been developed as discussed in Section 3.8. However, inherent difficulties in the generation process are the assumption of the phase distribution between the various single frequency waves and the duration of the record. Therefore, signals that match the same spectrum may look different and, more importantly, may lead to different structural responses. A closer fit between the spectrum of the generated signal and that of the design spectrum should be sought in the vicinity of the structural fundamental period. It should also be recog-

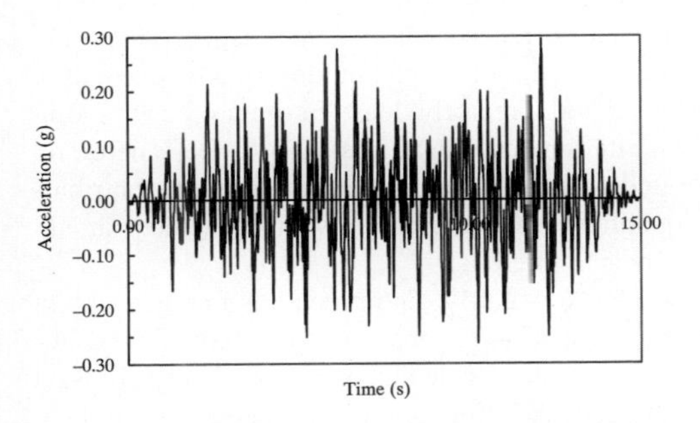

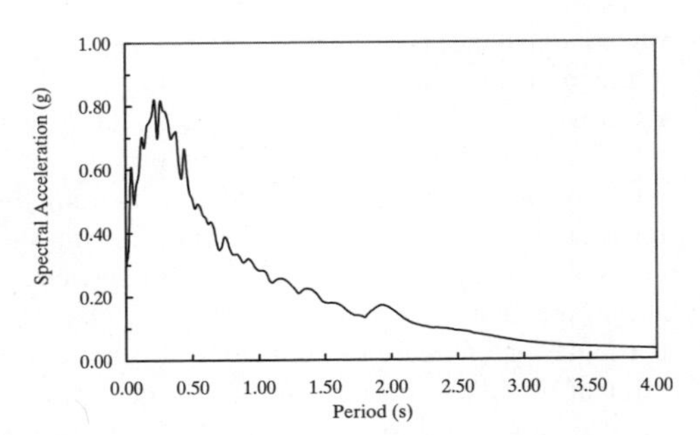

Figure 3.18 Acceleration artificial record (*left*) matched to a code spectrum (*right*)

nized that artificial records do, on average, exhibit a larger number of cycles than natural records and hence they may impose unrealistic seismic demand for inelastic structural systems as discussed in Section 3.6. The main reason for the over-conservatism in spectrum-compatible artificial signals emanates from the spectrum itself. Uniform hazard and code spectra represent many different magnitude–distance–soil triads. Having a single record representing tens of feasible scenarios may lead to over-conservatism of the ensuing artificial motion.

3.5.3 Records Based on Mathematical Formulations

Considerable advances in earthquake geophysics and wave propagation modelling have resulted in the development of complex formulations for the generation of earthquake-like signals. The latter provide an alternative to statistical treatments of observational data, i.e. attenuation relationships presented in Sections 3.3 and 3.4.3.1. The purpose of this research activity is twofold From the seismological point of view, parametric variation in source, path and site characteristics coupled with comparisons with measured strong motion, shed further light on the influence of the seismological environment on ground motion. From the earthquake engineering perspective, once such models have been developed and calibrated, even for a limited extent, they may be used to generate input motions in areas of the world where natural records do not exist. A review of models for generation of simulated ground motion is beyond the scope of this book. Valuable comments on the most widely used and developed models can be found in Bard *et al.* (1995). Following the seismic waves from source to site, mathematical formulations for ground-motion generation model the source, path and site response. They purport to represent low- and high-frequency near-source effects, various focal mechanisms, fault slip velocity and displacement, directivity, crustal propagation, soil and topography, all in three dimensions. The further development of such approaches is constrained by three specific factors, namely:

(i) Availability of detailed geophysical data on the area of interest;
(ii) Existence of observational data for validation;
(iii) Computing power for spatial simulation.

Whereas giant strides are achieved continuously on the third factor, the articulation of mathematical formulations for the generation of strong ground motion is severely hindered by the former two. There are a few documented cases of *a posteriori* success (e.g. Gariel *et al.*, 1990 for the Imperial Valley earthquake of 15 October 1979). However, even in the case of the southern California seismo-tectonic environment that is very well studied and documented, the model failed to represent, with an acceptable degree of accuracy, the response at a number of stations. The failure was attributed to site effects, which are still far from being well represented.

Whereas continuing efforts towards the refinement and calibration of this approach are totally warranted, to balance the statistical approach extensively used in current seismic hazard practice, it is unable at the moment to provide definitive guidance on the characteristics of ground motion where none exists. Even when a full array of geophysical data is available, where precisely does the rupture occur and in which direction it propagates will still only be known after an earthquake has taken place. Therefore, these methods provide insight and an opportunity for understanding the influence of different contributing parameters on strong-motion characteristics in a qualitative manner. Records, thus generated should be used with caution and with the full knowledge of their limitations.

An interesting approach was developed by Afra and Pecker (2001). The process is initiated with a particular natural earthquake record, the phase distribution of which is observed and recorded for subsequent use. The Fourier amplitude spectrum of the record is then calculated and iteratively adjusted to fit a target spectrum. During this process, the phase distribution of the original record is kept

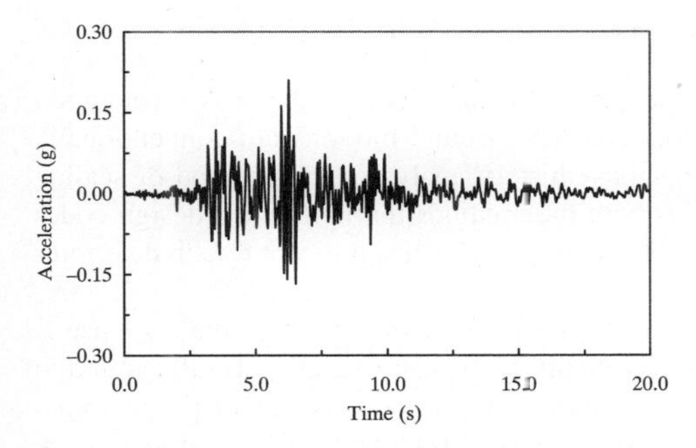

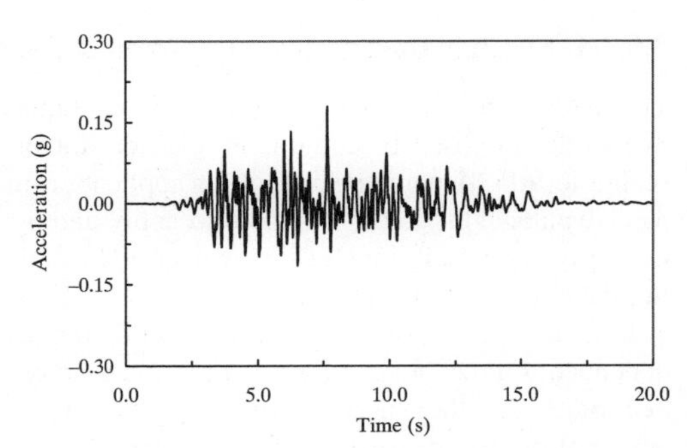

Figure 3.19 Natural record (*left*) at Westmorland Brawley, N45W (California, 1981) and relative artificial record generated (*right*)

unchanged. An inverse Fourier transformation is then applied to arrive at a non-stationary signal, alongside multiplication with an envelope function dependent on the type of motion sought. For near-source records, a pulse-like function with an exponential decay is used to scale the records, while a trapezoidal function is used for large distant earthquake records representation. The starting record for such an application is shown in Figure 3.19 along with the artificial record resulting from the procedure described herein. Whereas the records are distinct, the artificial record carries many of the modulation characteristics of its originator. This procedure has no physical justification. However, it is superior to the procedure of generating spectrum-compatible white noise signals, since the phase distribution in the former is taken from a natural earthquake recording. This is shown vividly by comparing Figure 3.18 and Figure 3.19. The former has too many peaks as compared to a natural record, while the latter looks like a natural earthquake record.

Other alternatives exist for generating ground motion-like signals either from source models or from other engineering criteria, such as spectrum matching. However, the above methods serve to highlight the expected features of natural and artificial input motion for structural analysis.

3.5.4 Scaling of Earthquake Records

The variation during response of the important features of earthquake structural response, which are stiffness, strength and ductility, discussed in Sections 2.3.1, 2.3.2 and 2.3.3 respectively, is highly dependent on ground-motion characteristics. Therefore, it is important when considering seismic response due to a number of earthquake motions, to ensure that the characteristics of each of these motions are similar. In this manner, the effects of other features, such as frequency content and duration of loading, can be assessed. Moreover, when evaluating the performance of a structure designed to code-prescribed seismic loads or a specific hazard level (or design scenario) by inspecting its response to actual ground motions, the seismic energy imparted to the structure by the imposed base accelerations should be comparable to that implied in the code design spectrum or the site-specific hazard definition. The various spectrum intensity scales suggested by researchers are presented and discussed hereafter. This is followed by an assessment of the validity of these techniques for structures of different characteristics, using a large number of earthquake records. Finally, based on the comparative accuracy of the various approaches and on practical considerations, a recommended scaling procedure for time history records is summarized. A complete, simple and effective framework of selection and scaling for structural application is thereby presented.

3.5.4.1 Scaling Based on Peak Ground Parameters

The loads acting on a structure during an earthquake are proportional to its instantaneous acceleration due to the imposed base motions. On account of this, recorded ground motions are conventionally scaled to a PGA value prior to their application in response history analysis. This method of scaling has advantages in that it is simple to apply and agrees with the methods through which design codes normally define seismic loads. However, PGV and PGD values also play a significant role in determining the severity of seismic response.

In general, it is possible to identify three ranges of structural periods within which the response is dependent on the values of ground-motion acceleration, velocity or displacement, as also illustrated in Section 3.4.4. Short-period structures (typically less than 0.5 second) are sensitive to peak ground acceleration, while structures of moderately long period (i.e. 0.5 to ~2.0 seconds) are sensitive to peak ground velocity. The response of structures of exceptionally long period (i.e. longer than 2–3 seconds) is likely to be more dependent on displacement. The dependence of intermediate period structures on velocity is directly recognized by some codes of practice, such as in the Japanese code (BSL, 2004), where peak ground velocity is used for the design of tall buildings. Therefore, it is appropriate to scale earthquake records in a manner that reflects the response periods of the structure under consideration. This has been confirmed by a number of studies on large suites of earthquake records (Chandler, 1991; Tso *et al.*, 1992) in which the degree of spectral dispersion is reduced in the low-period range when acceleration scaling is applied and a similar reduction is observed at longer periods with velocity scaling. This observation applies equally to the important parameter of the ductility demand imposed on structural systems when the additional effects of varying yield strength must be considered.

While replacing PGA with PGV for intermediate period structures offers an improvement over using acceleration for the full range of period, it is still not sufficiently accurate. It was shown by several researchers (e.g. Nau and Hall, 1984; Matsumura, 1992) that PGA and PGV are not always adequate measures over the wide range of frequencies, since they are based on a single point of the response spectrum. Response velocity spectra of various ground motions normalized by PGV may be remarkably different. Furthermore, existing codes of practice require seismic loading to be uniquely defined in terms of PGA, with an associated acceleration response spectrum. This implies that scaling ground motion to a common peak velocity would disturb the equivalence between the time-history record and the design spectrum.

3.5.4.2 Scaling Based on Spectrum Intensity

In view of the above, spectrum intensity, rather than a peak ground parameter, may be utilized as a reference for scaling earthquake records. This scaling procedure assumes that the seismic energy imparted by the scaled earthquake record is equal to that implied in the design spectrum of the adopted seismic code. Several spectrum intensity scales have been suggested in the literature, the most pertinent of which are discussed hereafter.

(i) Housner Spectrum Intensity

The spectrum intensity proposed by Housner (1952) has received considerable attention in conjunction with scaling procedures for earthquake records. Using the assumption that elastic response spectra may be used to estimate the energy available to cause damage, it was suggested that the velocity spectrum could provide a measure of the severity of structural response. The intensity of shaking of an earthquake at a given site was represented by the spectrum intensity SI_{H}, defined as the area under the elastic velocity spectrum, between the periods 0.1 and 2.5 seconds:

$$SI_{\mathrm{H}} = \int_{0.1}^{2.5} S_{\mathrm{v}}(T, \xi)\, \mathrm{d}T \qquad (3.30)$$

where S_v is the velocity spectrum curve, T is the period of vibration and ξ is the damping coefficient.

The justification given by Housner (1952) for the selected integration limits is that they encompass a range of typical periods of vibration of structures. Hence, the above definition of spectrum intensity may be considered an overall measure of the capability of an earthquake to excite a population of structures with response periods between 0.1 and 2.5 seconds.

Although the use of Housner's spectrum intensity for scaling purposes may be considered an effective overall scaling procedure, it ignores important parameters of response such as yield period, period elongation associated with structural damage and energy distribution in the frequency domain. It therefore requires modification to account for the inelastic dynamic characteristics of the anticipated response of the structure. Several proposals to modify the integration limits proposed by Housner (1952) have been suggested, some of which are discussed below.

(ii) Intensity Scales of Nau and Hall

Nau and Hall (1984) conducted a study on scaling methods for earthquake response spectra. Scaling factors considered were based on ground-motion data and on elastic response. Factors derived from ground-motion data included the integrals of the square of acceleration, velocity and displacement, and the associated root square, mean square and root mean square. Factors based on elastic response included spectrum intensity and mean Fourier amplitude. The effectiveness of the scaling parameters was assessed in terms of the average coefficient of variation of the pseudo-velocity spectra for elastic and bilinear hysteretic SDOF systems, computed for an ensemble of 12 earthquake records.

To reduce the dispersion in results, a three-parameter system of spectrum intensities computed within low-, medium- and high-frequency regions was proposed. This system accounts for the sensitivity of the response to acceleration, velocity or displacement and is given by:

$$SI_a = \int_{0.03}^{0.19} S_v(T, \xi)\, dT \quad \text{for } 0.12\,\text{s} < T \leq 0.5\,\text{s} \tag{3.31.1}$$

$$SI_v = \int_{0.29}^{2.00} S_v(T, \xi)\, dT \quad \text{for } 0.5\,\text{s} < T \leq 5.0\,\text{s} \tag{3.31.2}$$

$$SI_d = \int_{4.17}^{12.50} S_v(T, \xi)\, dT \quad \text{for } 5.0\,\text{s} < T \leq 14.09\,\text{s} \tag{3.31.3}$$

where SI_a is the spectrum intensity in the acceleration region, SI_v is the spectrum intensity in the velocity region and SI_d is the spectrum intensity in the displacement region. It is noteworthy that in the original paper by Nau and Hall (1984), the units were cycle per second (cps) and frequency rather than period was used. Conversion and rounding off were undertaken in the above equations (3.31.1) to (3.31.3).

Although, the intensity scale of Nau and Hall (1984) follows well-established observations, it has some shortcomings. The number of records considered was relatively small (12 ground motions), while the variation of PGA/PGV ratio within the selected records was inadequate to reflect the importance of this ratio in assessing the overall characteristics of the frequency content. More importantly, the calibrations performed were primarily based on the dispersion of the pseudo-velocity spectra for the elastic case. Also, it was not shown whether or not the reduction in the dispersion of the pseudo-velocity spectra would effectively result in a reduction in the dispersion of displacement ductility demands. In this context, it is important to note that the assessment of seismic response based on elastic response-related parameters does not necessarily translate to structural damage. Economically designed earthquake-resistant structures are expected to perform in the inelastic range under the design earthquake. Hence, parameters of inelastic response, such as displacement ductility demand, appear to be of prime importance in calibrating spectrum intensities.

(iii) Matsumura Spectrum Intensities

Matsumura (1992) conducted a parametric study on the intensity measures of strong motions and their correlation with structural damage. Four intensity measures of ground motion, namely PGA, PGV, SI_M and V_e, were examined by evaluating the inelastic response of SDOF systems. Both SI_M and V_e are intensity measures suggested by Matsumura. SI_M is referred to as the 'Matsumura Spectrum Intensity' and is defined as the mean spectral velocity between T_y and $2T_y$, where T_y is the period corresponding to yield of a SDOF structure with critical damping ratio ξ of 5%. On the other hand, V_e is the mean equivalent velocity converted from the input energy E_i, between T_y and $2T_y$, with conversion given by:

$$V_e = \sqrt{\frac{2E_i}{m}} \tag{3.32}$$

where m is the mass of the SDOF system.

The adopted period interval, i.e. T_y to $2T_y$, is based on the assumption that the response of the structure in the inelastic range will be in the domain defined by its yield and twice its yield period. The bounds of the period range were found by studying its correlation with SI_M and V_e for a ductility factor μ equal to 2.0.

Based on the results obtained for four Californian and eight Japanese records, Matsumura confirmed that PGA and PGV are well correlated to damage in short- and long-period structures, respectively. However, they are not reliable measures of intensity for other frequency ranges. It was also observed that V_e and SI_M are measures of intensity that correlated well with damage for a wide range of frequencies. The correlation coefficient for V_e was found to be slightly higher than that for SI_M. It should be noted that in practice SI_M, as in the case of all spectrum intensity scales, can be directly used to define a scaling procedure since the pseudo-velocity spectrum could be derived from the acceleration spectrum provided by the adopted seismic code. Conversely, V_e cannot be directly applied due to the fact that the input energy E_i is not currently specified in seismic design code provisions.

Although the study of Matsumura (1992) showed that SI_M is an adequate scaling parameter, a comparison of the effectiveness of Housner's spectrum intensity with that of the four parameters included in his study was not performed. In addition, the post-yield stiffness ratio considered in the study was 0.5, which appears to be very high and not representative of typical earthquake-resistant structures with significant ductility demand.

(iv) Comparisons and Recommended Scaling Procedures

Martinez-Rueda (1997) carried out a preliminary evaluation by comparing the performance of the Housner's intensity with the three-parameter system of spectrum intensities proposed by Nau and Hall (1984). This indicated that the three-parameter system does not result in an improvement of the correlation with displacement ductility demand. The spectrum intensity of Nau and Hall appeared to be marginally less stable, particularly for short-period structures. However, there was an improvement in the velocity region when using Nau and Hall procedure. In addition, Housner's intensity involves a single parameter for all period ranges and hence is simpler to use in practice. The discussion given below focuses mainly on the comparison between the intensity scales of Housner SI_H and Matsumura SI_M as well as a third intensity scale SI_{yh} suggested by Martinez-Rueda (1997).

The spectrum intensity scales were represented as average spectrum velocities for damping ratios $\xi = 0.05$, such that Housner average spectrum intensity is given by:

$$\overline{SI_H} = \frac{1}{2.4}\int_{0.1}^{2.5} S_v(T, 0.05)\,dT \tag{3.33.1}$$

and Matsumura average spectrum intensity is given by:

$$\overline{SI_M} = \frac{1}{T_y}\int_{T_y}^{2T_y} S_v(T, 0.05)dT \tag{3.33.2}$$

Martinez-Rueda (1997) also suggested changing the second integration limit of Matsumura to T_h, which represents the hardening period of the structure. This was based on the assumption that the ground-motion frequencies contributing to the failure of the structure are contained within the period interval of T_y to T_h. Using these integration limits, average spectrum intensity may be represented as:

$$\overline{SI_{yh}} = \frac{1}{T_h - T_y}\int_{T_y}^{T_h} S_v(T, 0.05)dT \tag{3.33.3}$$

In order to assess the effectiveness of the spectrum intensities considered, a large number of inelastic time-history analyses were performed for a wide range of parameters of seismic input and dynamic response using more than 100 earthquake records. The structural parameters were the yield force ratio C_y and the hardening parameter α. A sample of results is given in Table 3.22. From the extensive results, it was concluded that the Matsumura definition is the most reliable regardless of the period range. Within the intermediate period range, Housner's intensity gives higher coefficient of correlation, but the improvements are not very significant. Martinez-Rueda (1997) gives recommended scaling procedures for each range of period, yield force ratio and hardening parameter. This is perhaps the most comprehensive study of practical application of spectrum intensity scaling. However, one main concern is that the spectra plotted for the yield force ratio do not control the ductility demand. Values in Table 3.22 show that the ductility demand imposed is up to 14. Hence, if the results are constrained to practical limits of ductility demand, i.e. up to about 6, the observations and conclusions may vary. Constant ductility spectra are considered more appropriate for such applications, since the practical yield force ratio is very wide, from structures that are not designed for seismic loading, to earthquake-resistant structures exhibiting high overstrength values.

A parametric study was conducted by Elnashai (1998) with 30 earthquake records using constant ductility spectra. The coefficients of variation (COVs) for elastic and inelastic spectral ordinates were calculated. Comparison between the elastic and inelastic spectra of scaled records and a target smoothed 'code-like' spectrum was also undertaken, alongside comparison of the former with the elastic and inelastic spectra of records compatible with the smoothed spectrum. A sample of the results is shown in Figure 3.20.

Table 3.22 Correlation coefficient for spectrum intensity scales (Yield force ratio, $C_y = 0.3$; hardening parameter, $\alpha = 0.1$).

	T_y (seconds)		
Intensity scale	0.4	1.4	2.4
$\overline{SI_H}$	0.84	0.91	0.70
$\overline{SI_M}$	0.93	0.84	0.92
$\overline{SI_{yh}}$	0.92	0.79	0.88
Intensity range (g-sec)	0.0–0.20	0.0–0.20	0.0–0.20
Ductility demand range	0.0–14.0	0.0–2.5	0.0–2.0

Key: $\overline{SI_H}$ = Average Housner spectrum intensity; $\overline{SI_M}$ = average Matsumura spectrum intensity; $\overline{SI_{yh}}$ = average Martinez-Rueda spectrum intensity; T_y = period at yielding

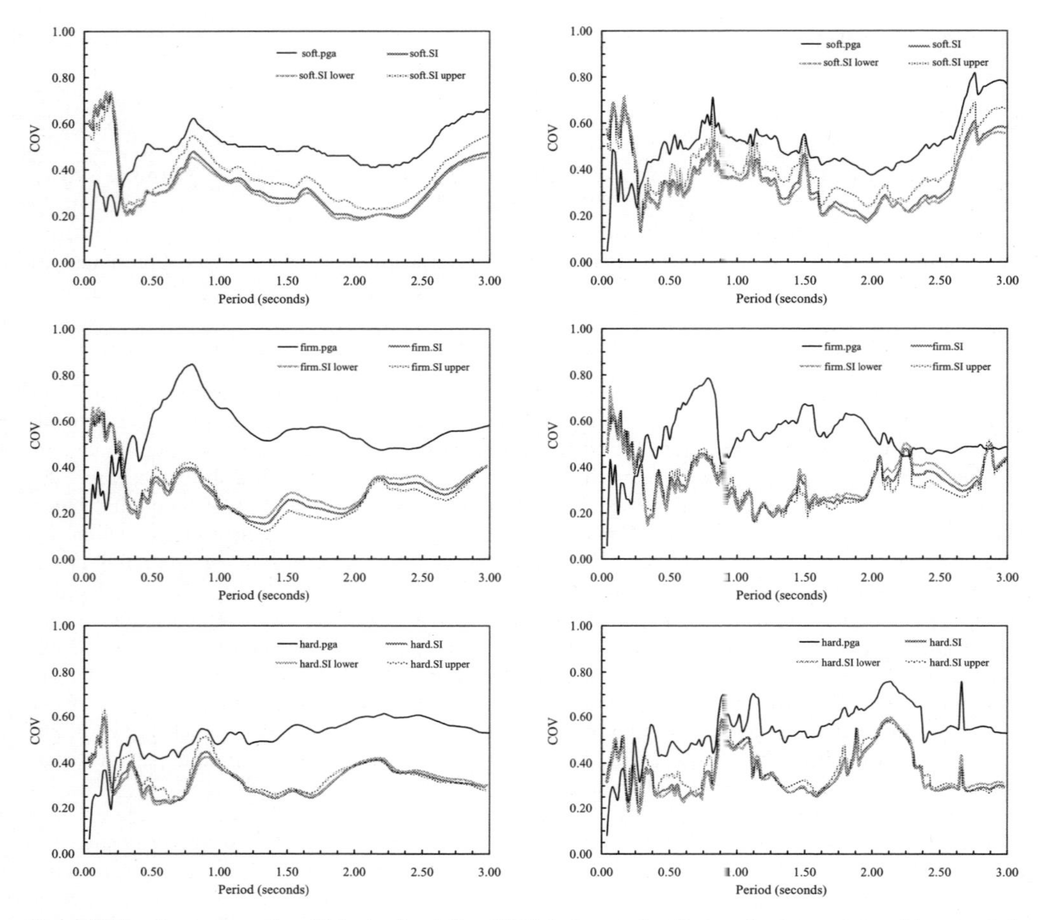

Figure 3.20 Comparison of coefficients of variation (COVs) of spectral ordinates of a strong-motion data set scaled by different spectrum scales and peak ground acceleration (PGA) for soft (*top*), firm (*middle*) and stiff soil condition (*bottom*): elastic (*left*) and inelastic (*right*) spectra (*courtesy* of Pacific Earthquake Engineering Research Center, USA)

The general observations from this study on scaling of earthquake records to minimize the coefficient of variation in structural response characteristics are summarized in Tables 3.23 to 3.25. The artificial records are signals derived to fit a target spectrum. This is only indicative since the results will vary when the target spectrum varies.

Based on the above study, it is recommended that Table 3.23 should be used to scale records in order to minimize the dispersion of results obtained from using a proposed set of natural records. Tables 3.24 and 3.25 should be consulted when the objective is to acquire a close match between results of analysis using spectrum-compatible and natural earthquake records.

Examination of Table 3.25 also provides physical interpretation of the results. Rock records are dominated by high-frequency waves. Hence, scaling them using the lower-period range, i.e. 0.1 to 1.5 seconds, provides a close match in the period range up to 1.0 second. The opposite is true for soft soil records that are best scaled using the upper-period range, i.e. 1.5 to 2.5 seconds. For scaling in the long-period range, e.g. 1 to 3 seconds, amplifications are low on average because of the rarity of records

Table 3.23 Coefficients of variation (COVs) for all records and scaling procedures in the elastic range.

Site condition	Period range (seconds)		
	0.01–0.3	0.3–1.0	1.0–3.0
Rock	0.25 (*acc.*)	0.30 (SI/SIL)	0.35 (SI/SIL/SIU)
Firm soil	0.30 (*acc.*)	0.30 (SIU)	0.25 (SIU)
Soft soil	0.30 (*acc.*)	0.35 (SI/SIL)	0.30 (SI/SIL)

Key: Average COV is approximate and serves only as an indication of the relative consistency of the scaling.

SI, SIL and SIU stand for scaling using the velocity spectrum for the ranges 0.1–2.5, 0.1–1.5 and 1.5–2.5 seconds. In the case of rock sites, the first period range is better represented by 0.01–0.2 seconds. For both rock and firm sites, the second period range is better represented by 0.2–1.1 and 0.3–1.2 seconds.

Table 3.24 Ratios of elastic spectral ordinates.

Site condition	Period range (seconds)		
	0.01–0.4	0.4–1.5	1.0–3.0
Rock	0.8 (*acc.*)	~1.0 (SI/SIL)	~1.0 (SI/SIL/SIU)
Firm soil	1.4 (SIU)	~1.0 (SIU)	~1.0 (SIU)
Soft soil	1.2 (SI)	~1.0 (SI)	~1.0 (SI/SIL)

Key: Ratio of spectral ordinates is the average (approximate) ratio between natural and artificial spectra in the given period range. A value greater than unity implies that the natural spectrum exceeds the artificial one.

Table 3.25 Ratios of inelastic spectral ordinates.

Site condition	Period range (seconds)	
	0.01–1.0	1.0–3.0
Rock	1.05 (SIL)	~1.0 (SIU)
Firm soil	1.1 (SIU)	~1.0 (SIU)
Soft soil	~1.0 (SI)	~1.0 (SIU)

Key: Ratio of spectral ordinates is the average (approximate) ratio between natural and artificial spectra in the given period range. A value greater than unity implies that the natural spectrum exceeds the artificial one.

with large amplifications in the long-period range. Therefore, they should be scaled using the long-period range to achieve parity with the artificial records that purport to represent a uniform hazard spectrum. Firm soil sites are somewhere in between rock and soft, hence, no clear trend is shown but fortuitously. For the firm and soft soil conditions, it was observed that the higher-ductility plots give a marginally closer match between artificial and natural inelastic spectra than the lower-ductility ones.

This was reversed in the case of rock records. However, the differences are not significant and therefore, are not worthy of further consideration.

In conclusion, where the periods of structures are not known *a priori*, use of Matsumura intensity for scaling is recommended. Housner's intensity scaling is effective for intermediate period ranges in its original form. Nau and Hall process provides improvement in the velocity-sensitive range of the response. The top and bottom ends of the spectrum may be used, as shown above, to improve the performance of the Housner's intensity scaling results. For short periods, scaling using acceleration is usually reliable.

Problem 3.8

Normalize the ground motions from the 1994 Northridge (Station 90056 Newhall – W. Pico Canyon Rd, component NORTHR/WPI046) and 1995 Kobe (Station Takarazuka, component KOBE/TAZ000) such that PGA = 0.5 g. The records can be downloaded from http://peer.berkeley.edu/smcat/. Compare the PGV and Housner spectral intensity of each ground motion using damping ratio $\xi = 0.05$. Comment on the results.

3.6 Duration and Number of Cycles of Earthquake Ground Motions

Earthquake strong motions are commonly characterized by the peak ground parameters and the response spectra illustrated in Section 3.4. However, the seismic behaviour of structural systems, especially those with stiffness and strength degradation is significantly affected by the duration and number of cycles of the ground motion (Jeong and Iwan, 1988). For ductile structures responding beyond their elastic limits, the magnitude of permanent deformations depends on how long the shaking is sustained.

Several analytical studies have demonstrated that the duration of the ground motion and the corresponding number of cycles are of importance for the assessment of low-cycle fatigue damage (e.g. Cosenza *et al.*, 1993; Mander *et al.*, 1995; Malhotra, 2002; Kunnath and Chai, 2004). For example, the onset of brittle rupture of longitudinal steel reinforcement bars in RC columns resulting from low-cycle fatigue depends on the number of inelastic load reversals. Similarly, the fracture and buckling of steel components and connections in moment-resisting frames and concentrically braced frames (e.g. the damage shown in Figures B.39 to B.42 of Appendix B) is influenced by ground-motion duration and loading history. It is, therefore, important to account for the effects of duration and number of cycles of earthquake records on the structure, especially when inelastic response history analyses are used. It is also important to ensure that the duration of shaking is consistent with the design scenario (Bommer and Acevedo, 2004). Existing seismic codes of practice do not provide guidelines on the selection of strong-motion records with adequate duration and number of cycles, probably due to the lack of generally accepted definitions of these parameters.

A number of definitions of strong-motion durations have been proposed by different researchers (e.g. Housner, 1965; Trifunac and Brady, 1975; Vanmarcke and Lai, 1980; Kawashima and Aizawa, 1989; Novikova and Trifunac, 1994, among others). Bommer and Martinez-Pereira (1999) reviewed some 30 different definitions based on earthquake records. Definitions of earthquake duration can be classified into three groups (Figure 3.21):

- *Bracketed duration*: defined as the total time D_b elapsed between the first and last excursions of a specified acceleration threshold a_0;
- *Uniform duration*: defined as the sum of the time intervals D_u during which the acceleration is greater than a given threshold a_0;
- *Significant duration*: defined as the time intervals D_s over which a portion of the total energy integral is accumulated.

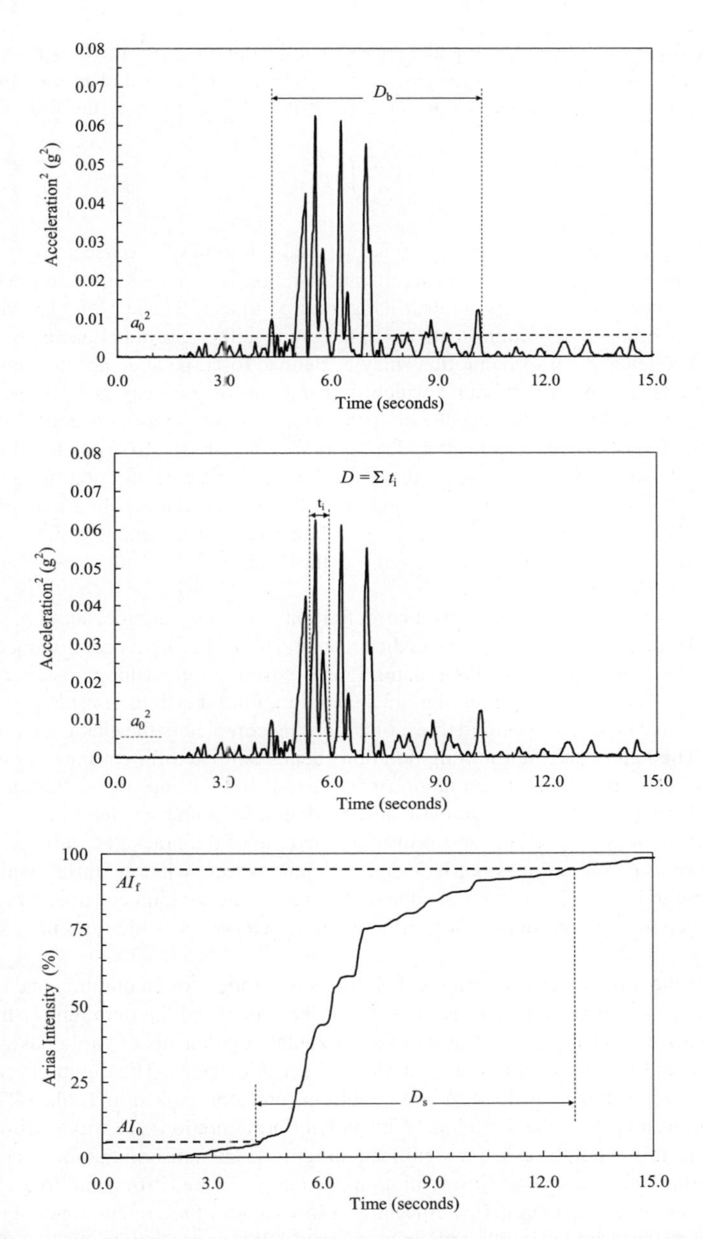

Figure 3.21 Definition of duration of earthquake strong motion: bracketed (*top*), uniform (*middle*) and significant (*bottom*) duration (*courtesy* of Pacific Earthquake Engineering Research Center, USA)

The accumulation of energy in earthquake records can be computed as the integral of the square of the ground acceleration, velocity or displacement. If the integral of the ground acceleration is employed then the quantity is related to the Arias intensity, *AI* (Arias, 1970), given by the following:

$$AI = \frac{\pi}{2g}\int_0^{t_r} a^2(t)\mathrm{d}t \tag{3.34}$$

where $a(t)$ is the acceleration time history, t_r is the total duration of the accelerogram and g is the acceleration due to gravity. Generally, the significant duration is assumed equal to the build-up of the Arias intensity between two arbitrary limits; this is referred to as a 'Husid plot' (Husid, 1969).

All the aforementioned definitions can be based upon either absolute or relative criteria. For example, the bracketed and uniform durations may be defined for a specified absolute level of threshold acceleration, or alternatively for a bound which is a fraction of the peak acceleration. By adopting absolute criteria, the values of the acceleration thresholds a_0 for bracketed and uniform durations can vary between 0.05 g and 0.10 g (Bolt, 1973). Trifunac and Brady (1975) and Dobry *et al.* (1978) assumed, as the absolute arbitrary limits in the definition of significant duration, the interval between the times at which 5% and 95% of the total integral of square of the acceleration is attained. Comprehensive studies by Somerville *et al.* (1997) have shown that more reliable estimates of earthquake significant durations can be derived by assuming the limits of 5% and 70% of the total Arias intensity.

It is unwise to rely upon a single, universal definition for strong-motion duration, since different definitions may be more or less appropriate in different situations. Furthermore, Bommer and Martinez-Pereira (1999) demonstrated that all three definitions of strong-motion durations as given above are flawed in some instances. For example, the bracketed duration takes into consideration only the first and the last peaks that cross the specified threshold, ignoring completely the characteristics of the strong shaking phase. The latter can result in long durations being estimated for earthquakes with small sub-events occurring after the main shock motion has passed. In addition, the definition can be rather unstable if low thresholds of acceleration are employed and, for some accelerograms, a change of the threshold, e.g. from 0.03 g to 0.02 g, can result in an increase of the bracketed duration by 20 seconds or more. The former definition does not include a continuous time window during which the shaking can be considered to be strong. On the other hand, the significant duration considers the characteristics of the entire accelerogram and defines a continuous time window in which the motion may be considered strong.

To investigate the differences between the definitions of strong-motion duration and their sensitivity to threshold values, Bommer and Martinez-Pereira (1999) calculated the duration for the set of accelerograms summarized in Table 3.26. The sample horizontal components of earthquakes included very strong, moderate and very weak motions, and multiple seismic events. The durations calculated using a few representative examples of the different definitions are summarized in Table 3.27.

The computed results show that both bracketed and uniform durations are very sensitive to acceleration thresholds a_0; this is not the case for significant durations. The uniform durations are characterized by low values for high thresholds. Strong-motion records generated from multiple ruptures exhibit unrealistic values of bracketed durations, especially if low values of a_0 are employed. For example, for the 23 November 1980 Irpinia (Italy) earthquake – a multiple seismic event – the bracketed durations are as high as 40–50 seconds if relatively low threshold values of accelerations, e.g. 0.05 g, are used and the second event is included in the strong shaking phase. When the threshold is raised from 0.05 g to 0.10 g, the duration is reduced by almost 35 seconds, i.e. about 70–80% lower. To identify the limits of the strong shaking sequence of the 1971 San Fernando earthquake (Table 3.26), relative criteria in the definition of durations should be used. The bracketed durations based on absolute thresholds of 0.05 g and 0.10 g result in values of zero for this earthquake ground motion, whereas non-zero durations are derived by adopting the definitions based on relative criteria.

Table 3.26 Sample earthquakes used for the evaluation of the strong-motion duration (*after* Bommer and Martinez-Pereira, 1999).

No.	Earthquake	Date	Station	Component	a_{max} (g)	*AI* (m/s)	t_r (s)
1	Imperial Valley	15 October 1979	Cerro Prieto	147°	0.166	1.145	64.0
2	Tabas	16 September 1978	Tabas	N16W	1.049	11.167	63.6
3	San Fernando	9 February 1971	Via Tejon PV	155°	0.042	0.034	70.4
4	Managua	23 December 1972	ESSO Refinery	E-W	0.368	1.613	45.9
5	Irpinia	23 November 1980	Sturno	E-W	0.319	1.497	60.48

Key: a_{max} = maximum ground acceleration; AI = Arias intensity; t_r = total duration

Table 3.27 Strong-motion durations for the sample earthquakes in Table 3.26 (in seconds).

Definition of duration	Duration of records				
	1	2	3	4	5
Bracketed (a_0 = 0.03 g)	47.6	46.6	1.9	20.8	50.0
Bracketed (a_0 = 0.05 g)	32.8	43.4	—	14.3	43.5
Bracketed (a_0 = 0.10 g)	19.2	42.6	—	8.9	9.1
Uniform (a_0 = 0.03 g)	15.3	24.8	0.1	10.0	12.2
Uniform (a_0 = 0.05 g)	8.3	18.4	—	6.5	7.2
Uniform (a_0 = 0.10 g)	1.6	10.2	—	2.9	2.6
Bracketed ($a_0 = 0.1a_{max}$)	61.7	42.5	63.2	17.9	49.9
Bracketed ($a_0 = 0.3a_{max}$)	32.8	38.7	32.8	8.9	9.2
Bracketed ($a_0 = 0.5a_{max}$)	21.6	3.7	4.7	4.5	7.9
Uniform ($a_0 = 0.1a_{max}$)	26.7	10.0	22.5	8.5	11.4
Uniform ($a_0 = 0.3a_{max}$)	8.4	1.7	2.9	2.5	2.8
Uniform ($a_0 = 0.5a_{max}$)	2.8	0.5	0.4	0.6	0.6
Significant (0–90% I_A)	29.0	20.7	51.8	10.3	20.7
Significant (5–95% I_A)	31.2	18.0	55.3	10.9	38.2

Comparisons between records with similar acceleration or energy indicate that significantly different effects on structural systems may occur. Whereas for equal accelerations, greater duration is generally more damaging, for equal energy, shorter duration presents a greater seismic hazard. An earthquake of short duration may not produce enough load reversals to damage a structure, even if the magnitude of the motion is high. A ground motion with moderate amplitude but long duration can, conversely, lead to substantial damage due to the resulting high number of load reversals.

From the seismological standpoint, the duration of strong ground motion is related to the time required for the release of accumulated strain energy by rupture along the fault. As the length, or the area, of fault rupture increases, the time required for rupture also increases. Consequently, the duration of ground shaking increases with earthquake magnitude. The duration of earthquakes of different magnitudes has been investigated by Chang and Krinitzsky (1977). Using 0.05 g threshold acceleration, they estimated the bracketed durations for soil and rock sites at short epicentral distances as given in Table 3.28.

Correlations between earthquake strong-motion duration and number of cycles have been calculated (Hancock and Bommer, 2005). The better correlation is with uniform duration since the latter effectively identifies the strong cycles of motion without including the rest of the record. To obtain close correlation, a threshold value of the acceleration a_0 equal to 0.10 g should be employed. The cycles-to-duration

Table 3.28 Typical earthquake bracketed durations at epicentral distances less than 10 km.

Magnitude	Duration (seconds)	
	Rock sites	Soil Sites
5.0	4	8
5.5	6	12
6.0	8	16
6.5	11	23
7.0	16	32
7.5	22	45
8.0	31	62
8.5	43	86

ratio decreases with the distance, as a result of the faster attenuation of high-frequency motion (reducing the number of cycles) and the separation of the cycles (increasing the duration) due to the different wave propagation velocities. Rock site motions produce higher ratios than soil sites. The durations on soil sites are marginally longer but due to the higher frequency of motions at rock sites, the number of cycles may be significantly higher than at soil sites. Additionally, the cycles-to-duration ratio decreases with magnitude, presumably the result of the greater proportion of long-period waves in ground motions produced by larger earthquakes. Long-period cycles, by definition, contribute more to the duration of the motion.

To account for the dependence of structural response on the number of load reversals, Kawashima and Aizawa (1989) developed acceleration response spectra taking account of the number of response cycles $\overline{S_a}(T,\xi,n)$. They were derived through a reduction factor $\eta(T,\xi,n)$ from conventional acceleration response spectra $S_a(T,\xi)$ as defined in Section 3.4.2:

$$\eta(T,\xi,n) = \frac{\overline{S_a}(T,\xi,n)}{S_a(T,\xi)} \tag{3.30.1}$$

with T being fundamental period of vibration, ξ the damping ratio and n number of cycles. The expression of $\eta(T,\xi,n)$, based on the analysis of 394 ground accelerations, is as follows:

$$\eta(T,\xi,n) = \frac{1}{1+\mathrm{e}(T,\xi)(n-1)} \tag{3.30.2}$$

where

$$\mathrm{e}(T,\xi) = \frac{80\,\xi}{60\,\xi+1} 0.0815\,T^{\,0.349} \tag{3.30.3}$$

If it is assumed that $\xi = 0.05$ and the number of cycles $n = 10$, then for a SDOF system with the fundamental period of vibration $T = 1.0$ second, the reduction factor η is approximately 0.2, i.e. the amplitude of response acceleration $\overline{S_a}(1.0, 0.05, 10)$ over which 10 acceleration reversals occur is only 20% of the peak acceleration $S_a(1.0,0.05)$.

Smooth cyclic demand spectra have been formulated by Malhotra (2002) by using the cumulative damage model for low-cycle fatigue of Coffin (1954) and Manson (1954). The results of the extensive parametric study showed that the number of cycles, and hence the cyclic demand, decreases as the damping ratios increase. Damping ratios ξ affect the cyclic demand spectra only for systems that are of intermediate stiffness.

3.7 Use of Earthquake Databases

Contrary to the situation in the 1970s and earlier, high-quality strong-motion data are freely available. Some sources of strong-motion data on the web are as follows:

(i) http://db.cosmos-eq.org: The web site of the Consortium of Organisations for Strong-Motion Observation Systems (COSMOS) allows access to a relational database of strong ground-motion parameters. Data are provided by the core members, i.e. US Geological Survey, California Geological Survey, US Army Corps of Engineers and US Bureau of Reclamation, for several earthquake-prone regions worldwide.
(ii) http://www.consrv.ca.gov/cgs/smip/: The California Strong-Motion Instrumentation Program (CSMIP) records the strong shaking, both free field and in structures during earthquakes for engineering use through a state-wide network of strong-motion instruments.
(iii) http://peer.berkeley.edu/smcat/: The Pacific Earthquake Engineering Research Center (PEER) database contains records of earthquakes publicly available from Federal, State, and private providers of strong-motion data. The database includes earthquake strong motions from several active regions worldwide.
(iv) http://www.isesd.cv.ic.ac.uk/esd: In this European Commission project site, acceleration time histories are archived as uncorrected and corrected records, together with the corresponding elastic response spectra. The acceleration time histories are from earthquakes in Europe and adjacent areas.
(v) http://www.k-net.bosai.go.jp: Kyoshin Net (K-NET) is a Japanese government project network, which avails of strong-motion data on the Internet. Such data are obtained from observatories deployed all over Japan.

There are many other sites, but the above provide a reasonable representation of the data freely available for downloading. Such records, if presented in their raw format, require baseline correction and filtering. Software for strong-motion processing is also abundant and is discussed below. Some of the above-mentioned earthquake databases provide response spectra online. For example, the PEER strong-motion site provides two types of search: earthquake or station characteristics and peak values can be selected or parameters for the response spectra specified as displayed in Figure 3.22. Moreover, earthquakes are archived on the basis of (i) source mechanism and distance, (ii) location and station, (iii) site classification, (iv) component, (v) date and time, and (vi) magnitude. Characteristics of low- and high-pass filters are also specified.

The results of the search for the 1995 Kobe earthquake are provided in Figure 3.23. Three components, i.e. two horizontal and one vertical, are available at 12 stations as shown in Figure 3.23. The source data were provided by the Conference on the Usage of Earthquakes (CUE), Railway Technical Research Institute of Tokyo. The characteristics of high-pass and low-pass filters are included. Peak ground values of acceleration (PGA), velocity (PGV) and displacement (PGD) are also provided. The acceleration, velocity and displacement time histories for the Kakogawa vertical component are provided in Figure 3.24, along with the 5% damping response spectra.

The ASCII files are also available for each earthquake components; thus, the strong motions can be used as input for structural assessment. Response spectra are also available in ASCII format for different values of damping.

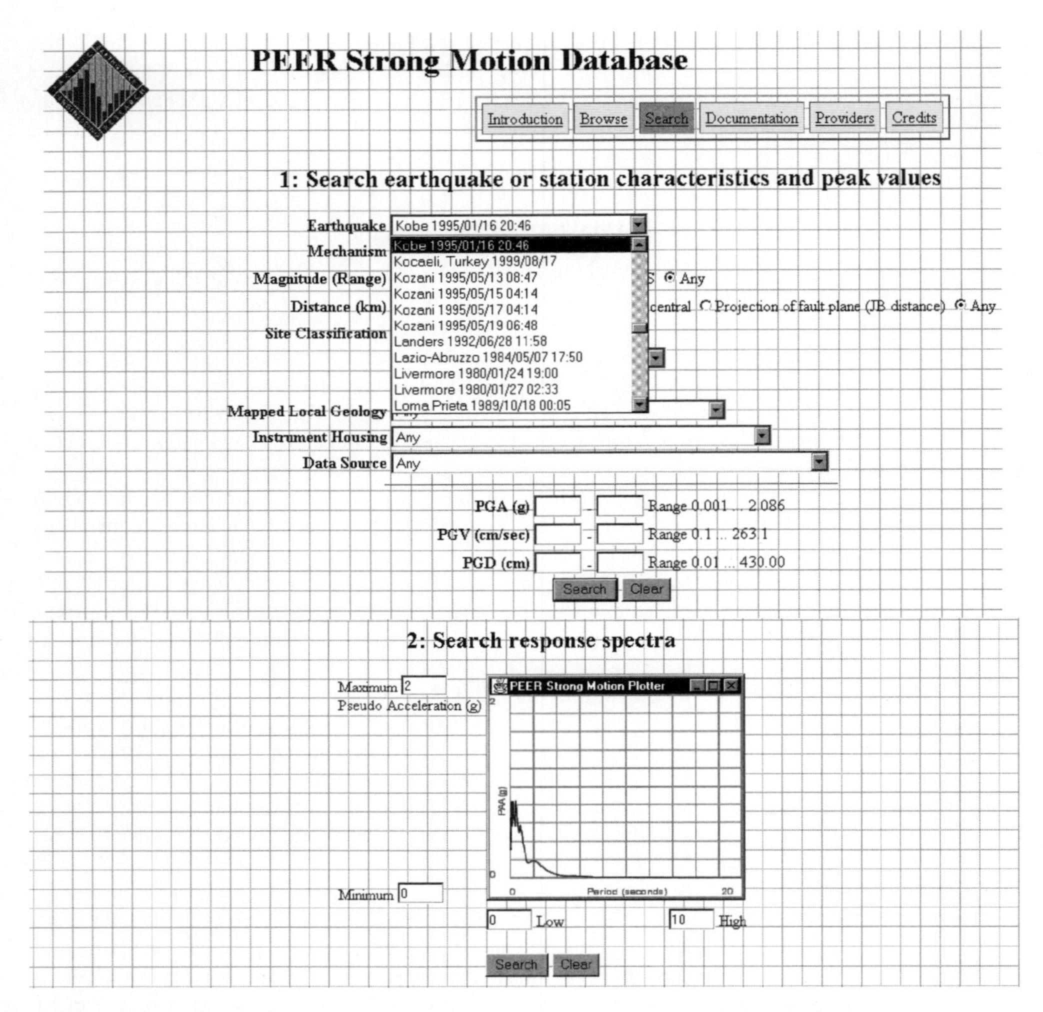

Figure 3.22 Search options from Pacific Earthquake Engineering Research Center (PEER) strong-motion database: selection of earthquake or station and peak values (*top*) or response spectra parameters (*bottom*) (*adapted from* Pacific Earthquake Engineeering Research Center, USA)

3.8 Software for Deriving Spectra and Generation of Ground-Motion Records

Several Windows®-based computer programs with user-friendly graphical interfaces are available to derive elastic and inelastic earthquake spectra. Similarly, artificial ground motion with spectra either matching or are compatible with a set of specified smooth response spectra can be generated with the aid of software packages (e.g. Gasparini and Vanmarcke, 1976). The use of artificially generated earthquakes can be useful when response history analyses are required and natural strong motions are scarce or non-existent. The software for deriving spectra and generation of artificial accelerograms is discussed hereafter.

PEER Strong Motion Database

Introduction | Browse | Search | Documentation | Providers | Credits

Query Results

Record ID	Earthquake	Station	Data Source	Record/Component	HP (Hz)	LP (Hz)	PGA (g)	PGV (cm/s)	PGD (cm)
P1038	Kobe 1995/01/16 20:46	0 FUK		KOBE/FUK-UP	0.05	null	0.01	1.7	0.67
P1038	Kobe 1995/01/16 20:46	0 FUK		KOBE/FUK000	0.05	null	0.034	4.3	1.28
P1038	Kobe 1995/01/16 20:46	0 FUK		KOBE/FUK090	0.05	null	0.042	5.3	2.08
P1040	Kobe 1995/01/16 20:46	0 HIK		KOBE/HIK-UP	0.05	null	0.039	3.3	0.92
P1040	Kobe 1995/01/16 20:46	0 HIK		KOBE/HIK000	0.05	null	0.141	15.6	3.08
P1040	Kobe 1995/01/16 20:46	0 HIK		KOBE/HIK090	0.05	null	0.148	15.4	1.96
P1041	Kobe 1995/01/16 20:46	0 Kakogawa	CUE	KOBE/KAK-UP	0.1	null	0.158	10.5	2.91
P1041	Kobe 1995/01/16 20:46	0 Kakogawa	CUE	KOBE/KAK000	0.1	null	0.251	18.7	5.83
P1041	Kobe 1995/01/16 20:46	0 Kakogawa	CUE	KOBE/KAK090	0.1	null	0.345	27.6	9.6
P1043	Kobe 1995/01/16 20:46	0 KJMA		KOBE/KJM-UP	0.05	null	0.343	38.3	10.29
P1043	Kobe 1995/01/16 20:46	0 KJMA		KOBE/KJM000	0.05	null	0.821	81.3	17.68
P1043	Kobe 1995/01/16 20:46	0 KJMA		KOBE/KJM090	0.05	null	0.599	74.3	19.95
P1045	Kobe 1995/01/16 20:46	0 MZH		KOBE/MZH-UP	0.05	null	0.041	2.5	2.0
P1045	Kobe 1995/01/16 20:46	0 MZH		KOBE/MZH000	0.05	null	0.07	4.4	1.54
P1045	Kobe 1995/01/16 20:46	0 MZH		KOBE/MZH090	0.05	null	0.052	4.7	1.87
P1046	Kobe 1995/01/16 20:46	0 Nishi-Akashi	CUE	KOBE/NIS-UP	0.1	23.0	0.371	17.3	5.63
P1046	Kobe 1995/01/16 20:46	0 Nishi-Akashi	CUE	KOBE/NIS000	0.1	23.0	0.509	37.3	9.52
P1046	Kobe 1995/01/16 20:46	0 Nishi-Akashi	CUE	KOBE/NIS090	0.1	23.0	0.503	36.6	11.26
P1047	Kobe 1995/01/16 20:46	0 OKA		KOBE/OKA-UP	0.05	null	0.038	2.5	1.68
P1047	Kobe 1995/01/16 20:46	0 OKA		KOBE/OKA000	0.05	null	0.081	4.8	2.12
P1047	Kobe 1995/01/16 20:46	0 OKA		KOBE/OKA090	0.05	null	0.059	3.2	1.62
P1048	Kobe 1995/01/16 20:46	0 OSAJ		KOBE/OSA-UP	0.05	null	0.064	7.5	3.73
P1048	Kobe 1995/01/16 20:46	0 OSAJ		KOBE/OSA000	0.05	null	0.079	18.3	9.26
P1048	Kobe 1995/01/16 20:46	0 OSAJ		KOBE/OSA090	0.05	null	0.064	17.0	8.03
P1054	Kobe 1995/01/16 20:46	0 Shin-Osaka	CUE	KOBE/SHI-UP	0.1	23.0	0.059	6.4	2.16
P1054	Kobe 1995/01/16 20:46	0 Shin-Osaka	CUE	KOBE/SHI000	0.1	23.0	0.243	37.8	8.54
P1054	Kobe 1995/01/16 20:46	0 Shin-Osaka	CUE	KOBE/SHI090	0.08	23.0	0.212	27.9	7.64
P1056	Kobe 1995/01/16 20:46	0 Takarazuka	CUE	KOBE/TAZ-UP	null	40.0	0.433	34.8	12.38
P1056	Kobe 1995/01/16 20:46	0 Takarazuka	CUE	KOBE/TAZ000	null	40.0	0.693	68.3	26.65
P1056	Kobe 1995/01/16 20:46	0 Takarazuka	CUE	KOBE/TAZ090	0.13	33.0	0.694	85.3	16.75
P1057	Kobe 1995/01/16 20:46	0 Takatori	CUE	KOBE/TAK-UP	0.2	null	0.272	16.0	4.47
P1057	Kobe 1995/01/16 20:46	0 Takatori	CUE	KOBE/TAK000	null	null	0.611	127.1	35.77
P1057	Kobe 1995/01/16 20:46	0 Takatori	CUE	KOBE/TAK090	null	null	0.616	120.7	32.72
P1058	Kobe 1995/01/16 20:46	0 TOT		KOBE/TOT-UP	0.05	null	0.015	1.3	0.8
P1058	Kobe 1995/01/16 20:46	0 TOT		KOBE/TOT000	0.05	null	0.076	10.9	3.71
P1058	Kobe 1995/01/16 20:46	0 TOT		KOBE/TOT090	0.05	null	0.075	7.6	4.58

Figure 3.23 Results of an earthquake search from the Pacific Earthquake Engineering Research Center (PEER) strong-motion database (*adapted from* Pacific Earthquake Engineeering Research Center, USA)

3.8.1 Derivation of Earthquake Spectra

The format of the computer programs used to derive earthquake spectra is based generally on a point-and-click interface that allows the user to navigate through the menus and to select analysis options. Response spectra are displayed on the screen and can be saved as ASCII files. Acceleration, velocity and displacements can be plotted versus either the periods or frequencies of the SDOF system. Moreover, plots may be copied to Windows® Clipboards and then used in other Windows® applications, such as Microsoft Excel® and Word®. Some of these computer programs can be downloaded freely

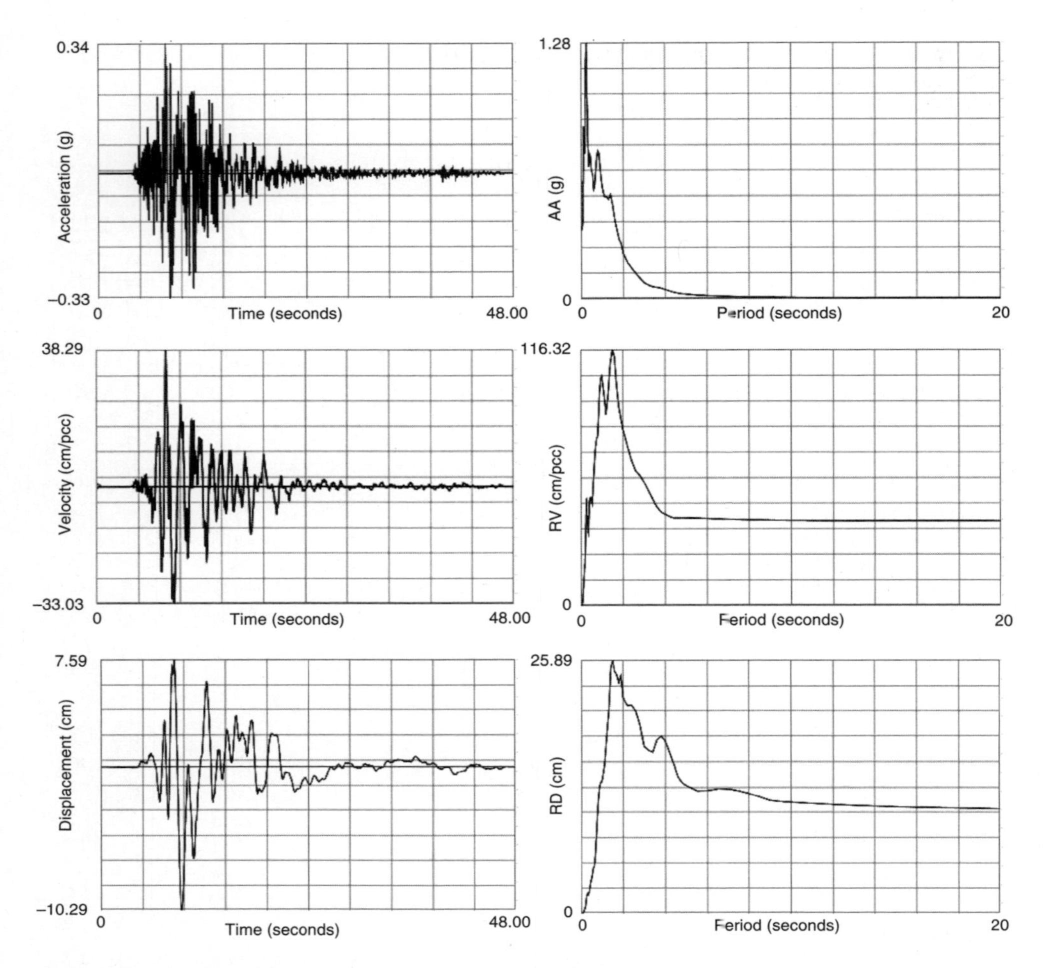

Figure 3.24 Time histories *(left)* and response spectra (*right*) computed automatically online from Pacific Earthquake Engineering Research Center (PEER) strong-motion database: acceleration (*top*), velocity (*middle*) and displacement (*bottom*) (*adapted from* Pacific Earthquake Engineeering Research Center, USA)

from Internet web sites such as the Utility Software for Earthquake Engineering (USEE) (available at http://mae.ce.uiuc.edu/software_and_tools/index.html) and Seismo-Signal (available at http://www.seismosoft.com/Downloads/SeismoSignal.htm). The former was developed at the Mid-America Earthquake Center in the University of Illinois at Urbana-Champaign (Inel *et al.*, 2001). The latter is the new release of the software used for strong-motion processing at Imperial College, London. In both programs, the user is guided through several data input screens.

The steps required to generate earthquake spectra can be summarized as follows:

(i) *Select the strong motion from a databank*: the input files should have standard extensions for USEE, i.e. '.mae', while there are no limitations for the format utilized by Seismo-Signal. Moreover, three types of base input accelerations are provided as default within the USEE library: recorded ground motions, synthetic motions and simple pulse waveforms. The user can define additional motions, which should, however, conform to the standard format described in the recorded ground-motion help topic. USEE recognizes the ground-motion data files only when they are located in the sub-directory with the other recorded ground motions. Strong-motion scaling factors should be specified along with their duration and time steps. Appropriate force-length units from a tab in the main menu should be selected for USEE.

(ii) *Select the load–deformation model and define the response spectra parameters*: three models are available in USEE, namely linear elastic, bilinear and stiffness-degrading systems. Seismo-Signal employs only two models: linear elastic and bilinear with hardening. The linear elastic model can be used when ground shaking intensities are relatively small and hence the SDOF remains elastic. In this case, the initial stiffness is sufficient to characterize the load–deformation curve. Generally, bilinear models are used for structures with stable and full hysteretic loops, for example, steel structures as discussed in Section 4.5.1. The parameters for the bilinear model are yield strength, initial stiffness and post-yield stiffness. The elastic-perfectly plastic model is a special case obtained using a zero post-yield stiffness. Similarly, the force–deformation curve of the stiffness-degrading model requires the specification of the yield strength, the initial stiffness and post-yield stiffness. Degrading models are generally used for concrete structures, which exhibit substantial degradation due to shear or bond deterioration, e.g. Section 4.5.2. Response spectra parameters can also be varied to include viscous damping, strength reduction factors and displacement ductility.

(iii) *View the computed earthquake response spectra*: The response spectra can be plotted as a function of the period, frequency or yield displacement. Spectral acceleration, velocity and displacements can be computed with Seismo-Signal. Peak displacements, ductility ratios, absolute accelerations as a percentage of the acceleration of gravity, and the ratios of the peak strength and the weight of the SDOF system are estimated with USEE. Five and six values of target ductility can be specified in USEE and Seismo-Signal, respectively; the maximum ductility ratio in Seismo-Signal is 10. The elastic spectra are plotted as default. Response data can be reported according to the user-specified output time step. A large number of steps are generally required to ensure accuracy of the solution.

It is instructive to note that the USEE computes the spectral response using the linear acceleration method or Newmark method (e.g. Chopra, 2002). Indeed, the parameters of the implicit integration scheme are set as $\alpha = 1/2$ and $\beta = 1/6$. Thus, the response acceleration varies linearly during the time step and the properties of the SDOF are assumed to be invariant. On the other hand, the integration parameters in Seismo-Signal can be selected by the user; the algorithm employed is still the Newmark method. To obtain unconditionally stable solutions, the parameters α and β of the implicit integration scheme should be assumed that $\alpha \geq 1/2$ and $\beta \geq 1/4\ (\alpha + 1/2)^2$ (Bathe, 1996).

USEE and Seismo-Signal possess additional utilities for the processing of ground motions. For example, Seismo-Signal can compute the fast Fourier transform for the accelerograms, the Arias intensity and the different types of duration (bracketed, uniform, significant and effective) presented in Section 3.6. In addition, Seismo-Signal includes a specific module to perform baseline correction and filtering. The former can be utilized to remove from the input motion spurious baseline trends, usually well noticeable in the displacement time history. Filtering can be employed to remove undesirable frequency components from a given record. In so doing, Seismo-Signal implements four filters: (i) low-pass filtering which suppresses frequencies that are higher than a user-defined cut-off frequency; (ii) high-pass filtering which allows frequencies that are higher than the cut-off frequency to pass through; (iii) band-pass filtering allows signals within a given frequency range bandwidth to pass through; and (iv) band-stop filtering suppresses signals within the given frequency range.

Online databases, such as those at the PEER Center or the European Strong-Motion Database presented in Section 3.7, provide actual strong-motion time histories and resulting spectra. The latter are generally computed in terms of acceleration, velocity and displacement for various damping values between 0.5% and 20% as a percentage of the critical.

Some general-purpose finite element programs for structural analysis possess internal routines to compute elastic and inelastic spectra. For example, they allow the user to evaluate elastic earthquake spectra by selecting the plot of the time history of the base fixed node of a cantilever beam to which the strong motion is applied. Spectra can thus be plotted in terms of acceleration, velocity and displacement for various values of damping.

Problem 3.9

Use Seismo-Signal or USEE to plot the acceleration, velocity and displacement response spectra for the 1999 Kocaeli (Turkey) earthquake (Station Ducze, horizontal component). Consider an elastic system with 5% viscous damping and an inelastic system with ductility of 4. For the latter system, assume two inelastic response models, namely linear elastic-perfectly plastic and linear elastic strain hardening. Use four values for the post-yield hardening of 1%, 5%, 10% and 20%. Comment on the results. For the inelastic spectra, use a viscous damping of 0.5%.

3.8.2 Generation of Ground-Motion Records

Statistically independent accelerograms can be simulated through SIMQKE-1 (Gasparini and Vanmarcke, 1976), which can be downloaded from the Internet at http://nisee.berkeley.edu/elibrary/Software/SIMQKE1ZIP. The source code is written in standard Fortran and the executable file can be obtained by using a standard compiler. This program generates earthquake strong motions whose response spectra either match or are compatible with a set of specified response spectra. Moreover, SIMQKE-1 performs baseline correlation on the generated motions to ensure zero final ground velocity. Response spectra are also calculated automatically. It is worth mentioning that the generation of the spectrum-compatible motion is based on the relationship between the response spectrum values, for a given damping, and the expected Fourier amplitudes of the earthquake motion. The latter is synthesized by superimposing sine and cosine components and pseudo-random phase angles. The stationary trace is then multiplied by a function representing the variation of ground-motion intensity with time; this function is user-specified. Furthermore, iterative adjustments of the spectral density ordinates may be required to improve the matching between the computed and the smooth target response spectra.

Ground-motion records can also be generated with the Simulink® of Matlab. This software has a user-friendly graphical interface along with online help and some examples for the processing of records.

Problem 3.10

Consider the design (5% damping) spectrum relative to a site of construction located, for example, in the western USA, expressed by equations (3.31.1) to (3.31.3). Use the software SIMQKE and Matlab (Simulink®) to determine spectrum-compatible earthquake records for the design spectrum.

$$0 \le T < T_1 \qquad S_a(T) = 0.40\, S_{a1} + 0.60 \frac{S_{a1}}{T_1} T \tag{3.31.1}$$

$$T_1 \le T < T_2 \qquad S_a(T) = S_{a2} \tag{3.31.2}$$

$$T \ge T_2 \qquad S_a(T) = \frac{S_{a2}}{T} \tag{3.31.3}$$

where T_1 = 0.166 second and T_2 = 0.830 second. The values of the spectral accelerations S_{a1} and S_{a2} are 1.207 g and 1.001 g, respectively.

References

Abrahamson, N.A. and Litehiser, J.J. (1989). Attenuation of vertical peak acceleration. *Bulletin of the Seismological Society of America*, 79(3), 549–580.

Afra, H. and Pecker, A. (2001). Stochastic characterization of seismic motion: Application to Eurocode 8 response spectra, *European Earthquake Engineering*, XV(2), 35–41.

Ambraseys, N.N. (1975). Trends in engineering seismology in Europe. *Proceedings of the 5th European Conference on Earthquake Engineering*, 3, 39–52.

Ambraseys, N.N. and Bommer, J.J. (1991). The attenuation of ground accelerations in Europe. *Earthquake Engineering and Structural Dynamics*, 20(12), 1179–1202.

Ambraseys, N.N. and Bommer, J.J. (1992). On the attenuation of ground accelerations in Europe. *Proceedings of the 10th World Conference on Earthquake Engineering*, 2, 675–678.

Ambraseys, N.N. and Bommer, J.J. (1995). Attenuation relationships for use in Europe: An overview. Proceedings of the 5th SECED Conference on European Seismic Design Practice, Balkema, Rotterdam, pp. 67–74.

Ambraseys, N.N. and Douglas, J. (2000). Reappraisal of the effect of vertical ground motions on response. Engineering Seismology and Earthquake Engineering, Research Report No. ESEE 00/4, Imperial College, London, UK.

Ambraseys, N.N. and Douglas, J. (2003). Near-field horizontal and vertical earthquake ground motions. *Soil Dynamics and Earthquake Engineering*, 23(1), 1–18.

Ambraseys, N.N. and Simpson, K.A. (1995). Prediction of vertical response spectra in Europe. Engineering Seismology and Earthquake Engineering, Research Report No. ESEE 95/1, Imperial College, London, UK.

Ambraseys, N.N. and Simpson, K.A. (1996). Prediction of vertical response spectra in Europe. *Earthquake Engineering and Structural Dynamics*, 25(4), 401–412.

Ambraseys, N.N., Simpson, K.A. and Bommer, J.J. (1996). Prediction of horizontal response spectra in Europe. *Earthquake Engineering and Structural Dynamics*, 25(4), 371–400.

Ambraseys, N.N., Douglas, J., Sarma, S.K. and Smit, P.M. (2005a). Equations for the estimation of strong ground motions from shallow crustal earthquakes using data from Europe and the Middle East: Horizontal peak ground acceleration and spectral acceleration. *Bulletin of Earthquake Engineering*, 3(1), 1–53.

Ambraseys, N.N., Douglas, J., Sarma, S.K. and Smit, P.M. (2005b). Equations for the estimation of strong ground motions from shallow crustal earthquakes using data from Europe and the Middle East: Vertical peak ground acceleration and spectral acceleration. *Bulletin of Earthquake Engineering*, 3(1), 55–73.

Aptikaev, F. and Kopnichev, J. (1980). Correlation between seismic vibration parameters and type of faulting. Proceedings of the 7th World Conference on Earthquake Engineering, Istanbul, Turkey, Vol. 1, pp. 107–110.

Arias, A. (1970). A measure of earthquake intensity. In *Seismic Design for Nuclear Power Plants*, R. Hansen, Ed., MIT Press, Cambridge, MA, USA, pp. 438–483.

Atkinson, G.M. and Boore, D.M. (1995). Ground motion relations for eastern North America. *Bulletin of the Seismological Society of America*, 85(1), 17–30.

Atkinson, G.M. and Boore, D.M., (1997). Some comparisons between recent ground-motion relations. *Seismological Research Letters*, 68(1), 24–40.

Atkinson, G.M. and Boore, D.M. (2006). Earthquake ground-motion prediction equations for eastern North America. *Bulletin of the Seismological Society of America*, 96(6), 2181–2205.

Bard, P.Y., Bouchon, M., Campillo, M. and Gariel, J.C. (1995). Numerical simulation of strong-motion using discrete wavenumber method: A review of main results. Recent Advances in Earthquake Engineering and Structural Dynamics. Proceedings of the Fifth AFPS/EERI Microzonation Conference, Nice, France.

Bathe, K.J. (1996). *Finite Element Procedures in Engineering Analysis*. Prentice Hall, Englewood Cliffs, NJ, USA.

Bolt, B.A. (1973). Duration of strong ground motions. *Proceedings of the 5th World Conference on Earthquake Engineering*, 1, 1304–1313.

Bolt, B.A. (1978). Fallacies in current ground motion prediction. Proceedings of the Second International Conference on Microzonation, San Francisco, CA, USA, Vol. 2, pp. 617–633.

Bolt, B.A. (1996). *From Earthquake Acceleration to Seismic Displacement: Fifth Mallet-Milne Lecture (Mallet-Milne lecture)*. John Wiley & Sons, Chichester, England.

Bolt, B.A. (1999). *Earthquakes*. 4th Edition, W.H. Freeman and Company, New York, NY, USA.

Bommer, J.J. and Acevedo, A.B. (2004). The use of real earthquake accelerograms as input to dynamic analysis. *Journal of Earthquake Engineering*, 8(1), 43–91.

Bommer, J.J. and Elnashai, A.S. (1999). Displacement spectra for seismic design. *Journal of Earthquake Engineering*, 3(4), 1–32.

Bommer, J.J. and Martinez-Pereira, A. (1999). The effective duration of earthquake strong-motion. *Journal of Earthquake Engineering*, 3(2), 127–172.

Bommer, J.J., Elnashai, A.S. and Weir, A.G. (2000). Compatible and acceleration displacement spectra for seismic design codes. *Proceedings of the Twelfth World Conference on Earthquake Engineering*, Auckland, New Zealand.

Boore, D.M. (1983). Stochastic simulation of high-frequency ground motions based on seismological models of the radiated spectra. *Bulletin of the Seismological Society of America*, 73(6, Part A), 1865–1894.

Boore, D.M. (2005). Erratum: Equations for estimating horizontal response spectra and peak acceleration from western North American earthquakes: A summary of recent work. *Seismological Research Letters*, 76(3), 368–369.

Boore, D.M. and Atkinson, G.M. (2007). Boore-Atkinson NGA ground motion relations for the geometric mean horizontal component of peak and spectral ground motion parameters. Pacific Earhquake Engineering Research Center, Report No. 2007/01, Berkeley, CA, USA.

Boore, D.M. and Joyner, W.B. (1982). The empirical prediction of ground motion. *Bulletin of the Seismological Society of America*, 72(6, Part B), S43–S60.

Boore, D.M., Joyner, W.B. and Fumal, T.E. (1993). Estimation of response spectra and peak accelerations from western North American earthquakes: An interim report. US Geological Survey Open-File Report, pp. 93–509.

Boore, D.M., Joyner, W.B. and Fumal, T.E. (1994). Estimation of response spectra and peak accelerations from western North American earthquakes: An interim report – Part 2. US Geological Survey Open-File Report, pp. 94–127.

Boore, D.M., Joyner, W.B. and Fumal, T.E. (1997). Equations for estimating horizontal response spectra and peak acceleration from western North American earthquakes: A summary of recent work. *Seismological Research Letters*, 68(1), 128–153.

Borzi, B. and Elnashai, A.S. (2000). Refined force reduction factors for seismic design. *Engineering Structures*, 22(10), 1244–1260.

Borzi, B., Elnashai, A.S., Faccioli, E., Calvi, G.M. and Bommer, J.J. (1998). Inelastic spectra and ductility-damping relationships for displacement-based seismic design. ESEE Research Report No. 98-4, Imperial College London, UK.

Borzi, B., Calvi, G.M., Elnashai, A.S., Faccioli, E. and Bommer, J.J. (2001). Inelastic spectra for displacement-based seismic design. *Journal of Soil Dynamics and Earthquake Engineering*, 21(1), 47–61.

Bozorgnia, Y. and Bertero, V.V. (2004). *Earthquake Engineering. From Engineering Seismology to Performance-Based Engineering*. CRC Press, Boca Raton, FL, USA.

Bozorgnia, Y., Niazi, M. and Campbell, K.W. (1994). Characteristics of the free-field vertical ground motion during the Northridge earthquake. *Earthquake Spectra*, 23(4), 515–525.

Bozorgnia, Y., Campbell, K.W. and Niazi, M. (2000). Observed spectral characteristics of vertical ground motion recorded during worldwide earthquakes from 1957 to 1995. Proceedings of the 12th World Conference on Earthquake Engineering, New Zealand, Paper No. 2671, CD-ROM.

Broderick, B.M. and Elnashai, A.S. (1996). Seismic response of composite frames, I: Response criteria and input motion. *Engineering Structures*, 18(9), 696–706.

Broderick, B.M., Elnashai, A.S., Ambraseys, N.N., Barr, J.M., Goodfellow, R.G. and Higazy, E.M. (1994). The Northridge (California) Earthquake of 17 January 1994: Observations, strong-motion and correlative response analysis. Engineering Seismology and Earthquake Engineering, Research Report No. ESEE 94/4, Imperial College, London, UK.

Building Standard Law of Japan (BSL). (2004). Building Research Institute, Tokyo, Japan.

Bycroft, G.N. (1960). White noise representation of earthquakes. *Journal of the Engineering Mechanics Division, ASCE*, 86(EM2), 1–16.

Campbell, K.W. (1985). Strong-motion attenuation relationships: A ten-year prospective. *Earthquake Spectra*, 1(4), 759–804.

Campbell, K.W. (1993). Empirical prediction of near-source ground motion from large earthquakes. Proceedings of the International Workshop on Earthquake Hazard and Large Dams in the Himalaya, New Delhi, India.

Campbell, K.W. (1997). Empirical near-source attenuation relationships for horizontal and vertical components of peak ground acceleration, peak ground velocity, and pseudo-absolute acceleration response spectra. *Seismological Research Letters*, 68(1), 154–179.

Campbell, K.W. and Bozorgnia, Y. (2003). Updated near-source ground-motion (attenuation) relations for the horizontal and vertical components of peak ground acceleration and acceleration response spectra. *Bulletin of the Seismological Society of America*, 93(1), 314–331.

Campbell, K.W. and Bozorgnia, Y. (2007). Campbell-Bozorgnia NGA ground motion relations for the geometric mean horizontal component of peak and spectral ground motion parameters. Pacific Earhquake Engineering Research Center, Report No. 2007/02, Berkeley, CA, USA.

Chandler, A.M. (1991). Evaluation of site-dependent spectra for earthquake-resistant design of structures in Europe and North America. *Proceedings of the Institution of Civil Engineers*, Part 1, 605–626.

Chang, F.K. and Krinitzsky, E.L. (1977). Duration, spectral content and predominant period of strong-motion earthquake records from western United States. Miscellaneous Paper 5-73-1, US Army Corps of Engineers Waterways Experiment Station, Vicksburg, MS, USA.

Chapman, M.C. (1999). On the use of elastic input energy for seismic hazard analysis. *Earthquake Spectra*, 15(4), 607–635.

Chopra, A.K. (2002). *Dynamics of Structures: Theory and Applications to Earthquake Engineering*. 2nd Edition, Prentice Hall College Division, Upper Saddle River, New Jersey, USA.

Chopra, A.K. and Goel, R.K. (1999). Capacity-demand-diagram methods based on inelastic design spectrum. *Earthquake Spectra*, 15(4), 637–656.

Clough, R.W. and Penzien, J. (1993). *Dynamics of Structures*. McGraw-Hill, New York, NY, USA.

JrCoffin, L.F.. (1954). A study of the effect of cyclic thermal stress in ductile metals. *Transactions of ASME*, 76, 931–950.

Collier, C.J. and Elnashai, A.S. (2001). A procedure for combining horizontal and vertical seismic action effects. *Journal of Earthquake Engineering*, 5(4), 521–539.

Coppersmith, K.S. and Youngs, R.R. (1990). Improved methods for seismic hazard analysis in the Western United States. *Proceedings of the 4th National Conference on Earthquake Engineering*, 1, 723–731.

Cornell, C.A. (1968). Engineering seismic risk analysis. *Bulletin of the Seismological Society of America*, 58(4), 1583–1606.

Cornell, C.A., Falayer, F., Hamburger, R.O. and Foutch, D.A. (2002). Probabilistic basis for 2000 SAC Federal Emergency Management Agency Steel Moment Frame Guidelines. *Journal of Structural Engineering, ASCE*, 128(4), 526–533.

Cosenza, E., Manfredi, G. and Ramasco, R. (1993). The use of damage functionals in earthquake engineering: A comparison between different methods. *Earthquake Engineering and Structural Dynamics*, 22(10), 855–868.

Crouse, C.B. (1991). Ground-motion attenuation equations for earthquakes on the Cascadia subduction zones. *Earthquake Spectra*, 7(2), 201–236.

Dahle, A., Bungum, H. and Kvamme, L. (1990). Attenuation models inferred from intraplate earthquake recordings. *Earthquake Engineering and Structural Dynamics*, 19(8), 1125–1141.

Deodatis, G. and Shinozouka, M. (1989). Simulation of seismic ground motion using stochastic waves. *Journal of Engineering Mechanics, ASCE*, 115(12), 2723–2737.

Dobry, R., Idriss, I.M. and Ng, E. (1978). Duration characteristics of horizontal components of strong-motion earthquake records. *Bulletin Seismological Society of America*, 68(5), 1487–1520.

Douglas, J. (2001). A comprehensive worldwide summary of strong-motion attenuation relationships for peak ground acceleration and spectral ordinates (1969 to 2000). *ESEE Research Report No. 01-1*, Department of Civil Engineering and Environmental Engineering, Imperial College, London, UK.

Douglas, J. (2002). Errata of and additions to ESEE Report No. 01-1 'A comprehensive worldwide summary of strong-motion attenuation relationships for peak ground acceleration and spectral ordinates (1969 to 2000)'. Department Research Report, Department of Civil Engineering and Environmental Engineering, Imperial College, London, UK.

Douglas, J. (2004). Ground motion estimation equations 1964 to 2003. Re-issue of ESEE Research Report No. 01-1 'A comprehensive worldwide summary of strong-motion attenuation relationships for peak ground acceleration and spectral ordinates (1969 to 2000)'. Technical Report No.04-001-SM. Department of Civil Engineering and Environmental Engineering, Imperial College, London, UK.

Douglas, J. (2006). Errata of and additions to 'Ground motion estimation equations 1964–2003'. ESEE Research Report No. 01-6, Department of Civil Engineering and Environmental Engineering, Imperial College, London, UK.

Elghadamsi, F.E. and Mohraz, B. (1987). Inelastic earthquake spectra. *Earthquake Engineering and Structural Dynamics*, 15(1), 91–104.

Elnashai, A.S. (1998). Use of real earthquake time-histories in analytical seismic assessment of structures. HM Nuclear Installations Directorate (NII) Report.

Elnashai, A.S. and Papazoglou, A.J. (1997). Procedure and spectra for analysis of RC structures subjected to strong vertical earthquake loads. *Journal of Earthquake Engineering*, 1(1), 121–155.

Elnashai, A.S., Bommer, J.J., Baron, I., Salama A.I. and Lee, D. (1995). Selected Engineering Seismology and Structural Engineering Studies of the Hyogo-ken Nanbu (Kobe, Japan) Earthquake of 17 January 1995. Engineering Seismology and Earthquake Engineering, Report No. ESEE/95-2, Imperial College, London, UK.

Esteva, L. and Villaverde, R. (1973). Seismic risk, design spectra and structural reliability. Proceedings of the 5th World Conference on Earthquake Engineering, Rome, Italy, Vol. 2, pp. 2586–2596.

Eurocode 8 (2004). Design provisions for earthquake resistance of structures. Part 1.3: General rules. Specific rules for various materials and elements. European Communities for Standardisation, Brussels, Belgium.

Fajfar, P. (1995). Elastic and inelastic spectra. *Proceedings of the 10th European Conference on Earthquake Engineering*, 2, 1169–1178.

Fajfar, P. (1998). Capacity spectrum method based on inelastic demand spectra. *Report EE-3/98*, IKPIR, Ljubljana, Slovenia.

Fernandez, J.A. (2007). Numerical Simulation of Earthquake Ground Motions in the Upper Mississippi Embayment. PhD Thesis, School of Civil and Environmental Engineering, Georgia Institute of Technology, Atlanta, GA, USA.

Fernandez, J.A. and Rix, G.J. (2006). Soil attenuation relationships and seismic hazard analyses in the Upper Mississippi Embayment. Proceedings of the 8th National Conference on Earthquake Engineering 8NCEE, San Francisco, CA, CD-ROM.

Frankel, A., Mueller, C., Barnhard, T., Perkins D., Leyendecker, E.V., Dickman, N., Hanson, S. and Hopper, M. (1996). National Seismic Hazard Maps: Documentation. OFR 96-532, US Geological Survey.

Free, M.W. (1996) The Attenuation of Earthquake Strong-Motion in Intraplate Regions. PhD Thesis, Imperial College, London, UK.

Freeman, S.A. (1998). The capacity spectrum method as a tool for seismic design. Proceedings of the 11th European Conference on Earthquake Engineering, Paris, France, CD-ROM.

Frohlich, C. and Apperson, K.D. (1992). Earthquake focal mechanisms, moment tensors, and the consistency of seismic activity near plate boundaries. *Tectonics*, 11(2), 279–296.

Fukushima, Y., Gariel, J.C. and Tanaka, R. (1995). Site-dependent attenuation relations of seismic motion parameters at depth using borehole data. *Bulletin of the Seismological Society of America*, 85(6), 1790–1804.

Gariel, J.C., Archuleta, R.J. and Bouchon, M. (1990). Rupture process of an earthquake with kilometric size fault inferred from near source records. *Bulletin of the Seismological Society of America*, 8(2), 870–888.

Gasparini, D.A. and Vanmarcke, E.H. (1976). Simulated earthquake motions compatible with prescribed response spectra. Department of Civil Engineering, Research Report R76-4, Massachusetts Institute of Technology, Cambridge, MA, USA.

Gutenberg, B. and Richter, C.F. (1954). *The Seismicity of the Earth.* Princeton University Press, Princeton, NJ, USA.

Hancock, J. and Bommer, J.J. (2005). The effective number of cycles of earthquake ground motion. *Earthquake Engineering and Structural Dynamics*, 34(6), 637–664.

Housner, G.W. (1952). Spectrum intensities of strong-motion earthquakes. Proceedings of Symposium on Earthquake and Blast Effects on Structures, EERI.

Housner, G.W. (1956). Limit design of structures to resist earthquakes. Proceedings of the 1st World Conference on Earthquake Engineering, 5-1 to 5-13, Berkeley, CA, USA.

Housner, G.W. (1959). Behavior of structures during earthquakes. *Journal of Engineering Mechanics Division, ASCE*, 85(EM-4), 109–129.

Housner, G.W. (1965). Intensity of ground motion shaking near the causative fault. Proceedings of the 3rd World Conference on Earthquake Engineering, Vol. 1, Auckland, New Zealand, pp. 81–94.

Hughes, T.J.R. (1987). *The Finite Element Method.* Prentice Hall, Englewood Cliffs, NJ, USA.

Husid, L.R. (1969). Characteristicas de terremotos. Analis general. *Revista del IDIEM, Santiago*, 8, 21–42.

Kalkan, E. and Gulkan, P. (2004). Empirical attenuation equations for vertical ground motion in Turkey. *Earthquake Spectra*, 20(3), 853–882.

Kamiyama, M. (1995). An attenuation model for the peak values of strong ground motions with emphasis on local soil effects. *Proceedings of the 1st International Conference on Earthquake Geotechnical Engineering*, 1, 579–585.

Kanai, K. (1983). *Engineering Seismology*. University of Tokyo Press, Tokyo, Japan.

Kawashima, K. and Aizawa, K. (1989). Bracketed and normalized durations of earthquake ground accelerations. *Earthquake Engineering and Structural Dynamics*, 18, 1041–1051.

Kawashima, K., Aizawa, K. and Takahashi, K. (1986). Attenuation of peak ground acceleration, velocity and displacement based on multiple regression analysis of Japanese strong-motion records. *Earthquake Engineering and Structural Dynamics*, 14(2), 199–215.

Kramer, S.L. (1996). *Geotechnical Earthquake Engineering*. Prentice Hall, Upper Saddle River, NJ, USA.

Krawinkler, H. and Nassar, A.A. (1992). Seismic design based on ductility and cumulative damage demand and capacities. In *Nonlinear Seismic Analysis and Design of Reinforced Concrete Buildings*, P. Fajfar and H. Krawinkler, Eds., Elsevier Applied Science, New York, NY, USA.

Kunnath, S.K. and Chai, Y.H. (2004). Cumulative damage-based inelastic cyclic demand spectrum. *Earthquake Engineering and Structural Dynamics*, 33(4), 499–520.

Idriss, I.M. (1978). Characteristics of earthquake ground motions. *Proceedings of the Geotechnical Engineering Division Specialty Conference: Earthquake Engineering and Soil Dynamics, ASCE*, III, 1151–1265.

Iervolino, I. and Cornell, C.A. (2005). Record selection for nonlinear seismic analysis of structures. *Earthquake Spectra*, 21(3), 685–713.

Inel, M., Bretz, E.M., Black, E.F. Ascheim, M.A. and Abrams, D.P. (2001). USEE 2001: Utility Software for Earthquake Engineering. Report and User's Manual. Mid America Earthquake Center Report, University of Illinois at Urbana-Champaign, IL, USA.

Iwasaki, T., Kawashima, K. and Saeki, M. (1980). Effects of seismic and geotechnical conditions on maximum ground accelerations and response spectra. *Proceedings of the 7th World Conference on Earthquake Engineering*, 2, 813–190.

Jennings, P.C. (1985). Ground motion parameters that influence structural damage. In *Strong-motion Simulation and Earthquake Engineering Applications – A Technological Assessment*, R.E. Scholl and J.L. King, Eds., Earthquake Engineering Research Institute, Publication No. 85-02.

Jeong, G.D. and Iwan, W.D. (1988). The effect of earthquake duration on the damage of structures. *Earthquake Engineering and Structural Dynamics*, 16, 1201–1211.

Johnson, R.A. (1973). An earthquake spectrum prediction technique. *Bulletin of the Seismological Society of America*, 63(4), 1255–1274.

Johnston, A.C., Coppersmith, K.J., Kanter, L.R. and Cornell, C.A. (1994). The earthquakes of stable continental regions: Volume 1 – Assessment of large earthquake potential. EPRI Report TR-10 2261-VI, Electric Power Research Institute, Palo Alto, CA, USA.

Joyner, W.B. and Boore, D.M. (1981). Peak horizontal acceleration and velocity from strong-motion records including records from the 1979 Imperial Valley, California earthquake. *Bulletin of the Seismological Society of America*, 71(6), 1479–1480.

Joyner, W.B. and Boore, D.M. (1988). Measurement, characterization, and prediction of strong-motion. Proceedings of Earthquake Engineering and Soil Dynamics II, Geotechnical Division, ASCE, pp. 43–102.

Joyner, W.B. and Boore, D.M. (1996). Recent developments in strong-motion attenuation relationships. Proceedings of the 28th Joint Meeting of the US-Japan Cooperative Program in Natural Resource Panel on Wind and Seismic Effects, pp. 101–116.

Lee, W.H.K., Kanamori, H., Jennings, P.C. and Kisslinger, C. (2003). *International Handbook of Earthquake and Engineering Seismology*. Academic Press, San Diego, CA, USA.

Mahin, S.A. and Bertero, V.V. (1981). An evaluation of inelastic seismic design spectra. *Journal of Structural Engineering, ASCE*, 107(9), 1777–1795.

Malhotra, P.K. (2002). Cyclic-demand spectrum. *Earthquake Engineering and Structural Dynamics*, 31(7), 1441–1457.

Mander, J., Peckan, G. and Chen, S. (1995). Low-cycle variable amplitude fatigue modelling of top-and-seat angle connections. *Engineering Journal, AISC*, 32(2), 54–63.

Manson, S.S. (1954). Behavior of materials under conditions of thermal stress. NACA TN 2933.

Martinez-Rueda, J.E. (1997). Energy Dissipation Devices for Seismic Upgrading of RC Structures. PhD Thesis, University of London, UK.

Matsumura, K. (1992). On the intensity measure of strong-motions related to structural failures. *Proceedings of the 10th World Conference on Earthquake Engineering*, 1, 375–380.

McGarr, A. (1984). Scaling of ground motion parameters, state of stress and focal depth. *Journal of Geophysical Research*, 89(B8), 6969–6979.

McGuire, R.K. (1978). Seismic ground motion parameter relations. *Journal of the Geotechnical Engineering Division, ASCE*, 104(GT4), 481–490.

Miranda, E. and Bertero, V.V. (1994). Evaluation of strength reduction factor for earthquake-resistance design. *Earthquake Spectra*, 10(2), 357–379.

Mohammadioun, G. (1994). Calculation of site-adapted reference spectra from the statistical analysis of an extensive strong-motion databank. Proceedings of the 10th European Conference on Earthquake Engineering, Vienna, Austria, Vol. 1, 177–181.

Naeim, F., Alimoradi, A. and Pezeshk, S. (2004). Selection and scaling of ground motion time histories for structural design using genetic algorithms. *Earthquake Spectra*, 20(2), 413–426.

Nau, J.M. and Hall, W.J. (1984). Scaling methods for earthquake response spectra. *Journal of Structural Engineering, ASCE*, 110(7), 1533–1548.

Newmark, N.M. and Hall, W.J. (1969). Seismic design criteria for nuclear reactor facilities. Proceedings of the 4th World Conference on Earthquake Engineering, Santiago, Chile, B-4, pp. 37–50.

Newmark, N.M. and Hall, W.J. (1982). Earthquake Spectra and Design, EERI Monograph Series, EERI, Oakland, CA, USA.

Newmark, N.M., Blume, J.A. and Kapur, K.K. (1973). Seismic design spectra for nuclear power plants. *Journal of Power Division*, 99(PO2), 287–303.

Novikova, E.I. and Trifunac, M.D. (1994). Duration of ground motion in terms of earthquake magnitude, epicentral distance, site conditions and site geometry. *Earthquake Engineering and Structural Dynamics*, 23(6), 1023–1043.

Nuclear Regulatory Commission (NRC) (2001). Technical Basis for Revision of Regulatory Guidance on Design Ground Motions: Hazard- and Risk-Consistent Ground Motion Spectra Guidelines. NUREG/CR-6728, Government Printing Office, Washington, DC, USA.

Papazoglou, A. and Elnashai, A.S. (1996). Analytical and field evidence of the damaging effect of vertical earthquake ground motion. *Earthquake Engineering and Structural Dynamics*, 25(10), 1109–1137.

Power, M., Chiou, B., Abrahamson, N. and Roblee, C. (2006). The Next Generation of Ground Motion Attenuation Models (NGA) project: An overview. Proceedings of the 8th National Conference on Earthquake Engineering 8NCEE, San Francisco, CA, USA, Paper No. 2022, CD-ROM.

Reiter, L. (1990). *Earthquake Hazard Analysis – Issues and Insights*. Columbia University Press, New York, NY, USA.

Rinaldis, D., Berardi, R., Theodulidis, N. and Margaris, B. (1998). Empirical predictive models based on a joint Italian and Greek strong-motion database: I. Peak ground acceleration and velocity. Proceedings of the 11th European Conference on Earthquake Engineering, Balkema, Rotterdam.

Saatcioglu, M. and Ozcebe, G. (1989). Response of reinforced concrete columns to simulated seismic loading. *ACI Structural Journal*, 86(1), 3–12.

Sadek, F., Mohraz, B. and Riley, M.A. (2000). Linear procedures for structures with velocity-dependent dampers. *Journal of Structural Engineering, ASCE*, 128(8), 887–895.

Sarma, S.K. and Srbulov, M. (1996). A simplified method for prediction of kinematic soil-foundation interaction effects on peak horizontal acceleration of a rigid foundation. *Earthquake Engineering and Structural Dynamics*, 25(8), 815–836.

Sarma, S.K. and Srbulov, M. (1998). A uniform estimation of some basic ground motion parameters. *Journal of Earthquake Engineering*, 2(2), 267–287.

Shinozouka, M. and Deodatis, G. (1988). Stochastic process models for earthquake ground motion. *Journal of Probabilistic Engineering Mechanics*, 3(3), 499–519.

Silva, W., Gregor, N. and Darragh, R. (2003). Development of Regional Hard Rock Attenuation Relations for Central and Eastern North America, Mid-Continent and Gulf Coast Areas. Pacific Engineering and Analysis, El Cerrito, CA, USA.

Skarlatoudis, A., Theodulidis, N., Papaioannou, C. and Roumelioti, Z. (2004). The dependence of peak horizontal acceleration on magnitude and distance for small magnitude earthquakes in Greece. Proceedings of Thirteenth World Conference on Earthquake Engineering, Paper No. 1857, CD-ROM.

Somerville, P.G., Smith, N.F., Graves, R.W. and Abrahamson, N.A. (1997). Modification of empirical strong-motion attenuation relations to include the amplitude and duration effects of rupture directivity. *Seismological Research Letters*, 68(1), 199–222.

Spudich, P., Fletcher, J.B., Hellweg, M., Boatwright, J., Sullivan, C., Joyner, W.B., Hanks, T.C., Boore, D.M., McGarr, A., Baker, L.M. and Lindh, A.G. (1997). SEA96 – A new predictive relation for earthquake ground motions in extensional tectonic regimes. *Seismological Research Letters*, 68(1), 190–198.

Spudich, P., Joyner, W.B., Lindh, A.G., Boore, D.M., Margaris, B.M. and Fletcher, J.B. (1999). SEA99 – A revised ground motion prediction relation for use in extensional tectonic regimes. *Bulletin of the Seismological Society of America*, 89(5), 1156–1170.

Stamatovska, S. (2002). A new azimuth dependent empirical strong-motion model for Vranchea subduction zone. Proceedings of the Twelfth European Conference on Earthquake Engineering. Paper No. 324, CD-ROM.

Takahashi, T., Kobayashi, S., Fukushima, Y., Zhao, J.X., Nakamura, H. and Somerville, P.G. (2000). A spectral attenuation model for Japan using strong-motion data base. Proceedings of the 6th International Conference on Seismic Zonation: Managing Earthquake Risk in the 21st Century, Earthquake Engineering Research Institute, Oakland, CA, USA.

Tolis, S.V. and Faccioli, E. (1999). Displacement design spectra. *Journal of Earthquake Engineering*, 3(1), 107–125.

Tong, H. and Katayama, T. (1988) Peak acceleration attenuation by eliminating the ill-effect of the correlation between magnitude and epicentral distance. *Proceedings of the 9th World Conference on Earthquake Engineering*, II, 349–354.

Trifunac, M.D. and Brady, A.G. (1975). A study of the duration of strong earthquake ground motion. *Bulletin Seismological Society of America*, 65(3), 581–626.

Trifunac, M.D. and Brady, A.G. (1976). Correlations of peak, velocity and displacement with earthquake magnitude, distance and site conditions. *Earthquake Engineering and Structural Dynamics*, 4(5), 455–471.

Tso, W.K., Zhu, T.J. and Heidebrecht, A.C. (1992). Engineering applications of ground motion a/v ratio. *Soil Dynamics and Earthquake Engineering*, 11(3), 133–144.

Uang, C.M. (1991). Establishing R (R_w) and C_d factors for buildings seismic provisions. *Journal of Structural Engineering, ASCE*, 117(1), 19–28.

Vanmarcke, E.H. and Lai, S.S. (1980). Strong-motion duration and RMS amplitude of earthquake records. *Bulletin Seismological Society of America*, 70(4), 1293–1307.

Vidic, T., Fajfar, P. and Fischinger, M. (1994). Consistent inelastic design spectra: Strength and displacement. *Earthquake Engineering and Structural Dynamics*, 23(5), 507–521.

Wesnousky, S., Scholz, C., Shimazaki, K. and Masuda, T. (1984). Integration of geological and seismological data for the analysis of seismic hazard: A case study of Japan. *Bulletin of the Seismological Society of America*, 74(2), 687–708.

Whittaker, A., Hart, G. and Rojahn, C. (1999). Seismic response modification factors. *Journal of Structural Engineering, ASCE*, 125(4), 438–444.

Youngs, R.R., Day, S.M. and Stevens, J.L. (1988). Near field ground motions on rock for large subduction earthquakes. Proceedings of Earthquake Engineering and Soil Dynamics II, Geotechnical Division, ASCE, pp. 445–462.

Youngs, R.R., Chiou, S.J., Silva, W.J. and Humphrey, J.R. (1997). Strong ground motion attenuation relationships for subduction zone earthquakes. *Seismological Research Letters*, 68(1), 58–73.

Zhu, T.J., Heidebrecht, A.C. and Tso, W.K. (1988). Effect of peak ground acceleration to velocity ratio on the ductility demand of inelastic systems. *Earthquake Engineering and Structural Dynamics*, 16(1), 63–79.

4

Response Evaluation

4.1 General

The objective from this chapter is to provide sufficient information for the effective modelling of reinforced concrete (RC) and steel structures, for the purposes of assessment and design under earthquake motion. Towards this end, the chapter starts with a conceptual framework that relates the various components required for effective inelastic analysis, namely models, input and method. The nature of commonly applied loads is explained, and load combinations are presented. This is followed by a detailed account of modelling issues on the material, section, member, connection and system levels, following the same hierarchical treatment used in Chapter 2. The backbone of the present chapter is an example of a three-dimensional structure employed to demonstrate many important issues that are described in the text. An RC building frame is also used for the project that serves as the capstone for the material presented in all four chapters. This closing chapter is the second of the two 'capacity' chapters, and is developed to complement the two 'demand' chapters, thus completing the set of tools that the writers intend to provide to graduate students and young practising engineers interested in earthquake structural engineering.

4.2 Conceptual Framework

Few issues of structural response to static loads remain unresolved. Therefore, further sophistication of procedures for static analyses and design is difficult to justify on the basis of either safety or economy. On the other hand, seismic loading and response of structures are far from being sufficiently understood. The analysis process, which leads to the evaluation of seismic actions and deformations, invokes knowledge from several sub-disciplines in engineering, such as engineering seismology, geotechnical engineering, structural analysis and computational mechanics. Therefore, to perform reliable seismic analyses of structural systems, a conceptual framework, such as that shown in Figure 4.1, is required. This framework includes the essential components summarized below:

- Ground motion and load modelling;
- Structural modelling;
- Foundation and soil modelling;
- Method of analysis;

Fundamentals of Earthquake Engineering Amr S. Elnashai and Luigi Di Sarno

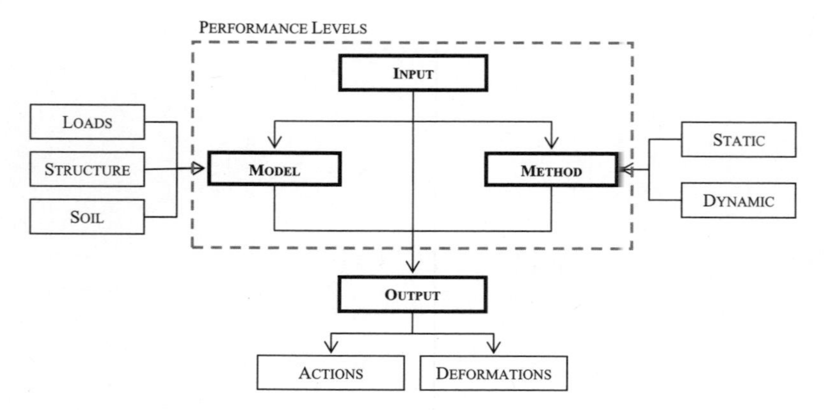

Figure 4.1 Conceptual framework for seismic analysis of structures

- Performance levels;
- Output for assessment.

The above components are discussed in subsequent sections, while noting that soil and foundation models are beyond the scope of this book. For the latter features, the reader may refer to the literature, (e.g. Wolf, 1994) for fundamentals, and to numerous applications (e.g. Monti *et al.*, 1996; Stewart *et al.*, 1999; Mylonakis *et al.*, 2001; Zhang and Makris, 2002; Sextos *et al.*, 2003a; Kwon and Elnashai, 2006; Elnashai and Kwon, 2007). As mentioned above, the topics discussed in this chapter are illustrated by an example application consisting of a three-storey irregular RC frame shown in Figure 4.2. This structure was comprehensively assessed within the European network on Seismic Performance Assessment and Rehabilitation (SPEAR). It features irregularities both in plan and elevation. It has heavily imbalanced stiffness in two orthogonal directions (x and y in Figure 4.2) as well as large eccentricity in plan and irregularity in elevation. The RC frame was designed according to modern design codes and with no seismic design provisions. It was constructed from weak concrete and smooth bars. Further details are given in Negro *et al.* (2004) and Jeong and Elnashai (2005).

4.3 Ground Motion and Load Modelling

Modelling of input quantities, such as ground motions and gravity loads, is a critical step in the earthquake response analysis of structures. Approaches used to model ground motions in structural assessment are visually summarized in Figure 3.1 and the various input motion forms are described in detail in Chapter 3. To perform dynamic analysis either response spectra or time histories of earthquake ground motion may be employed. For equivalent static analysis, as depicted in seismic codes, design spectra are utilized to estimate lateral forces as discussed in Section 4.6.2.1. Spectral representations are also used for advanced inelastic static analyses, e.g. adaptive pushover analysis presented in Section 4.6.2.2. Table 4.1 summarizes commonly used methods of analysis alongside the appropriate input representations and range of structural applications.

A number of time histories are specified in seismic design standards for dynamic analysis, usually three to seven records in each principal direction of structural response (e.g. Bazzurro and Cornell, 1994; Dymiotis *et al.*, 1999). For long-span structures, such as major bridges, it may be necessary to employ asynchronous earthquake ground motions at the base supports to account for the spatial and temporal variability of the input (e.g. Burdette and Elnashai, 2008; Burdette *et al.*, 2008). Asynchronous

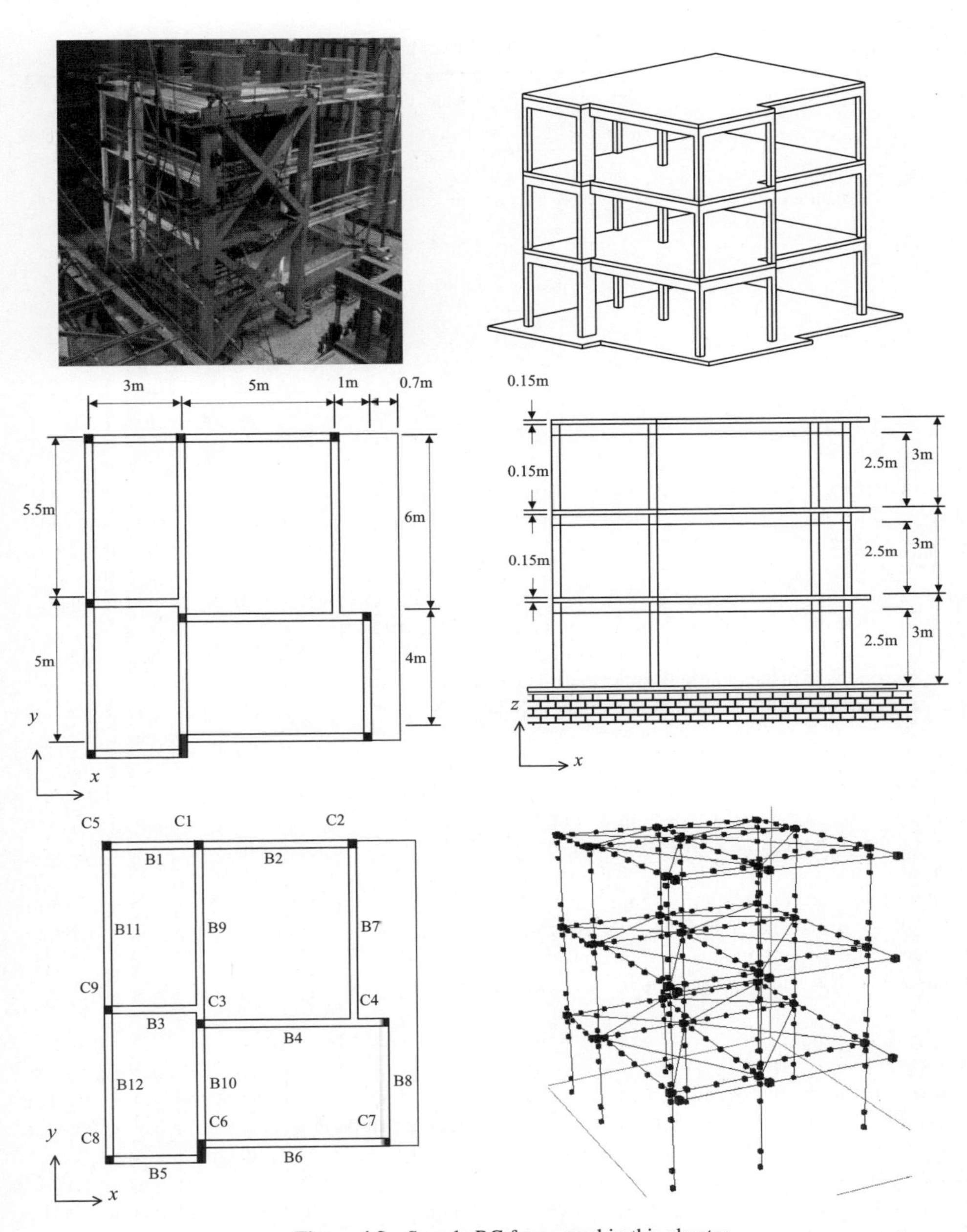

Figure 4.2 Sample RC frame used in this chapter

or non-coherent response analyses may be undertaken in the frequency or time domains. Further details on frequency-domain analysis are available in Clough and Penzien (1993) and Zerva (2008); meanwhile many researchers have developed analytical approaches for asynchronous analysis using response history representation (e.g. Price and Eberhard, 1998; Tzanetos *et al.*, 2000; Sextos *et al.*, 2003b).

Table 4.1 Typical methods of analyses and relative earthquake input representations.

Method	Analysis type	Reference (*Section*)	Representation	Application
Dynamic	Multi-modal spectral	4.6.1.1	Spectrum	Irregular structures
	Response history	4.6.1.2	Time history	Irregular, highly inelastic and important structures
	Incremental dynamic	4.6.1.3	Time history	Irregular, highly inelastic and important structures
Static	Equivalent static	4.6.2.1	Fixed	Regular and ordinary structures
	Conventional pushover	4.6.2.2	Fixed	Regular and important structures
	Adaptive pushover	4.6.2.2	Spectrum	Irregular and important structures

Several types of loads may be applied to a structure during its lifetime. These include primarily dead and live actions. Dead loads may be modelled reliably through Gaussian distributions and exhibit low coefficients of variation. Live loads exhibit higher variability and their statistical representation depends significantly on the type of live load considered. In addition, when an earthquake hits a structure, it is unlikely that all live loads would be at their respective maximum value. Therefore, load combination models are an important part of the definition of actions on structures.

Dead loads considered in static and dynamic analyses are due to the own weight of the structure as well as partitions, finishes and any other permanent fixtures. Live loads, on the other hand, are non-permanent and represent the use and occupancy. Seismic codes provide characteristic values of design loads.

Lateral loads, such as wind and earthquakes, occur only occasionally. Seismic loads are generated by the mass of the structure when accelerated by earthquake ground motion. As such, these loads are a function of the characteristics of both earthquake and structure. When calculating seismic loads, the weight of the structure does not correspond to the full dead and live load, since this would be over-conservative in view of the low probability of an earthquake occurring while the structure is at maximum live load. Furthermore, some live loads may not be rigidly fixed to the supporting system and do not necessarily move in phase with the rest of the structure. Therefore, it is appropriate to define percentages of dead and live loads when considering the tributary seismic weight W_{EQ} and corresponding mass M_{EQ}. The latter approach is implemented in seismic codes as follows:

$$W_{EQ} = p_1 \, \mathrm{DL} + p_2 \, \mathrm{LL} \tag{4.1}$$

where W_{EQ} is the seismic weight, p_1 and p_2 are percentages of the dead and live loads (DL and LL), respectively. It is often recommended by codes to assume p_1 as unity and p_2 as varying between about 0.15 and 0.3. Seismic codes may also recommend the use of different p_2 values for the roof level in buildings or ignore some types of live loads. The tributary seismic weight at each storey of the analytical model of the RC frame in Figure 4.2 is calculated as the sum of the total dead loads due to the self-weight of the structure and 30% of live loads on the slab; i.e. p_1 and p_2 in equation (4.1) are equal to 1.0 and 0.3, respectively. For multi-storey buildings, the evaluation of the term p_1DL in equation (4.1) should include the weight of the floor system, finishes, partitions, beams and columns one-half storey above and below a floor for fixed-base rigid foundations. In the sample SPEAR structure shown in Figure 4.2, the value of p_1 DL accounts for the self-weight of the structure (0.5 kN/m^2), estimated from the weight per unit volume of RC and finishes.

The mass M_{EQ} accelerated by earthquake motion is calculated by dividing W_{EQ}, in equation (4.1), by the gravitational acceleration g. Once the mass M_{EQ} is defined, gravity and seismic loads should be combined as described in Section 4.4.

4.4 Seismic Load Combinations

Dead, live and earthquake loads should be combined to perform response analysis of structural systems utilizing the methods presented in Section 4.6. Loads acting on structures during earthquakes are generally combined as follows:

$$L = \gamma_D \, DL + \gamma_L \, LL + \gamma_E \, EQ \tag{4.2}$$

where L is the total load, γ are load factors for dead loads DL (subscript D), live loads LL (subscript L) and for earthquakes EQ (subscript E). Different notations are used in different codes. Usually more than one load combination is used in analysis and design. The values of γ-factors and the number of different combinations depend on the limit states employed to assess structural performance. The symbol '+' in equation (4.2) means 'to be combined with' because the most adverse condition should be considered in the analysis, i.e. equation (4.2) is an algebraic expression for L.

Load factors γ in equation (4.2) account for the variability in the values of load; lower factors are used for loads that are unlikely to vary significantly from the specified characteristic value. In particular, γ_L factors are generally 20–30% greater than γ_D because LLs exhibit higher uncertainty than DLs (Nowak and Collins, 2000).

Combination coefficients Ψ_L are often used to account for the likelihood of certain live loads not being present over the whole structure during the occurrence of the earthquake. Equation (4.2) may be thus modified as follows:

$$L = \gamma_D \, DL + \Psi_L \, \gamma_L \, LL + \gamma_E \, EQ \tag{4.3}$$

where values of Ψ_L are less than unity. These Ψ_L-factors may also account for a reduced participation of masses in the motion due to the non-rigid connection between a structure and its contents, as mentioned in Section 4.3. Each code has its own system of factors, which are either computed by stochastic analyses or based on historical values derived by successful application experience. Expressions for Ψ_L-factors are given, for example, in ISO (1998) and JCSS (2001), for different probability distributions. Values of Ψ_L for residential buildings may vary between 0.20 and 0.70. Mixing of factors from different codes is therefore not allowed.

The design spectrum (e.g. from Section 3.4.5) may be multiplied by the importance factor γ_I, which is used both for equivalent static and dynamic (multi-modal spectral) analyses. Equation (4.3) is consequently modified as follows, to include γ_I:

$$L = \gamma_D \, DL + \Psi_L \, \gamma_L \, LL + \gamma_I \, \gamma_E \, EQ \tag{4.4}$$

The rationale for multiplying the seismic loads EQ by γ_I is that the return period of the design earthquake for critical structures is longer than normal-use structures. Alternatively, earthquake spectra for different return periods may be used in the analysis. Caution should be exercised to avoid duplicating and accumulating safety features.

Commonly used load combinations for building structures are summarized in Table 4.2. Earthquakes should be combined with snow loads only if the latter are severe. A portion of the snow load is included in seismic combinations because a significant amount of ice can build up on roofs. Seismic codes may

Table 4.2 Typical load combinations for building structures.

Load type	Load combinations	
	General case	Snow district
Permanent	$\gamma_D \cdot DL + \Psi_L \cdot \gamma_L \cdot LL$	$\gamma_D \cdot DL + \Psi_L \cdot \gamma_L \cdot LL + \Psi_S \cdot \gamma_S \cdot SL_1$
Transient	$\gamma_D \cdot DL + \Psi_L \cdot \gamma_L \cdot LL + \Psi_S \cdot \gamma_S \cdot SL_2$ $\gamma_D \cdot DL + \Psi_L \cdot \gamma_L \cdot LL + \Psi_W \cdot \gamma_W \cdot WL_1$ $\gamma_D \cdot DL + \Psi_L \cdot \gamma_L \cdot LL + \gamma_I \cdot \gamma_E \cdot EQ$	$\gamma_D \cdot DL + \Psi_L \cdot \gamma_L \cdot LL + \Psi_S \cdot \gamma_S \cdot SL_2$ $\gamma_D \cdot DL + \Psi_L \cdot \gamma_L \cdot LL + + \Psi_S \cdot \gamma_S \cdot SL_3 + \Psi_W \cdot \gamma_W \cdot WL_2$ $\gamma_D \cdot DL + \Psi_L \cdot \gamma_L \cdot LL + + \Psi_S \cdot \gamma_S \cdot SL_3 + \gamma_I \cdot \gamma_E \cdot EQ$

Key: SL_1, SL_2 and SL_3 = heavy snow load; WL_1 = strong wind load; WL_2 = wind load.

also require load combinations to prevent specific failure modes, e.g. global overturning. In such cases, the most conservative estimates should be considered for the load factors γ and Ψ_L.

A load combination such as equation (4.4) is employed to calculate the design value of the load L for the RC frame in Figure 4.2. In particular, it is assumed that the building is residential, which implies that for live loads LL, the factor $\Psi_L \, \gamma_L$ is 0.3, while $\gamma_I \, \gamma_L$ is 1.0 for seismic loads EQ.

Earthquake effects in two orthogonal horizontal directions should be combined. The following relationship is applicable for structures analysed using equivalent static methods or multi-modal spectral analysis:

$$(EQ)_i = \alpha_i \, EQ_T + \beta_i \, EQ_L \tag{4.5}$$

where EQ_T and EQ_L are the transverse and longitudinal earthquake actions, respectively, i is a counter for the number of combinations to be considered. The factors α_i and β_i account for the probability that both EQ_T and EQ_L are simultaneously acting on the structure with their maximum intensity. It is noteworthy that the orthogonal components of ground motion are not necessarily in phase. It is commonly assumed that i is 2, and that α_i and β_i are 1.00 and 0.30, respectively. Therefore, equation (4.5) becomes:

$$(EQ)_1 = 1.00 \, EQ_T + 0.30 \, EQ_L \tag{4.6.1}$$

$$(EQ)_2 = 0.30 \, EQ_T + 1.00 \, EQ_L \tag{4.6.2}$$

It is often required, as mentioned in Sections 3.4.6 and 3.4.7, to estimate the effects of the vertical component of earthquake ground motion. The latter should be combined with two orthogonal horizontal components: EQ_T and EQ_L. Seismic codes specify the proportion of each earthquake component to be used in load combinations for equivalent static methods or multi-modal spectral analyses. They are given by:

$$(EQ)_i = \alpha_i \, EQ_T + \beta_i \, EQ_L + \lambda_i \, EQ_v \tag{4.7}$$

where EQ_v is vertical earthquakes and λ_i has the same definition and values of α_i and β_i given above. Thus, the different seismic load cases can be assumed as given below:

$$(EQ)_1 = 1.00 \, EQ_T + 0.30 \, EQ_L + 0.30 \, EQ_V \tag{4.8.1}$$

$$(EQ)_2 = 0.30 \, EQ_T + 1.00 \, EQ_L + 0.30 \, EQ_V \tag{4.8.2}$$

$$(EQ)_3 = 0.30\ EQ_T + 0.30\ EQ_L + 1.00\ EQ_V \tag{4.8.3}$$

A more detailed procedure for combining transverse, longitudinal and vertical components of ground motion has been proposed by Collier and Elnashai (2001); the procedure takes into account the distance from the source and recommends that within a few kilometres from the source, peak vertical and horizontal actions should be directly combined without scaling.

For structures analysed in the time domain, the response may be calculated using natural earthquake input motions applied in two or three directions simultaneously.

4.5 Structural Modelling

Structural models are idealizations of the prototype and are intended to simulate the response characteristics of systems discussed in Section 2.3. Three levels of modelling are generally used for earthquake response analysis (Figure 4.3). These are summarized below in ascending order of complexity and accuracy:

i. *Substitute* (or *equivalent SDOF*) *models*: the structure is idealized as an equivalent single-degree of freedom (SDOF) system or 'substitute system'. Four parameters are needed to define the substitute system: effective mass M_{eff}, effective height H_{eff}, effective stiffness k_{eff} and effective damping ξ_{eff}. The height H_{eff} defines the location of the equivalent or effective mass M_{eff} of the substitute system. The equivalence used to estimate k_{eff} and ξ_{eff} assumes that the displacement of the original structure is the same as that of the substitute model. This approach is frequently used for inelastic structures (Gulkan and Sozen, 1974; Shibata and Sozen, 1976; Kowalski

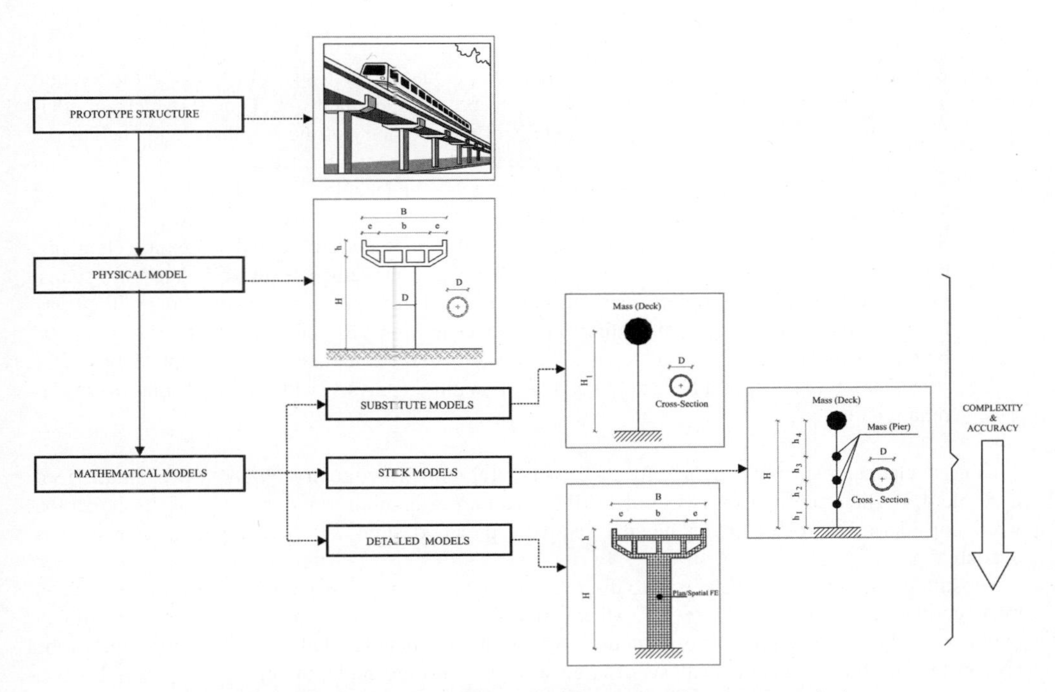

Figure 4.3 Levels of structural modelling for earthquake response analysis

et al., 1995; Priestley, 2003). For inelastic systems, the effective stiffness k_{eff} may be assumed as the secant stiffness at some given displacement, while ξ_{eff}, which is utilized to quantify the energy dissipation, is assumed as the equivalent viscous damping. The value of the secant stiffness is estimated using the definition illustrated in Section 2.3.1. The effective or equivalent damping ξ_{eff} is computed from equation (2.23) in Chapter 2. Relationships between damping ξ_{eff} and translational ductility μ_Δ for SDOFs depend significantly on the hysteretic action–deformation characteristics, e.g. elastic-perfectly plastic, hardening and softening behaviour (Section 3.4.4). It can be utilized for spectral and response history dynamic analyses described in Sections 4.6.1.1 and 4.6.1.2. Substitute models are inadequate to assess local response of structures, although they are effective for global analyses.

ii. *Stick models*: these consist of multi-degree of freedom (MDOF) systems in which each element idealizes a number of members of the prototype structure. In multi-storey building frames, each storey is modelled by a single line of finite elements (FE) representing the deformational characteristics of all columns and their interaction with beams. For three-dimensional models, the stick element relates the shear forces along two horizontal orthogonal directions and the storey torque to the corresponding inter-storey translations and rotations, respectively. The lateral stiffness of each equivalent stick element is the stiffness of the frame comprising columns connected to beams. For dynamic analysis, the mass of each floor is concentrated at the nodes representing the centroid of the slab. Lumping both mass and stiffness at a limited number of nodes and pairs of nodes leads to a significant reduction in the size of the problem to be solved. This issue is further discussed in Section 4.5.4. Distributed masses are seldom employed for stick models. They are used, for example, to simulate the response of structural walls. Shear beam elements are also utilized as stick elements for multi-storey frames employing members where shear deformation cannot be ignored. Stick models are suitable for sensitivity analyses to assess the effects of various design parameters, such as beam-to-column strength ratio and the degree of irregularity along the height. Conversely, they cannot be used to evaluate the distribution of ductility demands and damage among the individual structural members.

iii. *Detailed models*: these include general FE idealizations in which structures are discretized into a large number of elements with section analysis or spatial elements in 2D or 3D. Such a modelling approach allows representation of details of the geometry of the members, and enables the description of the history of stresses and strains at fibres along the length or across the section dimensions. Provided that the problem size remains manageable, detailed models also provide global response quantities and the relationships between local and global response. In the detailed modelling approach, beams and columns of frames are represented by flexural elements, braces by truss elements, and shear and core walls by 2D elements, such as plates and shells. For accurate evaluation of deformations and member forces, three-dimensional modelling may be required. Its use is essential to study stress concentrations, local damage patterns or interface behaviour between different materials. However, spatial FE models are often cumbersome for large structures, especially when inelastic dynamic analysis with large displacements is required.

Generic characteristics of the three levels of structural modelling mentioned above are summarized in Table 4.3. Their comparison is useful for the selection of an appropriate method of discretization while considering the objective of the analysis, the accuracy desired and the computational resources available.

Substitute and detailed models used to discretize structural systems may be described as macro- and micro-models as shown in Figure 4.4. Stick models constitute an intermediate group and employ member-level representations. Hybrid models, e.g. combining detailed and stick elements, can also be used especially for the seismic analysis of large structures. For example, the upper deck of multi-span bridges, which is expected to remain elastic, is often discretized using beam elements, while fine FE

Table 4.3 Comparisons between different levels of structural models (*relative measure*).

Model type	Discretization type	Tridimensional effects	Structure prototype	Analysis target	Complexity/ Accuracy	Computational demand
Substitute	SDOF	Usually not accommodated	Primarily regular structures	Global response	Low	Low
Stick	MDOF	Accommodated	All types of structure	Global response	Medium	Medium
Detailed	MDOF	Accommodated	All types of structure	Local and Global response	High	High

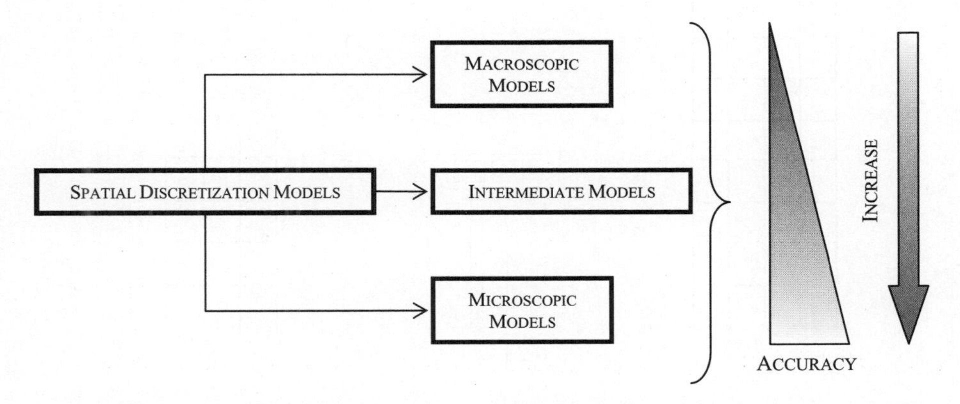

Figure 4.4 Classification of models for spatial discretization of structural systems

meshes are utilized for the piers, where inelasticity is expected. For buildings, detailed models are often used to idealize the frame of the superstructure, while stick models are used for foundations. Where walls and cores exist, there are possibilities of modelling them using 2D or even 3D continuum elements to detect the spread of inelasticity.

Problem 4.1

Consider the medium-rise building and the suspension bridge shown in Figure 4.5. Sketch suitable models for the analysis of the two structures. What type of analysis should be used and why? If vertical and horizontal components of ground motion should be accounted for, how would they be combined?

Problem 4.2

Sketch suitable models for the three structural systems shown in Figure 4.6 and justify the selection of models.

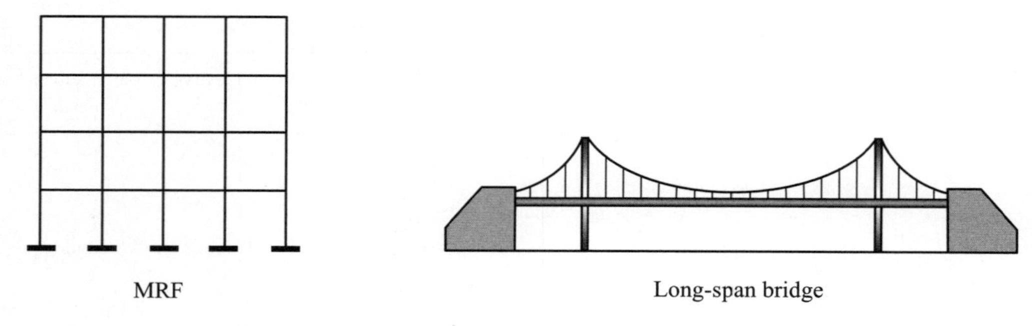

Figure 4.5 Multi-storey building (*left*) and suspension bridge (*right*)

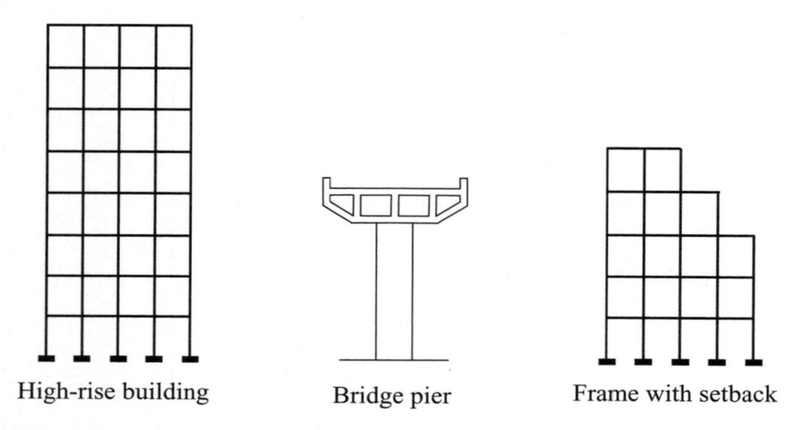

Figure 4.6 Regular multi-storey building (*left*), bridge pier (*middle*) and irregular building (*right*)

4.5.1 Materials

(i) Metals

Under monotonic loading, metals are modelled by simple uniaxial constitutive relationships in FE software packages. Linear elastic models (LEMs) can be used for structural systems which are not expected to experience inelastic deformations. These LEMs may be thus utilized for elastic static and dynamic analyses. In such cases, two model parameters are specified: Young's modulus E and Poisson's ratio ν. For mild steel, it may be assumed that $E = 205{,}000$–$210{,}000\,\text{MPa}$ and $\nu = 0.30$–0.35. The shear modulus G is derived from the relationships of linear elastic continua, i.e. $G = E/[2 \cdot (1 + \nu)]$. On the other hand, uniaxial elastic-plastic models (EPMs) may be employed to perform inelastic analysis of structures. These material models are based on the theory of plastic flow. Therefore pre- and post-yield constitutive models are required along with a yield criterion. The most common EPMs employed in inelastic analysis are as follows:

- Linear elastic-perfectly plastic (LEPP);
- Linear elastic-plastic with strain hardening (LESH);
- Linear elastic-plastic with non-linear hardening (LENLH);
- Power laws: e.g. Ramberg–Osgood (RO) and Menegotto–Pinto (MP) models.

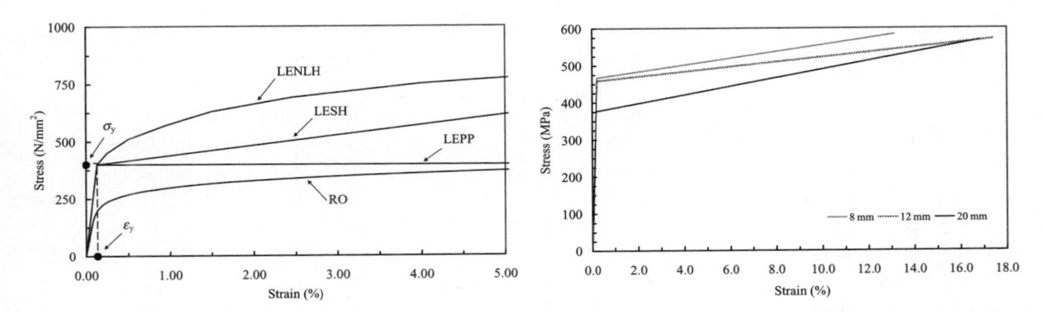

Figure 4.7 Uniaxial models for stress–strain relationships: comparison between different models (*left*) and linear elastic strain-hardening model used in the assessment of the sample SPEAR frame for different bar diameters (*right*) *Key*: LENLH = linear elastic-plastic with non-linear hardening; LESH = linear elastic-plastic with strain hardening; LEPP = linear elastic-perfectly plastic; RO = Ramberg-Osgood

Table 4.4 Comparison between uniaxial models.

Properties	Model types			
	Linear elastic-perfectly plastic	Linear elastic-plastic strain hardening	Linear elastic-plastic non-linear hardening	Power Law (Ramberg–Osgood model)
Constitutive equations*	$\sigma = E\varepsilon$ $\sigma = E(\varepsilon - \lambda)$	$\sigma = E\varepsilon$ $\sigma = \sigma_y + E_t\left(\varepsilon - \frac{\sigma_y}{E}\right)$	$\sigma = E\varepsilon$ $\sigma = k\varepsilon^n$	$\varepsilon = \frac{\sigma}{E} + a\left(\frac{\sigma}{b}\right)^n$
Yield criterion	$\sigma = \sigma_y$	$\sigma = \sigma_y$	$\sigma = \sigma_y$	$\sigma = \sigma_{y,0.2}$
Hardening type	Null	Linear	Non-linear	Non-linear
Pros and Cons	Easy to implement No spreading of plasticity Not suitable for stress-controlled inelastic analysis Suitable for mild steel	Easy to implement Spreading of plasticity Suitable for stress-controlled inelastic analysis Suitable for mild steel	Easy to implement Gradual spreading of plasticity Experimental data for calibration Suitable for mild steel	Implementation more time-consuming Suitable for high inelasticity (alloy metals) Experimental data for calibration

Key: * = the first equation holds for $\sigma \leq \sigma_y$ and the second for $\sigma > \sigma_y$; λ = non-zero scalar; *a*, *b* and *n* are material constants that can be computed through laboratory tests.

Comparisons between the above models are provided in Figure 4.7; the pros and cons of employing each model are summarized in Table 4.4. Bilinear models are simple to implement and computationally efficient. Nevertheless, they grossly misrepresent the plastic hinge lengths (between 5% and nearly 50%), as explained in Elnashai and Elghazouli (1993) and Elnashai and Izzuddin (1993). In the capacity-design framework, inaccuracy in determining the spread of inelasticity in dissipative zones renders the prediction of global structural performance less reliable.

Elastic-plastic models may exhibit kinematic, isotropic or mixed strain hardening. Values ranging between 1.0% and 3.0% of the elastic stiffness are frequently used for the strain-hardening stiffness when modelling mild steels using LESH. Figure 4.7 shows the linear elastic strain-hardening model used for the assessment of the SPEAR frame. The steel properties used to plot the bilinear curves in the figure are based on material test results carried out for reinforcement bars. The steel reinforcement includes three different diameters ϕ, namely 8 mm, 12 mm and 20 mm. The strain hardening of the bars vary between 3.2% (ϕ = 12 mm) and 5.6% (ϕ = 20 mm); these values are, however, higher than the average observed in the literature.

On the other hand, metal alloys show significant non-linear strain hardening and may be simulated using LENLH or RO models. The RO provides an accurate description of the response of metals such as aluminium alloys and stainless steels with high non-linearity at low stress and without a well-defined yield point (Di Sarno and Elnashai, 2003).

Analytical expressions for LEPP, LESH and LENLH are given in explicit form; the equations provide stresses σ as a function of strains ε. In addition, explicit equations for modelling metals can be formulated in terms of strains as a function of stresses. Inelastic response models described explicitly as stress-versus-strains are frequently used in FE analysis of structures. The most successful in this category of models is the MP model (Menegotto and Pinto, 1973).

The EPMs presented above can be used in FE computer programs as monotonic envelope curves in tension and compression to describe the inelastic behaviour of metals. For cyclic loading, a backbone curve is required alongside rules for unloading and reloading, to account for hysteretic energy dissipation, stiffness and strength degradation. Simplified uniaxial models, such as those in Figure 4.7, are computationally efficient alternatives to more elaborate multiaxial models. This efficiency is particularly important in cases where several iterations on the stress–strain relationship are required for each load step at each fibre of sections and members. Simplified uniaxial EPMs exhibit, however, three main shortcomings because of their inability to accommodate the following response features:

- Presence of a horizontal yield plateau followed by a strain-hardening zone;
- Reduction in strain-hardening slope with the increase in strain amplitude;
- Experimentally observed cyclic hardening and softening.

Under such conditions, more complex models would be more appropriate than the models above. Comparisons between experimental and analytical results computed using simple EPM and significantly more complex multi-surface plasticity models were undertaken (Elnashai and Izzuddin, 1993). For moderate levels of ductility demand, in the range of displacement ductility of 2 to 4, EPM models, though less accurate than multi-surface plasticity formulations, are adequate for seismic assessment. For higher levels of ductility, use of models that account for non-linear hardening and softening, as well as mean stress relaxation, is necessary.

(ii) Reinforced Concrete

Reliable models to fully describe the material behaviour of RC structures (Figure 4.8) up to and beyond the ultimate capacity require the following:

- Non-linear stress–strain relationship;
- Fracture and failure surface;
- Post fracture and failure;
- Model for reinforcing bars;
- Bond-slip and interface characterization.

The first two items are straightforward. However, there are a number of models in the literature for both compression and tension from which to choose. These may be broadly subdivided into microscopic and macroscopic models. Such models can express either uniaxial or multiaxial stress states. By contrast, non-linear stress–strain models currently used for the seismic shear response of RC structures are still complex and rarely utilized in common earthquake response analyses of systems; such models are still the subject of active research (e.g. Collins, 1978; Vecchio and Collins, 1982; Collins *et al.*, 1996; Gerin and Adebar, 2004). The third item is the most controversial. Different assumptions may yield

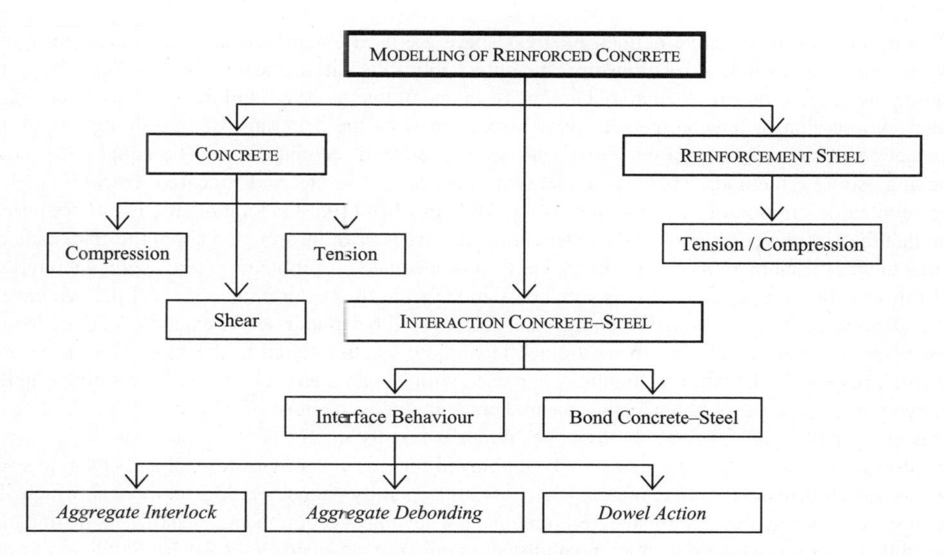

Figure 4.8 Material modelling of reinforced concrete (without torsion)

widely varying results, and calibration is often impossible. The fourth item was covered in the previous section. Bond-slip and interface characterization can be ignored, unless reliable data are used to calibrate models for interface behaviour between steel and concrete. Comprehensive formulations for the simulation of the behaviour of concrete are available in the literature (e.g. Chen, 1982, among many others).

Macroscopic models have been extensively used with varying degrees of success. These models include:

- Linear elastic models with tension cut-off (LEMs);
- Non-linear elastic models (NLEMs);
- Elastic-plastic models (EPMs).

Stress–strain relationships for LEMs and NLEMs are invariably represented by empirical formulae fitted to a given set of experimental results. Expressions for these models are presented in the form of total stress–strain equations. Young's modulus E, at the origin E_0 or as tangent value E_t, stresses, crushing f_c and cracking f_t, and Poisson's modulus ν are often required for isotropic formulations. As such, these models cannot be reliably adapted to variable amplitude cyclic loading conditions, required for earthquake response analysis. The extension of these models to cyclic response involves even more assumptions and empirical formulations. The main advantages of LEMs and NELMs are that they are easy to implement and duplicate fairly well test results to which they have been calibrated. On the other hand, EPMs are based on the well-developed and concise theory of incremental plasticity; hence cyclic loading can be readily accommodated. They can be easily implemented in FE programs. Obtaining realistic hysteretic loops might, however, not be as straightforward. The main objection to the use of these models is that there is no evidence that concrete flows plastically under stress. Nevertheless, non-linear hardening formulations describe the stress–strain relationship very closely, where the first 'yield' point is coincident with the departure from linearity and the remaining ascending and descending regions are modelled by piecewise linear or non-linear hardening.

Failure criteria for concrete defined in stress space are used extensively in structural applications. The non-linear constitutive relationships are only applicable within the failure envelope. Once the failure criteria are satisfied, the material suffers from one of two modes of failure: crushing or cracking, under compression or tension, respectively. In compression, the material loses its strength upon the satisfaction of the given criterion. For tensile failure, several techniques may be adopted to model cracking using FE methods. These fall under two categories: 'discrete' and 'smeared' cracking models. The detailed description of these methods is available in Chen (1982) among others. It suffices here to note that the discrete crack model is a better choice if there is prior information about the crack patterns so that the mesh can be refined along crack paths. It is adequate for problems where aggregate interlock and dowel action are significant. Discrete crack models are the better choice if local behaviour, e.g. local stresses and crack sizes, is more important than overall behaviour, as for example load–deflection curves. On the other hand, the smeared crack technique is better suited to the assessment of overall structural response. The latter technique, when used with isoparametric elements, is versatile, efficient and economical. As a consequence, the smeared crack model is the most suitable choice for earthquake engineering applications where the global behaviour is usually the focus of the analysis.

Multi-linear elastic-plastic models may be employed for RC and composite steel and concrete structures as further discussed in Section 4.5.3.1. These are generally phenomenological models, which also take into account the presence of steel reinforcement. They may accommodate the stiffness degradation caused by the onset of concrete cracking and steel yielding (e.g. Takeda *et al.*, 1970; Saiidi and Sozen, 1979; Ibarra *et al.*, 2005, among others). Such models employ generally uniaxial formulations.

Most of the models described above consider strain–stress relationships for concrete under compressive loads with an envelope curve, which matches the material response obtained under increasing loads. If the stress is decreased, an unloading curve will be traced. Increasing the stress forces the material along a reloading branch of the material response curve. Figure 4.9 shows the most commonly used uniaxial relationship for concrete under monotonic and cyclic loading, which is the model by Mander *et al.* (1988). The latter can be utilized to simulate the behaviour of both confined or 'core' and unconfined or 'cover' concrete in cross sections modelled by fibres as illustrated in Section 4.5.2.

A simplified non-linear concrete model, which may simulate constant active confinement, is implemented in Zeus-NL (Elnashai *et al.*, 2003); this model is derived from Mander's formulation. A constant confining pressure is assumed, taking into account the maximum transverse pressure generated by

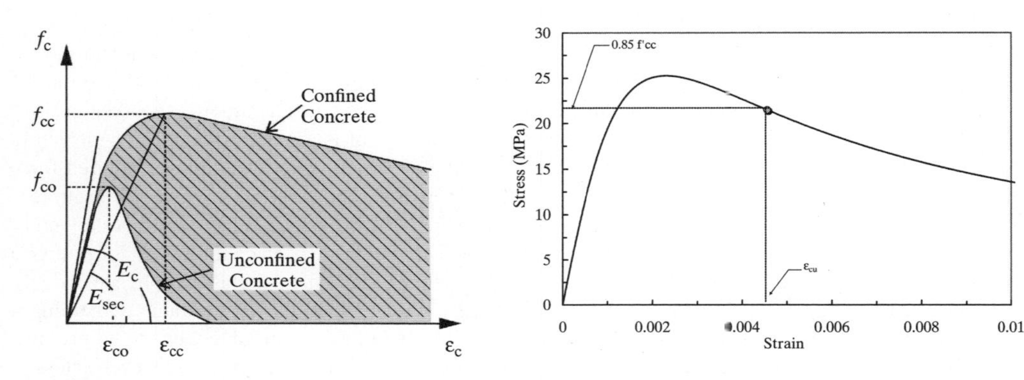

Figure 4.9 Uniaxial models for concrete: Mander *et al.* (1988) (*left*) and constant confinement Zeus-NL model used in the assessment of the sample SPEAR frame (*right*)

confining steel. This is introduced in the model as a constant confinement factor $k > 1.0$, which is used to scale up the stress–strain relationship throughout the entire strain range. The confinement factor k is expressed as the ratio of confined concrete strength f'_{cc} to plain concrete strength f'_c. The simplified model in Zeus-NL can be used to simulate the response of concrete under monotonic and earthquake loads; its cyclic rules enable the prediction of cyclic degradation of strength and stiffness and ensure numerical stability under large displacements analysis (Martinez-Rueda and Elnashai, 1997). Four parameters are required to fully define the model. These are compressive strength f_c, tensile strength f_t, crushing strain ε_c and a confinement factor k. Figure 4.9 shows the stress–strain relationship of the constant confinement concrete model with $k = 1.01$ used in the assessment of the SPEAR frame. The low value of k is due to the limited amount of transverse reinforcement used for the members of the assessed RC building structure.

One of the main shortcomings of this 'active confinement' approach is that it couples the peak of the concrete constitutive curve to the yield point of the steel reinforcement, regardless of spacing and bar diameter. This implies that even at low levels of axial stress, the material is fully confined as if the steel is at the point of yield. Clearly, this is not the case. A passive confinement model was developed by Madas and Elnashai (1992a) and extensive analyses have shown that the peak of concrete and steel strength is not reached at the same time. Passive confinement models are therefore preferable to their active confinement counterparts.

Problem 4.3

Consider the cross section in Figure 4.10, which is representative of corner composite RC and steel columns in multi-storey frames. Which of the material models described in Section 4.5.1 would you select for the column to perform non-linear dynamic analysis? Comment on your answer. Assume that the column is part of a frame with high ductility.

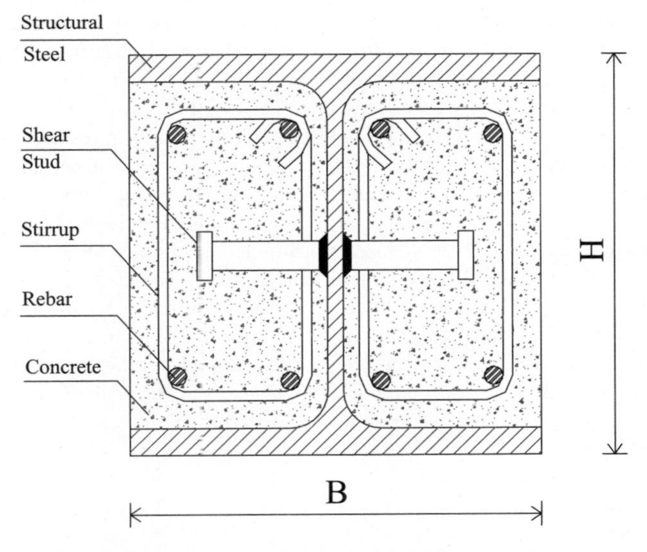

Figure 4.10 Steel and concrete composite section

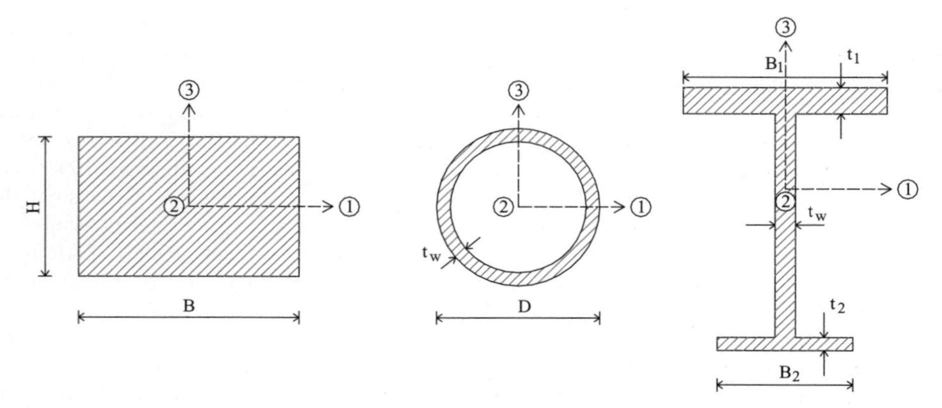

Figure 4.11 Typical sections implemented in the non-linear finite element code Zeus-NL

4.5.2 Sections

Cross sections are modelled by defining geometric and material properties of their components. Comprehensive section libraries are generally included in FE programs for different materials of construction as shown, for example, in Figure 4.11. It suffices to define only certain sectional dimensions, e.g. width and height for rectangular sections, because the remaining properties, such as cross-section area A, effective areas for shear (A_{vy} and A_{vz}), flexural (I_y and I_z) and torsional (J) moments of inertia, are computed automatically. Sections with non-standard shapes can be modelled by defining manually all geometric and mechanical properties. The number of properties required is a function of the problem under consideration. For example, to analyse plane structures, it is required to specify A, A_{vz} and I_z, where z is the axis perpendicular to the plane.

Depending on the type of section employed, e.g. steel, RC or composite, a different number of material properties is specified. These are a function of the model adopted for the material(s) as illustrated in Section 4.5.1. For RC and composite sections, it is necessary to specify the area and location of steel reinforcing bars, which are located with reference to the local axes of the section. These axes generally coincide with the section principal directions, e.g. axes labelled '1' and '3' in the sections shown in Figure 4.11. The correct orientation of local axes should always be thoroughly checked before running analyses as it is a common source of errors in FE modelling of structures. Sections with different moments of inertia about the principal directions (e.g. sections with a T-shape in Figure 4.11) may be oriented incorrectly. Modern software packages employ user-friendly graphical environments that render checks at the sectional level easy to perform. For example, Figure 4.12 shows how the section local axes 1-2-3 relate to the global axes X-Y-Z in beam-elements implemented in Zeus-NL (Elnashai *et al.*, 2003). Nodes n1 and n2 are the end nodes of the element. The element local 2-axis lies on the line defined by them, i.e. n1-n2. However, n1 and n2 give no information for the T-section in the figure and its orientation. Node (3) is thus required to define the (local) 2-3 plane and can be a non-structural node. It is possible, and advisable, to use one non-structural node as the third node for all the beam elements that lie on the same plane of the model; this is also shown for the sample SPEAR frame in Figure 4.12, where non-structural nodes are indicated by dark markers.

Three basic formulations can be used for section analysis with FEs:

- Fibre (or filament) models;
- Phenomenological (or mathematical) models;
- Mechanical (or sectional spring) models.

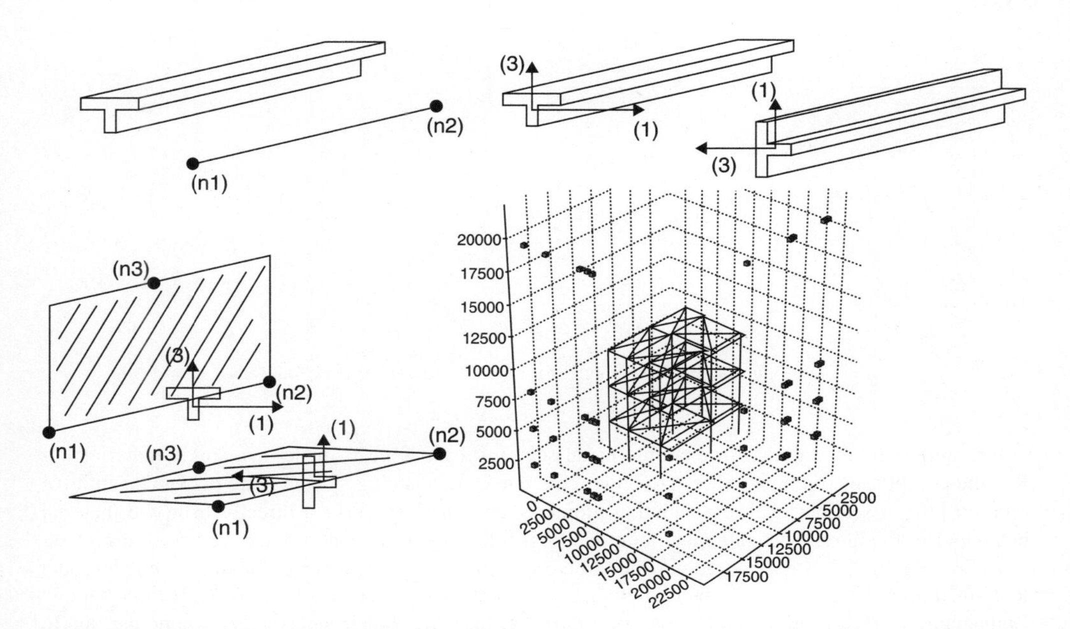

Figure 4.12 Typical local axis orientation for beam-column elements in Zeus-NL: beam with a T-section to be modelled (*top-left*), two possible orientations of the T-section of the beam (*top-right*), the correct position of node n3 for modelling the orientations (*bottom-left*) and location of non-structural nodes in the finite element model of the SPEAR frame (*bottom-right*)

Fibres form the basis of distributed inelasticity models. The latter are the most reliable formulations to predict the earthquake response of structural systems. Nevertheless, they may be time-consuming in practical applications for the analysis of the inelastic behaviour of large structures. Phenomenological models are employed to simulate the monotonic and hysteretic response of cross sections under different loading conditions. These are efficient models but often require tedious calibrations of the large set of parameters utilized to describe, for example, moment–rotation relationships of steel, RC or composite sections. Phenomenological models are generally used for lumped inelasticity discretizations of structural systems. In mechanical models, cross sections are idealized through a discrete number of springs. These models combine some features of the fibre method for the construction of the section stiffness matrix with the basic principles of lumped inelasticity models.

In fibre-based formulations, the area A of the section is divided into finite regions (or fibres), e.g. a rectangular grid of lines parallel to cross-sectional principal axes for 2D analysis. Each fibre is characterized by two geometric quantities, these are its location in the local reference system of the section, defined as 'monitoring point', and the fibre area dA. Figure 4.13 shows typical subdivision in fibres for RC sections. The number of fibres is dependent on the type of section, the target of the analysis and the degree of accuracy sought. Refined subdivisions with a large number of monitoring points increase exponentially the computations required in the analysis. For rectangular sections, the number of fibres can be determined by subdividing width B and height H in segments of length equal to 1/10 of B and H, respectively. At least two lines of fibres should be used for flanges and webs in standard metal profiles. Fibre-based discretizations of RC rectangular and T-shaped sections of the SPEAR frame in Figure 4.2 employ 200 monitoring points. In 3D, subdivision into fibres along two orthogonal section axes results in the definition of smaller areas with distances of the monitoring points to both axes, as discussed further below.

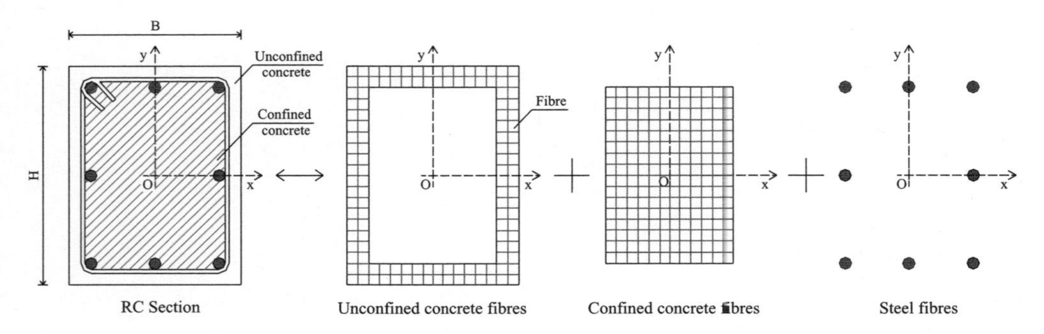

Figure 4.13 Fibre model for a reinforced concrete section

In homogeneous sections, for example metal sections, each fibre is made of a single material; for RC and composite structures, fibres are either concrete or reinforcing steel (Figure 4.13). Constitutive relationships at section level, e.g. bending moment-curvature, are derived directly from the material behaviour of the fibre. Adequate material modelling is thus a key element of the fibre-based discretization. For example, in steel and composite structures, fibre discretization may be employed to model local buckling of flanges and webs by defining adequate stress–strain relationships (e.g. Elnashai and Elghazouli, 1993; Broderick and Elnashai, 1996). On the other hand, inelasticity of concrete due to cracking, crushing and post-peak softening can be easily accommodated by fibre models. Tension-stiffening may be implemented either by modifying the stress–strain relationship of the concrete in tension after cracking or by adapting the cyclic inelastic constitutive model for reinforcing bars. In so doing, the presence of discrete concrete cracking, the contribution of concrete in tension between the cracks, and the bond-slip behaviour is accounted for in a smeared fashion. In addition, adopting fibre models for RC and composite sections enables the use of constitutive relationships for unconfined and confined concrete as shown Figure 4.13. Partially confined concrete can also be simulated in composite sections by combining adequate material model and fibre discretizations. For the modelling of the SPEAR frame, the decomposition of the RC cross section in Figure 4.13 is assumed for both rectangular and T-shaped sections. Material models in Figures 4.7 and 4.9 are associated with concrete and steel fibres, respectively; for unconfined concrete the confinement factor $k = 1.0$, while $k = 1.01$ is used for the sparsely confined concrete of the core of the sections.

Cross-sectional subdivision in fibres of RC sections of the three-dimensional SPEAR building is an efficient tool to quantify biaxial bending moments in the columns of the frame subjected to bidirectional earthquake records. Conversely, for 2D frame models, member sections can be discretized into strips or fibres normal to the plane of the structure, as the bending moment is uniaxial.

To obtain reasonable estimates of fundamental periods, displacements and distribution of lateral force using elastic analysis (static, eigenvalues and dynamic), stiffness properties for RC sections should account for cracking. Effective stiffnesses may be defined for axial loads, flexure, shear and torsion. For frame analysis, when applying capacity design rules, the reduction in the flexural stiffness of the beams should be higher than that of the columns. Several values of the effective stiffness have been suggested in the literature (e.g. Elnashai and Mwafy, 2002, among others).

Problem 4.4

Which of the section discretizations shown in Figure 4.14 would you select to perform three-dimensional analysis of a multi-storey steel frame? Justify your selection. What are the other options available to the analyst/designer to assess the three-dimensional response of structures? Compare alternative options in terms of computational efficiency and accuracy.

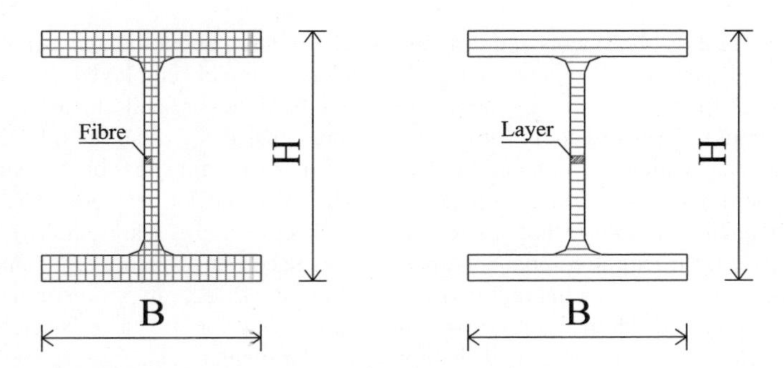

Figure 4.14 Discretization of cross sections: fibre (*left*) and layer (*right*) model

4.5.3 Components and Systems for Structural Modelling

The evaluation of seismic response of structures by FE analysis requires that system components, such as beam-columns, braces, slabs and walls are modelled using discrete elements spanning between nodes. Below is a summary of commonly used elements for earthquake response analysis:

i. *Beams*: Beam elements are used for both planar and spatial models of structures to represent beams and columns. In 2D analysis, they possess 3 DOFs per node: 2 displacements and a rotation about the axis perpendicular to the plane of the structure. Axial loads, bending moments and shears are the internal actions. Three-dimensional beams have 6 DOFs per node, these are 3 displacements and 3 rotations. Internal actions include axial load, 2 shear forces, 2 bending moments and torsion. Beam elements are commonly implemented in FE programs as two-noded elements. Higher-order formulations are, however, also available (e.g. Cook *et al.*, 2002). The higher-order beams include one or two additional nodes at intermediate locations along the element length. The number of nodes of beam elements depends on the types of polynomial used as shape functions. Euler or Timoshenko formulations are used depending on the geometric characteristics of the beam element. For deep and stocky members (aspect ratios less than ~3–4), shear deformations are significant and it is more accurate to employ Timoshenko's beam theory. Similarly, torsional deformations should be accounted for when employing elements with open sections. Cubic elastic-plastic 3D beam-column elements in Zeus-NL (Elnashai *et al.*, 2003) may be used to model RC frame members of the SPEAR building in Figure 4.2. This cubic element was formulated by Izzuddin (1991) and is used for detailed inelastic modelling, making use of the uniaxial inelastic material models for steel and concrete described in Section 4.5.1. It accounts for the spread of inelasticity along the member length and across the section depth. Geometrical non-linearities are accommodated using a Eulerian formulation. Beam elements can also be used to model structural components subjected only to axial forces, such as braces and cables. These elements are known as 'rods' or 'bars'. In some FE computer programs, rods are derived from beam elements by releasing all but the axial DOFs;

ii. *Plates and shells:* These include triangular and rectangular elements. Isoparametric formulations permit quadrilateral elements to have non-rectangular shapes and mild curvature or irregular geometry. The number of nodes of plate and shell elements depends on the types of polynomial used as shape functions, as in the case of beams. Lower- and higher-order formulations have been implemented for these FEs (e.g. Cook *et al.*, 2002). Following the general plate theory, plate elements are assumed to have 3 DOFs per node: translation perpendicular to the plate and rotations about two perpendicular axes in the plane of the plate. The DOFs per node for shells are usually 5; 3 displacements and 2 rotations. The rotation about the axis perpendicular to the element mid-plane is not represented in the kinematic models employed for shells. However, in some FE computer programs, the kinematic model used for shell elements may employ an artefact sixth degree of freedom (drilling) to ensure consistency of all the 6 DOFs in space. In-plane elements, which are frequently used where either constant stress or strain can be assumed through the thickness, are special plate elements with 2 DOFs per node. These are in-plane displacements along two orthogonal directions. Plates and shells are employed in FE elastic analysis of 2D structural components that may displace out-of-plane, e.g. diaphragms, walls, arches and vaults. An essential requirement is to limit the aspect ratio of the finite elements used for the geometric discretization. For instance, narrow and long rectangular elements (e.g. $L/B > \sim 6$) should be avoided, as should shallow triangular elements.

iii. *Solid elements:* These are 3D elements which are directly related to rectangular plane elements. Each node has 3 DOFs: 3 displacements along orthogonal directions. Bricks, i.e. rectangular solid elements, may be linear or quadratic. Linear elements have 8 nodes and 24 DOFs, while higher-order quadratic elements have 20 nodes and 60 DOFs. As for plates and shells, meshes employing solid elements with excessive aspect ratios (e.g. $L/B > \sim 6$) should be avoided as they affect the accuracy of the results. Bricks are rarely used in earthquake response analysis of framed structures since they require intensive computational effort, especially if inelasticity is taken into account. Solid elements are frequently used to assess local stress analysis in system sub-assemblages under seismic loads, such as in the case of beam-to-column connections or at the base of columns.

The characteristics of the aforementioned FEs are compared in Table 4.5. In all the FEs discussed above, the integration of displacement interpolation functions over the length, area or volume is usually performed numerically by Gauss quadrature. This is an efficient method that locates few sampling points within each element and assigns weights so as to minimize integration errors when the integrand in the expression of the stiffness matrix is a general polynomial. The Gauss–Lobatto formulation was also proposed for practical earthquake response analyses due to its efficiency and accuracy. In this formulation, the integration points are at the nodes, as well as along the length of the elements (Spacone *et al.*, 1996a). This may result in a viable integration scheme especially when the curvature at the end of the element is included in the computations or for elements with spreading of inelasticity along their length (e.g. Spacone *et al.*, 1996b; Cordova and Deierlein; 2005; Berry 2006, among others).

Table 4.5 Types of finite elements used in analysis of structures under earthquakes.

Element	Geometry	Nodes		Limitations
		Nodes	DOF per node	
Beam	1D	2*	3 (planar) 6 (spatial)	Some types of inelasticity in RC and composite structures. Limited application for complex geometries
Plate and Shell	2D	4*	2 (planar) 3 or 5** (spatial)	Several types of inelasticity in RC, masonry and composite structures
Solid	3D	8*	3	General applications. Time-consuming for inelastic analysis

* Higher-order formulations are also available (additional nodes located mid-side and/or at centre-element).
** Shell elements with drilling degree of freedom (6 DOFs per node) are also used in computer programs.

4.5.3.1 Beams and Columns

Beam elements are the most widely used FEs for earthquake engineering applications. For linear elastic analyses, standard prismatic beams with cubic displacement variation along the element length are frequently utilized. For inelastic analyses, two alternative models are available:

- *Lumped* (or *concentrated*) *inelasticity models*;
- *Spread* (or *distributed*) *inelasticity models*.

Inelasticity in ductile framed structures is concentrated at or near member ends. In lumped inelasticity models, the element response is thus generally represented by zero-length plastic hinges, also referred to as 'point hinges', located at member ends. The point hinges are inelastic springs. The stiffness matrix of the member is computed from the stiffness of the single or multiple springs. For elements with inelastic flexural behaviour, the inflection point is usually assumed to be at member mid-length and the inelastic flexural deformations in each half of the beam are lumped in a rotational spring at the end of the member. Figure 4.15 shows a typical lumped model for members governed by bending moments. Concentrated inelasticity models may be utilized to describe complex hysteretic behaviour by the selection of appropriate constitutive relationships for the end springs.

Typical force–displacement models utilized for inelastic springs are provided in Table 4.6. Such models can accommodate cyclic stiffness degradation in flexure and shear, pinching and fixed-end rotations at the beam–column joint interface due to bar pull-out. An extensive discussion of mathematical functions appropriate for inelastic spring models for RC may be found in CEB (1996). For steel components, the Ramberg–Osgood model is generally sufficiently accurate (Bruneau *et al.*, 1998).

The main advantage of the lumped inelasticity models presented above is the simplicity of their formulation. Notwithstanding, these models oversimplify certain aspects of hysteretic behaviour, which may render them of limited applicability. They are more retrospective than predictive (i.e. the behaviour is needed before the analysis is performed). Conversely, distributed inelasticity models provide a more accurate description of the hysteretic behaviour but at the expense of high computational demands. In the distributed inelasticity models, material non-linearity can take place at any section and the element response is estimated by weighted integrations of the sectional stiffness. Integrals needed to update the tangent stiffness matrix of member are evaluated numerically at selected points, referred to as 'Gauss points'. Generally only two or three Gauss points are used to monitor the member response. Consequently, only sections at these monitoring locations or 'slices' contribute to the member stiffness. Sampling Gauss points in isoparametric beam elements of length L are located at $L/2\left(1-1/\sqrt{3}\right)$, i.e.

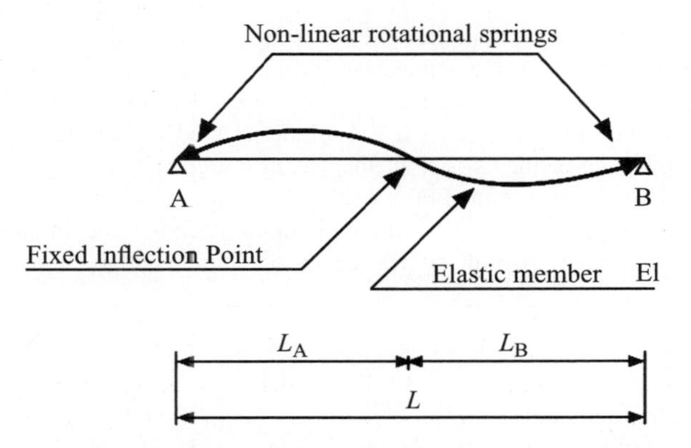

Figure 4.15 Lumped plasticity model for elements with inelastic flexural behaviour

Table 4.6 Common hysteretic relationships for inelastic springs in lumped models.

Model type	Reference	Sketch
Bilinear with axial interaction	Takayanagi and Schnobrich (1979)	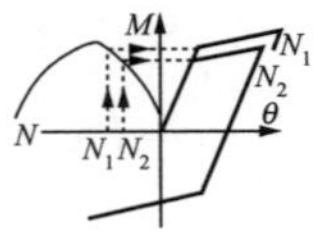
Stiffness degrading	Clough and Johnston (1966)	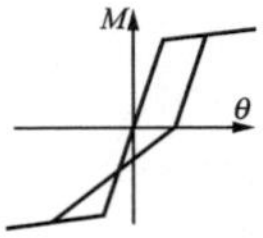
Stiffness degrading with strength deterioration	Saiidi and Sozen (1979)	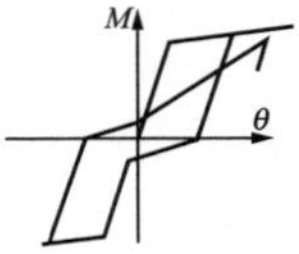
Takeda hysteretic model	Takeda *et al.* (1970)	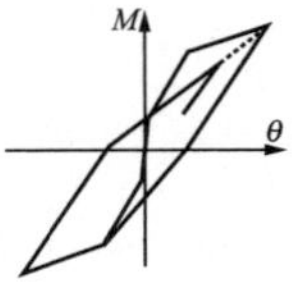
Ramberg–Osgood model	Park *et al.* (1987)	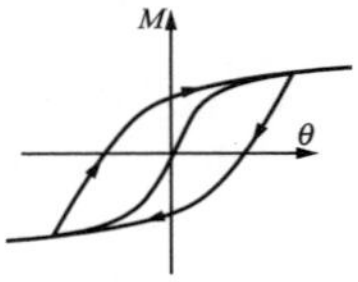

at about $0.21\,L$, from each node. It is of importance, when performing inelastic analyses, to ascertain that zones where inelasticity is expected, i.e. critical zones, contain at least two Gauss points. This can be achieved by either automatic procedure based on adaptive analysis (Izzuddin and Elnashai, 1993; Karayannis *et al.*, 1994) or by starting the analysis with refined meshes in critical zones in the FE structural model. As discussed before, the spread-of-inelasticity approach is versatile for earthquake analysis of structures with different materials of construction. Monitored sections are decomposed into fibres and constitutive relationships are defined at the centre of each infinitesimal area as illustrated in Section 4.5.2. The sectional stiffness is derived from the integral of all contributions of fibres. In turn, the integral along the element length of the stiffness of the selected slices leads to the member stiffness. Fibre discretization of beam elements is implemented in the computer program Zeus-NL (Elnashai *et al.*, 2003). Figure 4.16 shows a refined FE model utilized for a plane frame extracted from the prototype of the SPEAR building. To quantify reliably inelastic action-effects in dissipative zones of the structure, refinement of FE mesh is utilized near beam-to-column and base connections.

Problem 4.5

Consider the two bridge piers in Figure 4.17. Five possible idealizations are given in Figure 4.18. Which is the most appropriate discretization, in terms of both accuracy and efficiency, to estimate the inelastic response of both bridge piers? Give comments and illustrate your answers with simple sketches.

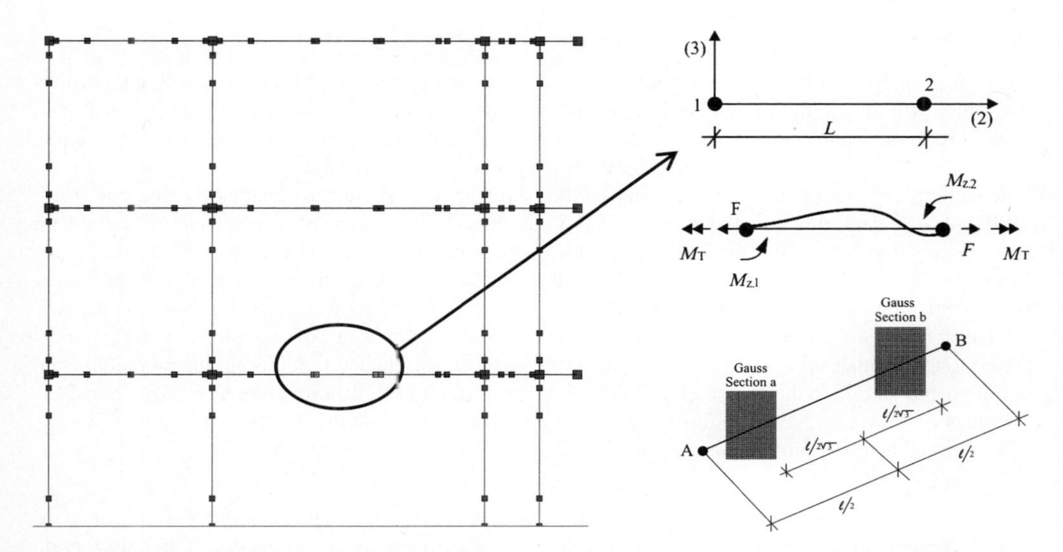

Figure 4.16 Refined finite element (FE) model for plane frame extracted from SPEAR structure: FE plane frame (*left*) and beam-column modelling (*right*)

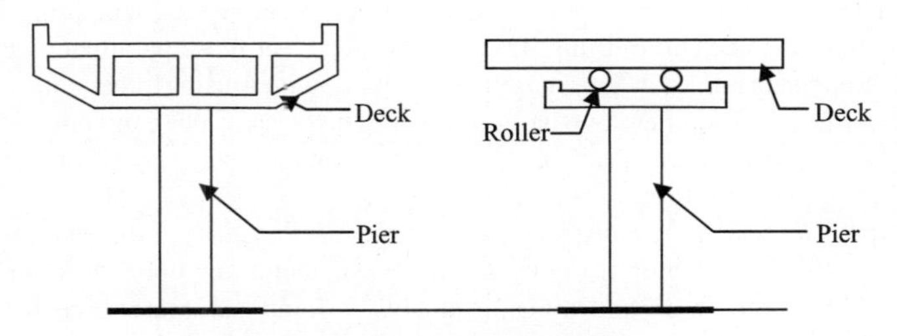

Figure 4.17 Bridge piers

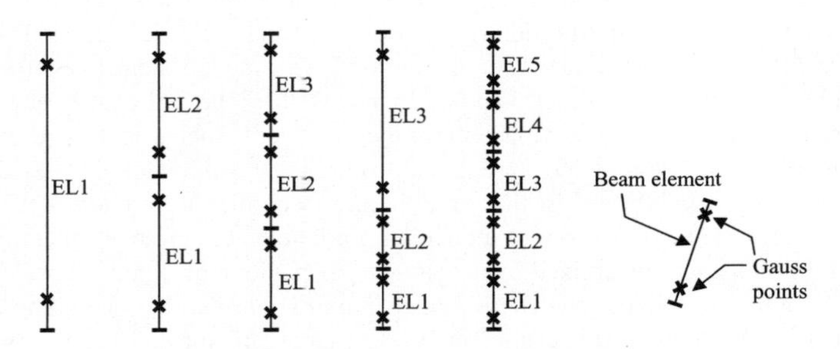

Figure 4.18 Alternative idealizations of bridge piers

4.5.3.2 Connections

Structural analysis of frames is generally performed using the centreline of beams and columns. Effects of joint stiffness, strength and ductility significantly affect the global structural response and should therefore be accounted for. In steel framed structures, all connections may be classified, in principle, as semi-rigid (e.g. Nader and Astaneh, 1989, among many others). Analytical modelling of semi-rigid connections is a key feature in assessing earthquake response of framed structures (Elnashai *et al.*, 1998). Using centrelines in the response analysis of steel sub-assemblages may lead to significant underestimation of the contribution of columns to storey drifts (Gupta and Krawinler, 2000). Furthermore, it is likely that using centrelines only leads to mislocating plastic hinges, thus areas of high ductility demand are not accurately identified.

The above discussion demonstrates that explicit modelling of joint response could be of great importance to avoid misleading results. Several models have been formulated to represent the monotonic and cyclic behaviour of connections in RC, steel and composite frames. A thorough review is given in CEB (1996) and in Mazzolani and Piluso (1996). Connection models may be divided into three groups as summarized below in order of increasing complexity for their representation:

i. *Phenomenological* (or *mathematical*) *models*: these are mathematical expressions fitted to monotonic envelopes or hysteretic rules for experimentally determined action–deformation curves. The response curve is given as an expression, generally in terms of moment–rotation M–θ or shear force–angular distortion V–γ. This approach can be used to fit virtually any shape of action–deformation response, but most have difficulties matching complicated hysteretic and degrading curves. In recent years, neural network representations have overcome the difficulties of matching the latter class of response (Yun *et al.*, 2007). However, polynomial, piece-wise linear and power-law functions are generally employed to match experimental data. Users of such models have to specify numerous empirical input parameters, which may require tedious iterative calibrations or curve fitting. It is difficult to select these parameters without access to appropriate experimental data or analytical results obtained using other more refined models. This is also true of neural network-based models. The class of phenomenological models in its entirety also suffers from the disadvantage that they cannot be employed for types of connection different from those used for their derivation.
ii. *Physical* (or *mechanical*) *models*: these consist of mechanical analogues (springs, bars, etc.), each representing one or more response mode or connection component. Physical models can be employed in both static and dynamic analyses. Input parameters are the geometric and mechanical properties of the components. In component-based models, the global response of the joint is computed as the aggregate of the stiffness and strength contribution of all components of the connection (e.g. Madas and Elnashai, 1992b). Alternatively, calibrations using laboratory tests or more refined approaches, e.g. FE discretization, may be adopted. Strength and stiffness deterioration and pinching in RC joints may be accommodated in physical models (e.g. Ghobarah and Biddah, 1999; Mitra and Lowes, 2007). Similarly, they can describe deterioration of response parameters due to plastic deformation of steel and inelastic behaviour of bolts and angles in steel and composite joints. The efficiency of these models stems from the simple representation of the connection geometry and mechanical behaviour. They may be used to predict the response of connections that are made up of the same components as those used in the derivation of the model, hence they have predictive features.
iii. *Finite element* (or *detailed*) *models*: these are the most complex in terms of development but also the most accurate models for joint modelling. Finite element meshes generally employing shells or solid elements are utilized to assess local behaviour of connections, e.g. beam-to-column and column-to-foundation. These models require considerable modelling and analysis efforts, which are prohibitive for seismic analyses of large structures. Their use is thus limited

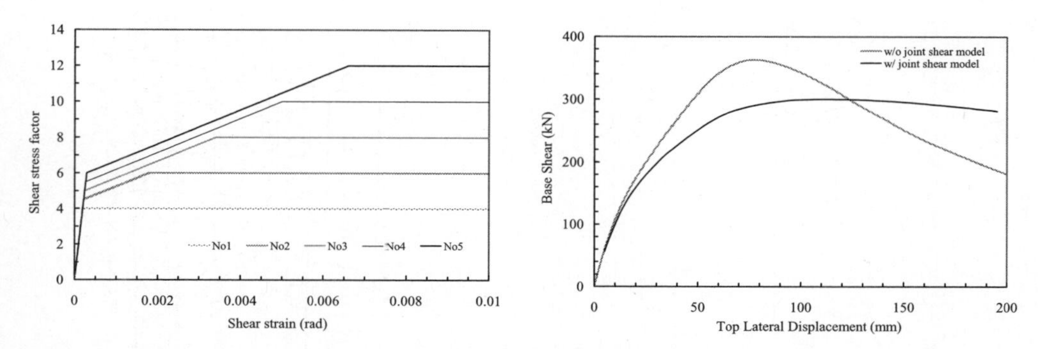

Figure 4.19 Beam-to-column joint modelling: trilinear shear strength–strain curves (*left*) and lateral global of the frame with and without joint modelling (*right*)
Key: No1 = knee joint; No2 = external joint without transverse beams; No3 = external joint with transverse beams; No4 = internal joint without transverse beams; No5 = internal joint without transverse beams

to special applications, e.g. fracture mechanics or when complex local three-dimensional stress–strain distributions need to be determined, e.g. in beam-to-column connection sub-assemblages (e.g. El-Tawil *et al.*, 2000, among others). They are, however, suitable for the derivation of moment–rotation relationships, which are then implemented as mathematical models (type *i* above).

Phenomenological and physical models are widely used in structural analysis to account for connection behaviour. They are utilized in planar or space frame analysis to model pin joints, inclined supports, elastic-plastic joint behaviour, soil-structure interaction and structural gaps through employing appropriate joint curves. Seven force–displacement curves are, for example, available for use with joint elements in the computer program Zeus-NL (Elnashai *et al.*, 2003): elastic linear, trilinear symmetrical and asymmetrical elastic-plastic curves, hysteretic shear model under constant axial load or with axial force variation, hysteretic flexure model under constant axial load or with axial force variation. Figure 4.19 shows trilinear symmetrical shear strength–strain relationships utilized to model beam-to-column joints of the SPEAR frame. In the same figure, the lateral response of the frame (pushover) with and without the joint modelling is provided, for the weak x-direction as per Figure 4.2. Shear joint relationships in Figure 4.19 require the definition of stiffness (initial and post-cracking) and strength (shear stress at cracking and at maximum capacity) parameters. The plot of the base shear-versus-lateral displacement curves in Figure 4.19 demonstrates that modelling of beam-to-column joints may lead to lower values of all global structural response characteristics discussed in Section 2.3, i.e. stiffness, strength and ductility.

4.5.3.3 Diaphragms

Framed structures are analysed using spatial FE discretizations with the common assumption that deck systems serve as rigid diaphragms between the vertical elements of the lateral load-resisting system. In-plane stiffness of bridge decks and floor slabs in buildings with regular plan layout are indeed very high compared to the lateral stiffness of the frame (Jain, 1983). However, the validity of this assumption should sometimes be checked, because the flexibility of diaphragms can significantly affect the three-dimensional dynamic behaviour of structural systems. This is especially the case when exceptionally shallow slabs are used or where the slab has large openings.

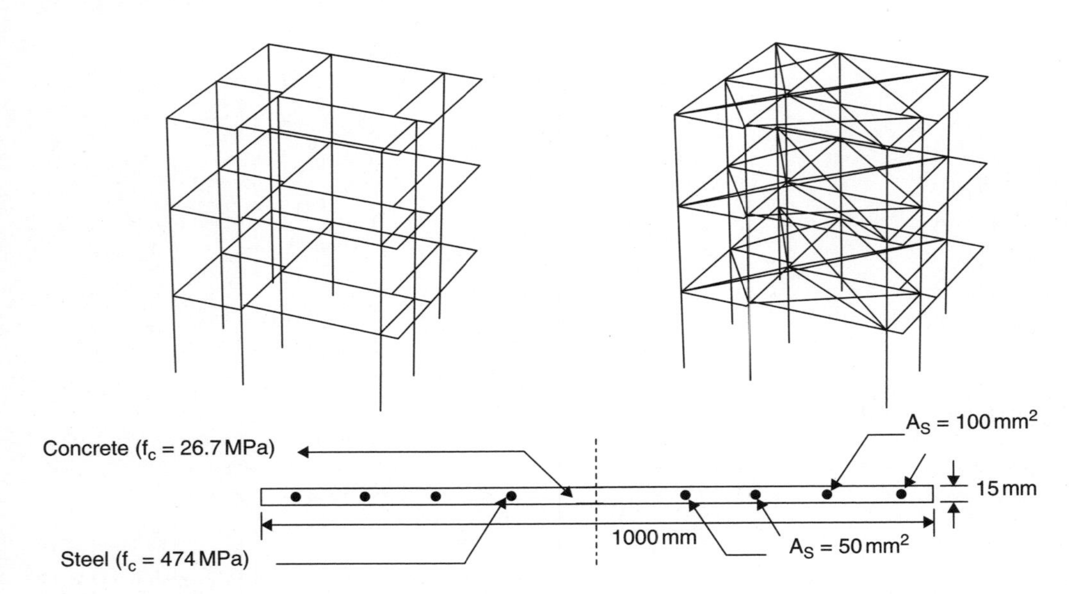

Figure 4.20 Modelling of rigid diaphragms in framed structures: 3D finite element model of SPEAR frame without (*top-left*) and with rigid (*top-right*) diaphragms and horizontally rigid member used to simulate the diaphragmatic action (*bottom*)

The assumption of rigid floors is economical in the analysis since several degrees of freedom can be condensed and the order of the stiffness matrices reduced. For example, the three-storey and two-bay irregular frame shown in Figure 4.20 possesses 162 DOFs: at each storey, the slab connects 9 nodes with 6 DOFs (three translations and three rotations).

If the three diaphragms of the SPEAR frame in Figure 4.20 are rigid, in-plane displacements of all nodes, defined as 'slave nodes', can be expressed, through principles of basic mechanics, as a function of the corresponding DOFs of a 'master node', at each storey level. Consequently, the total number of DOFs is reduced from 162 to 90. These include two translations (in the plane of the diaphragm) and one rotation (about the z-axis) for the master nodes and one translation (along z-axis) and two rotations for each slave node. Master nodes are generally assumed to coincide with the mass centre of the diaphragm.

In several commercial software packages, in-plane rigidity of horizontal diaphragms is modelled automatically, e.g. by means of diaphragm constraints or constraint equations. In these cases, master nodes possess 3 DOFs at each storey level: the translations along x- and y-direction and a rotation about the z-axis. The three out-of-plane DOFs are often not kinematically constrained. Rotational DOFs about x- and y-axes can be neglected. Alternatively, diaphragm effects can be simulated by inserting horizontally rigid members in the slab (e.g. Jeong and Elnashai, 2005). In the sample frame in Figure 4.2, in order to model rigid diaphragms, each corner of the slabs is diagonally connected to the opposite corner. The dimensions and reinforcement of connecting RC members shown in Figure 4.20 are such that the additional members do not provide duplicate stiffness to the flexural resistance of beams. The contribution of slabs to flexural stiffness of beams is already modelled by effective width of T-beam models. The thickness of the connecting elements is determined by iteration such that the contribution of connecting elements to the vertical stiffness is negligible, while the contribution to the horizontal stiffness remains significant. Figure 4.21 shows the angle of torsion at all corners of slabs that are located at column points for the first floor of the SPEAR frame. The analytical model with additional connecting

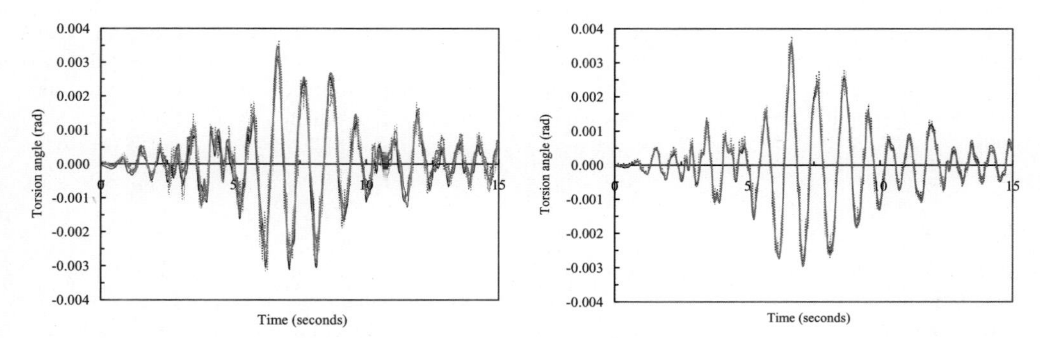

Figure 4.21 Angle of torsion time history at all columns in the first floor of the full-scale frame in Figure 4.2 (Montenegro 1979 – Herceg Novi, 0.2 g, bidirectional loading

members shows the same angle of torsion at every corner, thus satisfying the assumption of an in-plane rigid floor.

Computer programs do not generally accommodate the modelling of rigid inclined decks, which are typically used as roofs. Inclined diaphragms can be discretized in a number of different ways. The most common models include equivalent one-dimensional elements, e.g. trusses or beams, and two-dimensional elements, e.g. shells. The choice of model for inclined rigid diaphragms depends on the degree of accuracy sought. One-dimensional elements are a compromise between accuracy and economy, especially for inelastic dynamic analyses as those illustrated in Sections 4.6.1.2 and 4.6.1.3. Evaluation of local stresses requires the use of shell elements. The mesh should be refined gradually towards openings and other regions where high stress gradients are expected.

4.5.3.4 Infills

Masonry and concrete block infills are frequently used for interior partitions and exterior walls in steel, RC and composite frames. Infills can either be isolated or connected to the bounding frame. In the latter case, the interaction between walls and frame should be taken into account. Infills can be discretized using one of two approaches:

- *Micro-* or *refined models*;
- *Macro-* or *simplified models*.

Micro-models generally utilize three types of FEs to discretize the frame, infills and the interaction between the two. One-dimensional elements are used to model the frame while walls are idealized using triangular and rectangular plate or shell elements. Interface elements are essential for the simulation of cracking, which may take place between the frame and infills. They also account for the likelihood of separation and describe friction conditions where contact remains. It is difficult to define accurately the boundary conditions at the interface between infill and frame, a region affected by materials, details and construction method. For instance, in reinforced masonry walls, if the beams of the bounding frame are supported by the masonry when cast *in situ*, the interaction is full. Conversely, when infills are built after the frame and a separation gap is used between them, the interaction is negligible. Micro-models can be employed to assess global and local response of frames with infills provided that the properties of the interface are clearly defined. Macro-models are an attractive alternative to detailed models because of computational simplicity and efficiency. Diagonal struts with appropriate mechanical characteristics may be used to simulate the presence of infills (e.g. Stafford-Smith, 1968; Mainstone, 1974;

Balendra and Huang, 2003; Erberik and Elnashai, 2004, among others). The width of the equivalent diagonal strut can be taken as 0.20 of the total distance along the diagonal of the infill between the two nodes at the beam-to-column intersection. Macro-models employing equivalent multi-struts and improved hysteretic relationships are available and should be considered (Chrysostomou *et al.*, 1992).

4.5.3.5 Frames

Structural models for frames may be 2D or 3D depending on the geometry of the system, boundary conditions and applied loads. For example, symmetric multi-storey buildings subjected to gravity loads and horizontal seismic forces are generally discretized as series of two-dimensional FE models. Under earthquake loads, it is unlikely that all members yield at the same time, because of the randomness of both demand and supply. Non-uniform yielding may indeed cause asymmetry in systems that appear to be symmetric. Three-dimensional models should be employed to capture biaxial bending effects, especially in the inelastic range. Moreover, for plan asymmetric structures and for buildings with flexible diaphragms, 3D models provide more accurate estimations of the earthquake response than 2D discretizations. Figure 4.22 shows the three-dimensional FE model employed for the irregular multi-storey SPEAR frame presented in Figure 4.2. Modelling of frame components, e.g. beam-columns, joints, and diaphragms, has been discussed in detail in Sections 4.5.3.1 to 4.5.3.4. Focus is placed herein on the assembly of the above components into a load-resisting system. The fundamental steps for the modelling of frames are outlined in Figure 4.23. The definition of the geometric model, mechanical properties and loads is of critical importance since it significantly affects the reliability and efficiency of the analysis. Further refinement in the FE models does not necessarily improve accuracy because of other over-riding factors such as material modelling inaccuracies. As a general rule, the simplest adequate model should be sought, and no further refinement should be employed.

The first step in the definition of the geometric model is the selection of node locations. The nodes used in FE models of frames are often classified into structural and non-structural. Structural nodes are necessary to define the connectivity of the elements utilized for the discretization, while non-structural nodes are employed for the orientation of member axes, as discussed in Section 4.5.2. In some commercial software packages, only structural nodes are explicitly defined and a default orientation is assigned to the element set of axes. Adequate FE models for frames should have nodes at the following locations:

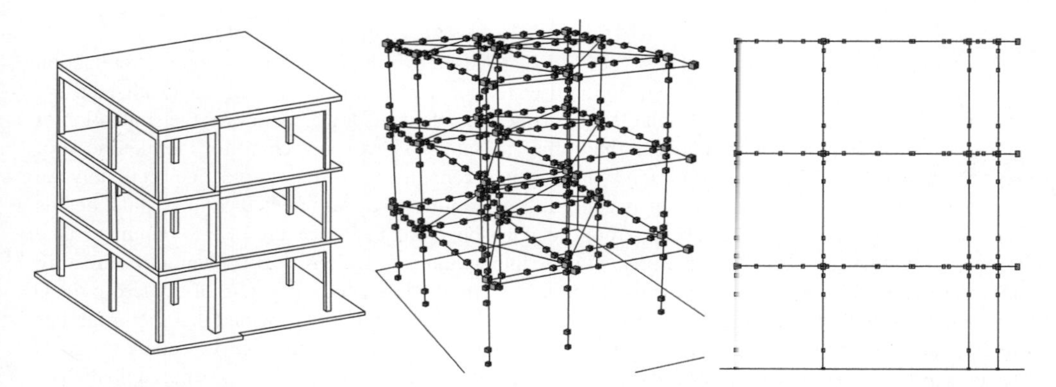

Figure 4.22 Detailed structural modelling of the SPEAR frame in Figure 4.2: full-scale prototype (*left*), 3D (*middle*) and plane (*right*) finite element spine models

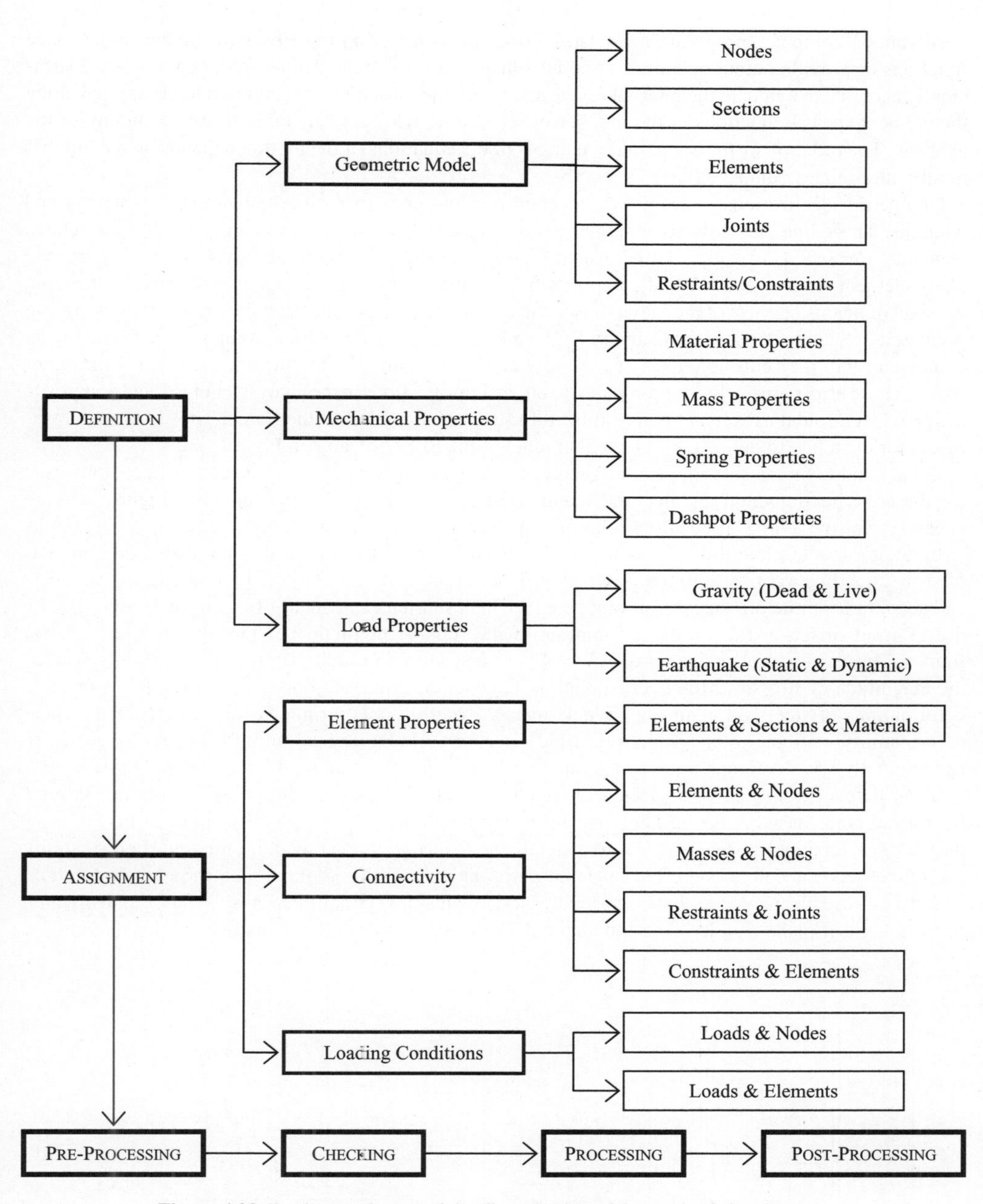

Figure 4.23 Fundamental steps of the discretization of frames by finite elements

- Intersections between beams and columns;
- Points where abrupt changes in the geometry occur;
- Free-standing ends of cantilever elements;
- Points of application of concentrated actions (forces and moments) or imposed deformations;
- Points where output quantities are required.

Advanced computer programs for seismic assessment may employ FEs with higher-order shape functions or provide output quantities at points other than beam ends. Higher-order elements are sufficiently accurate to evaluate the internal forces and moments caused by concentrated loads applied along the element span. Reductions in the number of structural nodes are beneficial for the economy of the analysis. The order of stiffness and mass matrices in the equations of the motion depends on the number of structural nodes connected by FEs.

Geometric idealizations of frames often employ centrelines to replace actual sizes of beams and columns. These line elements are assumed to be coincident with the centroidal axes of the discretized elements. For metal structures with constant section members, the centroidal axis is uniquely defined along element lengths. For RC and composite frames, the sectional properties vary, for example, along the span of beams because of the inevitable occurrence of cracking of the concrete at low stresses, as also discussed in Section 2.3.1.1. For ordinary RC framed buildings, having straight lines in the frame model coincident with the centroidal axes of the gross concrete sections of the members can result in minor errors. These errors are negligible compared to other simplifications in the analysis. For RC and composite members, centroidal axes may be thus utilized for the 'spine model' of the frame. This assumption was made, for example, to define the refined 3D model of the SPEAR frame in Figure 4.22.

Beams and columns seldom have aligned centroidal axes because of the different dimensions used for the cross sections and architectural features. In these cases, offsets as shown in Figure 4.24 are necessary for modelling of beam-to-column joints for the sample frame.

In addition, centreline dimensions are used as clear spans for beams and columns in spine models of structures. In several FE programs, a rigid-offset option is available to model non-aligned members and account for finite dimensions of joint panels, which shorten clear spans of beams and columns. The use of rigid offsets may, however, compromise the reliable prediction of the dynamic response of frames. These offsets introduce unrealistic stiffness concentrations and may result in local increases of the magnitude of stresses. Moreover, spread of diagonal cracking and bond-slips in RC joints or large deformations of steel and composite panel zones at beam-to-column intersections increase the shear deformability of frame joints. Therefore, rigid offsets should be utilized only for beams connected to exceptionally wide columns, like for example element 'C6' in Figure 4.24, or connected to walls.

Shear deformations can be neglected for slender members. However, for deep beams and columns, FE formulations based on Timoshenko beam theory should be employed as discussed earlier. In ordinary frames, torsional stiffness GJ is also generally neglected. For open sections, torsional deformability should be accounted for. The effect of axial deformations in beam-columns on the global structural response is usually neglected but in special cases. Beams are monolithic with the floor slabs, which possess high in-plane rigidity. The influence of axial stiffness of columns increases with the number

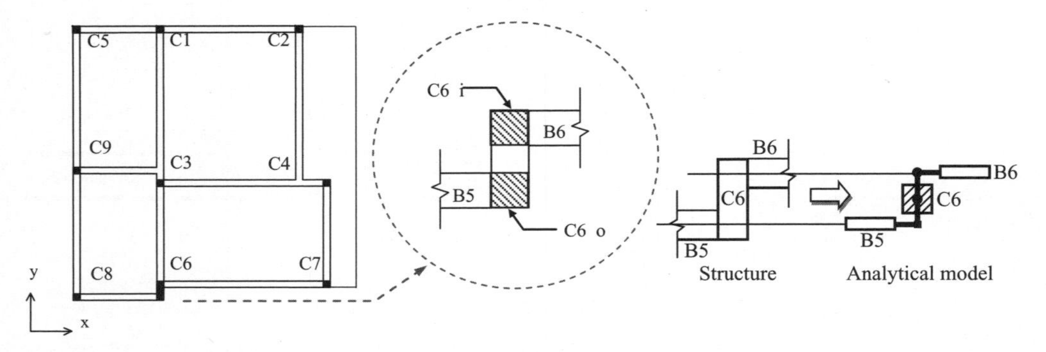

Figure 4.24 Sample SPEAR frame with in-plane beam offset (*left*), modelling of corners (*middle*), and intersections of beams and columns (*right*)

of storeys and when beams with large flexural stiffness are used. Axial deformations of columns should also be considered for high-rise buildings, i.e. structures with height-to-width ratios greater than 5 (e.g. Balendra, 1993; Taranath, 1998).

To assess the inelastic earthquake response of RC and composite frames, concrete slabs are rarely modelled by shell elements. The in-plane rigidity can be simulated by equivalent strut-and-ties or beams, as discussed in Section 4.5.3.3. Floor beam sections may employ equivalent sections to take into account the presence of concrete slabs. Effective flange widths b_{eff} are utilized for L- and T-shaped sections in RC frames or for steel wide-flange sections in composite constructions to simulate the contribution of floor slabs to beam stiffness and strength. The effective flange width b_{eff} used by Jeong and Elnashai (2005) for the seismic performance assessment of the SPEAR frame is the beam width plus 7% of the clear span of the beam on either side of the web. It is noted that effective beam widths are a function of deflection; the effective width increases with increasing deflection.

To define accurately the spreading of inelasticity in frames, it is required to use refined meshes at the locations of expected inelastic straining. The latter locations may however not be known *a priori*. Preliminary elastic analyses can be carried out to identify expected locations and drive the design of the FE mesh. The FE models in Figures 4.16 and 4.22 employ sufficiently refined meshes to identify plastic hinges and quantify the inelastic deformation capacity of the sample RC frame.

In the evaluation of the global structural response, geometric non-linearities, e.g. P-Δ effects should be included. The influence of geometric non-linearities increases as the deformations of a ductile structure progress into the inelastic region and may cause significant reduction in the global lateral load resistance. These effects may be included in the incremental form of the governing equilibrium equations of the structural system. Typical formulations for geometric non-linearity include geometric stiffness matrices; their coefficients change during the load process in compliance with the deformed shape of the geometric model of the frame. The cubic elastic-plastic 3D beam-column element formulated by Izzuddin and Elnashai (1993) and shown in Figure 4.16 for the modelling of the sample frame can be reliably employed to simulate large displacements of framed structures under inelastic dynamic loads, such as earthquake ground motions. Higher-order polynomials have also been used for beam-column elements to reduce the number of elements required to capture curvatures in frame members (Izzuddin and Elnashai, 1993).

4.5.3.6 Structural Walls

Structural walls are effective lateral action-resisting systems for medium-to-high-rise buildings as further discussed in Appendix A. Steel walls employ diagonal braces or steel and composite panels in bounding frames. The modelling of equivalent braces has been outlined in Section 4.5.3.3, while methods to discretize non-structural walls (infills) have been provided in Section 4.5.3.4. This section focuses on models for the response of structural walls.

Structural walls are generally systems with constant cross section along the height and may include openings for windows and doors, a feature that may significantly influence the modelling approach (Figure 4.25). The purpose of the analysis, i.e. evaluation of local or global response, also dictates the selection of refined meshes or simplified 'equivalent' systems.

Walls without openings may be represented by two-dimensional elements, e.g. shells or membrane elements, as presented in Section 4.5.3. The size of the adopted FEs should be proportional to the dimensions of the structural systems, e.g. plate/shell length equal to 1/5–1/10 of the width of the wall. Refined meshes should be employed for zones where high stress and strain gradients are expected, such as points of application of loads and sharp corners. This geometric idealization is suitable for the assessment of local distribution of actions and deformations within the wall. Moreover, the spread of cracking and reinforcement slippage may be modelled for RC walls, while plate buckling may be investigated for metal and composite systems.

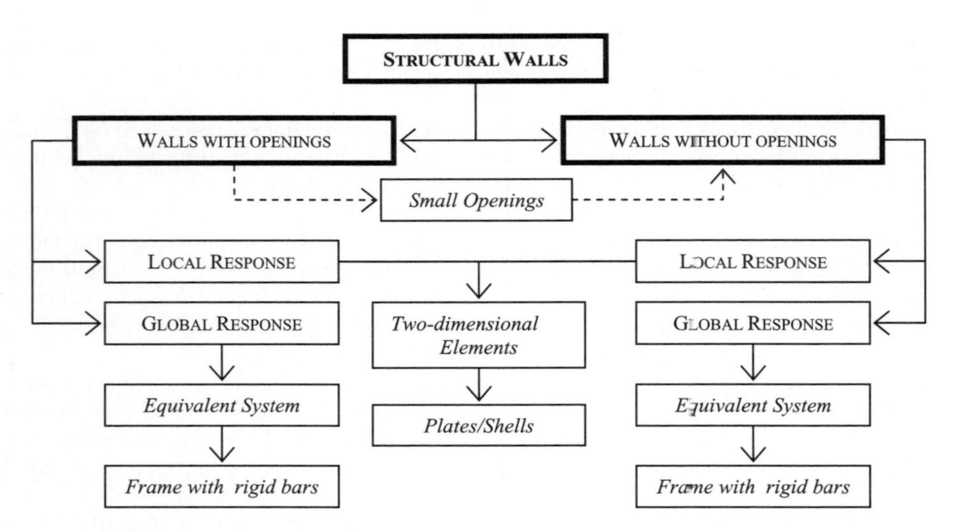

Figure 4.25 Analytical models to perform seismic assessment of structural walls

The global response parameters of walls without openings may be computed by assuming equivalent columns at the vertical centreline of the wall. The sectional properties of the columns are those of the wall. In some computer programs, the effective shear area A_v and shear modulus G are computed automatically; these quantities account for the shear deformability. The equivalent column is a simplified model, which is sufficiently accurate for walls with aspect ratios $H/B > 3$–4, where H is the height and B the width. The equivalent column approach is extensively used for the analysis of three-dimensional models of buildings with frames and wall systems. It, however, does not provide information on the local response of walls, the detailed distribution of deformation or the failure modes.

Openings in structural walls are often arranged in a regular pattern; their size can vary with respect to the length of the wall. In the case of small openings, the system can be considered equivalent to walls without openings, with appropriate reduction in stiffness and strength. For systems with large windows and doors, two options are available (Figure 4.26). To assess local effects in structural walls, plate or shell elements may be used. Refined meshes should be placed close to the openings as shown in Figure 4.26 to adequately describe the stress flow.

Equivalent frame discretizations are simple and effective for models of wall systems with large widths (Figure 4.26). Rigid end bars are introduced to simulate the high stiffness of the joint panel zones. As for the equivalent column elements described above, these models do not capture local effects. They are used for both elastic and inelastic analyses of dual systems, especially for three-dimensional modelling, because of their high efficiency.

In dual systems which employ refined two-dimensional models for walls, the modelling of connections between frames and plate elements is often a critical issue. Bending moments should be transferred from the beam-columns to the plates. However, the latter two-dimensional FEs do not possess rotational stiffness. In these cases, plates should be replaced by shell elements. Alternatively, additional beam elements are connected to the plates to transfer distributed forces equivalent to the beam bending moment among several nodes. Connecting one-dimensional FEs can be either orthogonal to the centroidal axis of the frame (T-shape connection) or penetrating within the wall. The system with T-shape connections can be used to estimate local effects at the interface between the beam of the frame and the wall, while the model with penetrating beams is appropriate for the global response at the frame–

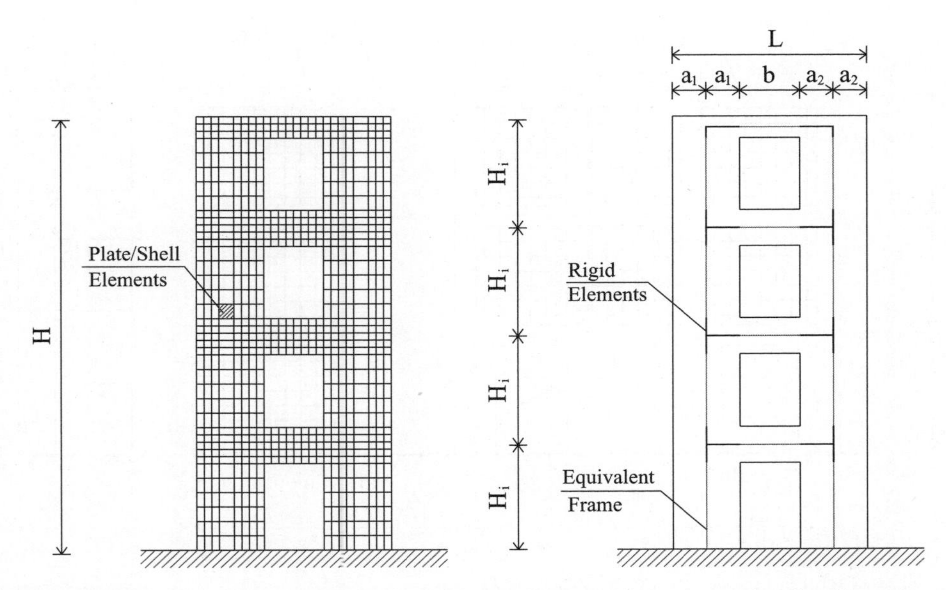

Figure 4.26 Refined (*left*) and simplified (*right*) finite element models for structural walls with openings

wall interface. Moreover, for walls with edge flanges, additional vertical one-dimensional FEs, i.e. beam elements, can be used at the interface between frame and the plate elements. These additional FEs resist the axial loads associated with the global overturning moments while the plates are subjected primarily to shear forces.

Several FE commercial programs do not allow connections between rigid diaphragms and walls discretized with two-dimensional elements. In this case, the connection may be achieved by duplicating the nodes of the wall located at each storey level. Slave nodes are connected to the master joint(s) of the floor introduced in Section 4.5.3.3. Rigid links are used between slave nodes and the corresponding nodes of the FEs on the wall. Alternatively, the nodes of plates or shells at the floor level are connected by means of rigid rods to the nodes of the diaphragm. It is customary to assume that framed structures are fully restrained against rotations at the base. For wall systems which possess large horizontal stiffness, this assumption is, however, less accurate than in the case of slender columns. Refined assessment requires including the different features of soil-foundation-structure interaction.

Problem 4.6

Define a finite element model for the hybrid system shown in Figure 4.27. Comment on your answer.

4.5.4 Masses

Both distributed and lumped mass representations are employed to model translational m_t and rotational m_r masses in earthquake response analysis. A mass matrix is a discrete representation of a continuous mass distribution assembled for the purposes of FE analysis. Mass matrices representing distributed

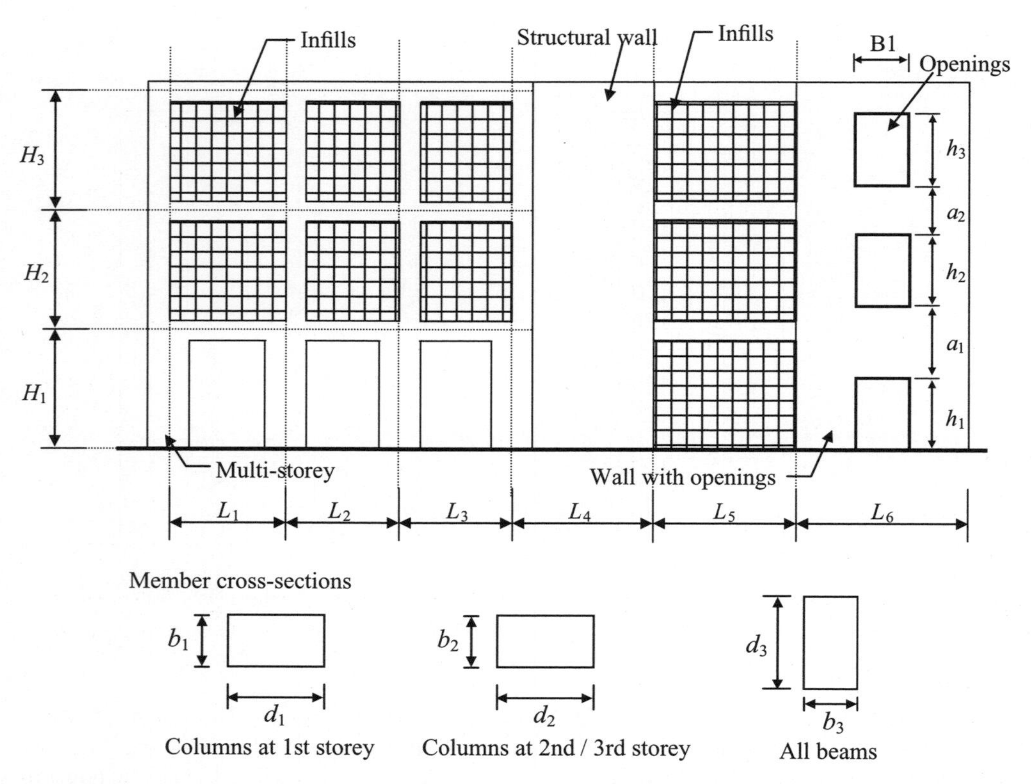

Figure 4.27 Hybrid lateral resisting system: partially and fully infilled framed and wall system
Note: $d_3 < d_2 < d_1$; shear deformations are negligible and shear capacities of all members and connections are expected to be far beyond shear demand

masses are characterized by two functions, which represent the values and the distribution of the mass within the structure. These functions are the mass density μ and shape function ϕ, representing value and distribution, respectively. If the displacement shape function is used to discretize the distributed masses, this is referred to as consistent mass representation. On the other hand, lumped mass matrices are obtained by placing masses (m_t and/or m_r) at nodes of the geometric idealization of the structure. Lumped mass matrices are matrices with the non-zero terms on the leading diagonal, while consistent matrices have non-zero terms that are off-diagonal in addition to the diagonal terms. As a result, lumped mass representations are computationally more efficient since they require less storage space and processing time. Advanced numerical techniques may, however, be implemented in FE computer programs to diagonalize consistent mass matrices.

Mass representations affect the evaluation of eigenvalues in modal analysis of structural systems. The computed natural frequencies tend to be upper bounds of exact frequencies from experiments or closed-form solutions. Consistent mass matrices are more accurate for flexural problems, such as beams and plates. On the other hand, lumped masses exhibit sufficient accuracy, especially for seismic analyses (e.g. Kim, 1993), and will therefore be discussed in detail hereafter.

Lumped mass representation is often used for bridge and building FE models. The number of lumped masses employed for a structural system depends on its geometry, loading conditions and type of geo-

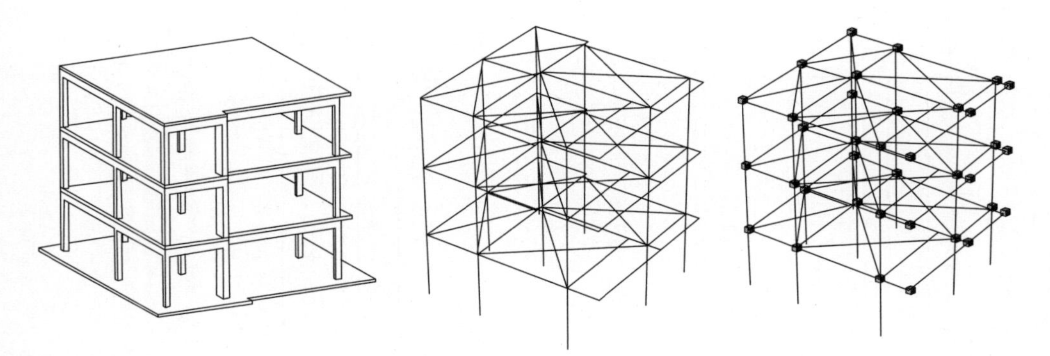

Figure 4.28 Mass modelling of RC frame in Figure 4.2: 3D structure (*left*), FE model without (*middle*) and with (*right*) masses
Key: Cubic markers indicate the location of lumped masses

metric discretization adopted. For multi-storey framed buildings, it is commonly assumed that structural masses are lumped at storey levels. Figure 4.28 shows the lumped mass modelling of the SPEAR frame first presented in Figure 4.2. The masses are applied at beam-to-column connections at all storeys. It is instructive to note that the actual fundamental period of the full-scale RC frame is higher than the values computed by eigenvalue analysis of the refined 3D-FE model shown in Figures 4.22 and 4.28. Variations between analytical and experimental periods may be caused by micro-cracks, which are inevitable in RC members. However, the smallest differences, on average less than 4% (Jeong and Elnashai, 2005), are found for the FE models that include rigid diaphragms and shear joint models presented in Sections 4.5.3.3 and 4.5.3.1, respectively.

In commercial software packages for earthquake analysis, the mass of each beam is automatically lumped at the member end nodes. Further, in FE modelling of framed structures with 'diaphragm constraints' and master nodes, which have been presented in Section 4.5.3.3, two translational masses and one rotational mass about the vertical axis are necessary for 3D dynamic analysis. The two translational masses are applied along the principal directions of the framed systems. For two-dimensional frame models, only one translational mass is required.

Bridge piers are modelled as 2D or 3D flexural systems. Most of the mass is associated with the deck. Bridge decks are often represented as linear elastic systems. The mass of the majority of bridge piers is a small proportion of the upper deck and can therefore be neglected or lumped with the deck mass. If the stresses in the pier are critically affected by its own vibration modes, or its mass is non-negligible, its mass is represented separate from the deck (Figure 4.29). When decks are sufficiently wide, their torsional modes may be important and an adequate mass representation along the width of the deck is necessary. Accurate representations of rotational inertia mass should be included particularly for large decks in single-pier bridges to account for higher mode effects that may be introduced by near-field earthquake strong motions.

Whenever vertical effects of earthquake ground motion (Sections 3.4.6 and 3.4.7) are of interest, the vertical translational mass has to be included in the FE model. It may also be necessary to model masses on the beams to capture the effects of their vibrations under vertical motion.

The description above has focused on the representation of structural masses for dynamic analyses. Non-structural masses may also be present in bridges and buildings, e.g. in the form of attached machinery, water tanks or other heavy electrical and mechanical equipment. These should be included in the model.

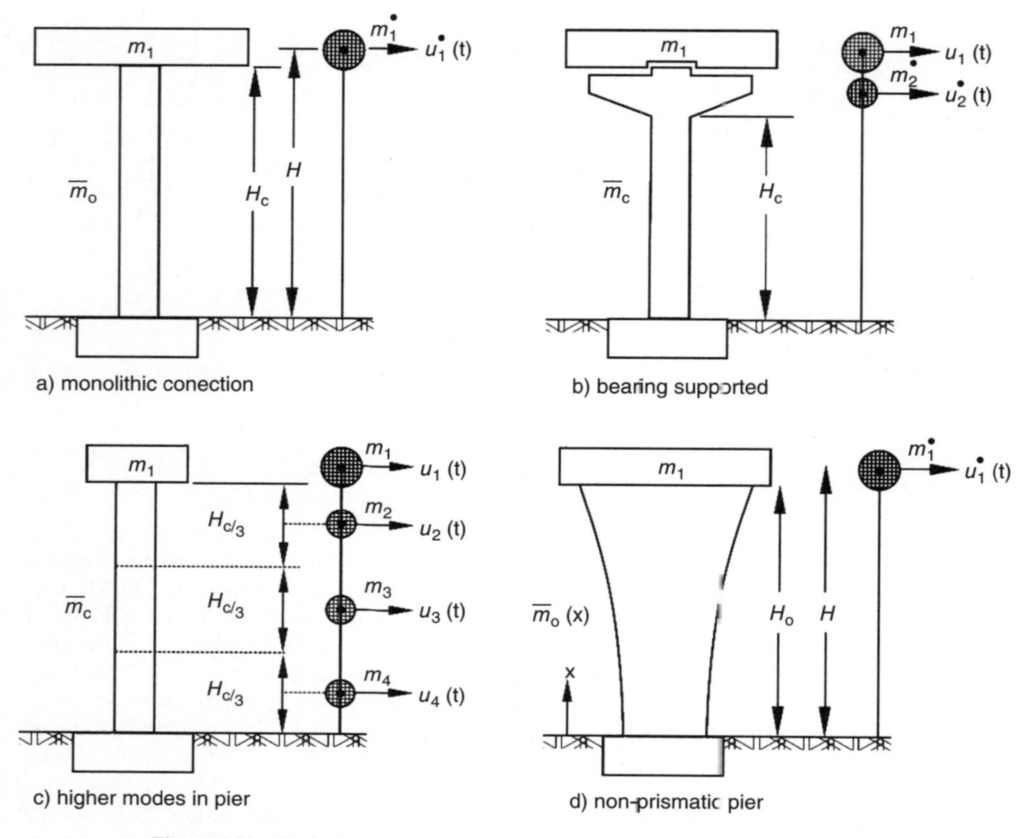

Figure 4.29 Typical mass modelling for bridge piers (*after* Priestley *et al.*, 1996)
Key: m = element total mass

Problem 4.7

Consider the multi-span bridge model in Figure 4.30. Draw the most suitable representation of masses for the cases of (i) horizontal motion only, and (ii) horizontal and vertical motion.

4.6 Methods of Analysis

The use of seismic analysis both in research and practice has increased substantially in recent years due to the proliferation of verified and user-friendly software and the availability of fast computers. This section presents an overview of the main methods of structural analysis used in earthquake engineering, summarized in Figure 4.31. The methods reviewed are grouped into static or dynamic methods, which are applied in elastic and inelastic response analysis. Dynamic analysis is the most natural approach towards the assessment of earthquake response, but is significantly more demanding than static analysis in terms of computational effort and interpretation of results.

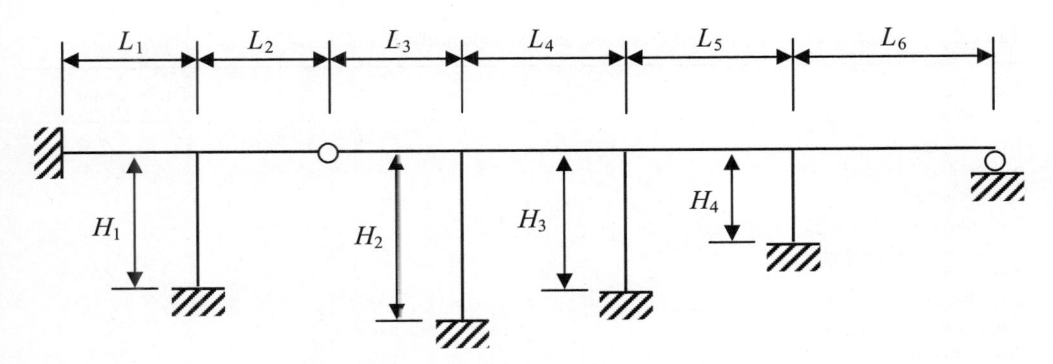

Figure 4.30 Multi-span bridge system

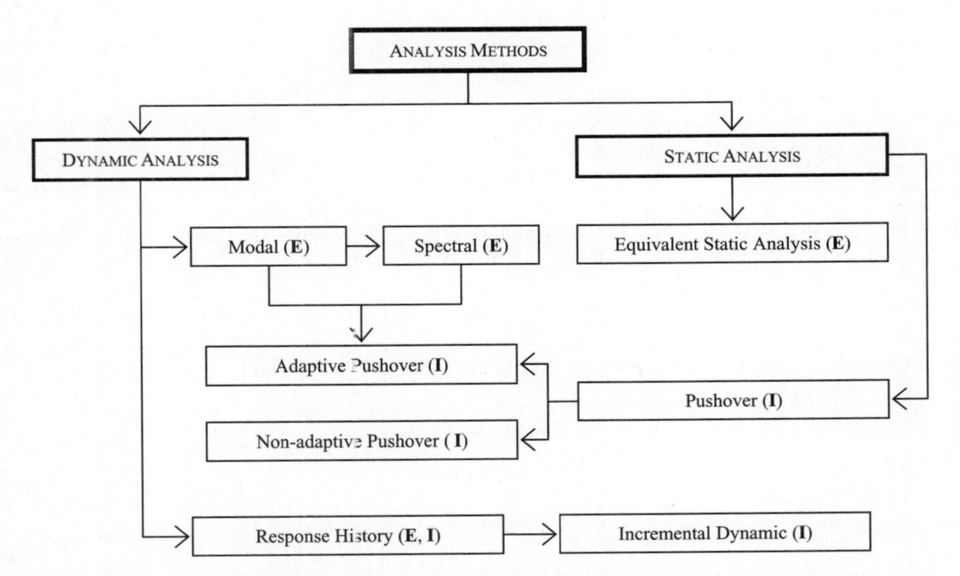

Figure 4.31 Common methods of structural analysis used in earthquake engineering
Key: **E** = elastic analysis; **I** = inelastic analysis

Table 4.7 lists the features of and requirements for static and dynamic analyses. It is clear that static approaches are less time-consuming but often reliable only for a limited class of structures, e.g. regular systems discussed in Appendix A, under normal strong ground motion, as discussed in Chapter 3. Inelastic large displacement response history is the most powerful tool of analysis. However, its potential accuracy and reliability are balanced by its complexity due to the selection of the seismic input as illustrated in Section 3.5.1.2, the structural modelling presented in Section 4.5 and the time-consuming computational schemes, which are discussed in Section 4.6.1.

Table 4.7 Comparisons of requirements for static and dynamic analyses.

Properties	Static analysis	Dynamic analysis
Detailed models	✓	✓
Stiffness and strength representation	✓	✓
Mass representation	✗*	✓
Damping representation	✗	✓
Additional operators	✗	✓
Input motion	✗	✓
Target displacement	✓	✗
Action distribution fixed	✓*	✗
Short analysis time	✓	✗

Key: ✓ = Yes; ✗ = No; * = not necessarily for adaptive pushover analysis.

4.6.1 Dynamic Analysis

The equation of equilibrium for a multi-degree of freedom (MDOF) system subjected to earthquake action is as follows:

$$\mathbf{F}_{\mathrm{I}}+\mathbf{F}_{\mathrm{D}}+\mathbf{F}_{\mathrm{R}}=\mathbf{F}_{\mathrm{E}} \tag{4.9.1}$$

where $\mathbf{F}_{\mathrm{I}}$ is the inertia force vector, $\mathbf{F}_{\mathrm{D}}$ the damping force vector, $\mathbf{F}_{\mathrm{R}}$ the vector of restoring forces and $\mathbf{F}_{\mathrm{E}}$ the vector of earthquake loads. Equation (4.9.1) may be expressed as:

$$\underline{\mathbf{M}}\,\mathbf{x}+\underline{\mathbf{C}}\,\mathbf{x}+\mathbf{F}_{\mathrm{R}}=-\underline{\mathbf{M}}\,\mathbf{I}\,\mathrm{x}_g \tag{4.9.2}$$

where the inertia, damping and earthquake forces are expressed, respectively, as:

$$\mathbf{F}_{\mathrm{I}}=\underline{\mathbf{M}}\,\mathbf{x} \tag{4.10.1}$$

$$\mathbf{F}_{\mathrm{D}}=\underline{\mathbf{C}}\,\mathbf{x} \tag{4.10.2}$$

$$\mathbf{F}_{\mathrm{E}}=-\underline{\mathbf{M}}\,\mathbf{I}\,\mathrm{x}_g \tag{4.10.3}$$

in which $\underline{\mathbf{M}}$ and $\underline{\mathbf{C}}$ are the mass and damping matrices, x_g the acceleration of the ground, $\mathbf{x}$ is the vector of (absolute) accelerations of the masses and $\mathbf{x}$ is the vector of velocity relative to the base of the structure, respectively. $\mathbf{I}$ is a vector of influence coefficients, i.e. the ith component represents the acceleration at the ith degree of freedom due to a unit ground acceleration at the base. For simple structural models with degrees of freedom corresponding to the horizontal displacements at storey level, $\mathbf{I}$ is a unity vector. In this case, it represents the rigid body acceleration of the structure due to a unit base acceleration. The use of MDOF lumped systems for dynamic analyses results in a diagonal mass matrix $\underline{\mathbf{M}}$ in which translational and rotational masses are located along the main diagonal, as stated in Section 4.5.4. Use of consistent mass representations leads to a fully populated mass matrix. If the MDOF system behaves linearly, the vector of the restoring forces in equation (4.9.1) can be expressed as follows:

$$\mathbf{F}_{\mathrm{R}}=\underline{\mathbf{K}}\,\mathbf{x} \tag{4.10.4}$$

in which $\underline{\mathbf{K}}$ is the stiffness matrix and $\mathbf{x}$ the vector of displacements.

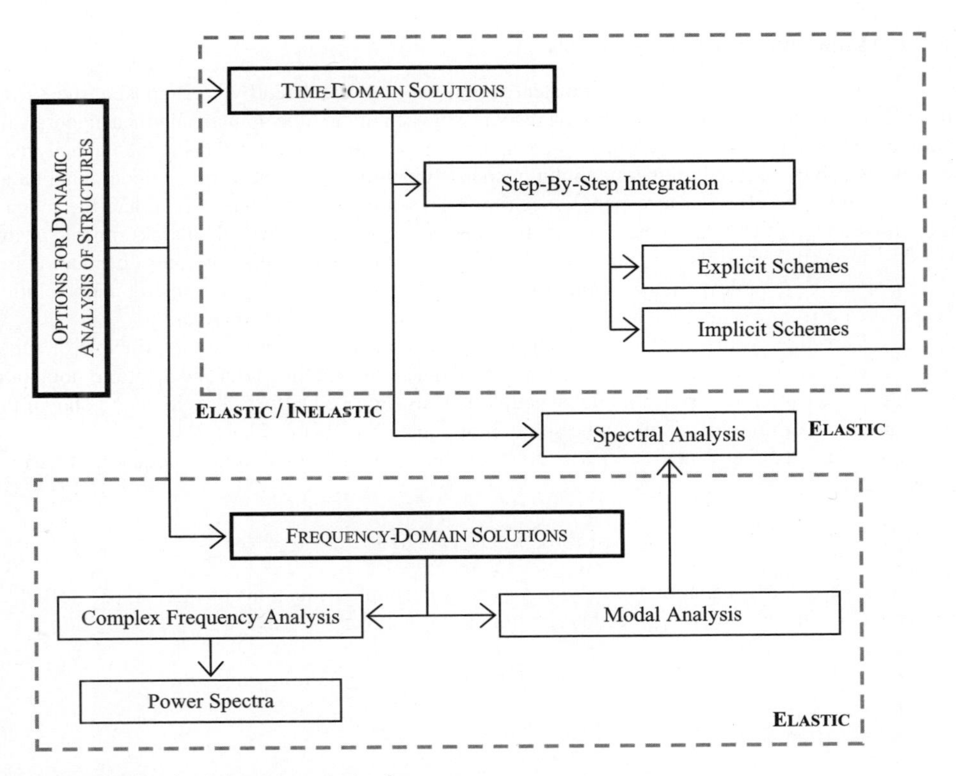

Figure 4.32 Methods of dynamic analysis of structures
Note: The nature of modal and modal-spectra analysis is considered herein as spanning between time and frequency domains

The matrix form of the dynamic equilibrium of motion given in equation (4.9.2) is identical to the equation of motion for single-degree of freedom (SDOF) systems given by equation (3.14) in Chapter 3. However, mass, damping and restoring forces (or stiffness for linearly elastic structures) for MDOF systems are expressed by matrices of coefficients representing the additional degrees of freedom.

Several methods of dynamic analysis of structures exist as shown in Figures 4.31 and 4.32. These methods can be employed either in the time or the frequency domain. The most commonly used methods for dynamic analysis of structures subjected to earthquake loads are modal, spectral and response history. These methods are presented hereafter. It is, however, beyond the scope of this book to provide a comprehensive discussion of numerical algorithms used for each method. The reader is referred to the existing extensive literature (e.g. Bathe, 1996, among many others). It is important to note the special nature of modal analysis. Modal decomposition of the coupled equations of motion leads to a number of equations describing the motion of individual modes in the time domain. Once the maxima of response quantities are evaluated without response history analysis, the method crosses the boundary between time domain, where the individual modes are described, and the frequency domain, where the maxima may be considered with no reference to their time of occurrence. It is the writer's opinion that modal analysis has time-domain as well as frequency-domain features, but this view is not universally accepted.

4.6.1.1 Modal and Spectral Analyses

The response of MDOF systems to a transient signal may be calculated by decomposing the system into series of SDOF systems, calculating the response of each in the time domain and then algebraically combining the response history to obtain the response of the MDOF system. This is modal analysis. If the analysis is only focused on the maximum response quantities, then the various modal maxima are calculated under the effect of a response spectrum representing the transient signal, and the maxima are combined to give an upper bound of the maximum response of the MDOF. This is modal spectral analysis, or spectral analysis for short. Both the above methods are applicable only to linear elastic systems, since they employ superposition. Modal analysis may be considered a time-domain solution, whereas it can be argued that modal-spectral analysis is a frequency-domain solution.

Two concepts are needed for the development of modal analysis. These are the principle of superposition and the convolution integral. Selection of earthquake spectra (input) and adequate combinations of modes are essential to perform modal spectral analysis. For a SDOF system, it can be shown that the displacement at time t is given by the solution of equation (3.14) in Chapter 3.

The coupled equation of motion for MDOF structures given in matrix form in equation (4.9.2) can be rewritten for linearly elastic systems as follows:

$$\underline{\mathbf{M}}\,\mathbf{x}+\underline{\mathbf{C}}\,\mathbf{x}+\underline{\mathbf{K}}\,\mathbf{x}=-\underline{\mathbf{M}}\,\mathbf{I}\,\mathrm{x}_g \tag{4.11}$$

By a change of basis, equation (4.11) yields a set of uncoupled equations of motion, each of which represents a SDOF system. The procedure is summarized below:

(a) Assume that the displacement vector can be expressed in the following form:

$$\mathbf{x}=\underline{\mathbf{\Phi}}\,\mathbf{Y}(t) \tag{4.12}$$

where $\underline{\mathbf{\Phi}}$ is the modal matrix and $\mathbf{Y}(t)$ is the vector of modal (or normal) coordinates. The modal matrix is non-singular positive and hence can be inverted. Note that the columns of the matrix $\underline{\mathbf{\Phi}}$, i.e. the modes of vibrations $\underline{\mathbf{\Phi}}_{\mathrm{i}}$, are not known at this stage;

(b) Formulate the eigenvalue problem for the MDOF system as follows:

$$\underline{\mathbf{K}}\,\mathbf{\Phi}_i=\omega_i^2\,\underline{\mathbf{M}}\,\mathbf{\Phi}_i \tag{4.13}$$

(c) Compute the N eigenvalues (or frequencies), and eigenvectors (or modes of vibration) from equation (4.13). This is a conventional eigenvalue analysis. Alternatively, Ritz vectors can also be employed, especially for complex structural systems, they provide more accurate results for the same number of modes computed through the eigenvectors. The mode with the lowest frequency is the fundamental mode and the corresponding frequency is the fundamental frequency of vibration. Once the frequencies are known, they can be substituted one at a time into the following equation:

$$\left(\underline{\mathbf{K}}-\omega^2\,\underline{\mathbf{M}}\right)\mathbf{x}=\mathbf{0} \tag{4.14}$$

which can be solved for the relative amplitudes of motion for each of the displacement components in the particular mode of vibration. The key characteristic of the mode shapes is that they are orthogonal with respect to the mass $\underline{\mathbf{M}}$ and stiffness $\underline{\mathbf{K}}$ matrices.

(d) Assume mode-proportional damping (i.e. total damping is the sum of the modal damping contributions), given by:

$$\mathbf{\Phi}_i^T\,\underline{\mathbf{C}}\,\mathbf{\Phi}_j=2\,\omega_i\,\xi_{\mathrm{i}}\,\delta_{ij} \tag{4.15}$$

In most FE codes, the 'mass and stiffness proportional' damping is used as an efficient technique of assembling a damping matrix without reference to the element contribution. If two modes only are involved, this is termed 'Rayleigh damping' and is given by the following expression:

$$\underline{\mathbf{C}} = \alpha\, \underline{\mathbf{M}} + \beta\, \underline{\mathbf{K}} \tag{4.16.1}$$

The parameters α and β can be evaluated if the damping ratio ξ_i is known for any two modes. Using the following relationship:

$$\alpha + \beta\, \omega_i^2 = 2\, \omega_i\, \xi_i \tag{4.16.2}$$

two simultaneous equations in α and β are derived for two known values of ξ_i. Consequently, the damping ratio ξ_i in any mode can be calculated as below:

$$\xi_i = \frac{\alpha + \beta\, \omega_j^2}{2\, \omega_j} \tag{4.16.3}$$

The above assumption is essential to retain the option of solving decoupled equations of motion. Since the mode shapes are orthogonal to $\underline{\mathbf{M}}$ and $\underline{\mathbf{K}}$, they are also orthogonal to the Rayleigh damping matrix.

(e) Formulate the equations of motion in terms of normal (or generalized) coordinates Y_i:

$$\mathrm{Y}_i + 2\, \xi_i \omega_i\, \mathrm{Y}_i + \omega_i^2\, \mathrm{Y}_i = -\Gamma_i\, \mathrm{x_g} \tag{4.17.1}$$

where the angular frequency ω_i for the ith mode is:

$$\omega_i = \sqrt{\frac{\hat{K}_i}{\hat{M}_i}} \tag{4.17.2}$$

in which $\hat{M}_i$ is the generalized mass given as follows:

$$\hat{M}_i = \mathbf{\Phi}_i^T\, \underline{\mathbf{M}}\, \mathbf{\Phi}_i \tag{4.17.3}$$

and $\hat{K}_i$ represents the generalized stiffness expressed by:

$$\hat{K}_i = \mathbf{\Phi}_i^T\, \underline{\mathbf{K}}\, \mathbf{\Phi}_i \tag{4.17.4}$$

The factor Γ_i is called the 'modal participation factor' and provides a measure of the degree to which the ith mode participates to the global dynamic response. This factor is as below:

$$\Gamma_i = \frac{L_i}{\hat{M}_i} \tag{4.17.5}$$

where:

$$L_i = \mathbf{\Phi}_i^T\, \underline{\mathbf{M}}\, \mathbf{I} \tag{4.17.6}$$

(f) Compute the solutions of the system of N uncoupled equations in normal coordinates given in equation (4.17.1). The response of the ith mode of vibration at any time t can be expressed by the convolution (Duhamel) integral in the form:

$$\mathrm{Y}_i(t) = \frac{L_i}{\hat{M}_i \omega_i} A_i(t) \tag{4.18}$$

where $A_i(t)$ is given by the solution of equation (3.14) in Chapter 3. Alternatively, the equation of motion can be solved numerically in the time or frequency domain. These approaches are known as 'direct integration method' and 'fast Fourier transform', respectively.

(g) Compute the total elastic restoring force as follows:

$$\mathbf{R} = \underline{\mathbf{K}}\, \underline{\mathbf{\Phi}}\, \mathbf{Y(t)} = \sum_{i=1}^{N} \frac{L_i}{\hat{M}_i} A_i(t) \underline{\mathbf{M}}\, \mathbf{\Phi}_i \tag{4.19}$$

(h) Compute the total seismic base shear V_B. It can be obtained by summing the effective earthquake forces over the height of the structure:

$$V_\mathrm{B} = \sum_{i=1}^{N} \frac{L_i^2}{\hat{M}_i} A_i(t) \tag{4.20}$$

(i) Compute the relative displacement with respect to the base of the structure corresponding to the ith mode of vibration:

$$\mathbf{x}_i = \underline{\mathbf{\Phi}}\, \mathbf{Y}_i\, (\mathbf{t}) = \frac{L_i}{\hat{M}_i} A_i(t) \mathbf{\Phi}_i \tag{4.21}$$

Equation (4.16.3) makes damping frequency-dependent. The procedure illustrated in (d) to compute ξ_i will usually over-damp the higher modes of vibration, thus affecting the reliability of results for high-rise structures or systems subjected to near-field earthquake ground motions. Proportional damping can be visualized as immersion of the structure in a non-physical fluid whose viscosity becomes infinite for rigid-body motion of the structure ($\omega = 0$). For higher frequency modes, viscosity acts to damp relative motion of the MDOF, with increasing effect as ω increases. Non-physical high-frequency vibrations, also known as 'noise', generated by numerical response simulation can be damped by the term $\beta \underline{\mathbf{K}}$.

The term $L_i^2 / \hat{M}_i$ in equation (4.20) is defined as the 'effective modal mass'. This quantity generally diminishes inversely with the order of modes. For example, in regular shear frame buildings, the fundamental mode accounts for up to 85–90% of the total mass. Therefore, summing the response for the first two to three modes will represent the MDOF system. On the other hand, slender long-span bridges usually respond in tens or even hundreds of modes, all of which will be required to achieve adequate representation of the MDOF. The sum of the modal masses is the total mass of the structure; i.e.:

$$\sum_{i=1}^{N} \frac{L_i^2}{\hat{M}_i} = \sum_{i=1}^{N} M_i \tag{4.22}$$

Equations (4.19) and (4.21) express the entire history of actions and deformations of MDOF structures. Lumped systems with N degrees of freedom possess N independent mode shapes. It is thus possible to express the deformed shape of the structure in terms of amplitudes of these shapes by treating them as generalized coordinates $\mathrm{Y}(t)$ as shown in equation (4.18).

In seismic analysis, the evaluation of maximum values of displacements and internal forces rather than their whole time history, is often the primary purpose, especially in design. Peak responses obtained for individual modes can be combined using statistical methods. The modal spectral (or spectral, or response spectrum) analysis estimates peak values of structural response by combining maximum modal contributions. These maxima are determined from earthquake response spectra for elastic SDOFs. The spectral analysis procedure is summarized in the following steps:

(a) Compute modes and frequencies of the MDOF by following steps (a) to (d) of the procedure for modal analysis given above.
(b) Compute for each mode the generalized mass $\hat{M}_i$ and the modal participation factor Γ_i from equations (4.17.3) and (4.17.5), respectively.
(c) Select an acceleration spectrum (e.g. as in Section 3.4.2).
(d) Compute the spectral accelerations S_{ai} corresponding to the periods T_i determined for each mode of vibration.
(e) Compute the maximum inertia forces for each mode. The vector of earthquake forces $\mathbf{F}_{\max,i}\ (t)$ for the ith mode is as follows:

$$\mathbf{F}_{\max,i}(t) = \underline{\mathbf{M}}\ \mathbf{\Phi}_i \frac{L_i}{\mathbf{\Phi}_i^T\ \underline{\mathbf{M}}\ \mathbf{\Phi}_i} S_{ai} \tag{4.23}$$

(f) Compute the maximum values of response parameters, e.g. actions (moments, shears, axial loads and torsion, if any) and deformations (displacements and rotations) discussed in Section 4.8. The response quantities can be determined from static analysis.
(g) Combine the quantities determined in step (f) for each mode to determine the total response parameters.

Decisions are needed for the number of modes to be combined and the combination method. The choice of number of modes to be combined has implications on both accuracy and economy of the procedure. In most cases of structural applications, two to three modes are sufficient, as mentioned earlier. The objective is to account for at least 85–90% of the total mass, which is achieved in regular structures with relative ease. In special structures, e.g. slender long-span bridges, reaching the minimum 85–90% limit may require combining tens or even hundreds of modes.

Various approximate formulae for superposition may be used in spectral analysis. The most commonly used methods are the square root of the sum of the squares (SRSS) and the complete quadratic combination (CQC). A reasonable safe upper bound on the overall response parameters is obtained by assuming that the response measures in the different modes are uncorrelated. For three-dimensional structures with a large number of almost similar periods of vibration, this assumption is not applicable.

In the SRSS, the total value of the response parameter E is given by:

$$E = \sqrt{\sum_{i=1}^{N} E_i^2} \tag{4.24}$$

If the difference between two modal frequencies is less than 10%, the SRSS may lead to underestimating the structural response. Notwithstanding, the SRSS combination approach secures a safe upper bound on global response quantities in most cases, as mentioned above. In some cases, local response parameters may not be a safe upper bound, due to the effect of higher modes on local quantities. The simplified code approach presented in Section 4.6.3 is indeed a simplification of the SRSS method, by just replacing the modal mass of the fundamental, or predominant, mode by the total mass.

When modes are closely spaced, a combination approach that includes cross-modal contributions is required, since the closely spaced modes are at least partially correlated. This procedure may be used for all structures; where cross-correlations are low or non-existent, the cross-coupling terms will be small or zero. The CQC is expressed as follows:

$$E = \sqrt{\sum_{i=1}^{N}\sum_{j=1}^{N} E_i E_j \, \rho_{ij}} \tag{4.25.1}$$

where ρ_{ij} is a cross-modal coefficient. This coefficient is generally expressed as a function of the modal frequencies and damping characteristics and, for equal modal damping i.e. $\xi_i = \xi_j = \xi$, is as follows (Der Kiureghian, 1980):

$$\rho_{ij} = \frac{8\xi^2(1+r)r^{3/2}}{\left(1-r^2\right)^2 + 4\xi^2 r(1+r)^2} \tag{4.25.2}$$

where $r = \omega_j/\omega_i$; the coefficient ρ_{ij} varies between 0 and 1 for $i = j$. If the modal frequencies of the MDOF are well separated, the off-diagonal terms tend to zero and the CQC method approaches the SRSS.

Estimates of the total value of the response parameter E obtained by CQC rule may be larger or smaller than the estimates provided by the SRSS rule (Chopra, 2002). Figure 4.33 shows the bending moment diagrams computed from response spectral analysis for a plane frame extracted from the sample SPEAR building in Figure 4.2. The modal combination rules discussed above, i.e. SRSS and CQC, are utilized. The damping value ξ used for analyses is 5%. SRSS and CQC provide values that are in good agreement. The modal analysis of the 3D frame shows that the frequencies of the system are not closely spaced; the minimum difference between two frequencies is greater than 10%. However, the sample SPEAR structure is a multi-storey building with asymmetric plan and hence the SRSS leads to reasonable estimates of response. The differences between the values computed through the CQC and SRSS are lower than 10%.

4.6.1.2 Response History Analysis

In contrast to the frequency-domain solutions presented in Section 4.6.1.1 (notwithstanding the special nature of modal analysis), the response of MDOF systems to a transient signal may be calculated by

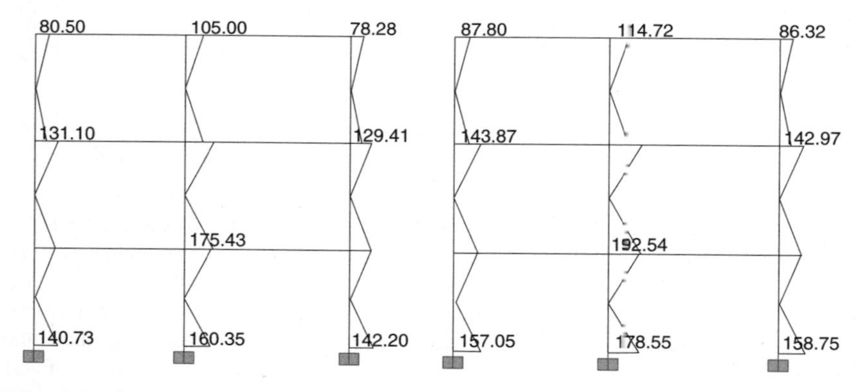

Figure 4.33 Bending moments (in kNm) computed through response spectral analyses using two different modal combinations for three-storey frame: square root of the sum of the squares (*left*) and complete quadratic combination (*right*)

time-stepping techniques where series of coupled equations of motion are solved as static equilibrium systems, but including inertia and damping effects. Time-stepping or response-history analysis is the most natural and intuitive approach. It, however, requires significantly more computing resources than modal and spectral methods. It is noteworthy that the individual modes equations of motion derived by decoupling in Section 4.6.1.1 may be solved in the time domain using the methods outlined in the current section, thus providing a link between frequency-domain and time-domain solutions.

When subjected to strong ground motion, structures generally undergo deformations in the inelastic range, as discussed in Section 2.3.3. Since their deformation is also relatively large, geometric non-linearity may be significant. Analysis of non-linear and inelastic systems subjected to seismic loads involves continuously changing temporal solution characteristics. This is due to changes in stiffness, and hence periods of vibration. To compute the response history of inelastic structures, it is necessary to integrate directly the coupled equations of dynamic equilibrium [given by equations (4.9.1) and (4.9.2)] as the principle of superposition is not applicable. Many numerical integration schemes are available in the literature. A review is provided by Dokainish and Subbaraj (1989), Subbaraj and Dokainish (1989) and Wood (1990). Time-marching schemes are either conditionally stable (explicit) or unconditionally stable (implicit). The response history is divided into time increments Δt and the structure subjected to a sequence of individual time-dependent force pulses $\Delta \mathbf{F}(t)$. During each Δt, the structure is assumed to be linear and elastic. Between intervals, the material and geometry components of the system stiffness matrix are modified to reflect the current state of deformation. The non-linear response is thus approximated by a series of piece-wise linear systems.

The steps required to perform response history analysis of MDOF structures subjected to seismic loads are as follows:

(a) Formulate the equation of motion for the discretized structure in incremental form as given below:

$$\underline{\mathbf{M}}\,\Delta \mathbf{x} + \underline{\mathbf{C}}\,\Delta \mathbf{x} + \underline{\mathbf{K}}_t(t)\Delta \mathbf{x} = \Delta \mathbf{F}(t) \tag{4.26}$$

where $\underline{\mathbf{K}}_t(t)$ is the stiffness matrix for the time increment beginning at time t, and $\Delta \mathbf{x}$ is the displacement increment during the time interval Δt.

(b) Integrate the incremental form given in equation (4.26) for each time step by using one of the numerical integration schemes available in the literature.

(c) Evaluate the increments of displacement, velocity and displacement at the given time step.

(d) Update the displacement, velocity and acceleration at the beginning of the interval to derive the corresponding quantities at the end of the time step interval.

(e) Evaluate stress states corresponding to the total displacements at the end of the given time step.

(f) Update the tangent stiffness matrix $\underline{\mathbf{K}}_t(t)$, if necessary.

The above steps show that the determination of the matrix $\underline{\mathbf{K}}_t(t)$ for each increment is the most demanding part in the response history analysis. All individual member stiffnesses are re-computed within each time increment and iteration within the time increment. This requires considerable computing resources for large structural systems.

The selection of the integration scheme to solve equation (4.26) and the value of integration operators have significant effects on the results. Manipulating algorithmic damping (intentionally) or falling victim to it (inadvertently) could lead to 50% or more variation in force response (Broderick *et al.*, 1994). The selection of damping parameters in the presence of hysteretic (material) damping is also a serious consideration that affects the results of the analysis.

In the modal and spectral analysis described in Section 4.6.1.1, the damping matrix $\underline{\mathbf{C}}$ in the equation of motion is often represented as a linear combination of the mass and stiffness matrices, e.g. equation (4.16.1). Similarly, for non-linear systems, $\underline{\mathbf{C}}$ can be assumed as a linear combination of the mass and stiffness of the initial elastic system. It was demonstrated that this assumption provides a reasonable approximation of the damping (Anderson and Gurfinkel, 1975). In addition, more refined formulations of the damping matrix are not as important in inelastic systems as it is for their elastic counterparts. In the inelastic range, the principal mechanism of energy dissipation is that due to irrecoverable deformations, as discussed in Section 2.3.3. The latter mechanism is accounted for by modelling the hysteretic behaviour of the materials. Even though the role of hysteresis in damping is prevalent, the selection of values of damping coefficients is of importance too, in both elastic and inelastic dynamic analyses (Broderick *et al.*, 1994). Critical damping calibrated on target values in the lower modes can over-damp the contribution of higher modes to the total response. This problem can be solved by adopting selectively dissipative numerical integration schemes, e.g. the Hilber–Hughes–Taylor α-integration scheme (Hilber *et al.*, 1977). By introducing intentional integration errors, observed as period elongation and amplitude decay, spurious mode contributions can be eliminated, thus improving the overall quality of the response calculations.

The application of time-stepping procedures to integrate the equations of motion of MDOF systems requires controlled values of the time interval Δt. Since higher modes are of short periods, small Δt permits accurate integration of higher modes. The higher modes are, however, poorly represented (from a finite element discretization viewpoint) in the dynamic structural response; thus, it is not necessary to use a time increment derived from the highest mode. Instead, a time increment sufficiently small to integrate the highest mode of interest should be utilized. For non-linear problems, where the reduction in stiffness may lead to the sudden inclusion of higher modes of vibration in the integrated response, it is recommended to employ the time integration algorithms developed by Hilber *et al.* (1977), which require the definition of three parameters, generally indicated as α, β and γ. Optimal solutions, in terms of accuracy, analytical stability and numerical damping, are obtained for values of $\beta = 0.25 \cdot (1 - \alpha)^2$ and $\gamma = 0.5 - \alpha$, with $-1/3 \le \alpha \le 0$. These algorithms, which exhibit low numerical damping for lower modes and high damping for higher (generally poorly represented) modes, are implemented in several advanced or commercial FE computer programs, e.g. the computer program Zeus-NL (Elnashai *et al.*, 2003).

It is noteworthy that modal and modal spectral analysis are primarily demand-oriented methods. In other words, they normally provide estimates of the demand imposed by an earthquake on the structural system investigated. They do not necessarily provide estimates of structural capacity, or 'supply'. Only response history analysis has the potential to be used in both 'demand' and 'supply' estimation. This is explored further below.

4.6.1.3 Incremental Dynamic Analysis

Incremental dynamic analysis (IDA), also termed dynamic pushover, is an analysis method that can be utilized to estimate structural capacity (or supply) under earthquake loading. It provides a continuous picture of the system response, from elasticity to yielding and finally to collapse. The rationale behind the IDA is derived by analogy with the incremental static analysis, or pushover, analysis which is discussed in Section 4.6.2.2. The concept of IDA is not new (*see*, for example Bertero, 1977; Nassar and Krawinkler, 1991). It has nevertheless more recently gained in popularity and wide use as a method to estimate the global capacity of structural systems (Vamvatsikos and Cornell, 2004).

The method constitutes subjecting a structural model to one or more ground-motion records, each scaled to multiple levels of intensity. Many dynamic analyses are undertaken and the response from these analyses is plotted versus the record intensity level. The resulting curves, termed IDA curves, give an indication of the system performance at all levels of excitation in a manner similar to the load–displacement curve from static pushover. The steps for obtaining a single earthquake record IDA are as follows:

(a) Define a suitable earthquake record consistent with the design scenario, as discussed in Section 3.5.

(b) Define a monotonic scaleable ground-motion intensity measure, e.g. the PGA, PGV, PGD or a combination (vector representation; e.g. Baker and Cornell, 2003).
(c) Define a damage measure or structural state variable, which could be force-based (maximum base shear, bending moment or axial load) or deformation-based (maximum storey drifts or member rotations) parameters. Energy-based quantities, such as ductility and/or hysteretic energy are also suitable damage indices.
(d) Define a set of scale factors to apply for the selected intensity measure in (b).
(e) Scale the sample record in (a) to generate a set of records that will test the structure throughout its response range, from elastic response to collapse.
(f) Perform response history analysis of the structural model subjected to the scaled accelerogram at the lowest intensity measure.
(g) Evaluate the damage measure in (c) corresponding to the scaled intensity measure in (b).
(h) Repeat steps (f) to (g) for all the scaled intensity measures.

The choice of a suitable intensity and damage measures in (b) and (c), respectively, depend on the purpose of the analysis and the system considered. For example, to assess structural damage of buildings, the maximum inter-storey drift $(d/h)_{max}$ is a reasonable choice since it is directly related to joint rotations and both local (or storey) and global (or system) collapse, as discussed in Sections 4.7 and 4.8.

Typical IDA curves for a multi-storey building are displayed in Figure 4.34. Four different structural responses are obtained for four different records. Common features of the analysis results plotted in Figure 4.34 are the initially linearly elastic branch (elastic stiffness for the given intensity and damage measure) and the flattening of the curves when, at the maximum value of intensity measure, the damage index tends towards very large values [cases (a) and (b) shown in Figure 4.34]. The final flat line indicates that the structure accumulates damage at increasingly higher rates. This is also referred to as 'dynamic instability'. The response curves in Figure 4.34 serve to demonstrate that a single-record IDA cannot fully define the behaviour of a structural system. The IDA is highly dependent upon the sample record. Therefore, a sufficient number of earthquake ground motions should be employed in the analysis. A multi-record IDA results in an IDA curve set, which can be analysed statistically.

Figure 4.35 compares the results computed from a number of incremental dynamic analyses and the response curve along x- and y-directions of the irregular SPEAR frame in Figure 4.2. The differences

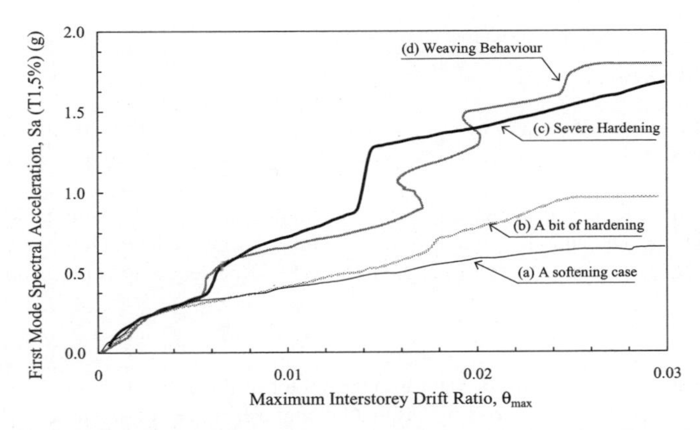

Figure 4.34 Typical incremental dynamic analysis curves (IDA curves) for a multi-storey building under four different earthquakes (*adapted from* Vamvatsikos and Cornell, 2002)

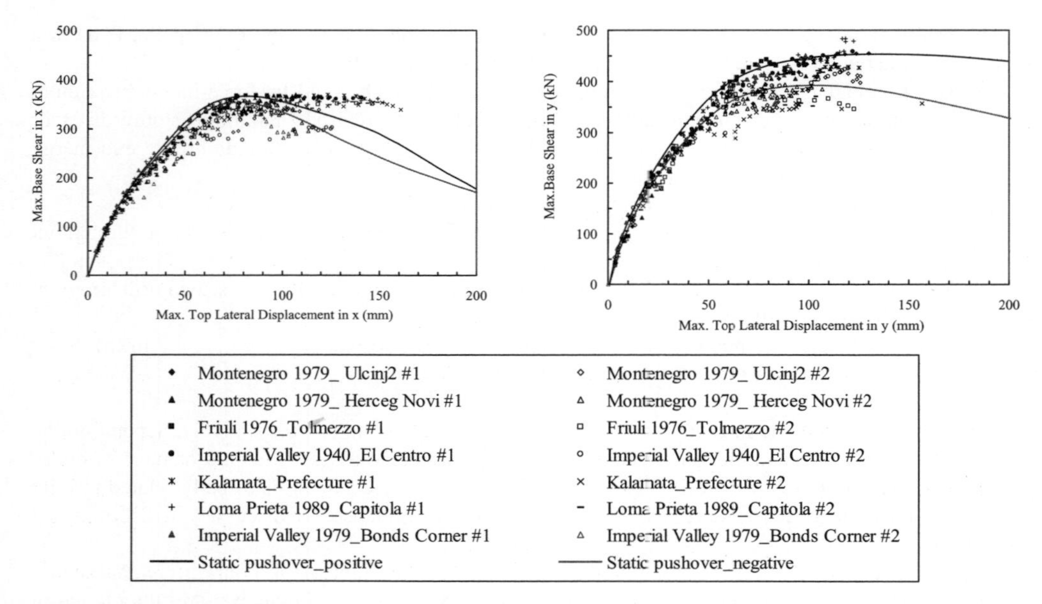

Figure 4.35 Incremental dynamic analysis for the irregular RC in Figure 4.2 (monitoring node@ C3): x-direction *(left)* and y-direction (*right*)

between the static pushover curve and maximum response points are mainly caused by structural irregularities of the assessed system.

The use of IDAs for earthquake engineering applications has several advantages (Vamvatsikos and Cornell, 2002). It provides a better understanding of the structural implications of rare ground-motion levels, which are also discussed in Section 4.7. Moreover, the IDA allows a thorough understanding of changes in the nature of the structural response as the intensity of the ground motion increases, e.g. changes in peak deformation patterns with height, onset of stiffness and strength degradation, and their patterns and magnitudes. It is also suitable to investigate the stability of all the above response features with changes in the input motion.

4.6.2 Static Analysis

Static methods are generally used to assess the capacity or 'supply' of the structural system in terms of actions and deformations at different limit states or performance objectives as those presented in Section 4.7.

Static analysis may be viewed as a special case of dynamic analysis when damping and inertia effects are zero or negligible. The equation of static equilibrium for a lumped MDOF system can be derived from equation (4.9.1) by setting inertia F_{I} and damping F_{D} forces equal to zero, leading to:

$$\mathbf{R} = \mathbf{F}(\mathbf{t}) \tag{4.27}$$

where $\mathbf{R}$ is the vector of restoring forces and $\mathbf{F}(\mathbf{t})$ the vector of the applied earthquake loads. The most commonly used static analysis methods in earthquake engineering are outlined below. Static methods can accommodate material inelasticity and geometric non-linearity. They, however, provide reliable results only for regular structural systems such as those discussed in Appendix A.

4.6.2.1 Equivalent Static Analysis

Equivalent static analysis (also referred to as equivalent lateral force, ELF method) is the simplest type of analysis that is used to assess the seismic response of structures. It is assumed that the behaviour is linear elastic (which corresponds to material linearity), while geometrical non-linearities, i.e. second-order (P-Δ) effects, can be accounted for implicitly. The horizontal loads considered equivalent to the earthquake forces are applied along the height of the structure and are combined with vertical (gravity) loads. Methods of structural analysis are used to solve the equilibrium equations for a MDOF system, e.g. equation (4.27) in which the vector of restoring forces can be assumed proportional to the vector of nodal displacements of the structure.

The critical issue in equation (4.27) is often the load magnitude and distribution. With regard to magnitude, the elastic forces are obtained from the mass of the structure and its predominant period of vibration, and the earthquake spectrum is scaled by a response modification factor, as discussed in Section 3.4.4. This factor is supposed to represent the ability of the structure to absorb energy by inelastic deformation and damage. With regard to load distribution, the most common is a code-type pattern corresponding to the predominant (usually fundamental) mode of vibration. For buildings, inverted triangular or parabolic load patterns are often used, depending on the period of the building. The magnitude of the force at each storey level is also calculated from the predominant mode shape. A triangular distribution provides a good approximation of horizontal forces for structures, which vibrate predominantly in the first mode, e.g. regular medium-rise building structures.

The steps required to assess structures by equivalent static analysis are summarized as follows:

(a) Assume a lateral load pattern distribution.
(b) Apply the gravity and horizontal loads defined in (a) in a single analysis.
(c) Evaluate displacements and hence internal forces.
(d) If scaled forces are used, the ensuing displacements also require scaling.

This method of assessment provides approximate estimates of the deformation of the structure up to the occurrence of significant inelasticity. It, however, ignores important response features, such as redistribution of internal forces, hysteretic effects, stiffness and strength degradation, and others.

4.6.2.2 Pushover Analysis

In this method, forcing functions, expressed either in terms of horizontal forces or displacements, are applied to the lateral action-resisting system. Static forces or displacements are distributed along the height of the structure so as to simulate the inertia forces or their effects. The forcing functions are increased in intensity and the pushover analysis (PA) terminates when the ultimate capacity corresponding to a set of ultimate limit states, as described in Section 4.7, is attained. These forcing functions correspond to one or more mode shapes. If the pattern of forcing function (loads or displacements) is kept constant throughout the analysis, the method is referred to as conventional pushover. If the pattern changes to account for variations in the mode shapes of the structure in the inelastic range, the method is referred to as adaptive pushover. There are variants in the literature, such as modal pushover and others. Further details are given below.

(i) Conventional Pushover Analysis

Conventional pushover is an inelastic static analysis method in which the idealized representation of the structure is subjected to constant gravity loads and to monotonically increasing lateral force or displacement pattern (also termed 'forcing function') of a constant shape. Because the structural model accounts directly for effects of both material inelasticity and geometric non-linearity, the PA is a capacity estimation method under a set of functions that represent inertial effects from the earthquake. This method is capable of shedding light on design weaknesses that elastic analysis cannot detect. For example in the equivalent static analysis presented in Section 4.6.2.1, or in the simplified code method in Section 4.6.3, weaknesses such as storey mechanisms cannot be readily detected.

The PA solution commonly utilizes an incremental-iterative solution of the static equilibrium equations. For a small load increment, the behaviour is assumed linear and equilibrium can be expressed in the form:

$$\underline{\mathbf{K}}\,\Delta\mathbf{x} = \Delta\mathbf{F} \tag{4.28.1}$$

which can be rewritten as:

$$\underline{\mathbf{K}}_{\mathrm{t}}\,\Delta\mathbf{x} + \mathbf{R}_{\mathrm{t}} = \mathbf{F} \tag{4.28.2}$$

where $\underline{\mathbf{K}}_{\mathrm{t}}$ is the tangent stiffness for the current load increment and $\mathbf{R}_{\mathrm{t}}$ the restoring forces at the beginning of the load increment. The forces $\mathbf{R}_{\mathrm{t}}$ can be expressed as:

$$\mathbf{R}_{\mathrm{t}} = \sum_{k=1}^{j-1} \underline{\mathbf{K}}_{\mathrm{t},k}\,\Delta\mathbf{u}_k \tag{4.28.3}$$

where j is the incremental step in the analysis.

During an increment, the resistance of the structure is evaluated from the internal equilibrium conditions and the tangent stiffness matrix $\underline{\mathbf{K}}_{\mathrm{t}}$ is updated when required by the iterative scheme adopted. The out-of-balance forces are re-applied until one or more convergence criteria are satisfied. At convergence, the tangent stiffness matrix is updated and another increment of displacements or forces is applied. The solution proceeds until a target displacement, which is associated with specific performance level (or limit state, as presented in Section 4.7), is reached or the program fails to converge. It is presumed that the program employed to carry out PA has been sufficiently verified so that the numerical, as opposed to structural, collapse is not operative (Elnashai, 2002). Internal forces and deformations computed at target displacements are used to quantify strength and deformation demands, which are, in turn, compared with available structural capacity. The results of pushover analysis (known as 'pushover or capacity curve') are often expressed in terms of global base shear V_{base} versus top lateral displacements δ_{top}. For multi-storey buildings, capacity curves can also be computed at each storey and presented as response curves consisting of storey shear V_i versus inter-storey drift ratio δ_i/h_{i}. Pushover curves for the three-storey irregular RC frames in Figure 4.2 are shown in Figure 4.36. The seismic response, expressed in terms of base shear versus the roof (or top) lateral displacement, was computed for different FE models of the frame system (termed Model #1 through Model #4 in the figure). Significant variations are observed by using structural discretization with and without rigid diaphragm and beam-to-column shear joint models, which are discussed in Sections 4.5.3.2 and 4.5.3.3. Since the direction of earthquake action that will cause collapse is not known, pushover curves were computed for push and pull, as shown in Figure 4.36. In asymmetric structures, these could be distinct and could even cause different failure modes. An example of such an effect is a steel frame diagonally braced in one direction. Its failure in the brace compression direction could be by brace buckling, an effect that will not occur when the single brace is under tension.

The critical parameters defining the characteristics of the conventional PA are the lateral load nature (forces or displacements), its distribution pattern along the height of the structure (triangular, uniform, etc.) and its magnitude. The number of applied load steps, and iterative strategy and convergence criteria, also play a significant role in the effectiveness and reliability of the analysis.

The PA is an analysis method more intuitive than mathematical (Krawinkler and Seneviratna, 1998; Elnashai, 2002). If a set of actions or deformations can be found such that a particular mode of vibra-

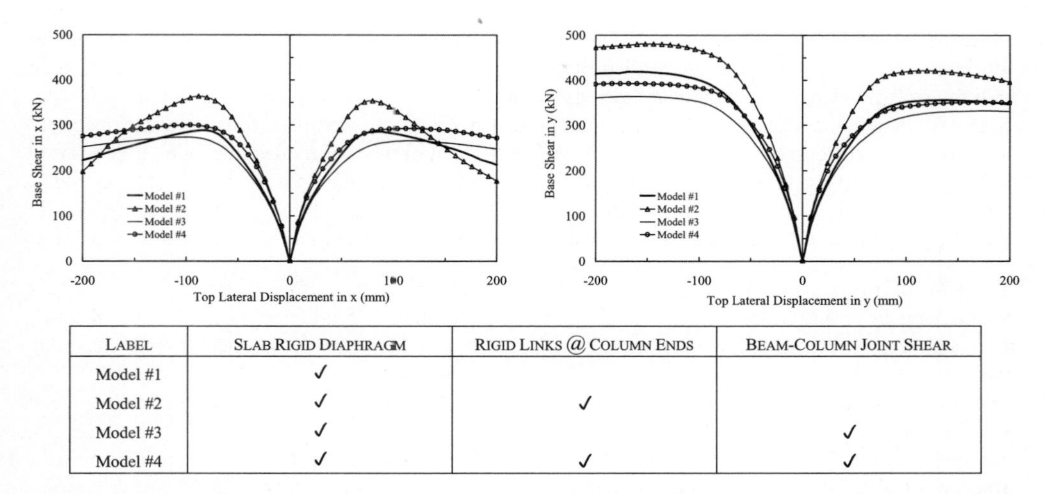

LABEL	SLAB RIGID DIAPHRAGM	RIGID LINKS @ COLUMN ENDS	BEAM-COLUMN JOINT SHEAR
Model #1	✓		
Model #2	✓	✓	
Model #3	✓		✓
Model #4	✓	✓	✓

Figure 4.36 Capacity curves for the sample RC frame in Figure 4.2: pushover along the x-direction (*left*) and y-direction (*right*)

tion or a set of modes is represented statically (i.e. single-mode and multi-mode conventional pushovers, respectively), then the results derived from response of the structure under a monotonically increasing vector of actions or deformations may replace results from dynamic analysis. There is also a variant of multi-mode pushover, where individual pushover curves are obtained for each load distribution, and thereafter combined to produce a multi-mode pushover curve that is superior to that obtained from conventional pushover with a load vector representative of a single or multiple mode (Chopra and Goel, 2002). Single-mode PAs are further discussed below.

The basic assumption of conventional PAs is that structural response is controlled by a single mode or a fixed ratio of modes. The steps required to perform PA are summarized in the following:

(a) Apply the gravity loads in a single step.
(b) Assume a lateral load pattern, either in terms of displacement shape $\boldsymbol{\Phi}$ or force vector $\mathbf{V}$.
(c) Select a controlling displacement node, e.g. the roof centre of mass for buildings.
(d) Determine the vertical distribution of lateral forces V_i ($= m_i\,\Phi_i$), if the displacement vector $\boldsymbol{\Phi}$ has been selected in (b). Conversely, determine the vertical displacement distribution Φ_i.
(e) Compute the incremental-iterative solution of the static equilibrium equations. This step is repeated until the target performance level, e.g. the target displacement of the roof centre of mass, is reached. The target displacement is intended to represent the maximum displacement likely to be experienced during the expected earthquake ground motion.
(f) For structures that are not symmetric about a plan perpendicular to the applied loads, the lateral load or displacement pattern should be applied in both positive and negative directions, e.g. in Figure 4.36, as discussed above.
(g) Determine the base shear V_{base}, top displacement δ_{top}, the storey shear V_i and storey drift δ_i.
(h) Plot the system (V_{base} versus δ_{top}) and the storey (V_i versus δ_i/h_i) pushover curves.

For both 2D and 3D analyses, at least two vertical distributions of lateral forces or displacements should be employed since the actual dynamic force distribution, which may be far from constant, is not known. The uniform pattern, which is proportional to the total mass at each floor, should be used along with the modal pattern. The latter can be the inverted triangular distribution, which is applicable when more than 85% of the total mass participates in the fundamental mode in the direction under consideration. Alternatively, a lateral distribution proportional to the storey inertia forces consistent with the storey shear distribution calculated by combination of modal responses, as illustrated in Sections 4.6.1.1 and 4.6.3, may be used. In so doing, it is necessary to first perform a response spectrum analysis as described in Section 4.6.1.1 including a sufficient number of modes to capture at least 85% to 90% of the total mass and use the appropriate design ground-motion spectrum. The choice of at least two load distributions along the main axis of the structure is a practical and viable solution to partly overcome the limitations of using a static analysis method to solve an inherently dynamic problem.

(ii) Adaptive Pushover Analysis

Adaptive pushover is a method by which possible changes to the distribution of inertial forces, as shown for example in Figure 4.37, can be taken into account during static analysis. As such, it responds to the main shortcoming of conventional pushover, where a constant forcing function has to be used. The time-invariant pattern of horizontal forces and displacements used in conventional pushover may indeed not reflect adequately the inelastic response characteristics of the structure (Elnashai, 2002). Several attempts at adapting the force distribution to the state of inelasticity are provided in the literature (e.g. Bracci *et al.*, 1997; Gupta and Kunnath, 2000). Consequently, a new method of analysis, referred to as 'adaptive pushover' was formulated.

The steps required to perform adaptive pushover analysis for structural systems are summarized as follows:

(a) Apply the gravity loads in a single step.
(b) Perform an eigenvalue analysis of the structure at the current stiffness state. The elastic stiffness can be used for the initial step. Eigenvalues and eigenvectors are computed.
(c) Determine the modal participation factors Γ_j for the jth mode using equation (4.17.5) in Section 4.6.1.1.
(d) Compute the modal storey forces at each floor level for the N modes deemed to satisfy mass participation of about 85–90% of the total mass. These forces $F_{i,j}$ are estimated at the ith level for the jth mode (being $1 \leq j \leq N$) as given below:

$$F_{i,j} = \Gamma_j M_i \Phi_{i,j} g \tag{4.29}$$

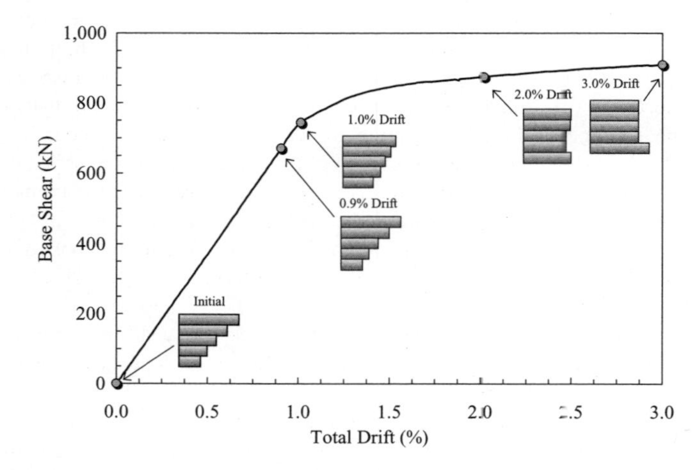

Figure 4.37 Changes of the distribution of inertial forces in a regular framed building (*adaptive force distribution*)

where M_i is the seismic mass of the ith level, g the acceleration of gravity.

(e) Perform a static pushover of the structure subjected to the storey forces computed in step (d) and corresponding to each mode independently.

(f) Estimate element (or local) and structure (or global) forces and displacements by means of SRSS combinations of each modal quantity for the kth step of analysis. Add the above quantities, i.e. forces and displacements, to the relevant quantity of the (k–1)th step.

(g) Compare the values established in step (f) to the limiting values for the specified performance goals at both local and global levels, as provided, for example, in Section 4.7. Return to step (b) until the target performance is achieved.

A variant on the above is empirically using the input motion spectrum to scale the modal contributions to the applied force or displacement vector, referred to as adaptive pushover with spectrum scaling. This procedure is not mathematically rigorous, but resembles the method of base shear calculation adopted in codes, and also used in modal spectral analysis. The lack of rigour results from the application of superposition to an inherently inelastic problem. Spectrum scaling provides an interesting angle, whereby the static capacity curve is no longer unique to a structure, but is a function of the input motion.

The steps of the adaptive pushover analysis utilizing the scaling of acceleration spectrum are as follows:

(a) Apply the gravity loads in a single step.

(b) Perform an eigenvalue analysis of the structure at the current stiffness state. The elastic stiffness can be used for the initial step Eigenvalues and eigenvectors are computed.

(c) Determine the modal participation factors Γ_j for the jth mode using equation (4.17.5) in Section 4.6.1.1.

(d) Compute the modal storey forces at each floor level for the N modes deemed to satisfy mass participation of at least 85–90% of the total mass. These forces $F_{i,j}$ are estimated at the ith level for the jth mode (being $1 \le j \le N$) as given below:

$$F_{i,j} = \Gamma_j W_i \bar{\Phi}_{i,j} S_{a,j}(T_j) g \tag{4.30}$$

where $S_{a,j}(T_j)$ is the spectral acceleration relative to the jth mode with period of vibration equal to T_j.

(e) Compute the modal base shears V_j as follows:

$$V_j = \sum_{i=1}^{N} F_{i,j} \tag{4.31}$$

where N is the number of stories.

(f) Combine the force determined in step (e). Use, for example, the SRSS combination rule as shown below:

$$V = \sqrt{\sum_{i=1}^{N} V_j^2} \tag{4.32}$$

(g) The storey modal base shears V_j computed in step (e) are uniformly scaled:

$$\bar{V}_j = S_n V_j \tag{4.33.1}$$

where the scaling factor S_n can be assumed as follows:

$$S_n = \frac{V_B}{N_s V} \tag{4.33.2}$$

in which V_B is the estimate of the total base shear of the structure and N_s the number of steps utilized to apply the base shear, e.g. $N_s = 100$.

(h) Perform a static pushover of the structure subjected to the scaled incremental storey forces computed in step (g) and corresponding to each mode independently. Different formulations can be used to describe the force variation, which is considered for the incremental updating of the force vector, during the pushover analysis (e.g. Bracci *et al.*, 1997; Gupta and Kunnath, 2000; Elnashai, 2002).

(i) Estimate element (or local) and structure (or global) forces and displacements by means of SRSS combinations of each modal quantity for the kth step of analysis. Add the above quantities, i.e. forces and displacements, to the relevant quantity of the (k–1)th step.

(j) Compare the values established in step (i) to the limiting values for the specified performance goals at both local and global levels, as provided, for example, in Section 4.7. Return to step (b) until the target performance is achieved.

In the above adaptive procedures there are, however, a number of controversial issues, e.g. issue of force distribution and updating. Research to refine adaptive pushover methods is still ongoing for both buildings (e.g. Antoniou and Pinho, 2004a,b; Chopra and Chintanapakdee, 2004; Goel and Chopra, 2004, among many others) and bridges (e.g. Aydinoglu, 2004; Kappos *et al.*, 2005, among others). Comparisons between conventional and adaptive pushover curves for regular and irregular structural systems are provided in Figure 4.38. The adaptive pushovers were performed by utilizing the scaling of acceleration spectrum. Two load patterns were employed for the conventional pushovers, i.e. uniform and triangular. The results of response history analyses are also included in Figure 4.38 as a benchmark.

It is observed from this simple comparison that the uniform distribution provides an upper bound of the lateral capacity in the inelastic range only for the regular model. In the case of irregular systems, the conventional PA is often inadequate to capture the dynamic behaviour, thus proving how misleading fixed patterns can be. In several cases, adaptive pushover is superior to the conventional variant, but this is by no means guaranteed. A wide-ranging comparison between conventional and adaptive pushover methods is available in Papanikolaou *et al.* (2006).

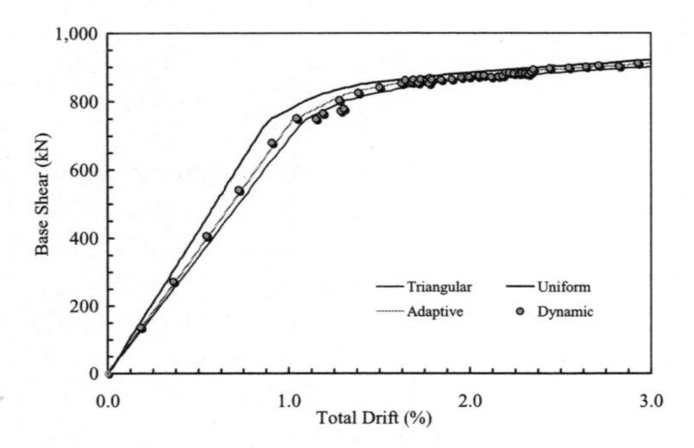

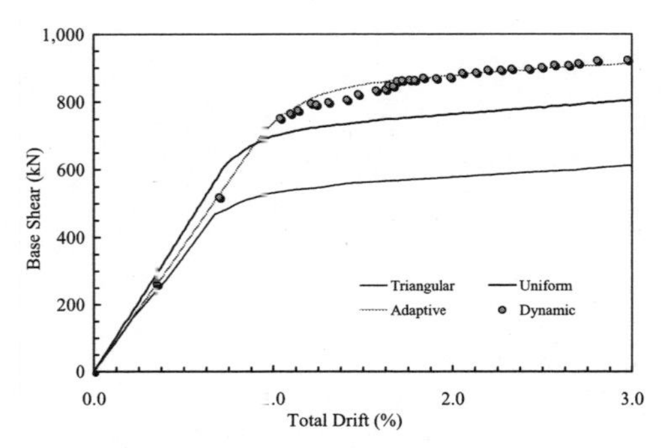

Figure 4.38 Conventional, adaptive and dynamic pushover curves for different structural models: regular (*left*) and irregular (*right*) systems

4.6.3 Simplified Code Method

The simplified code method is intended to replace dynamic earthquake loading by equivalent static loads acting horizontally. Such method is also referred to as 'equivalent lateral force method', as mentioned in Section 4.6.1. The equivalent static load is expressed as a percentage of the total seismic weight of the structure $W_{EQ,t}$. The basis of the method lies in modal decomposition of the response of MDOF systems, described in Section 4.6.1.1. Noting that the first mode modal mass is less than the total mass (indicated as $M_{EQ,t}$ consistent with the definitions in Section 4.3), the use of the latter in the expression of the fundamental mode contribution will result (with some exceptions in force distribution) in a safe upper bound on dynamic actions and their effects. The total horizontal force or base shear V_B acting on a structure is expressed as the product of the structural mass and the earthquake-induced acceleration. The maximum base shear is given by:

$$V_B = C\, W_{EQ,t} \tag{4.34}$$

where the total seismic weight $W_{EQ,t}$ includes the total dead loads and part of live loads. The contribution of the live loads depends on the type of structure, as discussed in detail in Section 4.3. The seismic weight $W_{EQ,t}$ can be computed from equation (4.1) as the sum of all $W_{EQ,i}$ corresponding to floor masses in buildings. The seismic base shear coefficient C is the main outcome sought in seismic codes. Since the effective weight of the fundamental mode $\overline{W}_i$ is about 70% to 80% of $W_{EQ,t}$, in regular structures (*see* Appendix A), equation (4.34) provides a value of V_B significantly larger than the first mode and approximately accounts for the base shear contributions of the higher modes. The effective modal weight $\overline{W}_i$ of the ith mode is given by:

$$\overline{W}_i = \frac{L_i^2}{\hat{M}_i} g \tag{4.35.1}$$

where $L_i^2/\hat{M}_i$ is the effective modal mass relative to the ith mode and g the acceleration of gravity. Note that:

$$\sum_{i=1}^{N} \overline{W}_i = \sum_{i=1}^{N} W_{EQ,i} = W_{EQ,t} \tag{4.35.2}$$

and

$$\sum_{i=1}^{N} \frac{L_i^2}{\hat{M}_i} = \sum_{i=1}^{N} \frac{W_{EQ,i}}{g} = \frac{W_{EQ,t}}{g} \tag{4.35.3}$$

in which N denotes the total number of modes of vibrations, determined through eigenvalue analysis as described in Section 4.6.1.1. The equivalent static load approach is, indeed, based on the modal analysis concept. Strictly speaking, the modal analysis is only applicable to structures with linearly elastic behaviour. However, the equivalent static load approach takes into account the ductility of the structure and hence is applicable to inelastic systems. This approach is not mathematically rigorous and is subject to the same criticism as the adaptive pushover method with spectral scaling.

Different codes attempt to estimate the value of seismic base shear coefficient C such that the obtained base shear V_B and its distribution over the structure represent a safe yet economical upper bound to the earthquake load. The evaluation of the seismic base shear coefficient is dependent mainly on the following parameters:

(i) Seismo-tectonic environment of the area;
(ii) Topography and soil condition of the site;
(iii) Dynamic characteristics of the structure;
(iv) Structural system, ductility and material used;
(v) Importance of the structure.

Whereas (iii) and (iv) above are clearly linked, they are treated separately in codes, with a degree of justification. The five parameters listed above are considered in different ways in seismic codes. A brief description of each parameter is given below:

(i) The 'Zone Factor' accounts for the anticipated seismic activity at the construction site. In this factor, the peak ground acceleration, obtained from seismic hazard studies, is given either directly (as a percentage of acceleration of gravity, g) or implicitly, as illustrated in Section 3.4.5.
(ii) The 'Site Factor' represents the effect of the different foundation materials on the strong-motion characteristics and the probability of high amplification or resonance due to the proximity of the period(s) of vibration of the site and the structure, as discussed in Section 1.3.2. The topology of the site is taken into account in only a small number of codes.
(iii) The 'Response Modification or Behaviour Factor' reflects the relative seismic performance of different structural systems, in terms of local/global ductility, redundancy and redistribution capability, and the predicted mode of failure (brittle or ductile). It is also referred to as the response reduction factor or, wrongly, the ductility factor. Different relationships between force reduction factor and ductility are reviewed in Section 3.4.4.
(iv) The 'Material Factor' reflects the ability of the structural material to dissipate energy and respond in a ductile manner. For instance, masonry is inherently less ductile than steel. In almost all cases, this factor is implicit in the behaviour factor described above.
(v) The 'Importance Factor' accounts for the importance of the building by decreasing the probability of damage or collapse for important, environmentally sensitive or exceptionally heavily populated structures; i.e. it depends on the occupancy of the structure (potential fatalities), its use (importance) and the consequence of its damage (environmental). Implicit in this parameter is the definition of an acceptable probability of being exceeded attached to the design ground acceleration. Higher probability is associated with less important structures, as stated in Sections 4.4 and 4.7. The design values of PGA specified in international building codes correspond frequently either to 10% of probability of being exceeded in 50 years (return period of about 475 years) or to 2% in 50 years (return period of about 2,475 years), especially in North America. However, more recently, acceleration response spectra are also provided for different return period and probability of exceedance corresponding to the various limit states to comply with.
(vi) The 'Design Spectrum', defined in Section 3.4.5, accounts for the coupling between structural periods of vibration and earthquake characteristics, as well as travel path such as attenuation and long-period amplification. The latter effects are expressed, in seismic codes and guidelines, by the corner periods T_1 and T_2 used to define the elastic response spectrum and the long-period exponent utilized to characterize the decay in the long-period range, as shown in Figure 4.39. In most codes, the spectrum is 'flattened' for periods shorter than T_1, to account for softening of short-period structures, which may lead to an increase in applied loads.

The above list is not exhaustive; each national code uses its own philosophy and notation and adopts a different format for the above-mentioned parameters. It provides, however, the fundamental ingredients to estimate the design base shear V_B from equation (4.34).

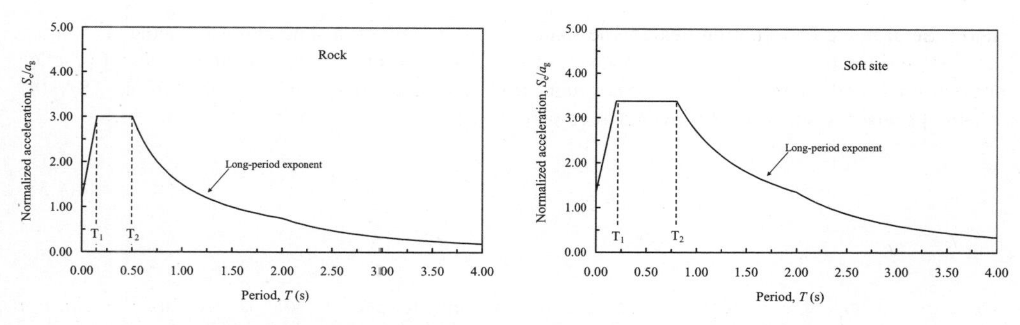

Figure 4.39 Standard smoothed spectra: rock site (*left*) and soft site (*right*)

The fundamental period of vibration T of a structure is essential to compute the base shear V_B. The importance of this dynamic parameter is twofold: the site-structure resonance and the design spectrum ordinate. Codes attempt to supply simplified, semi-empirical expressions for period estimation as a function of height, material, system and number of storeys. These expressions are also calibrated using regression analyses of data derived from system identification procedures (e.g. Goel and Chopra, 1997, 1998).

A reasonable evaluation of the fundamental period for a multi-storey structure requires calculations involving the mechanical properties of the members of the lateral resisting system. Clearly, for new structures these calculations cannot be carried out until the system is designed. It is customary, however, to check the period determined empirically through code-based formulae by using Rayleigh's method, which provides the following expression:

$$T = 2\pi \sqrt{\left[\sum_{i=1}^{N} W_i\, \delta_i^2\right] \bigg/ \left[g \sum_{i=1}^{N} F_i\, \delta_i\right]} \tag{4.36}$$

where W_i is the storey weight, F_i the force applied at the ith storey and δ_i the corresponding lateral displacement. Equation (4.36) is a simple application of the 'self-weight method'.

The distribution of seismic loads along the building height depends mainly on mass and stiffness distributions and the building configuration in plan and elevation as also discussed in Section 2.3.1.2 and Appendix A. The contribution of higher modes in the dynamic response of the structure also affects the load distribution. Codes attempt to supply a simplified method for load distributions based only on mass distribution and storey heights. A common expression for the seismic lateral force F_i at the ith storey of a building structure is given as:

$$F_i = V_B \frac{W_i H_i}{\sum_{j=1}^{N} W_j H_j} \tag{4.37}$$

where N is the total number of storeys, W_i and W_j are the seismic weight of the i-th and j-th storeys, respectively; they can be computed using equation (4.1). Similarly, H_i and H_j are the heights from ground level to the ith and jth level, respectively. Equation (4.37) provides a triangular distribution over the height for uniform mass and stiffness and is thus suitable for low-rise regular structures ($T \leq$ 0.5 second), for which the fundamental mode of vibration departs little from a straight line. For long-period structures, the influence of higher modes can be significant. In high-rise regular structures, the fundamental mode of vibration lies approximately between a straight line and a parabola (Di Sarno, 2002). Other force distributions are, however, also adopted by international seismic codes of practice or other published recommendations to account for higher mode effects. For example, in the USA, the seismic provisions utilize the following force pattern (FEMA 450, 2004):

$$F_i = V_B \frac{W_i \, H_i^k}{\sum_{j=1}^{N} W_j \, H_j^k} \tag{4.38}$$

where the power-law exponent k is related to the fundamental period T of the structural system. For short-period structures, such as those with $T \leq 0.5$ second, $k = 1$ and equations (4.37) and (4.38) are equivalent. For long-period structures, e.g. $T \geq 2.5$ seconds, it may be assumed that $k = 2$, while for structures with a period between 0.5 and 2.5 seconds, k should be conservatively set equal to 2 or determined by linear interpolation between 1 and 2.

Torsion is a serious problem in structures that have a non-coincident centre of mass and stiffness as illustrated in Appendix A. The geometric eccentricity between these two centres should be considered in the analyses of irregular or torsion-deformable structures. In addition to the geometric eccentricity, codes specify a value of eccentricity (referred to as 'accidental eccentricity') to account for uncertainty in the calculation of the actual centre of mass and stiffness. Codes define a minimum value of eccentricity as a ratio of the building dimension normal to the direction of the ground motion. The value of accidental eccentricity is frequently taken as 5%.

The steps required to perform the analysis based on the simplified code procedure are summarized in the following:

(a) Select the design earthquake spectrum. It is generally an elastic site-specific spectrum given in terms of PGA and for a specified value of structural damping ξ. Many codes provide spectra for $\xi = 5\%$, as illustrated in Section 3.4.5. This value can be considered adequate for RC and composite structures. For steel structures, values of ξ equal to 2–3% should be employed. It is possible to modify spectral ordinates by using the η-factor given below (Bommer *et al.*, 2000):

$$\eta = \sqrt{\frac{10}{5+\xi}} \geq 0.55 \tag{4.39}$$

where ξ is the viscous damping ratio, expressed in percent.

(b) Select the structural lateral force-resisting system, e.g. among those presented in Appendix A, material of construction and hence select the response modification factor from the values provided in the code.
(c) Scale the design spectrum by using the force reduction factor selected in (b).
(d) Estimate the fundamental period of vibration of the structure T Semi-empirical formulae or Rayleigh's method can be used.
(e) Compute the spectral acceleration corresponding to the fundamental period T, the assumed value of structural damping ξ and level of ductility (force reduction factor).
(f) Define the importance factor of the structure.
(g) Compute the seismic weight $W_{EQ,t}$. For buildings, it is sufficient to determine the weight of each floor in compliance with the rules provided in Section 4.3.
(h) Estimate the seismic coefficient C and hence compute the design base shear V_B from equation (4.34).
(i) Distribute the total seismic shear V_B computed in step (h) over the main axis of the structure in compliance with the relationship either in equation (4.37) or in equation (4.38).

(j) Perform a static structural analysis to evaluate the response quantities, as those described in Section 4.8. For 3D models it is necessary to apply the static horizontal distribution of forces along the principal directions of the plan layout fulfilling the combination rules outlined in Section 4.4.
(k) Scale the horizontal displacements computed in (j) by using an amplification factor, which is often assumed equal to the force reduction factor in (b), or a proportion of it. The estimated displacements are those generated by earthquake loading.

Seismic design codes allow the use of the equivalent lateral force procedure for relatively regular structures with fundamental periods not greater than 1.5–2.0 seconds. For irregular or long-period structures, more refined dynamic analyses such as modal spectral or inelastic response history analysis should be used. Recently, to permit performance-based assessment of structural systems, inelastic static pushovers and incremental dynamic analyses have also been recommended in seismic codes worldwide. Table 4.8 summarizes commonly used methods of analysis included in international seismic codes and their range of applicability.

In code application of modal spectral analysis, design spectra are scaled by using the all-embracing force reduction factor (R or q), and the elastic response of the structural system is computed using the response spectrum analysis, as detailed in Section 4.6.1.1. All significant modes are required in the combined response. This condition is satisfied by having the total effective modal mass included in the analysis equal to at least 85% to 90% of the total mass of the structure. Although calibrated for buildings, the latter rule is useful also for bridge structures. As with the static load procedure, 5% accidental eccentricity should be included in spatial pushover analysis. Methods of structural analysis presented in Sections 4.6.1, 4.6.2 and 4.6.3 are compared in Table 4.9, which complements Table 4.1 in Section 4.3. The comparison is expressed in terms of type of input, material inelasticity and geometric non-linearity, and accuracy.

Methods of analysis are often a compromise between accuracy and complexity. The simplest method that provides the desired information with reasonable accuracy is usually the preferred method. Unfortunately, the ideal solution is seldom available. For example, as shown in Table 4.9, inelastic dynamic response analysis is the most accurate and realistic method for seismic assessment because it can accommodate material inelasticity and geometric non-linearity. It is, however, also highly complex and time-consuming. Conversely, the equivalent static analysis is very simple to use but could be rather poor in accuracy. Its applicability is limited to the subclass of regular and short-period structures. Whereas inelastic static (or pushover) analysis is currently used extensively in the design office, its dynamic counterpart remains a challenge. To bridge the gap between research and application, it is necessary to provide training to civil engineers in advanced structural dynamics and inelastic behaviour, and create more user-friendly advanced software for inelastic dynamic analysis.

Table 4.8 Methods of analysis implemented in seismic design codes and their applicability.

Type of structure	Static analysis		Dynamic analysis		
	ELF	NSP	RSA	LRH	NRH
Regular	✓	✓	✓	✓	✓
Irregular	✗	✗	✓	✓	✓

Key: ELF = equivalent lateral force; NSP = non-linear static pushover; RSA = response spectrum analysis; LRH = linear response history; NRH = non-linear response history; ✓ = applicable; ✗ = not applicable.

Table 4.9 Comparisons between different types of analyses.

Analysis	Non-linearity type	Input	Mechanical	Geometric	Accuracy**
Equivalent static	Static	Horizontal force distribution		✓	•
Conventional pushover	Static	Horizontal force/ displacement distribution#	✓	✓	••
Modal	Dynamic	n.a.		✓	••
Spectral	Dynamic	Spectrum		✓	••
Response history	Dynamic	Earthquake	✓*	✓	•••
Adaptive pushover	Dynamic	Spectrum	✓	✓	••
Incremental dynamic	Dynamic	Earthquake	✓	✓	••
Simplified code	Static/Dynamic	Force distribution/ Spectrum		✓	•

Key: n.a. = not applicable; ✓ = applicable; • = low; •• = medium; ••• = high; * = not applicable for linear time history; ** = accuracy expressed as a relative measure; # Spectrum if used, force/displacement patterns are derived from simplified code method.

4.7 Performance Levels and Objectives

Whereas many recent publications refer to Performance-Based Engineering as a new trend in earthquake engineering, it is indeed not new at all. Limit states (LSs), a condition at which structural components and systems cease to perform their intended function (Wen *et al.*, 2004), and limit state design have been used since the early 70s (e.g. Baker and Heyman, 1969; Allen, 1975). Performance objectives are defined by limit states, which may or may not be structural, since the use of a structure can be impeded by non-structural issues.

In a broader socio-economic context, LSs may be related to repair costs (e.g. expressed as a percentage of replacement value) that are in excess of a desired amount, opportunity losses, or morbidity and mortality, as also discussed in Section 2.2.6. In structural earthquake engineering, narrowly defined structural LSs may be either strength, deformation or energy-related. This applies to performance levels of structural, non-structural and content systems (Miranda and Aslani, 2003; Taghavi and Miranda, 2003). This book deals primarily with structural LSs; the latter are related to the fundamental structural response quantities defined in Sections 2.3.1 to 2.3.3. However, the framework proposed herein is applicable also to non-structural components and contents. The definition of performance levels (PLs) is of importance for seismic assessment of structures. Numerous analytical approaches based on multiple LSs have been presented in the literature (e.g. Bozorgnia and Bertero, 2004). The seemingly different approaches exhibit common features. The proposed LS-based frameworks can accommodate the randomness and uncertainty that are inevitably present in the process of seismic performance assessment of structural systems. Current LS approaches used in earthquake engineering rely upon principles of probabilistic analysis to handle uncertainty and randomness (Cornell *et al.*, 2002; Wen *et al.*, 2004). Performance assessment employing a three-level LS format, as shown in Table 2.1, is the most suitable means of assessing the earthquake response of structural systems. These levels may be defined using terminology intended to be comprehensible to stakeholders and risk managers, as well as engineers and social scientists. An example of typical performance levels is provided in Table 4.10. The recommended LSs, i.e. serviceability, damage control and collapse prevention are intrinsically related to stiffness, strength and ductility, respectively. A performance objective is the association of a certain level of seismic action with LSs, as also shown in Figure 4.40.

Table 4.10 Recommended three-level format for limit states of structural systems.

Performance level		Performance objectives		Performance criteria		Seismic hazard	
Limit state	Structural characteristics	Engineering	Socio-economic	Type	Prob. exceedance (in %)	Type	Prob. event (in %)
Serviceability	Stiffness	Non-structural damage	Operational	Near elastic response	P_{c1} in N_{c1} years	Frequent	P_{s1} in N_{s1} years
Damage control	Strength	Moderate structural damage	Limited economic loss	Limited inelastic response	P_{c2} in N_{c2} years	Occasional	P_{s2} in N_{s2} years
Collapse prevention	Ductility	Severe structural damage	Life loss prevention	Large inelastic response	P_{c3} in N_{c3} years	Rare	P_{s3} in N_{s3} years

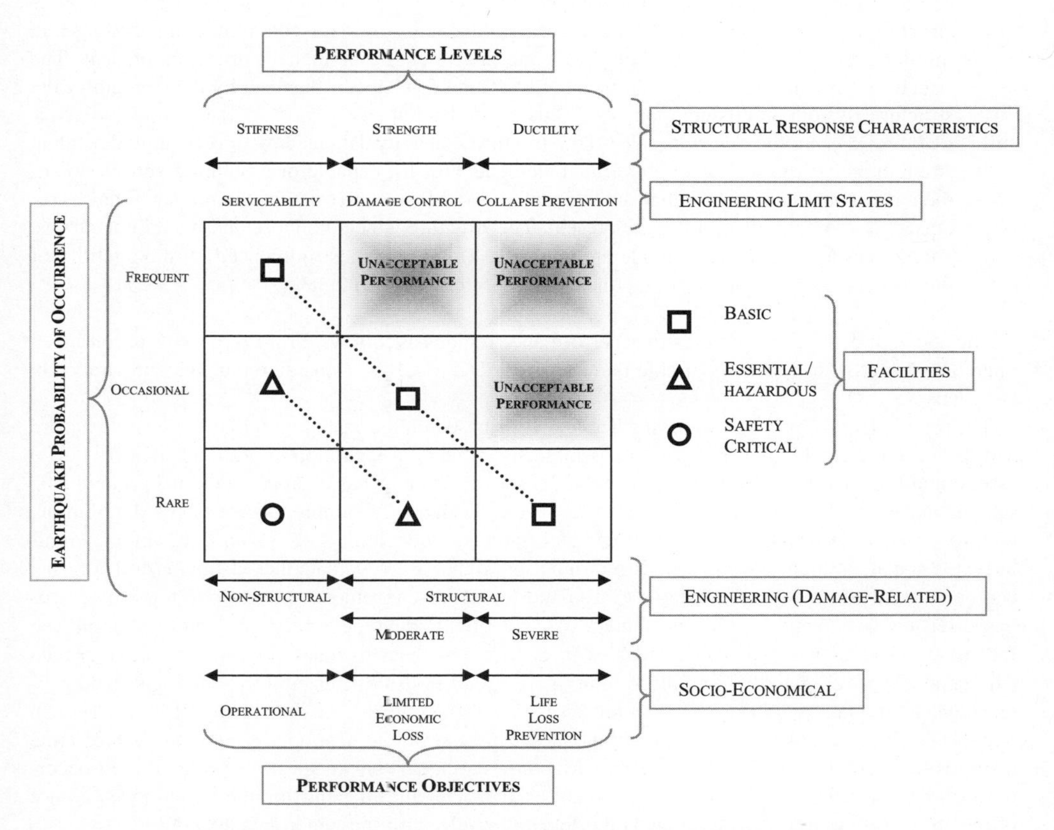

Figure 4.40 Conceptual correlation matrix to identify performance levels

The terminology used for the performance objectives in Table 4.10 signifies that engineers have been traditionally concerned with the effects caused by earthquakes on the 'system' (or structure), while risk managers and decision- and policy-makers are interested in the effects on the 'service' and 'life loss'. The objective of modern earthquake risk management includes both engineering and socio-economic objectives (referred to as the 'holistic approach'), i.e. it is aimed at controlling the risk to socio-economically acceptable levels. The achievement of a given performance objective is specified in Table 4.10 in statistical terms, i.e. probability P_{ci} (in percentage) of exceeding the performance objective for an earthquake with a given probability P_{si} of occurrence (in percentage) in N years. The reference periods of time (in years) for the probabilities P_{ci} and P_{si} are not necessarily the same; in general, also $N_{ci} \neq N_{si}$.

The LSs for structural systems indicated in Table 4.10 may be defined as follows:

i. *Serviceability limit state (SL)*: The structure is only slightly damaged. Structural elements have not reached significant yielding and have retained their strength and stiffness. Non-structural components such as partitions and infills may show some minor cracking that could, however, be economically repaired or even masked. No permanent drifts are present. This LS is most influenced by the stiffness of the structural system and its components discussed in Section 2.3.1.
ii. *Damage control limit state (DC)*: The structure is significantly damaged, but still retains considerable strength and stiffness. Vertical elements are capable of sustaining gravity loads, hence the structure is far from collapse. Non-structural components are damaged, although partitions and infills have not failed out-of-plane. Moderate and tolerable permanent drifts are present. The structure is repairable but at a non-trivial cost. This LS is most influenced by the strength of the structural system and its components discussed in Section 2.3.2.
iii. *Collapse prevention limit state (CP)*: The structure is heavily damaged, with very limited residual strength and stiffness. Although vertical elements are still capable of sustaining vertical loads, their resistance cannot be relied upon indefinitely. Most non-structural components have collapsed. Large permanent drifts are present. The structure is near collapse and would not survive another earthquake, even of moderate intensity. This LS is most influenced by the ductility of the structural system and its components discussed in Section 2.3.3.

The above three LS format yields four performance regions as follows. From zero to SL is continued operation, from SL to DC is repairable damage, from DC to CP is irreparable damage and above CP is collapse.

The probabilistic approach underlying Table 4.10 is based on the assumption that the seismic hazard and the structure can be treated separately, which is a common assumption in probabilistic earthquake assessment. The definition of the probability of the seismic event is based on conventional probabilistic seismic hazard analyses formulated by Cornell (1968). Using, for example, the Poisson's distribution, the probability of earthquake occurrence can be estimated by equation (3.1) given in Chapter 3. Similarly, the specification of the earthquake return period, required to evaluate the seismic hazard at a site, can be calculated employing equations (3.4.1) and (3.4.2) of the same chapter. Modern performance-based seismic design and assessment standards state explicitly the type of seismic event to be employed for structural assessment, as discussed also in Section 4.3. Hazard maps for different peak ground parameters, spectral quantities and mean return periods have been estimated by seismologists and geotechnical earthquake engineers for the most seismically active regions worldwide as illustrated in Section 1.1. Such maps are updated continuously as new seismological data become available. Time histories for ground motion have also been derived for earthquakes with specified probability of occurrence in the reference time window (N_{si} in Table 4.10). The latter depends on the importance and use of the structural system. For buildings and bridges, the selected time window is generally $N_{s1} = N_{s2} = N_{s3} = 50$ years and the following values are assumed as return periods T_{Ri} in Table 4.10:

- *Serviceability limit state:* $T_{R1} \approx 75$ years, corresponding to a probability of exceedance of 50% in 50 years;
- *Damage control limit state:* $T_{R2} \approx 475$ years, corresponding to a probability of exceedance of 10% in 50 years;
- *Collapse prevention limit state:* $T_{R3} \approx 2{,}475$ years, corresponding to a probability of exceedance of 2% in 50 years.

The values specified above for P_{si} and N_{si} are still controversial and are handled differently in seismic codes and recommendations worldwide for new and existing structures. This is also true with regard to the probability of exceedance of performance criteria P_{ci} and relative time windows N_{ci} for buildings and bridges. The common principle of the different existing LS variants is that the performance objectives increase, i.e. light damage is expected, for earthquake ground motions with high probability of occurrence (or frequent seismic event) or for an important structure or hazardous occupancy, e.g. health care centres, fire stations and power plants. Conversely, less critical buildings or temporary structures would be expected to suffer extensive and irreparable damage when subjected to a major earthquake (or rare seismic event). This approach is compliant with the γ_I-factor in equation (4.3) and implemented in the simplified code method illustrated in Section 4.6.3. Figure 4.40 provides the conceptual correlation matrix involving earthquake probability, structural performance levels and building occupancy. The matrix is based on the requirements outlined in Table 4.10. According to the proposed framework, the triad of response parameters discussed in Section 2.3, i.e. stiffness, strength and ductility take centre stage in the earthquake assessment of structures. In the event of a small or medium earthquake, a structure should exhibit adequate stiffness to ensure that non-structural damage is minimized, thus complying with the serviceability or operational performance target. Sufficient strength to control structural damage under intermediate events is also required to ensure the achievement of the damage control target, associated with limited economic loss. Finally, when the structure is subjected to a large earthquake, its ductility plays a critical part in guaranteeing that the structure can deform with significant loss of strength, but no collapse, thus safeguarding human lives and fulfilling the third performance target.

Performance levels expressed qualitatively in Table 4.10 should be established on the basis of thorough assessment of local (material, section, member and connection) and global (system) response parameters. These could be strength- or deformation-based as further discussed in Section 4.8. The definition of number, type and threshold values of LSs for earthquake engineering assessment depends on the material of construction and the structural system. Table 4.11 provides a correlation between

Table 4.11 Correlation of engineering limit states and performance levels.

Engineering limit states	Performance levels		
	Serviceability	Damage control	Collapse prevention
Cracking	■	■	■
First yielding	■	■	■
Spalling		■	■
Plastification		■	■
Local buckling		■	■
Crushing			■
Fracture/Fatigue			■
Global buckling			■
Residual drift			■

Key: Limit states to check are in grey boxes.

Table 4.12 Typical values of inter-storey drifts for the seismic performance assessment of framed structures.

Performance level	Damage type	Seismic hazard		Inter-storey drift (d/h)
Limit State	Level	Type	Prob. event (in %)	Values (in %)
Serviceability	Non-structural	Frequent	50% in 50 years	$0.2 < d/h < 0.5$
Damage control	Moderate structural	Occasional	10% in 50 years	$0.5 < d/h < 1.5$
Collapse prevention	Severe structural	Rare	2% in 50 years	$1.5 < d/h < 3.0$

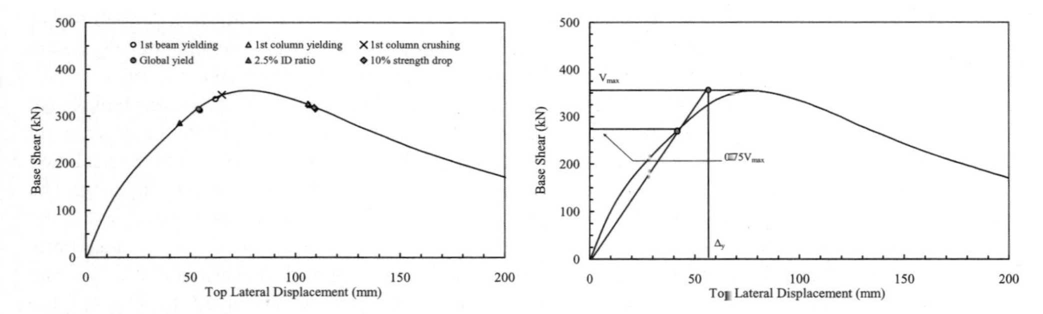

Figure 4.41 Limit states of the three-storey RC irregular frame (positive x-direction) in Figure 4.2: response curve and evaluation of global yielding

engineering LSs and PLs mentioned above. Some LSs, for example, crushing, plastification and buckling, may belong to either 'damage control' or 'collapse prevention' depending on their severity within the structural system; the classification in Table 4.11 is, therefore, indicative rather than definitive.

For practical applications, structural assessment should be based on values of measurable physical parameters that can be associated with engineering limit states and damage states. Comprehensive reviews of typical response parameters for seismic structural assessment, i.e. damage parameters and indices, and their values for different LSs, were given by Williams and Sexsmith (1995), Kappos (1997) and Ghobarah *et al.* (1999). Since earthquake-induced damage of building and bridge structures is generally related to inelastic deformations, deformation-based damage indices are more appropriate than force-based ones. Modern displacement-based seismic design guidelines provide values that rely primarily on inter-storey drifts d/h to assess the performance of structural systems. For bridge systems, maximum values of the lateral drift of the piers are generally recommended. Guiding values for the assessment of PLs at different seismic hazard levels are provided in Table 4.12 for buildings. For ductile multi-storey MRFs, the values of d/h may also be used as an indicator of the flexural rotational capacity θ of members (beams and columns) and connections, i.e. $\theta \approx d/h$ (Krawinkler *et al.*, 2003). These values are, however, not universally accepted.

An example of evaluation of LSs for an RC building is shown in Figure 4.41, depicting the response curve along positive x-direction of the sample frame in Figure 4.2.

The response curve in Figure 4.41 is obtained by conventional pushover analysis using a force pattern proportional to the first mode of vibration, as presented in Section 4.6.2.2. The assessment is performed in terms of both local and global LSs. The local LSs include member yielding and column crushing, while the global LSs are the onset of global yield, inter-storey drift ID equal to 2.5% and 10% strength drop. For the structural response curve in Figure 4.41, member yielding is conservatively defined as reaching the yield strain in longitudinal reinforcing steel, i.e. $\varepsilon_y = 0.002$. Column crushing is defined

Table 4.13 Local and global limit states relative to the response curve in Figure 4.41 of the sample SPEAR frame.

Local limit states		Top displacement		Global limit states		Top displacement	
		(mm)	% of Height			(mm)	% of height
Yield	1st Beam Yielding	62	0.69	Yield	Global yield	54	0.60
	1st Column Yielding	45	0.50				
Collapse	1st Column Failure	65	0.72	Collapse	2.5% ID ratio	106	1.18
					10% strength drop	109	1.21

Key: ID = inter-storey drift.

as corresponding to the extreme fibre of core concrete reaching its crushing strain, i.e. $\varepsilon_{cu} = 0.003$. The above LSs monitored for the sample frame are consistent with the framework provided in Figure 4.40 and with the 'engineering limit states' in Table 4.11. The value of inter-storey drift of ID = 2.5% corresponds to the LS of 'collapse prevention' in Table 4.12. Since the yield point is not clear in the plot of base shear-versus-top displacement of Figure 4.41, the proposal by Park (1988), presented in Section 2.3.3, is utilized to detect the 'global yield'. An idealized elastic-plastic system is used to define the yield point in the global response of the structure. The yield displacement is therefore based on the idealized elastic-plastic system with reduced stiffness, which is evaluated as the secant stiffness at 75% of the ultimate strength. Lateral displacements of the top of the frame corresponding to local and global LSs are summarized in Table 4.13.

The values in Table 4.13 may be employed to compute the frame global ductility μ by using equation (2.16) of Chapter 2. In so doing, it is observed that the sample structure has low ductility, i.e. $1.0 \leq \mu \leq 2.0$. Values of μ may be, however, overestimated if only global limit states are considered. For the sample frame, local limit states result in reliable damage quantification. It is, therefore, essential, when performing seismic structural assessment to compute LSs as both local and global response quantities. Exact damage assessment of the irregular SPEAR frame can be achieved only by using dynamic response history analysis, since pushover curves cannot reflect the effects of soft storey and torsion on member level damage using conventional damage assessment. Nevertheless, the LSs presented in Table 4.13 alongside the response curve in Figure 4.41 are useful guidelines for a quick and brief assessment of the capacity of the structure.

4.8 Output for Assessment

Previous sections of this chapter have focused on the definition of the load input, modelling issues, methods of structural analysis and limit states that may be used to assess the earthquake response of structures. The evaluation of seismic performance requires the selection of appropriate output quantities or response indicators. Commonly used indicators are summarized in Figure 4.42. Output quantities are subdivided into actions (stresses and their resultants) and deformations (strains and their resultants). Local and global indicators are used for accurate and reliable assessment of seismic response. In general, local output parameters are required primarily to detect potential damage localization and to evaluate the attainment of threshold values of stress and strain in fibres at different performance levels, such as yielding, cracking, crushing and buckling, as shown in Tables 4.11 and 4.13. On the other hand, global response indicators are used to estimate the fundamental structural response characteristics presented in Section 2.3. The evaluation of local and global parameters depends upon assumptions made regarding the level of discretization adopted for the structure. Substitute and stick models illustrated in Section

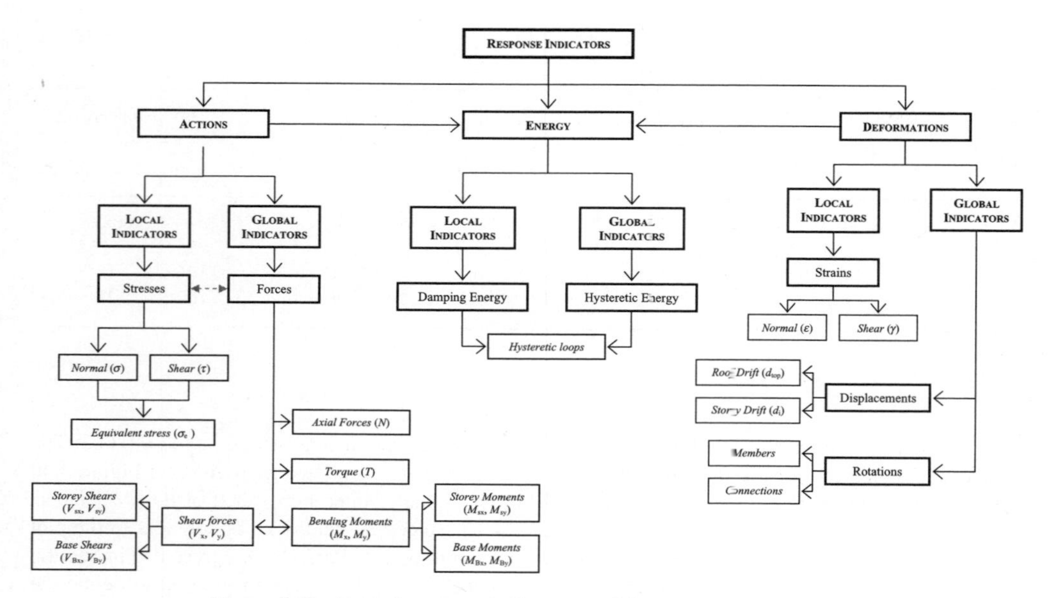

Figure 4.42 Typical response indicators used for structural assessment

4.4 provide global parameters. Conversely, refined models based on fibre formulations described in Sections 4.5.2 and 4.5.3, do not have any restriction on the response indicators. Global response parameters, e.g. lateral drift at the top of bridge piers, are more important than local quantities, such as section curvatures, in assessing member failure. Storey accelerations are also used to evaluate the seismic performance of structures, especially for base isolated structures, tall buildings and building contents. Human comfort criteria for high-rise structures subjected to environmental loads, e.g. earthquakes and winds, are expressed primarily in terms of acceleration (Di Sarno, 2002). The performance of tall buildings subjected to low-magnitude earthquakes can be reliably quantified by monitoring roof and storey accelerations.

Hysteresis loops are useful for action and deformation assessment, as well as energy absorbed and dissipated. Hysteresis loops are plots defining the structural response of components, connections or systems under load reversals. They are frequently expressed in terms of bending moment versus rotations, base or storey forces versus lateral displacements, and axial forces versus axial displacements. They are useful because they indicate the occurrence of stiffness and strength degradation at different structural resolutions, e.g. section, member, sub-assemblage and system.

Estimations of seismic behaviour may also be derived from the energy balance between seismic input and energy absorbed (Akiyama, 1985). During earthquake ground motions seismic energy is transferred to structural systems. Part of this energy is stored in the structure as kinetic and strain energy, and the rest, for inelastic systems, is dissipated through damping and hysteretic behaviour. Damping and hysteretic energy are response indicators, which are generally used to assess inelastic deformations in building and bridge systems.

4.8.1 Actions

Output for actions may be represented by local or global parameters. Local actions generally include stress and strain outputs at Gauss points within FEs of the discretized system. Normal stresses σ, shear

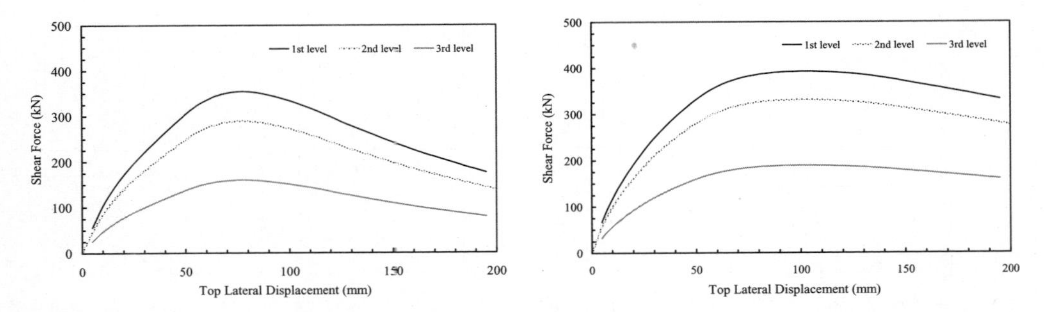

Figure 4.43 Storey pushover curves: positive X-direction (*left*) and Y-direction (*right*) of the sample frame in Figure 4.2 (top displacement at the centre column C3)

stresses τ or their combination (equivalent stresses, σ_{eq}) can be determined depending on the system geometry, discretization adopted and type of applied load. Equivalent stresses can be computed from 3D elasticity (e.g. Chen and Saleeb, 1982). On the other hand, global indicators correspond to internal actions, e.g. axial forces, bending moments, shear forces and torque, if any. Two bending moments M_x and M_y, e.g. about principal axes (x and y) of the member cross sections, and the shear forces, V_y and V_x, respectively, should be taken into account when performing three-dimensional analyses. In planar systems, output internal forces include only axial forces N, moment (M_x or M_y) and shear force (V_y or V_x). In framed structures, bending moments and shears are frequently monitored at each storey level (referred to as storey moments and shear forces) and at the base (known as base moment and base shear). These actions result from the contributions of all members at the storey level. Base and storey shear forces and moments may also be used to detect the occurrence of both local and global LSs presented in Section 4.7. For example, in the response curve of Figure 4.41, the seismic base shear is employed to characterize the global yield and the 10% drop in lateral strength. Similarly, the plots of the storey shear versus top lateral displacement (storey response curves) in Figure 4.43 show that the formation of a weak storey can be assessed by observing the change of storey shear during the pushover analysis.

Assuming that lateral force does not increase as the displacement increases, weak storey behaviour occurs when the capacity curve shows a descending branch as displayed in Figure 4.43. The ground storey loses its strength ahead of the second or third storey failure. Therefore, the failure occurs at ground floor. Because failure of ground storey indicates total loss of strength for the whole structure, monitoring that storey behaviour is, for the sample frame, a critically important measure of limit state attainment for the entire building.

4.8.2 Deformations

Deformation parameters provide a better indicator of damage of structures subjected to earthquakes than actions do. Local and global indicators may be used to assess system performance. Normal and shear strains, ε and γ respectively, can be obtained only from detailed geometric discretizations of the structure based on fibre models, described in Sections 4.5.2 and 4.5.3. Strain values are used to ascertain the likelihood of local buckling in steel or composite sections and buckling of reinforcement bars in RC members. For metal plates, it is necessary to determine shear strains γ to establish the occurrence of shear yielding and buckling. Evaluations of normal strains ε for section fibres are essential to monitor the curvatures in RC, steel and composite framed structures. Normal strains are frequently employed to determine also the occurrence of LSs, such as steel yield and concrete crushing. For example, the

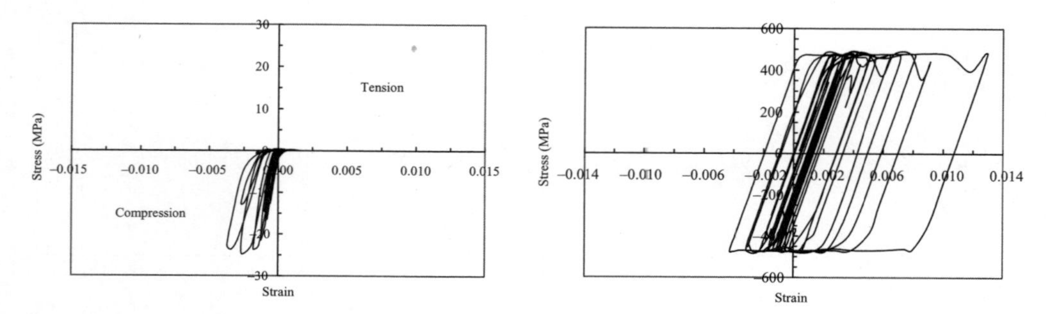

Figure 4.44 Time history and hysteretic response of normal strains within RC sections discretized through fibre elements in the model frame of Figure 4.2: confined concrete (*left*) and steel rebars (*right*)

response history of ε depicted in Figure 4.44 for both confined concrete and steel rebars can be used to monitor member yielding and column crushing in RC frames. Similarly, when performing inelastic static analyses, e.g. pushovers discussed in Section 4.6.2.2, the evaluation of axial strains in reinforcing steel (ε_y) and core concrete fibres (ε_{cu}) can be utilized to detect the occurrence of yielding and crushing. In Figure 4.44, member yielding is the LS corresponding to the onset of yielding strain $\varepsilon_y = 0.002$ in reinforcement steel fibres, while column crushing is attained if the extreme fibre of core concrete reaches its crushing strain $\varepsilon_{cu} = 0.003$.

On the other hand, global response deformational parameters, such as inter-storey drifts, may be used to determine the occurrence of different damage states as discussed in Section 4.7. Widely used values of inter-storey drifts for the seismic performance assessment of framed structures are given in Table 4.12. Excessive inter-storey drifts are indicators of structural failure, such as weak storeys. Figure 4.45 shows inter-storey drift response histories computed by Jeong and Elnashai (2005) using inelastic dynamic analyses for the irregular RC full-scale frame in Figure 4.2.

The sample structure fails under the September 1986 Kalamata earthquake, for a value of PGA equal to 0.20 g. The deformed shape of the building, which is also included in the figure, confirms the occurrence of the failure by weak storey, at the first floor. The latter can also be predicted from the results of the pushover curves in Figure 4.43 and from the distribution of plastic hinge formation at peak base shear for the sample frame shown in Figure 4.46. At maximum base shear, all the columns of the first storey exhibit plastic hinging at both ends.

For ductile multi-storey frames, e.g. with weak-beam strong-column, storey drifts are proportional to beam rotations, as also discussed in Sections 2.3.3 and 4.7. Shear deformations of beam-to-column connections significantly contribute to horizontal drifts. Moreover, ductility demand-to-capacity ratios at member levels should also be checked to prevent brittle failure modes. Therefore, beam, column and connection rotations should always be monitored, as should axial deformations in diagonal braces. Flexural curvatures and rotations do not account for shear effects in conventional frame analysis.

When assessing bridges with squat piers, shear effects can be monitored through global response indicators, such as displacement at the top of piers, which account for the contribution of both shear and flexure.

Final Project

The RC building shown in Figure 4.47 (Fardis, 1994) is to be constructed close to an active fault. Table 4.14 provides the dimension of the cross sections of the structural members. The characteristic concrete strength is 30 N/mm^2 and the characteristic yield strength is 420 N/mm^2 for both longitudinal and transverse steel.

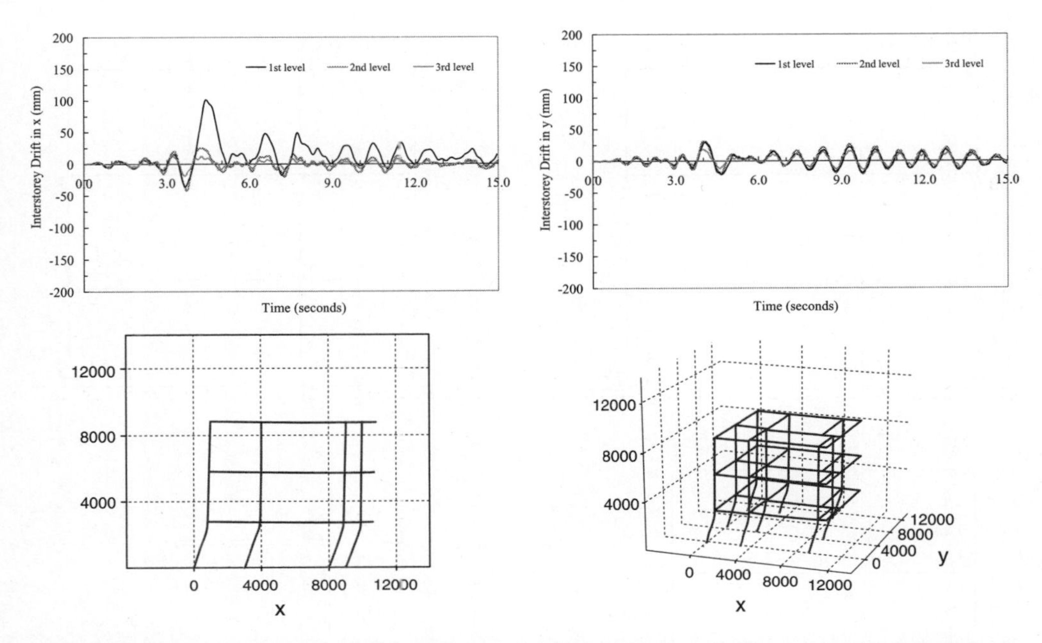

Figure 4.45 Inter-storey drift time histories (*top*) at column C3 of the frame in Figure 4.2 and deformed shapes (*bottom*) at 4.64 seconds of the 1986 Kalamata earthquake, (0.2 g, Kalamata-Prefecture, bidirectional loading)

Table 4.14 Member cross sections of the sample structure (in centimetres).

Columns			Beams $b \times h$			Slabs
Internal	External	Cut-off	X-dir. (1st floor)	X-dir. (2nd–8th floors)	Z-dir.	14
70 × 70	60 × 60	50 × 30	30 × 80	30 × 60	30 × 60	

The construction site is at an epicentral distance of 8.0 km from a thrust fault. A seismic hazard assessment for the site was carried out and a design earthquake with magnitude $M_w = 7.65$ and focal depth of 7.0 km was obtained. A number of borings drilled at the site indicated that the subsoil is rock with a shear wave velocity of 800 m/s.

The seismic hazard assessment recommended the following attenuation relationship to derive the peak ground acceleration (PGA) of the site:

$$\log(PGA) = -0.105 + 0.229(M_w - 6) - 0.778\log(R) + 0.162\, G_A + 0.251\, G_B \tag{4.40}$$

where the value of PGA is in g. The coefficients G_A and G_B can be obtained from Table 4.15 as a function of the soil shear wave velocity. The focal distance R should be computed using the following relationship:

$$R = \sqrt{d^2 + h^2} \tag{4.41}$$

where d is the epicentral distance and h the focal depth (in km).

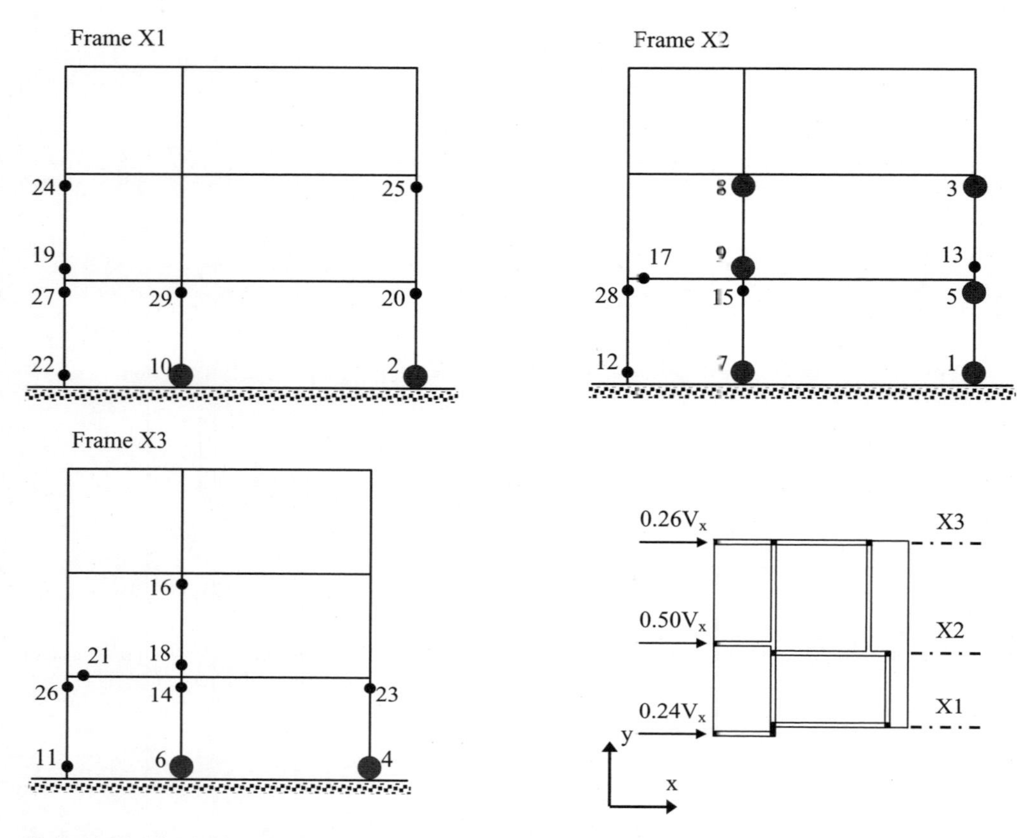

Figure 4.46 Plastic hinge formation at peak base shear (positive x-direction) of the three-storey irregular frame in Figure 4.2

Key: Formation orders of plastic hinges are represented by numbers and larger circles represent plastic hinges formed at early stages (from the 1st to 10th)

Table 4.15 Values of coefficients G in equation (4.40).

Soil type	Shear wave velocity (m/s)	G_A	G_B
Class A	$v_s > 750$	0	0
Class B	$360 < v_s \leq 750$	1	0
Class C	$v_s \leq 360$	0	1

The elastic acceleration response spectrum derived in the seismic hazard assessment is shown in Figure 4.48. A response modification factor of 8.0 and the PGA calculated from equation (4.40) should be used to scale the elastic spectrum given in Figure 4.48 and derive the design spectrum.

The distributed loads on beams are summarized in Table 4.16. Twenty-five percent of the live loads should be considered in all seismic design calculations. Do not account for any other reduction of live loads. The concentrated loads to apply at the beam-column connections are provided in Table 4.17.

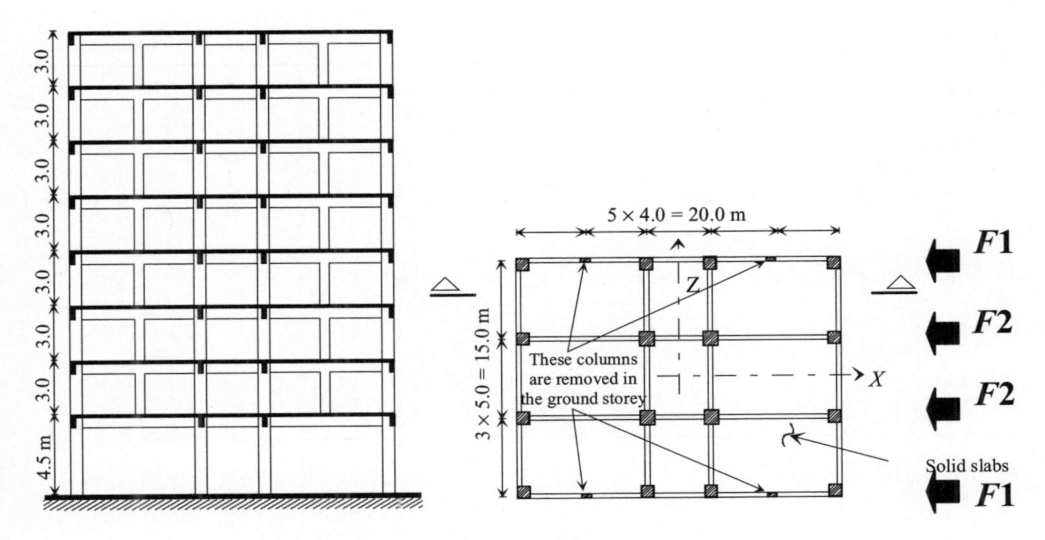

Figure 4.47 Sectional elevation and plane of the building

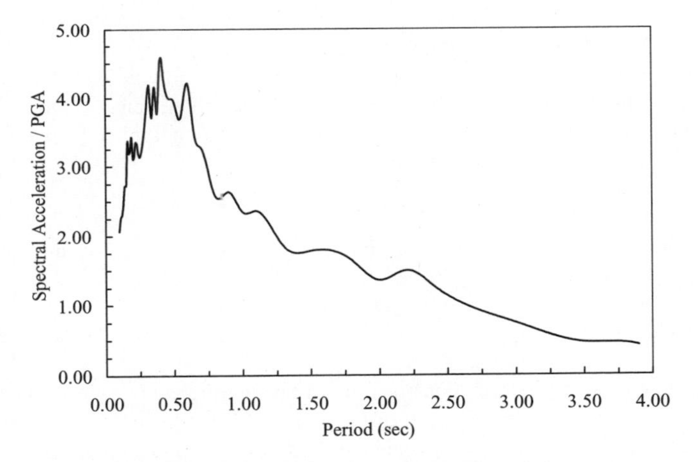

Figure 4.48 Recommended elastic spectrum (5% critical damping)

The elastic period of the structure may be estimated by the following:

$$T = 0.073(h_{\text{tot}})^{3/4} \tag{4.42}$$

where h_{tot} is the total height (in metres) from the foundation level. The design base shear (V) and the storey seismic forces (F_i) can be estimated by employing the relationships given below:

$$V = C\,W \tag{4.43}$$

Table 4.16 Distributed loads on beams.

Framing system	Dead load (kN/m)	Live load (kN/m)
External frames (F1)	20	10
Internal frames (F2)	30	15

Table 4.17 Concentrated loads at the beam-column connections.

Framing system	Load at external col. (kN)	Load at internal col. (kN)	Load at cut-off col. (kN)
External frames (F1)	80	100	10
Internal frames (F2)	125	180	n.a.

Key: n.a. = not applicable.

$$F_i = \frac{h_x\, W_x}{\sum_{1}^{n} W_i\, h_i} V \tag{4.44}$$

where W is the total load (dead load and 25% of the live load); h_i and h_x are the height from the foundation level to floor i and x; W_i and W_x are the portion of the total gravity load W located at level i or x; and n is the total number of stories. The seismic base shear coefficient C is the spectral response acceleration (expressed in 'g') obtained from the 'design' spectrum multiplied by the importance factor (I), which should be taken equal to 1.1 for this building.

It is required to:

1. Model the two lateral resisting systems in the X-direction (F1 and F2) using any finite element program and distribute the gravity loads on the two frames.
2. Calculate the actions of frame F1 from gravity loads.
3. Calculate the total base shear and distribute it along the height for the two lateral resisting systems F1 and F2.
4. Estimate the actions and deformations of frame F1 using the equivalent static force procedure. Modulus of elasticity of concrete, E_c, is 26 KN/mm^2 and Young's modulus of steel E_s is 200 KN/mm^2. Use 50% and 70% of the un-cracked stiffness of beams and columns, respectively, to estimate the effective flexural stiffness.
5. Estimate the periods of vibration and plot the first three mode shapes of frame F1.
6. Estimate the actions and deformations of frame F1 using the response spectrum analysis procedure. The response modification factor and the design PGA should be used to scale the elastic spectrum given in Figure 4.48 to obtain the design spectrum employed for the structural analysis.
7. Use the earthquake record relative to the horizontal component of the Loma Prieta earthquake (Northern California at Saratoga 'Aloha Ave.', USA, 1989). Scale the record to the PGA derived from the attenuation relationship for the construction site given in equation (4.40). Perform elastic response history analysis for frame F1 using the scaled record. Modern seismic codes allow for a reduction in base shear demand from elastic response history analysis by using the response modification factor (q- or R-factor).

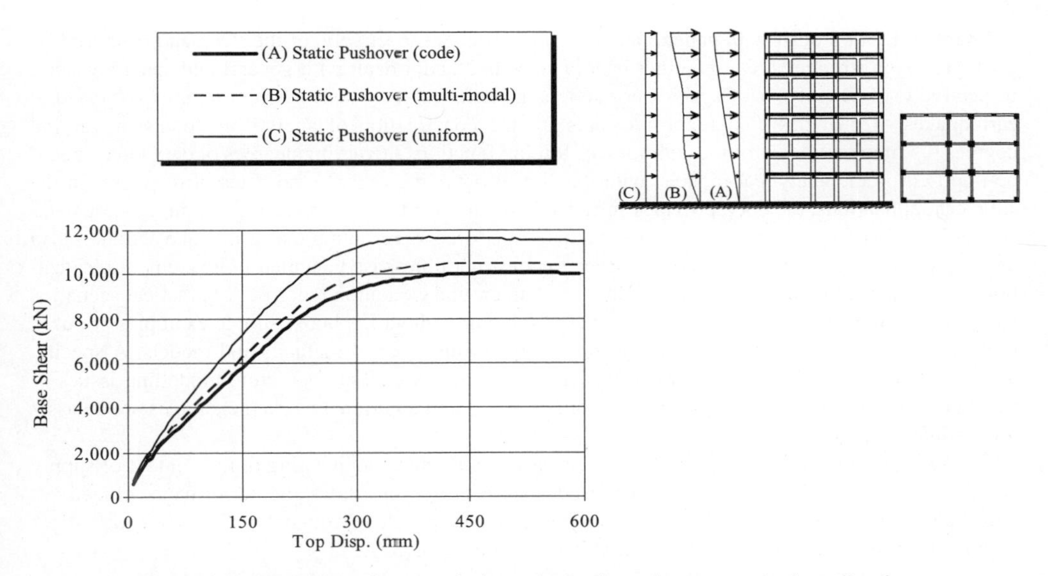

Figure 4.49 Inelastic pushover analysis results for the entire structure in the x-direction

8. Compare the results of different elastic analysis procedures.
9. Inelastic pushover analyses were conducted using Zeus-NL (Elnashai *et al.*, 2003) for the structure using the following lateral force distributions:
   i. Inverted triangular load (code pattern)
   ii. Lateral load distributions calculated from combinations of the first three modes of vibration (multi-modal pattern)
   iii. Uniform lateral load distribution

   The results from these analyses are provided in Figure 4.49. Comment on the results obtained from the lateral force patterns considered in the inelastic static analyses. Compare the ultimate strength of the building estimated from the inverted triangular load distribution and the design lateral force. Comment on the difference between the actual and the design strengths.

4.9 Concluding Remarks

Earthquakes continue to exact a heavy toll on vulnerable communities and this is unlikely to change significantly in the short term. Nonetheless, it is incumbent on the science and engineering communities to spare no effort to improve earthquake assessment, mitigation, response and recovery methods, and to educate society at large of the perils of earthquakes and of means to reduce their impact. There is a very large body of literature on the subject of earthquakes in general, and on earthquake engineering in particular. It is hoped that this education-oriented text will prove to be a valuable addition to the existing literature.

The book has covered structural aspects of earthquake engineering pertinent to buildings and bridges from the source of the earthquake to the actions and deformations required to dimension the structure. It is therefore a Source-to-Structure text when viewed from a Source-to-Society perspective. It stops where codes start; hence it is a code-independent text that focuses on fundamentals.

Chapters 1 and 3 dealt with general and specific aspects, respectively, of the 'Demand' imposed by earthquakes on structures. On the other hand, Chapters 2 and 4 dealt with general and specific issues, respectively, of 'Supply' or 'Capacity' for action and deformation resistance of structures subjected to earthquakes. It provides comprehensive tools for the construction of analytical models of varying degrees of complexity and for the definition of forcing functions of varying degrees of detail that should be imposed on the analytical models, with the aim of obtaining reliable estimates of response. In the simplest applications, the book provides sufficient guidance for the construction of a lumped-parameter single-degree-of-freedom structure subjected to a static force representing the earthquake action. It also provides, in the same level of detail, guidance to construct fibre-based detailed finite element idealizations of complex structures subjected to time-varying ground excitation records. All practical scenarios between the above two extremes are also catered for. Throughout the book, simple examples are used to provide guidance on the application of the described fundamental methods and models. A specific building is used as a threading example that links the various load and structure modelling issues of Chapter 4, the closing chapter. The set of summary slides and additional worked examples are an asset for graduate education.

The authors hope that their approach for discussing earthquake engineering in the context of supply and demand, projected in triads of return periods, engineering limit states and performance objectives, will appeal to both educators and graduate students as it has appealed through the years to a wide variety of students at the University of Illinois at Urbana-Champaign, USA.

References

Akiyama, H. (1985). *Earthquake-Resistant Limit-State Design for Buildings*. University of Tokyo Press, Tokyo, Japan.

Allen, D.E. (1975). Limit states design: A probabilistic study. *Canadian Journal of Civil Engineering*, 2(1), 36–49.

Anderson, J.C. and Gurfinkel, G. (1975). Seismic behaviour of framed tubes. *Earthquake Engineering and Structural Dynamics*, 4(2), 145–162.

Antoniou, S. and Pinho, R. (2004a). Advantages and limitations of adaptive and non-adaptive force-based pushover procedures. *Journal of Earthquake Engineering*, 8(4), 497–522.

Antoniou, S. and Pinho, R. (2004b). Development and verification of a displacement-based adaptive pushover procedure. *Journal of Earthquake Engineering*, 8(5), 643–661.

Aydinoglu, M.N. (2004). An improved pushover procedure for engineering practice: incremental response spectrum analysis (IRSA). Proceedings of the Workshop "Performance-based seismic design. Concepts and implementation." PEER Report No.2004/05, pp. 345–356.

Baker, J.W. and Cornell, C.A. (2003). Uncertainty specification and propagation for loss estimation using FOSM methods. Pacific Earthquake Engineering Research Center, PEER Report 2003/07, College of Engineering, University of California, Berkeley, CA, USA.

Baker, J.F. and Heyman, J. (1969). *Plastic Design of Frames 1: Fundamentals*. Cambridge University Press, Cambridge, England.

Balendra, T. (1993). *Vibration of Buildings to Wind and Earthquake Loads*. Springer-Verlag, London, UK.

Balendra, T. and Huang, X. (2003). Overstrength and ductility factors for steel frames designed according to BS 5950. *Journal of Structural Engineering, ASCE*, 129(8), 1019–1035.

Bathe, K.J. (1996). Finite Element Procedures in Engineering Analysis. Prentice Hall, Englewood Cliffs, NJ, USA.

Bazzurro, P. and Cornell, C.A. (1994). Seismic Hazard Analysis of Nonlinear Structures I: Methodology. *Journal of Structural Engineering, ASCE*, 120(11), 3320–3344.

Berry, M.P. (2006). Performance Modeling Strategies for Modern Reinforced Bridge Columns. PhD Thesis, Department of Civil and Environmental Engineering, University of Washington, Washington, DC, USA.

Bertero, V.V. (1977). Strength and deformation capacities of buildings under extreme environments. In *Structural Engineering and Structural Mechanics*, K.S. Pister, Ed., Prentice-Hall, Englewood Cliffs, NJ, USA, pp. 211–215.

Bommer, J.J., Elnashai, A.S. and Weir, A.G. (2000). Compatible acceleration and displacement spectra for seismic design codes. Proceedings of the 12th World Conference on Earthquake Engineering, Auckland, New Zealand, Paper No. 207, CD-ROM.

Bozorgnia, Y. and Bertero, V.V. (2004). *Earthquake Engineering. From Engineering Seismology to Performance-Based Engineering*. CRC Press, Boca Raton, FL, USA.

Bracci, J.M., Kunnath, S.K. and Reinhorn, A.M. (1997). Seismic performance and retrofit evaluation of RC structures. *Journal of Structural Engineering, ASCE*, 123(1), 3–10.

Broderick, B.M. and Elnashai, A.S. (1996). Seismic response of composite frames. I – Response criteria and input motions. *Engineering Structures*, 18(9), 696–706.

Broderick, B.M., Elnashai, A.S. and Izzuddin, B.A. (1994). Observations on the effect of numerical dissipation on the nonlinear dynamic response of structural systems. *Engineering Structures*, 16(1), 51–62.

Bruneau, M., Uang, C.M. and Whittaker, A. (1998). *Ductile Design of Steel Structures*. McGraw-Hill, New York, NY, USA.

Burdette, N.J. and Elnashai, A.S. (2008). The effect of asynchronous earthquake motion on complex bridges. Part II: Results and implication on assessment. *Journal of Bridge Engineering, ASCE*, 13(2), 166–172.

Burdette, N.J., Elnashai, A.S., Lupoi, A. and Sextos, A.G. (2008). Effect of asynchronous earthquake motion on complex bridges. Part I: Methodology and input motion. *Journal of Bridge Engineering, ASCE*, 13(2), 158–167.

Chen, W.F. (1982). *Plasticity in Reinforced Concrete*. McGraw-Hill, New York, NY, USA.

Chen, W.F. and Saleeb, A.F. (1982). *Constitute Equations for Engineering Materials: Elasticity and Modelling*. Vol. 1, John Wiley & Sons, New York, NY, USA.

Chopra, A.K. (2002). *Dynamics of Structures: Theory and Applications to Earthquake Engineering*. 2nd Edition, Prentice Hall College Division, Upper Saddle River, New Jersey, USA.

Chopra, A.K. and Chintanapakdee, C. (2004). Evaluation of modal and FEMA pushover analyses: Vertically 'regular' and irregular generic frames. *Earthquake Spectra*, 20(1), 255–271.

Chopra, A.K. and Goel, R.K. (2002). A modal pushover analysis procedure for estimating seismic demands for buildings. *Earthquake Engineering and Structural Dynamics*, 31(3), 561–582.

Chrysostomou, C.Z., Gergely, P. and Abel, J.F. (1992). Nonlinear seismic response of infilled steel frames. Proceedings of the 10th World Conference on Earthquake Engineering, Madrid, Spain, Vol. 8, pp. 4435–4437.

Clough, R.W. and Johnston, S. (1966). Effect of stiffness degradation on earthquake ductility requirements. Transactions of Japan Earthquake Engineering Symposium, Tokyo, Japan, pp. 195–198.

Clough, R.W. and Penzien, J. (1993). *Dynamics of Structures*. McGraw-Hill, New York, NY, USA.

Collier, C.J. and Elnashai, A.S. (2001). A procedure for combining horizontal and vertical seismic action effects. *Journal of Earthquake Engineering*, 5(4), 521–539.

Collins, M.P. (1978). Towards a rational theory for RC members in shear. *ASCE Proceedings*, 104(ST4), 649–666.

Collins, M.P., Mitchell, D., Adebar, P. and Vecchio, F.J. (1996). A general shear design method. *ACI Structural Journal*, 93(1), 36–45.

Comite Euro-International Du Beton (CEB) (1996). *RC Frames Under Earthquake Loading. State of the Art Report*. Thomas Telford, London, UK.

Cook, R.D., Malkus, D.S., Plesha, M.E. and Witt, R.J. (2002). Concepts and Applications of Finite Element Analysis. 4th Edition, John Wiley & Sons, New York, NY, USA.

Cordova, P.P. and Deierlein, G.G. (2005). Validation of the seismic performance of composite frames: full- scale testing, analytical modeling and seismic design. The John A. Blume Earthquake Engineering Research Center, Report No. 155, Department of Civil and Environmental Engineering, Stanford University, Stanford, CA, USA.

Cornell, C.A. (1968). Engineering seismic risk analysis. *Bulletin of the Seismological Society of America*, 58(4), 1583–1606.

Cornell, C.A., Jalayer, F., Hamburger, R.O. and Foutch, D.A. (2002). Probabilistic basis for 2000 SAC Federal Emergency Management Agency steel moment frame guidelines. *Journal of Structural Engineering, ASCE*, 128(4), 526–533.

Der Kiureghian, A. (1980). Probabilistic modal combination for earthquake loading. Proceedings of the 7th World Conference on Earthquake Engineering, Istanbul, Turkey, Vol. 6, pp. 729–736.

Di Sarno, L. (2002). Dynamic Response of Tall Framed Buildings Under Environmental Actions. PhD Thesis, University of Salerno, Salerno, Italy.

Di Sarno, L. and Elnashai, A.S. (2003). special metals for Seismic Retrofitting of Steel and Composite Buildings. *Journal of Progress in Structural Engineering and Materials*, 5(2), 60–76.

Dokainish, M.A. and Subbaraj, K. (1989). A survey of direct time-integration methods in computational structural dynamics – I. Explicit methods. *Computers and Structures*, 32(6), 1371–1386.

Dymiotis, C., Kappos, A.J. and Chryssanthopoulos, M.K. (1999). Seismic reliability of RC frames with uncertain drift and member capacity. *Journal of Structural Engineering, ASCE*, 125(9), 1038–1047.

Elnashai, A.S. (2002). Do we really need inelastic dynamic analysis? *Journal of Earthquake Engineering*, 6(1), 123–130.

Elnashai, A.S. and Elghazouli, A.Y. (1993). Performance of composite steel/concrete members under earthquake loading. Part I: Analytical model. *Earthquake Engineering and Structural Dynamics*, 22(4), 315–345.

Elnashai, A.S. and Izzuddin, B.A. (1993). Modelling of material non-linearities in steel structures subjected to transient dynamic loading. *Earthquake Engineering and Structural Dynamics*, 22(6), 509–532.

Elnashai, A.S. and Kwon, O.S. (2007). Distributed analytical modelling of bridges with soil-foundation-structure interaction. Proceedings of the 1st US-Italy Bridge Workshop, Pavia, Italy.

Elnashai, A.S. and Mwafy, A.M. (2002). Calibration of force reduction factors of RC buildings. *Journal of Earthquake Engineering*, 6(2), 239–273.

Elnashai, A.S., Elghazouli, A.Y. and Danesh, A.F.A. (1998). Response of semirigid steel frames to cyclic and earthquake loads. *Journal of Structural Engineering, ASCE*, 124(8), 857–867.

Elnashai, A.S., Papanikolaou, V. and Lee, D. (2003). Zeus NL – A System for Inelastic Analysis of Structures. Mid-America Earthquake Center, CD-Release, No.03-0, University of Illinois at Urbana-Champaign, IL, USA.

El-Tawil, S., Mikesell, T. and Kunnath, S.K. (2000). Effect of local details and yield ration on behavior of FR steel connections. *Journal of Structural Engineering, ASCE*, 126(1), 79–87.

Erberik, M.A. and Elnashai, A.S. (2004). Fragility analysis of flat-slab structures. *Engineering Structures*, 26(7), 937–948.

Fardis, M.N. (1994). Analysis and design of reinforced concrete buildings according to Eurocodes 2 and 8. Configuration 3, 5 and 6, Reports on Prenormative Research in Support of Eurocode 8.

Federal Emergency Management Agency (FEMA) (2004). NEHRP recommended provisions for seismic regulations for new buildings and other structures, Part 1 – Provisions. Report No. FEMA 450, Washington, DC, USA.

Gerin, M. and Adebar, P. (2004). Accounting for shear in seismic analysis of concrete structures. Proceedings of the 13th World Conference on Earthquake Engineering, Vancouver, Canada, Paper no. 1747, CD-ROM.

Ghobarah, A. and Biddah, A. (1999). Dynamic analysis of reinforced concrete frames including joint shear deformation. *Engineering Structures*, 21(11), 971–987.

Ghobarah, A., Abou-Elfath, H. and Biddah, A. (1999). Response-based damage assessment of structures. *Earthquake Engineering and Structural Dynamics*, 58(1), 79–104.

Goel, R.K. and Chopra, A.K. (1997). Period formulas for moment-resisting frame building. *Journal of Structural Engineering, ASCE*, 123(11), 1454–1461.

Goel, R.K. and Chopra, A.K. (1998). Period formulas for concrete shear wall buildings. *Journal of Structural Engineering, ASCE*, 124(4), 426–433.

Goel, R.K. and Chopra, A.K. (2004). Evaluation of modal and FEMA pushover analyses: SAC buildings. *Earthquake Spectra*, 20(1), 225–254.

Gulkan, P. and Sozen, M.A. (1974). Inelastic responses of reinforced concrete structures to earthquake motions. *ACI Journal*, 71(12), 604–610.

Gupta, A. and Krawinler, H. (2000). Estimation of seismic drift demands for frame structures. *Earthquake Engineering and Structural Dynamics*, 29(9), 1287–1305.

Gupta, B. and Kunnath, S.K. (2000). Adaptive spectra-based pushover procedure for seismic evaluation of structures. *Earthquake Spectra*, 16(2), 367–391.

Hilber, H.M., Hughes, T.J.R. and Taylor, R.L. (1977). Improved numerical dissipation for time integration algorithms in structural dynamics. *Earthquake Engineering and Structural Dynamics*, 5(3), 283–292.

Kappos, A.J. (1997). Seismic damage indices for R/C buildings: Evaluation of concepts and procedures. *Progress in Structural Engineering and Materials*, 1(1), 78–87.

Kappos, A.S., Paraskeva, T.S. and Sextos, A.G. (2005). Modal pushover analysis as a means for the seismic assessment of bridge structures. Proceedings of the 4th European Workshop the Seismic Behaviour of Irregular and Complex Structures, Thessaloniki, Greece, Paper No. 49.

Karayannis, C.G., Izzuddin, B.A. and Elnashai, A.S. (1994). Application of adaptive analysis to reinforced concrete frames. *Journal of Structural Engineering, ASCE*, 120(10), 2935–2957.

Kim, K.O. (1993). A review of mass matrices for eigenproblems. *Computers & Structures*, 46(6), 1041–1048.

Kowalski, M.J., Priestley, M.J.N. and MacRae, G.A. (1995). Displacement-based design of RC bridge columns in seismic regions. *Earthquake Engineering and Structural Dynamics*, 24(12), 1623–1643.

Krawinkler, H. and Seneviratna, G.D.P.K. (1998). Pros and cons of a pushover analysis of seismic performance evaluation. *Engineering Structures*, 20(4–6), 452–464.

Krawinkler, H., Medina, R. and Alavi, B. (2003). Seismic drift and ductility demands and their dependance on ground motions. *Engineering Structures*, 25(5), 637–653.

Kwon, O.H. and Elnashai, A.S. (2006). Analytical seismic assessment of highway bridges with soil-structure interaction. Proceedings of the 4th International Conference on Earthquake Engineering, Taipei, Taiwan, Paper No. 142.

Ibarra, L.F, Medina, R.A. and Krawinkler, H. (2005). Hysteretic models that incorporate strength and stiffness deterioration. *Earthquake Engineering and Structural Dynamics*, 34(12), 1489–1511.

International Organization for Standardization (ISO) (1998). General Principles on Reliability for Structures. Document N. ISO/FDIS 2394, Final Draft.

Izzuddin, B.A. (1991). Nonlinear Dynamic Analysis of Framed Structures. PhD Thesis, Department of Civil and Environmental Engineering, Imperial College, London, UK.

Izzuddin, B.A. and Elnashai, A.S. (1993). Eulerian formulation for large-displacement analysis of space frames. *Journal of Engineering Mechanics, ASCE*, 119(3), 549–569.

Jain, S.K. (1983). Analytical models for the dynamics of buildings. Earthquake Engineering Research Laboratory, Report No. EERL 83-02, California Institute of Technology, Pasadena, CA, USA.

Jeong, S.H. and Elnashai, A.S. (2005). Analytical assessment of an irregular RC frame for full-scale pseudo-dynamic testing. Part I: Analytical model verification. *Journal of Earthquake Engineering*, 9(1), 95–128.

Joint Committee on Structural Safety (JCSS) (2001). *Assessment of Existing Structures*. JCSS-RILEM Publication, Cachan, France.

Madas, P.J. and Elnashai, A.S. (1992a). A new passive confinement model for transient analysis of reinforced concrete structures. *Earthquake Engineering and Structural Dynamics*, 21(5), 409–431.

Madas, P.J. and Elnashai, A.S. (1992b). A component-based model for beam-column connections. Proceedings of the 10th World Conference on Earthquake Engineering, Spain, Vol. 8, pp. 4495–4499.

Mainstone, R.J. (1974). Supplementary note on the stiffness and strength of infilled frames. Current Paper CP 13/74, Building Research Station Garston, Watford, UK.

Mander, J.B., Priestley, M.J.N. and Park, R. (1988). Theoretical stress-strain model for confined concrete. *Journal of Structural Engineering, ASCE*, 114(8), 1804–1826.

Martinez-Rueda, J.E. and Elnashai, A.S. (1997). Confined concrete model under cyclic load. *Journal of Materials and Structures*, 30(4), 139–147.

Mazzolani, F.M. and Piluso, V. (1996). *Theory and Design of Seismic Resistant Steel Frames*. E&FN Spon, London, UK.

Menegotto, M. and Pinto, P.E. (1973). Method of analysis for cyclically loaded RC plane frames including changes in geometry and non-elastic behaviour of elements under combined normal force and bending. Preliminary Report IABSE, 13, pp. 15–22.

Miranda, E. and Aslani, H. (2003). Building-specific loss estimation methodology. PEER Report No. 2003-03, Pacific Earthquake Engineering Research Center, University of California at Berkeley, Berkeley, CA, USA.

Mitra, N. and Lowes, L.N. (2007). Evaluation, calibration and verification of a reinforced concrete beam-column joint model. *Journal of Structural Engineering, ASCE*, 133(1), 105–120.

Monti, G., Nuti, C. and Pinto, P.E. (1996). Nonlinear response of bridges under multisupport excitation. *Journal of Structural Engineering, ASCE*, 122(10), 1146–1159.

Mylonakis, G., Papastamatiou, D., Psycharis, J. and Mahmoud, K. (2001). Simplified modeling of bridge response on soft soil to nonuniform seismic excitation. *Journal of Bridge Engineering, ASCE*, 6(6), 587–597.

Nader, M.N. and Astaneh, A.A. (1989). Experimental studies if a single storey steel structure with fixed, semi-rigid and flexible connections. Earthquake Engineering Research Center, Report No. EERC/89-15, University of California, Berkeley, CA, USA.

Nassar, A.A. and Krawinkler, H. (1991). Seismic demands for SDOF and MDOF systems. The John Blume Earthquake Engineering Center, Report No. 95, Stanford University, Stanford, CA, USA.

Negro, P., Mola, E., Molina, F.J. and Magonette, G.E. (2004). Full-scale PSD testing of a torsionally unbalanced three-storey non-seismic RC frame. Proceedings of the 13th World Conference on Earthquake Engineering, Vancouver, British Columbia, Canada, CD-ROM.

Nowak, A.S. and Collins, K.R. (2000). *Reliability of Structures*. McGraw-Hill Higher Education, New York, USA.

Papanikolaou, V.K., Elnashai, A.S. and Pareva, J.F. (2006). Evaluation of conventional and adaptive pushover analysis II: Comparative results. *Journal of Earthquake Engineering*, 10(1), 127–151.

Park, R. (1988). Ductility evaluation from laboratory and analytical testing. Proceedings of the 9th World Conference on Earthquake Engineering, Tokyo-Kyoto, Japan, Vol. VIII, pp. 605–616.

Park, Y.J., Reinhorn, A.M. and Kunnath, S.K. (1987). IDARC: Inelastic damage analysis of reinforced concrete frame-shear-wall structures. National Center for Earthquake Engineering Research, Report No. NCEER-87-0008, Buffalo, NY, USA.

Price, T.E. and Eberhard, M.O. (1998). Effects of spatially varying ground motions on short bridges. *Journal of Structural Engineering, ASCE*, 124(8), 948–955.

Priestley, M.J.N. (2003). *Myths and Fallacies in Earthquake Engineering Revisited. The Mallet Milne Lecture*. IUSS Press, Pavia, Italy.

Priestley, M.J.N., Seible, F. and Calvi, G.M. (1996). *Seismic Design and Retrofit of Bridges*. John Wiley & Sons, New York, NY, USA.

Saiidi, M. and Sozen, M.A. (1979). Simple and complex models for nonlinear seismic response of reinforced concrete structures. Department of Civil Engineering, Structural Research Series 465, University of Illinois at Urbana-Champaign, IL, USA.

Sextos, A.G., Pitilakis, K.D. and Kappos, A.J. (2003a). Inelastic dynamic analysis of RC bridges accounting for spatial variability of ground motion, site effects and soil-structure interaction phenomena. Part 1: Methodology and analytical tools. *Earthquake Engineering and Structural Dynamics*, 32(4) 607–627.

Sextos, A.G., Pitilakis, K.D. and Kappos, A.J. (2003b). Inelastic dynamic analysis of RC bridges accounting for spatial variability of ground motion, site effects and soil-structure interaction phenomena. Part 2: Parametric study. *Earthquake Engineering and Structural Dynamics*, 32(4) 629–652.

Shibata, A. and Sozen, M.A. (1976). Substitute structure method for seismic design in reinforced concrete. *Journal of Structural Division, ASCE*, 102(ST1), 1–18.

Spacone, E., Filippou, F.C. and Taucer, F.F. (1996a). Fibre beam-column model for non-linear analysis of R/C frames. Part I. Formulation. *Earthquake Engineering and Structural Dynamics*, 25(7), 711–725.

Spacone, E., Filippou, F.C. and Taucer, F.F. (1996b). Fibre beam-column model for non-linear analysis of R/C frames. Part II. Applications. *Earthquake Engineering and Structural Dynamics*, 25(7), 727–742.

Stafford-Smith, B. (1968). Model test results of vertical and horizontal loading of infilled frames. *Journal of American Concrete Institute*, 65, 618–624.

Stewart, J.P., Fenves, G.L. and Seed, R.B. (1999). Seismic soil-structure interaction in buildings. I: Analytical methods. *Journal of Geotechnical and Geoenvironmental Engineering, ASCE*, 125(1), 26–37.

Subbaraj, K. and Dokainish, M.A. (1989). A survey of direct time-integration methods in computational structural dynamics – I. Implicit methods. *Computers and Structures*, 32(6), 1387–1401.

Taghavi, S. and Miranda, E. (2003). Response assessment of nonstructural elements. PEER Report No. 2003-04, Pacific Earthquake Engineering Research Center, Richmond, CA, USA.

Takayanagi, T. and Schnobrich, W. (1979). Non-linear analysis of coupled walls. *Earthquake Engineering and Structural Dynamics*, 7(1), 1–22.

Takeda, T., Sozen, M.A. and Nielson, N.N. (1970). Reinforced concrete response to simulated earthquake. *Journal of Structural Division, ASCE*, 96(ST12), 2557–2573.

Taranath, B.S. (1998). *Steel, Concrete and Composite Design of Tall Buildings*. McGraw-Hill, New York, NY, USA.
Tzanetos, N., Elnashai, A.S., Hamdan, F.H. and Antoniou, S. (2000). Inelastic dynamic response of RC bridges subjected to spatial non-synchronous earthquake motion. *Advances in Structural Engineering*, 3(3), 191–214.
Vamvatsikos, D. and Cornell, A.C. (2002). Incremental dynamic analysis. *Earthquake Engineering and Structural Dynamics*, 31(3), 491–514.
Vamvatsikos, D. and Cornell, A.C. (2004). Applied incremental dynamic analysis. *Earthquake Spectra*, 20(2), 523–553.
Vecchio, F.J. and Collins, M.P. (1982). The response of the reinforced concrete to in-plane shear and normal stresses. Publication No. 82-03, Department of Civil Engineering, University of Toronto, Toronto, Canada.
Wen, Y.K., Ellingwood, B.R. and Bracci, J. (2004). Vulnerability function framework for consequence-based engineering. MAE Center Project DS-4 Report, Department of Civil and Environmental Engineering, University of Illinois, Urbana, IL, USA.
Williams, M.S. and Sexsmith, R.G. (1995). Seismic damage indices for concrete structures A state of the art review. *Earthquake Spectra*, 11(2), 319–349.
Wolf, J.P. (1994). *Foundation Vibration Analysis Using Simple Physical Models*. Prentice-Hall, Englewood Cliffs, NJ, USA.
Wood, W.L. (1990). *Practical Time-Stepping Schemes*. Clarendon Press, Oxford, UK.
Yun, G.J., Ghaboussi, J. and Elnashai, A.S. (2007). Development of neural network based hysteretic models for steel beam-column connections through self-learning simulation. *Journal of Earthquake Engineering*, 11(3), 453–467.
Zerva, A. (2008). *Seismic Ground Motion Spatial Variability: Modeling and Engineering Applications. Advances in Engineering Series*. CRC Press, Boca Raton, FL, USA.
Zhang, J. and Makris, N. (2002). Seismic response analysis of highway overcrossings including soil-structure interaction. *Earthquake Engineering and Structural Dynamics*, 31(11), 1967–1991.

Appendix A

Structural Configurations and Systems for Effective Earthquake Resistance

A.1 Structural Configurations

Configuration plays an important role in the seismic performance of structures subjected to earthquake actions. Post-earthquake reconnaissance has pointed towards the observation that buildings with irregular configurations are more vulnerable than their regular counterparts. There are several reasons for this observed poor structural performance of irregular structures. Concentrations of inelastic demand are likely to occur in zones of geometrical discontinuities and/or mass and stiffness irregularities. If the available ductility is limited, failure is initiated, thus possibly leading to collapse. Unexpected load paths and overstress of components can cause significant adverse effects. To prevent unfavourable failure modes, adequate 'conceptual design' is required at an early stage. In addition, thorough assessment of the structural configuration is vital to achieve adequate seismic performance.

Structural configuration has two fundamental aspects: the overall form and the type of lateral resisting system employed. The impact of structural configuration, in plan and elevation, on seismic performance depends upon:

(i) *Size:* as the absolute size of the structure increases, the range of cost-efficient configurations and systems is reduced. For example, while standardized simple and symmetrical shapes are generally used for high-rise buildings, more options are available for low- to medium-rise structures. The same is also true in bridge engineering where very long spans (>600–800 m) impose the use of suspension cables. Size may also dictate the choice of specific materials of construction. For example, high-rise structures may require high-strength concrete (e.g. Laogan and Elnashai, 1999; Aoyama, 2001, among others).

(ii) *Proportion:* earthquake response of a structure depends on its relative proportions rather than absolute size. Low slenderness in plan and elevation is beneficial. Reduced elevation slenderness minimizes overturning effects. For buildings, the ratio of the height (H) to the smallest depth (B) should not exceed 4–5 (Dowrick, 1987). This figure is exceeded by far in modern tall buildings worldwide, which exhibit H/B of 10–15 (CTBUH, 1995). Multi-storey structures may also employ narrow shapes. In this case, the slenderness ratio is critical. Large aspect ratios in plan

Fundamentals of Earthquake Engineering Amr S. Elnashai and Luigi Di Sarno

render torsional effects more likely to occur. Asynchronous motions at the foundation of building structures may also be caused by high width-to-depth ratios.

(iii) *Distribution and concentration:* vertical and plan distribution of stiffness and mass is important to achieve adequate seismic performance. In tall and slender buildings, lateral deformability reduces the earthquake-induced forces. Problems related to deflection control may arise, however, in earthquake and wind response of high-rise structures. Low-rise buildings should be flexible to reduce the shear forces due to ground motions. Tall buildings should be stiff to control the lateral deformations. Seismic motions are multi-dimensional, thus structures need to be able to resist the imposed loads and deformations in any direction. Adequate distributions of structural systems to resist loads (vertical and lateral) can prevent concentrations of inelastic demands. Structural elements can be arranged in orthogonal directions to ensure similar stiffness and resistance characteristics in both main directions, i.e. they should possess bidirectional resistance and stiffness.

(iv) *Perimeter resistance:* torsional motion tends to stress lateral resisting systems non-uniformly. High earthquake-induced torsional moments can be withstood by lateral resisting components located along the perimeter of the structure as displayed in Figure A.1. Perimeter columns and walls create, for instance, structural configurations with high rigidity and strength (also referred to as 'torsional stiffness and resistance'). The location in plan of systems for earthquake resistance significantly influences the dynamic response. The higher the radius of gyration of the plan layout of the structure, the higher the lever arm to resist overturning moments. In framed systems, the bending stiffness is significantly affected by the layout of columns in plan and elevation. Frames employing perimeter columns possess high bending stiffness and resistance; this is also true for frame-wall systems.

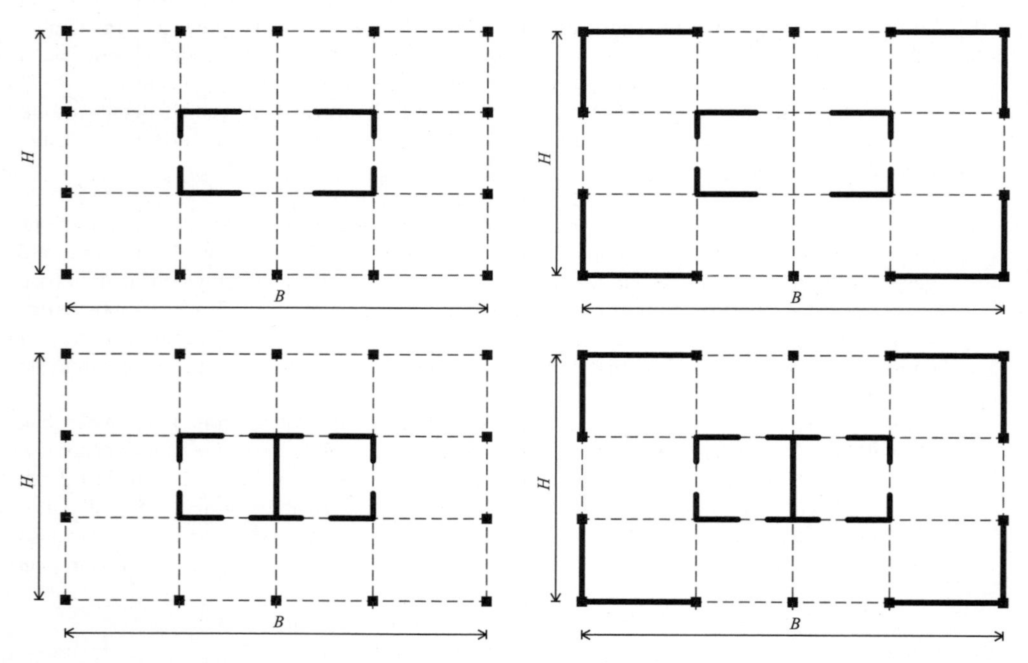

Figure A.1 Configurations with different perimeter resistance: low (*left*) and high (*right*) torsional resistance

The importance of structural configuration in earthquake response has been recognized and implemented by codes of practice and design guidance documents worldwide. To achieve adequate performance, these standards and guidelines provide basic principles for 'conceptual design', which are summarized below:

(i) *Simplicity:* consists of clear and direct paths for vertical and horizontal forces due to the combination of gravity and earthquake loading. Its fulfilment gives rise to reliable predictions of seismic behaviour. Compact, convex and closed shapes perform better than complex, concave and open sections. In addition, dimensioning, detailing and construction of simple structures are often more cost-effective than for complex structural systems.

(ii) *Uniformity:* implies even distribution of structural elements in plan and elevation, allowing for smooth and direct transmission of the inertial forces generated by the masses of structural and non-structural components height-wise. Concentrations of stresses or large ductility demands cause premature collapse. It may be necessary to subdivide the entire building into independent units by using seismic joints. Uniform distributions of mass, strength and stiffness eliminate large eccentricities between the centre of mass and that of stiffness. Torsion generates undesirable effects in the earthquake response of structures.

(iii) *Symmetry:* symmetrical or quasi-symmetrical structural layouts, well distributed in-plan, are a viable solution for the achievement of uniformity. Structural symmetry means that the centre of mass and centre of resistance are located at, or close to, the same point. Eccentricity produces torsion and stress concentrations. Symmetry is important in both directions in plan and elevations. The use of evenly distributed structural elements allows more favourable redistribution of action effects within the entire structure. Symmetry combined with simplicity is beneficial for earthquake response but architectural constraints sometimes make this difficult to achieve. Symmetrical shapes, which employ offset cores, cause undesirable torsional effects. Shapes with re-entrant corners can be symmetrical, but lack compactness.

(iv) *Redundancy:* this is a measure of the degree of indeterminacy and reliability of structural systems. Redundancy primarily arises from the capacity of structures to provide an alternative loading path after any component failure. The quantification of this system property in framed structures can be carried out through the 'redundancy index' (Bertero and Bertero, 1999). This index is defined as the number of critical (or inelastic) regions of the structural systems that dissipate significant amounts of hysteretic energy (or dissipative regions). In frames, adequate redundancy is achieved by ensuring that the number of beam plastic hinges is high, e.g. at all beam ends. Redundancy can be significantly affected by the configuration of the structure; it also depends on the connection behaviour. For example, for buildings under biaxial and torsional motions, redundant framed systems employing ductile connections exhibit adequate seismic performance (Wen and Song, 2003).

(v) *Bidirectional resistance and stiffness:* lateral resisting elements and systems arranged in an orthogonal in-plan pattern provide similar resistance and stiffness characteristics in the principal directions of the structure. High horizontal stiffness is effective in limiting excessive displacements that may lead to instabilities (e.g. due to P-Δ effects) or to extensive structural and non-structural damage.

(vi) *Torsional resistance and stiffness:* adequate torsional stiffness and resistance is necessary to reduce torsional motions which tend to stress the structural elements non-uniformly. In this respect, arrangements in which the main elements resisting the seismic actions are distributed close to the periphery of the building present clear advantages. Structures with compact and convex layouts exhibit high torsional stiffness and resistance. Inelastic demands on joints due to torsion are high. These structural components are generally weak-links in the load path for gravity and earthquake loads as illustrated in Section 2.3.2.2: they should possess adequate stiffness, strength and ductility.

Table A.1 Attributes and benefits of optimal structural configurations.

Attributes	Benefits
Low width-to-depth ratio	Low torsional effects
Low height-to-base ratio	Low overturning effects
Similar storey heights	Elimination of weak/soft storeys
Short spans	Low unit stress and deformation
Symmetrical plan shape	Elimination/reduction of torsion
Uniform plan/elevation stiffness	Elimination of stress concentrations
Uniform plan/elevation resistance	Elimination of stress concentrations
Uniform plan/elevation ductility	High energy dissipation
Perimeter lateral resisting systems	High torsional resistance potential
Redundancy	High plastic redistribution

(vii) *Diaphragm behaviour at storey level:* floor and roof systems act as horizontal diaphragms in building structures. These collect and transmit inertia forces to the vertical elements of lateral resistant systems, i.e. columns and structural walls. They also ensure that vertical components act together under gravity and seismic loads. Diaphragm action is especially relevant in cases of complex and non-uniform layouts of vertical structural systems, or where systems with different horizontal deformation characteristics are used together (as in dual or mixed systems). High in-plane stiffness and resistance is required to ensure adequate seismic response of storey diaphragms.

(viii) *Adequate foundation:* stiff and resistant foundations and their connections with the superstructure ensure that the whole structure is subjected to uniform seismic excitation. Rigid, box-type or cellular foundations, containing a foundation slab and a cover slab are adequate for structures composed of a discrete number of structural walls, which differ in width and stiffness. Buildings with isolated foundation elements – footings or piles – should utilize a foundation slab or tie beams between these elements in both main directions.

Ideal structural configurations for earthquake-resistant design should possess the attributes listed in Table A.1. Major benefits that can be achieved are also given in the table.

Features in Table A.1 can be utilized to classify structural configurations as 'regular' or 'irregular'. Regular structures are those employing the attributes in Table A.1. These systems generally show adequate seismic performance; regularity is thus necessary but not sufficient under earthquake loading. Detailing is as important as regularity. Although expressed in a qualitative rather than quantitative manner, Table A.1 provides simple guidelines that can be used in conceptual structural seismic design.

The physical significance of structural regularity is intuitive but its quantitative definition is often very difficult. Structures may have plan and elevation irregularities as illustrated in Figure A.2; these depend on geometry, lateral stiffness and strength distributions, mass ratios along the height, mass-resistance eccentricity and discontinuity in diaphragm stiffness. Regular structures are likely to exhibit uniform energy distribution, hence uniform damage distribution under earthquake actions.

Irregularities are commonly associated with geometrical properties, such as size and shape. However, buildings with irregular plans and elevations may employ regular structural systems to resist vertical and lateral loads. Criteria to identify irregularities exist and it is often possible to estimate them (e.g. Arnold and Reitherman, 1982). Torsion increases as a function of the eccentricity between centres of mass C_M and rigidity C_R, as discussed in Section 2.3.1.2. The distance between C_M and C_R can be used

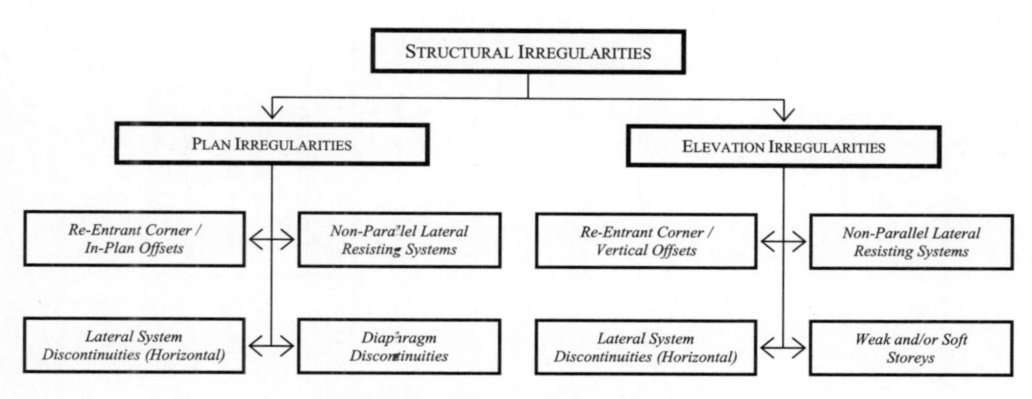

Figure A.2 Typical structural irregularities

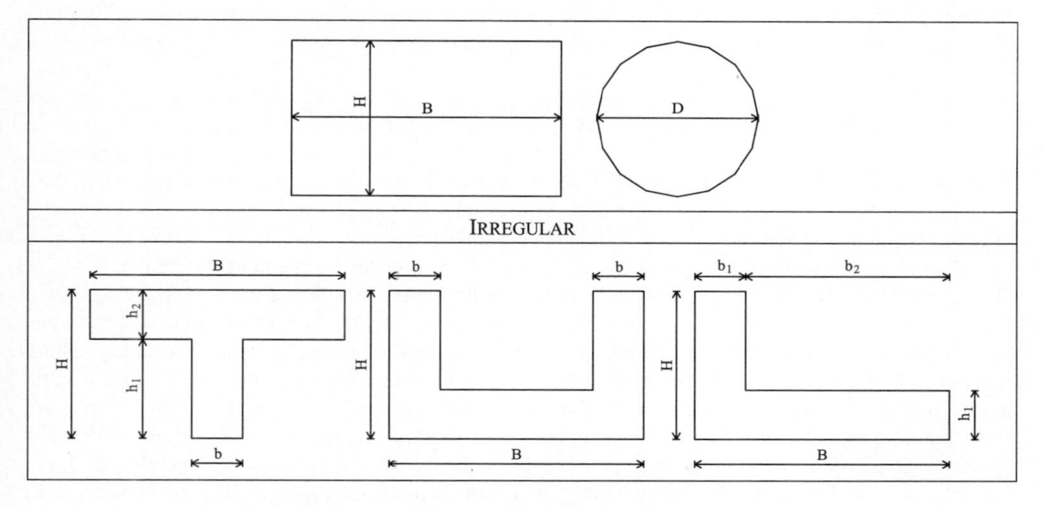

Figure A.3 Common regular and irregular shapes for plan layout

to quantify torsional effects. Criteria for regular structures are outlined hereafter for plan and elevation, respectively.

A.1.1 Plan Regularity

Structures with regular plan configurations are compact, i.e. described by polygonal convex lines. Square, rectangular and circular shapes are compact. Square or rectangular configurations with minor re-entrant corners can still be considered regular. Large re-entrant corners creating crucifix forms give rise to irregular configurations (Figure A.3). The dynamic response of the wings (also termed 'multi-mass structures') generally differs from that of the structure as a whole. Multi-mass structures are highly vulnerable at connections between wings. Relative displacements cause severe damage at the intersection of various blocks; torsional effects are likely to occur. Other plan configurations with geometrical

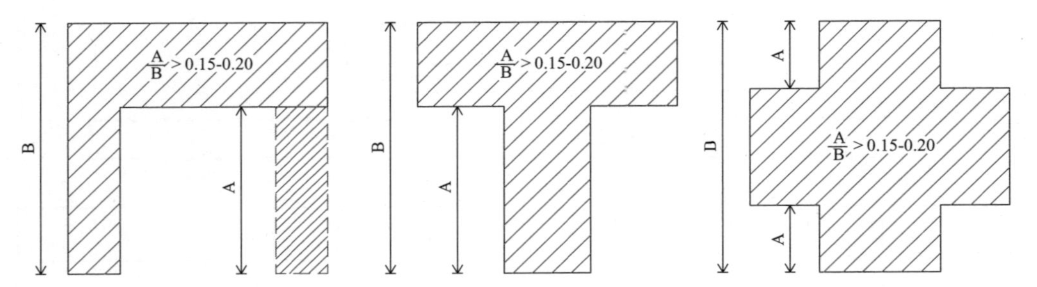

Figure A.4 Typical limits for plan irregularities (*adapted from* FEMA 450, 2004)

symmetry, e.g. I- and H-shapes, are also irregular because of the response of the wings. Plan irregularities depend upon the size of setbacks, i.e. re-entrant corners and edge recesses. Limitations for the setbacks can be expressed as a function of their geometry. For example, for L-, T- and X-sections, the following limitation can be used:

$$\frac{A}{B} > 0.15 \div 0.20 \tag{A.1}$$

where A and B are the length and the depth of the re-entrance, respectively, as shown in Figure A.4. Equation (A.1) provides the limitation included in seismic design recommendations in the USA (e.g. FEMA 450, 2004). Alternatively, regularity in plan may be assumed if, for each setback, the area between the outline of the floor and a convex polygonal line enveloping the floor does not exceed 5% of the total area. This criterion is adopted in European design practice (e.g. Eurocode 8, 2004).

A building structure may have a symmetrical geometric shape without re-entrant corners and wings but can still be classified as irregular in plan, since the distribution of mass or vertical seismic resisting elements may be asymmetric. Torsional effects due to earthquake motions can occur even when static centres of mass C_M and resistance C_R coincide. For example, ground-motion waves acting at an angle to the building axis also cause torsion, as may crack and yield in a non-symmetrical fashion. Additionally, these effects can magnify torsion due to eccentricity between the static centres. Generally speaking, buildings having an eccentricity between the static centre of mass and the static centre of resistance in excess of 10% of the building dimension perpendicular to the direction of the seismic force are considered irregular. Quantitative criteria for torsional effects are often provided in a few modern international seismic codes.

Structures with symmetric and compact shapes but employing plan discontinuity for lateral resisting systems are not regular. Typical examples are three-sided buildings that experience high torsional effects under earthquake loading (Ambrose and Vergun, 1999). Several failures have been observed in past earthquakes for these structures, which are utilized mainly, but not exclusively, for low- to medium-rise constructions. Architectural reasons generally impose arrangements of plan layout with steel or reinforced concrete (RC) frames and walls located along three sides of the perimeter (Figure A.5). In commercial buildings, the necessity for large openings for shop windows on the facade may lead to the use of three-sided buildings of this type. Continuity in plan between lateral resisting systems is essential for clear and continuous load paths.

Core-type buildings with the vertical seismic-force-resisting system concentrated near the centre tend to behave poorly during earthquakes. Better performance has been observed when vertical components are distributed near the perimeter of the building. This may, however, cause instability due to torsion. Eccentric locations of rigid cores for external lifts and stairwells also generate undesirable torsional

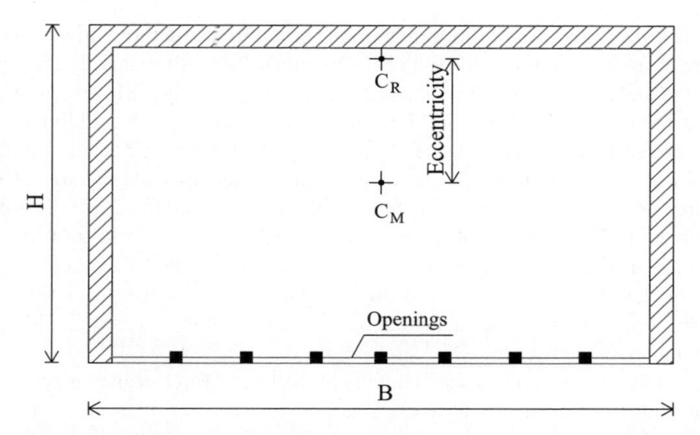

Figure A.5 Irregularities due to plan discontinuity for lateral resisting systems (*three-sided buildings*)
Key: C_M = centre of mass; C_R = centre of rigidity

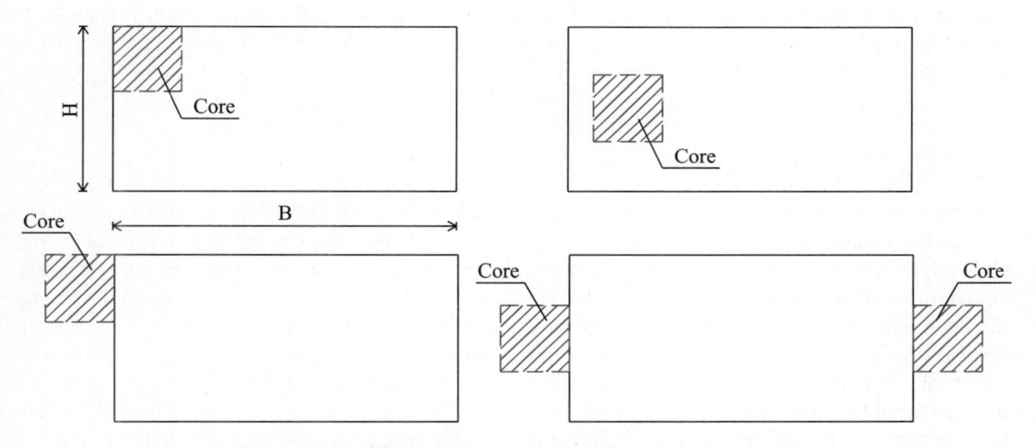

Figure A.6 Plan irregularities due to unfavourable core location

effects (Figure A.6). For example, external access towers, which are meant to be used during seismic events, often fail in their function because they experienced large rotations or collapse.

Diaphragm action is another requirement for plan regularity. Relative stiffness and strength of floors and bracing systems are critical for earthquake response. Floor systems with high stiffness and strength ensure adequate distribution of seismic actions among vertical structural elements. Where discontinuities in the lateral force resistance path exist, the structure is no longer regular. Significant differences in stiffness between portions of diaphragms may cause a change in the distribution of seismic forces to the vertical components and create torsional forces.

Building structures with large aspect ratios in plan are susceptible to incoherent earthquake motion (also referred to as 'out-of-phase effects'). Different foundation materials may generate amplification of the dynamic response in different parts of the building. The higher the aspect ratio, the higher the likelihood of incoherence effects as illustrated in Section 1.3.3. These effects depend on whether foundation systems, as well as superstructures, are continuous or not. The probability of having similar live loads in large structures is inversely proportional to the size of the structure (Nowak and Collins, 2000). Therefore, the plan aspect ratio should be not greater than 2–3. Alternatively, the structure may be subdivided into independently responding parts by using seismic joints. Movement gaps are relatively easy to construct for bridge structures but are often highly unreliable in buildings. Separation joints should be large enough to accommodate lateral displacements between adjacent buildings and to avoid pounding, as discussed in Section A.1.2. Out-of-phase movements dictate the size of the gap between adjacent structures. As a rule of thumb, the separation(s) can be assumed as 1/100 of the maximum height (*H*) of the adjacent structures, in metres. Separation joints can help to mitigate unfavourable seismic effects on multi-mass structures. It should, however, be noted that they can have disastrous effects because of gas entrapment during post-earthquake fires. Debris from severely damaged or partially collapsed upper storeys can also fall in separation joints. These should be sealed, where possible, to prevent such occurrences.

Irregularities in plan arise when vertical elements of the lateral force-resisting system are not parallel to or symmetric with major orthogonal axes. Shapes with sharp corners are unsuitable for seismic resistance because of the high probability of torsional forces under earthquake motions. Wedge-shape plans have large eccentricity between centre of mass and centre of rigidity. In addition, different relative stiffnesses between narrower and wider perimeter sides exacerbate torsional effects.

Discontinuities in horizontal and vertical lateral resistant systems are an additional source of irregularity in plan. Out-of-plane offsets of vertical elements, for example, may impose significant demands on structural components of earthquake-resistant structures. Extensive damage may be caused by these offsets; they should not be employed in seismic areas.

A.1.2 Elevation Regularity

Structural systems can be characterized by several types of irregularities in elevation depending on their geometrical configuration and mechanical properties along the height. For example, asymmetrical geometry with respect to the vertical axis can cause vertical irregularities in buildings. Structures with setbacks, i.e. with re-entrant corners along the height are irregular. Setbacks often introduce stiffness and strength discontinuities in lateral force-resisting systems. High inelastic demands are concentrated in zones of vertical offsets. Damage is likely to occur in these 'notch regions' during earthquakes. Unfavourable effects due to setbacks depend on the relative proportions and absolute size of the system. Pyramid and inverted pendulum configurations are extreme examples of vertical setbacks (Figure A.7), but they do not have corners. The pyramid is the optimal geometric shape for earthquake resistance. The bulk of the mass is located near the ground and its plan density is extremely high. By contrast, the inverted pendulum (or inverted setback) has low resistance to overturning and unfavourable location of the mass at the top of the structure. In both cases, the absence of re-entrance prevents concentration of inelastic demand at corners along the height. Inverted pendulum structures also have low redundancy and overstrength, and concentrate their inelastic behaviour at their bases. They exhibit substantially lower energy dissipation capacity compared to pyramidal shapes, as well as to several other lateral resisting systems.

The aspect ratio of the building in elevation affects the overturning moment exerted on the foundations. Very slender structures suffer from higher mode contributions, which can cause damage at intermediate storeys. In addition, structures with higher mode effects exhibit a complex dynamic response that necessitates the use of more elaborate seismic force calculation methods. Therefore, buildings and structures employing low aspect ratios (*H*/*B*) are considered regular.

Post-earthquake observations have shown that a considerable percentage of damaged structures suffer from non-orthogonal and non-coaxial member axes. Axes with offsets are found either in plan or elevation in several RC multi-storey buildings. All beams and columns should therefore have the same axes with no offset between adjacent members. Examples of members with offsets are shown in Figure A.8. Large variations in size between connected members undermine the uniformity of load paths

Figure A.7 Extreme examples of setbacks: traditional pyramid in Egypt (*left*), modern US pyramids in Indianapolis (*middle*) and an inverted pyramid in Dallas (*right*)

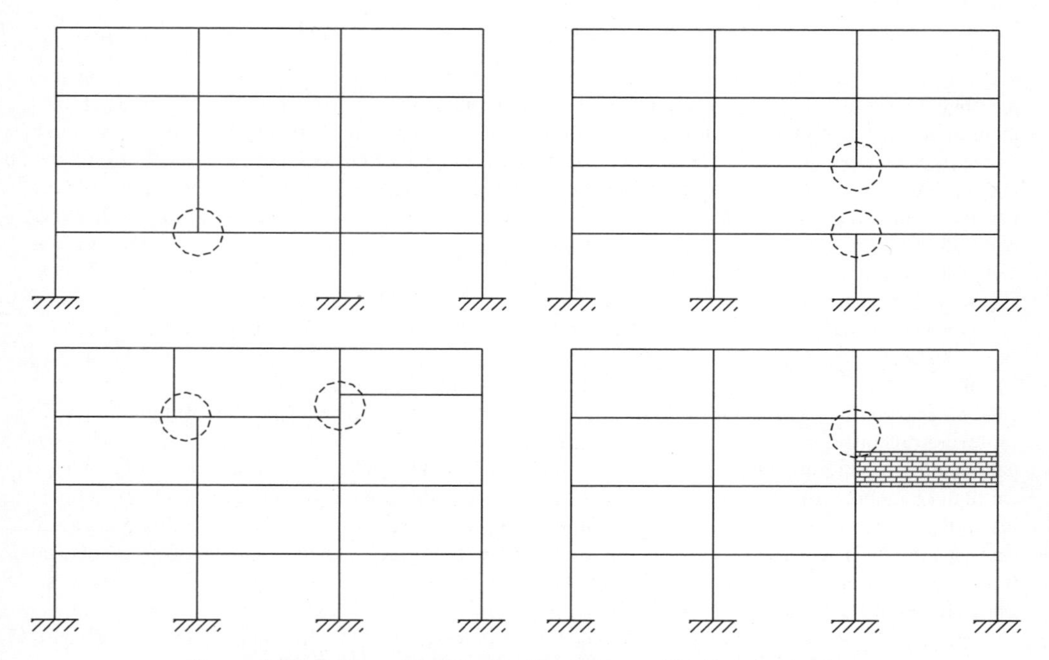

Figure A.8 Unfavourable discontinuities and axes with offsets in elevation
Key: Circles indicate areas of concern

(as discussed in Section 2.3.2.2) and cause large stress concentrations in connected structural elements. This problem frequently occurs in RC frames when large flat beams frame into columns along their weak axes. In this case, the difference between the widths of beams and columns may endanger the uniform transfer of flexural, shear and axial actions between the connected members. This type of detail also generates high stress concentrations at beam-to-column joints.

Beams or columns supported on beams should be avoided since the imposed local demand, especially in torsion and shear, is very difficult to accommodate. In V-, inverted V- and K-braced frames, braces do not intersect beam-to-column joints at both ends. Brace-to-beam and brace-to-column connections are located along the spans and heights, respectively, as shown in Figure A.9. Short columns generated in K-braced frames give rise to unfavourable failure modes.

Under horizontal seismic forces, the compression brace buckles and its load-bearing capacity is reduced dramatically. Tests carried out by Hassan and Goel (1991) on steel structures showed that the

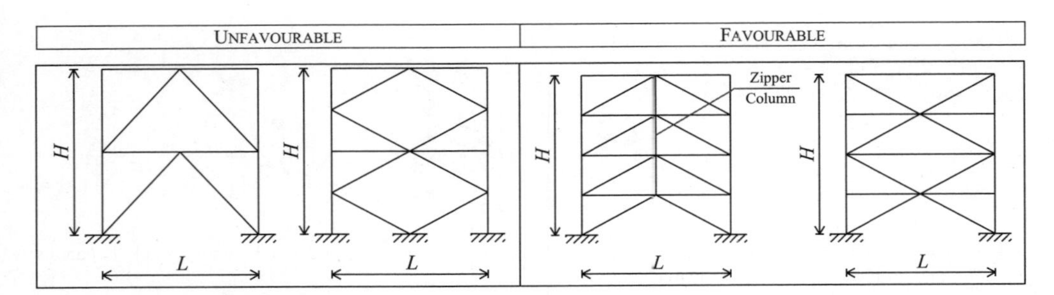

Figure A.9 Unfavourable and favourable brace arrangements for inverted V-bracing (*left*) and K-bracing (*right*)

post-buckling residual strength is typically 30% of the initial compressive strength. By contrast, forces increase in the brace in tension up to the yielding point. The net result is an unbalanced vertical force on the intersecting beam. The unfavourable effects of the unbalanced force can be mitigated by employing configuration of braces as, for example, those given in Figure A.9. The presence of zipper columns has been found to be very effective for systems with V- or inverted V-braces (Khatib *et al.*, 1988; Yang *et al.*, 2006).

Vertical configuration irregularities affect structural response at the various floor levels in buildings; elastic and inelastic demands at these levels are significantly different from uniform distributions. If there are abrupt changes in stiffness-strength or mass irregularities in elevation, high demand concentrations will ensue, as indicated in Figure A.10.

Soft storeys occur in buildings whenever the stiffness of a storey to resist lateral demands is significantly less than that of adjacent storeys. This is because structural systems with this configuration tend to develop inelastic behaviour at the most vulnerable storey. As a result, significant changes in load paths and deformation patterns arise (*see* Sections 2.3.1.2 and 2.3.2.2). Soft storeys experience large earthquake-induced displacements and, in turn, cause extensive damage and even collapse. Global instability in many multi-storey buildings is initiated by soft storeys. Similarly, if two or more adjacent storeys exhibit large variations in strength, this leads to the effect known as 'weak storey'. Soft and weak storeys often occur simultaneously due to the close relationship between strength and stiffness, as discussed in Sections 2.3.1.2 and 2.3.2.2.

Typically soft and weak storeys are located at the ground floor of buildings and are caused by excessive inter-storey heights due, for example, to large shop windows or garages, changes in stiffness and strength above the first floor and discontinuity in lateral resisting systems. Braces in framed structures and shear walls are often interrupted at the ground floor for architectural reasons. Soft storeys are also generated by components which are non-structural. For example, infills create unexpected bracing actions (also known as 'stiffening') of upper floors in buildings; large concentrations of inelastic demand are thus imposed at ground floor level. Clearly, irregular distributions of infills along the building height can cause unfavourable failure modes, which are not necessarily localized at the base. Bracing due to partially infilled frames, mezzanines and hillside sites may lead to short column effects, which are highly unfavourable (*see* Section 2.3.1.2). Infills are frequently made of heavy masonry or RC panels such that non-uniform arrangements affect the mass distribution (mass irregularity in elevation). As a general rule, differences of more than 20–25% in mass or stiffness and strength between consecutive floors can cause unfavourable failure modes. This not only infers that column dimensions should be reduced with caution, but also suggests the necessity for restrictions on linkages between adjacent buildings (such as walkways) as well as on setbacks.

Vertical continuity for earthquake-resisting systems is essential for regularity in elevation. Cores, structural walls, frames with or without braces, and all other lateral resisting structures should run

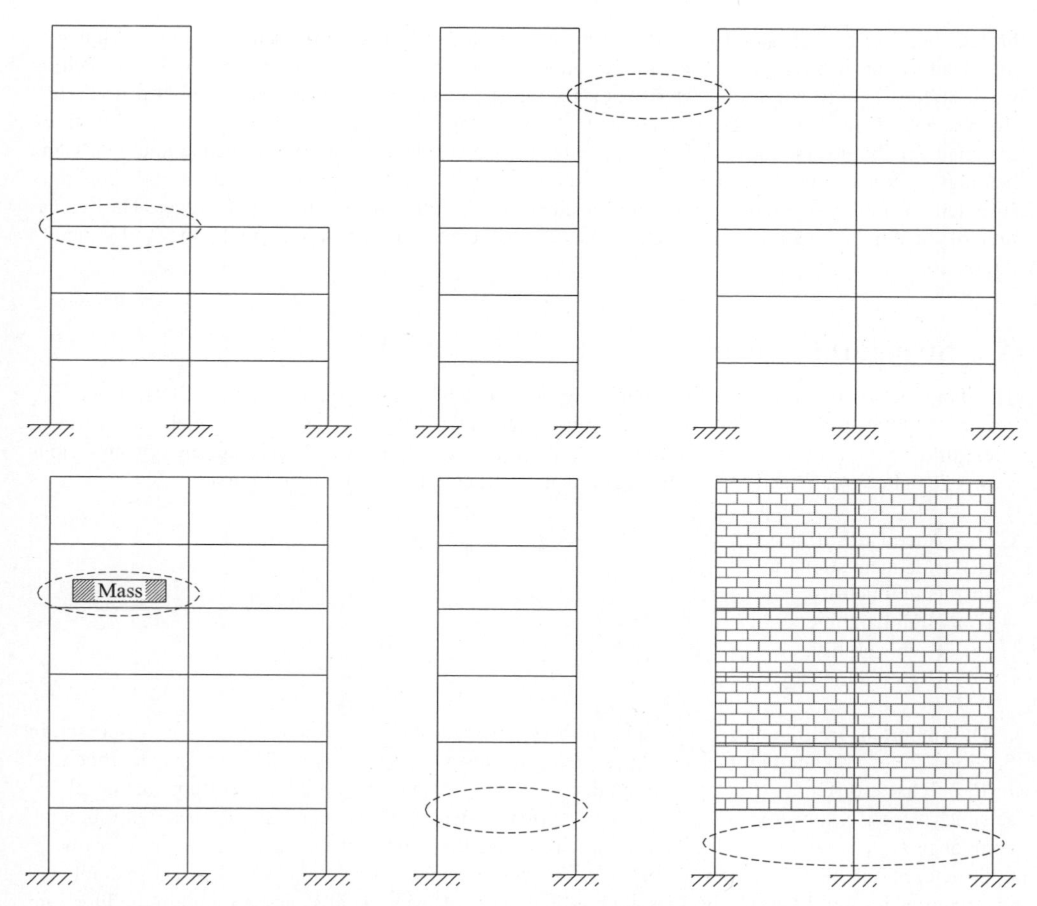

Figure A.10 Irregularities in elevation due to non-uniform stiffness, strength and mass distribution
Key: Circles indicate areas of concern

without interruption from their foundations to the top of the building. This ensures a clear and continuous load path and prevents concentrations of high ductility demands.

Vertical and plan layouts should be selected such that damage of adjacent structures is avoided. Excessive drifts during earthquakes may cause damage to proximate systems or between different wings of the same structure (also referred to as 'pounding'). Structural damage for pounding (also termed 'hammering' or 'battering') is induced by high momentum transferred between colliding structures. Pounding is a major cause of damage to buildings in cities located in seismic regions. It may occur in multi-mass structures, which employ structural components with very different relative stiffness. Typical examples are buildings with RC infilled frames for lower storeys and steel bare frames for the upper storeys. In this case, the less stiff structural systems and attachments (penthouses, roof tanks) move considerably with respect to rigid parts; high differential movements may cause pounding. Bell towers in historical buildings typically suffer damage caused by this type of pounding. Out-of-phase vibrations and separation joints inadequate to accommodate large drifts have caused extensive damage worldwide in past earthquakes (e.g. Kasai *et al.*, 1996). Examples of effects of pounding in building structures are shown in Figures B.17 and B.18. Pounding may also cause damage in multiple-frame bridges because of their low lateral stiffness and restrainer stiffness. Moderate-to-strong earthquakes may lead to out-of-phase motion of bridge frames due to the variability of ground motion, travelling wave effects and structural characteristics affecting the dynamic response, especially stiffness (DesRoches and Fenves, 1997; DesRoches and Muthukumar, 2002). Inertial forces may exceed those assumed for the design earthquake and in order to prevent damage at column bents, abutments and bearings, adequate lateral stiffness of piers and restrainers is essential (Kim *et al.*, 2000). It is generally sufficient to employ separation joints between adjacent buildings and in multi-span bridges. The estimation of such joints requires thorough assessment of the seismic response including soil-structure interaction.

A.2 Structural Systems

The dynamic behaviour of structures under earthquake actions is dependent upon the lateral resisting system employed. Construction materials and structural configurations differ widely in stiffness, strength and ductility; thus, different systems deform, resist actions and dissipate energy in various ways. To achieve satisfactory seismic performance, structural systems should possess:

(i) Adequate stiffness;
(ii) Adequate strength;
(iii) High ductility;
(iv) High damping;
(v) High stability;
(vi) High redundancy.

The importance of the above attributes in the seismic response of structures has been discussed in Section 2.3. Several lateral force-resisting systems, however, possess only a few of the above properties. In these cases, different structural components or systems may be combined to improve the global seismic response. For example, dual (or hybrid) systems, which combine frames with bracing components such as structural walls, are more effective than either of the components on their own.

Structures suitable for earthquake resistance include horizontal and vertical systems. Those employed in structures for buildings and bridges are outlined below. Design details can be found in the literature for the various construction materials (e.g. Dowrick, 1987; Paulay and Priestley, 1992; Priestley *et al.*, 1996; Foliente, 1997; Bruneau *et al.*, 1998).

A.2.1 Horizontal Systems

Horizontal bracing in buildings and bridges is provided by floor and deck framing systems (also known as 'horizontal diaphragms'), respectively. Floor and deck systems have two functions. They carry gravity loads and transfer them to vertical structural elements as described in Section A.2.2. They also collect and distribute inertial forces among lateral load-resisting components. Force-resisting mechanisms in horizontal diaphragms are very complex because of the interaction between in-plane and out-of-plane behaviour. Figure A.11 compares the structural response of rigid and flexible diaphragms under horizontal loads for a simple box system. If the in-plane stiffness of the floor is high (*rigid diaphragm*), horizontal actions (F in the figure) are distributed to vertical elements in proportion to their relative stiffness, as also illustrated in Section 2.3.1.2. Floor deformations are negligible compared to those of vertical resisting systems. Conversely, a flexible diaphragm distributes horizontal inertial actions to vertical components as a series of simple supported beams spanning between these components. In this case, the action distribution is governed by equilibrium conditions. Floor deflections may exceed those of lateral resisting systems. Diaphragms, especially in existing buildings, are neither perfectly rigid nor

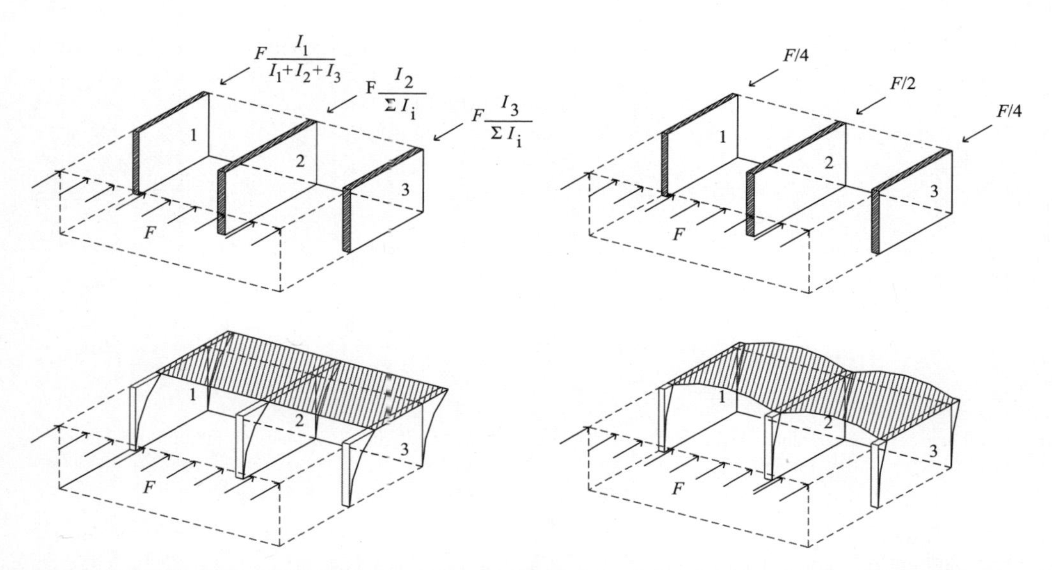

Figure A.11 Diaphragm behaviour under horizontal loads: shear distribution (*top*) for rigid (*left*) and flexible (*right*) diaphragms and lateral displacements (*below*)

Table A.2 Behaviour of horizontal diaphragms for different depth-to-span (aspect) ratios.

Depth-to-span ratios	Behaviour	Contribution to response	
		Shear	Bending
1	Deep beam	✓	
2	Deep-to-stiff beam	✓	
4	Stiff beam	✓	✓
10	Flexible beam		✓

completely flexible; they possess intermediate behaviour (referred to as 'semi-rigid diaphragms'). Continuity between lateral resisting systems and floor decks is essential for action distribution and in-plane rigid displacements.

Structural response of floor and deck diaphragms under lateral loads depends upon the materials of construction as well as their geometry, e.g. depth-to-span ratios (also quoted as 'aspect ratios'). Concrete slabs possess adequate in-plane rigidity despite showing high out-of-plane flexibility. Metal and wood decks may, on the other hand, exhibit low in-plane stiffness. Depth-to-span ratios depend on the lateral force-resisting systems and are correlated to the material of construction. For common bays of RC frames for buildings, which are in the range of 4–6 m, adequate thicknesses of concrete slabs to achieve diaphragm actions are 4–6 cm. Rough estimates for depth-to-span ratios are provided in Table A.2. Requirements for minimum aspect ratios are also given by seismic codes of practice.

Diaphragms behave in-plane as horizontal continuous beams supported by vertical lateral resisting systems (also referred to as 'beam analogy'). The deck or slab is the web of the beam carrying the shear and the perimeter spandrel or wall is the flange of the beam resisting bending (Figure A.12). Seismic forces may be considered uniform distributed loads for stiff floors, as also shown in Figure

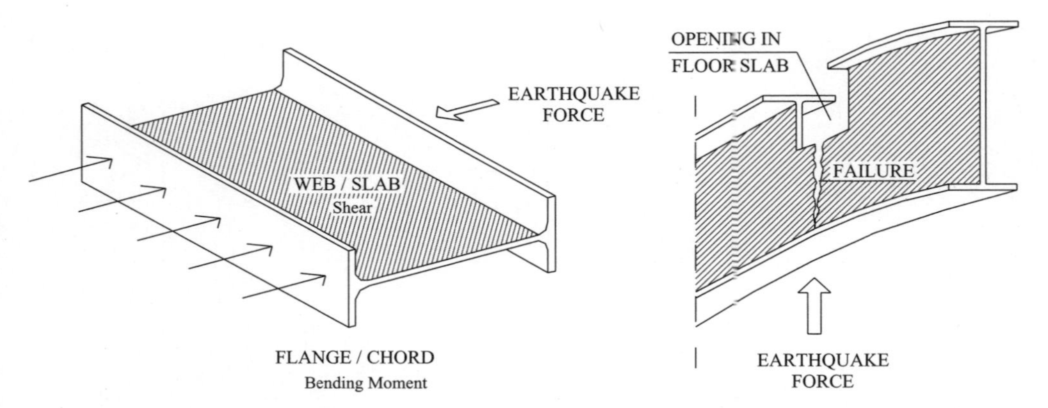

Figure A.12 Beam analogy for horizontal diaphragms: load distribution (*left*) and common failure (*right*)

A.11. Diaphragms should possess adequate shear and bending resistance to withstand in-plane seismic loads and out-of-plane gravity loads.

Inappropriate locations of large openings, due for example to stairs or elevator cores, can create problems similar to those openings in the web of a beam (also known as 'notch effects'). These openings significantly reduce the diaphragm action and can lead to failure. Reinforcement around the weakened regions helps to redistribute the actions in the slab around the opening.

A.2.2 Vertical Systems

Structural and non-structural damage under earthquakes is caused by inadequate stiffness and/or strength of vertical components of lateral structural systems used for buildings, bridges and other types of construction. Vertical components may also fail because of insufficiency or absence of ductility. To achieve satisfactory seismic performance, vertical components of lateral resisting systems should comply with the structural requirements discussed in Section A.1. Seismic behaviour depends on materials of construction, system configurations and failure modes.

Earthquake resistance can be achieved through a wide range of vertical systems, which can range from free-standing columns to complex three-dimensional framed tubes and/or cores. Figure A.13 shows basic structural systems, which have been ranked according to their lateral stiffness. Columns are the simplest structural elements with lateral stiffness and strength. The relationship between applied actions and lateral deformations depends on their geometric and mechanical properties, as discussed in Section 2.3.1.2.

The deformed shape of columns is generally characterized by double curvature, thus inelastic demand can be concentrated at both ends. Frames show higher stiffness, strength and ductility than free-standing columns because of their deflected shape. Frame behaviour significantly depends on the relative rigidity of structural members (beams and columns) and connections (beam-to-columns and base columns). Frames with diagonal braces exhibit higher lateral stiffness and strength than moment frames; the ductility of braced systems is generally endangered by the occurrence of member (diagonal) buckling. Moment frames can be stiffened by infill panels. Infilled frames exhibit higher stiffness, strength and ductility than bare frames. Under lateral seismic loads, infills behave like one diagonal compression brace. Infill panels are often made of brittle materials, such as masonry or concrete, which crack due to their low tensile strength. Lateral stiffness of braced and infilled frames can be enhanced by employing structural walls. These elements usually exhibit high in-plane stiffness and resistance; their ductility

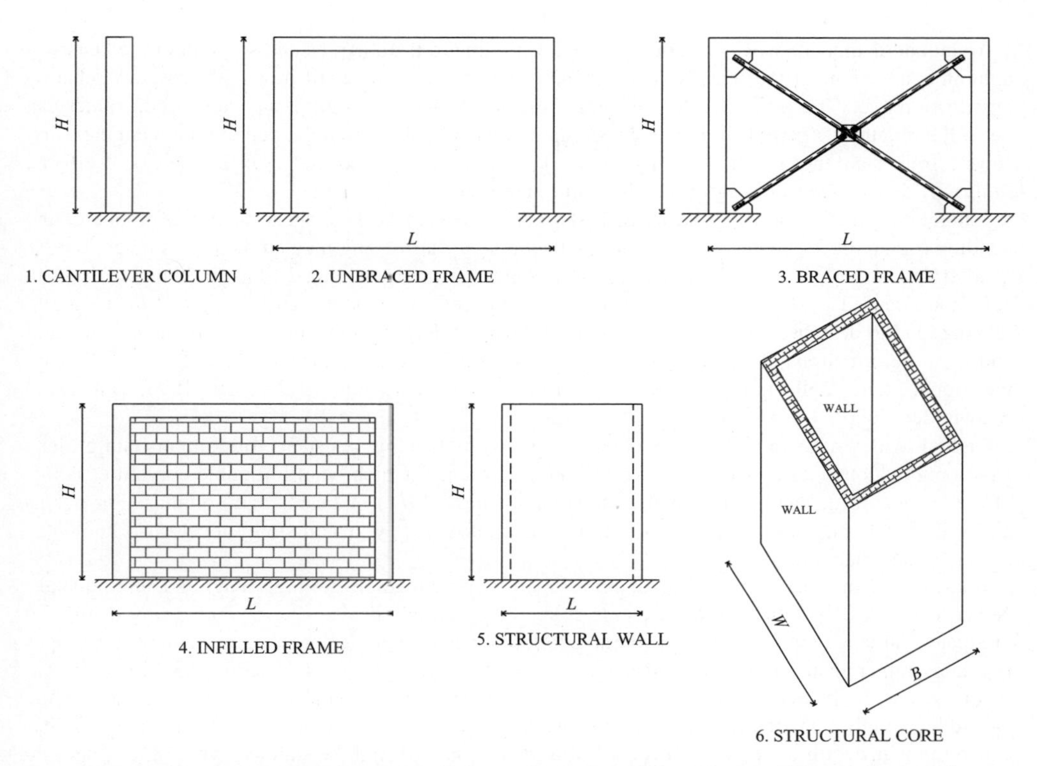

Figure A.13 Basic structural systems with increasing lateral stiffness (*from top left to bottom right*)

depends primarily on the detailing of the foundation connection and their shape. Walls can be arranged to form rigid core systems. The latter possess high resistance but, as for structural walls, their inelastic behaviour can be impaired by seismic details with low ductility.

Basic elements shown in Figure A.13 are used for vertical systems of buildings and bridges. Typical lateral load-resisting systems include the following:

(i) *Moment-Resisting Frames*
(ii) *Braced Frames*
(iii) *Structural Walls*
(iv) *Hybrid Systems*
(v) *Tube Systems*

Moment-resisting frames can dissipate a large amount of energy, but they often suffer from large lateral displacements. Conversely, braced frames possess high lateral stiffness but relatively low deformation capacity. Hybrid and tube systems generally exhibit adequate structural performance. The seismic response characteristics of the above structural systems under horizontal forces are discussed below.

(i) Moment-Resisting Frames

Moment-resisting frames (MRFs) are structural systems consisting of beams, columns and joints. These systems are frequently used as structural skeletons in RC, steel and composite buildings and bridges.

Metal and composite MRFs can be classified according to the stiffness and strength of the beam-to-column connections or the sensitivity to second-order effects. Where stiffness is the response characteristic employed, frames can be 'rigid' or 'semi-rigid'. Where, in turn, resistance is used, frames can be 'full strength' or 'partial strength'; the strength is quantified through the bending moment capacity. 'Sway' frames are those with lateral stiffness inadequate to prevent secondary effects, e.g. P-Δ effects; in turn, if these effects are negligible, the frames are described as 'non-sway'.

Lateral deformation of MRFs is caused by two components: shear (or shear racking component) and bending (or cantilever bending component) deflections. Shear racking may amount to as much as 80% of total deflection, while bending component accounts for the remaining 20%. Unacceptable drifts in MRFs are caused by shear racking due to the bending of columns and girders. Storey drifts due to racking tend to decrease with height; by contrast, bending deformations increase with height. Flexural and shear deformations of beam-to-column and base-column connections at ground floor can increase the global lateral deflections in MRFs (e.g. Elnashai and Dowling, 1991; CEB, 1996; Gupta and Krawinkler, 2000).

Lateral load resistance of MRFs is chiefly provided by bending resistance of columns and girders. Horizontal actions (storey shears) generate shear in columns which bend in double curvature. Points of contra-flexure depend on the relative flexural stiffness (EI/L) of beams and columns. For beams stiffer than columns, the point of contra-flexure is approximately at mid-storey level (also known as 'shear frame behaviour').

Framed systems generally possess ductile response under medium- to high-magnitude earthquakes. Nevertheless, MRFs often suffer excessive lateral deformations, e.g. storey and roof drifts. To dissipate a large amount of energy, favourable failure modes are global mechanisms with plastic hinges in beams rather than columns – referred to as 'beam-sway' – (see also Section 2.3.3.2). Different requirements can be adopted for the design of dissipative zones in MRFs, depending on the seismic hazard level of the construction site. For example, framed systems employing details with high ductility, e.g. rotational ductilities $\mu_\phi > 8$–10, can be used in zones of high seismicity, where high inelastic demands are expected. Conversely, frames with very low values of sectional or member ductilities, e.g. $\mu_\phi \approx 1.5$–2.0, can be utilized in zones of low seismicity.

Moment-resisting frames are cost-effective for buildings up to about 30 storeys (Balendra, 1993). Framed systems with semi-rigid connections are, however, conveniently used for low- to medium-rise structures, up to six to nine storeys (Di Sarno, 2002). Drift control, i.e. stiffness requirements, necessitates the use of deep beams and columns, which becomes highly uneconomical as the height increases. For taller buildings, MRFs can be used in combination with braced frames and structural walls. Moment frames employ either configurations with several lines of frames along principal directions of the building or layouts with perimeter MRFs. Perimeter frames are very common especially in US design practice; these layouts lead to tube systems, which are discussed in the next sections. In bridges, frames are generally employed when the width of the upper deck is high and/or the piers are very slender and may not withstand high seismic overturning moments.

(ii) Braced Frames

Braced frames (BFs) are lateral force-resisting systems which consist of beams, columns, diagonal braces and joints. Many brace configurations may be efficiently employed to withstand earthquake loads. Braced frames are often grouped into two categories, i.e. concentrically braced frames (CBFs) and eccentrically braced frames (EBFs), depending on the layout of the diagonals employed. Knee-braced frames (KBFs) have also been found to be efficient for new and existing multi-storey buildings (Balendra *et al.*, 1991, 1997). The most common bracing configurations for CBFs and EBFs are provided in Figure A.14. Configurations for BFs with V-, inverted V- and K-braces are not included; their use should be avoided, especially in regions of high seismicity, because of unfavourable dynamic structural performance as discussed in Section 2.3.3.3.

In CBFs, beams, columns and braces intersect at a point. The entire system acts as a vertical cantilever truss. Beams and braces constitute the web of the truss while columns form the chords. The lateral

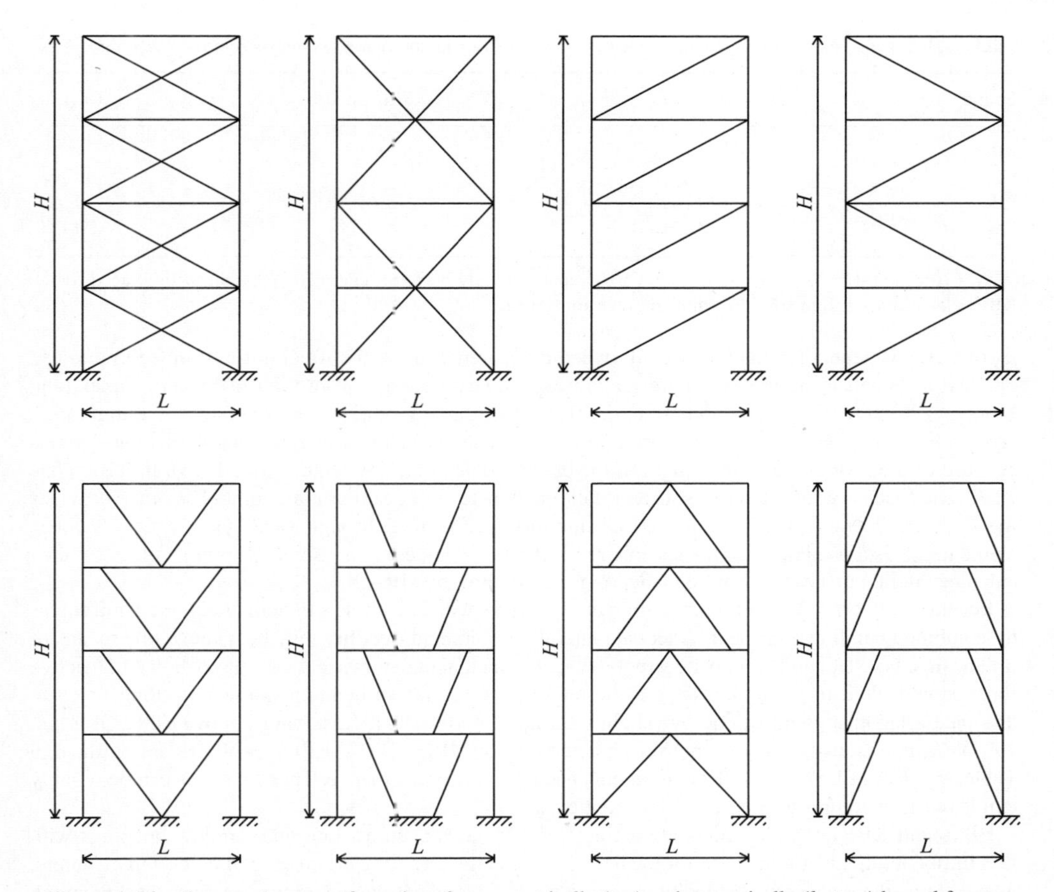

Figure A.14 Common brace configurations for concentrically (*top*) and eccentrically (*bottom*) braced frames

stiffness and strength of CBFs are well above that of MRFs. Lateral deflection modes depend upon the vertical slenderness of braced bays. In low-rise (squat) BFs, shear deflections are predominant, while high-rise (slender) BFs displace primarily in flexural modes. Shear deformations are caused by the elongation of braces and shortening of beams. Conversely, flexural deformations are generated by shortening and elongation in exterior columns.

In CBFs, internal actions are primarily transmitted through axial actions, either compression or tension. Under horizontal seismic forces, beams are in compression while braces are either in tension or compression. By reversing the load direction, beams are in tension. Lateral strength of CBFs depends on the capacity of braces, beams and columns.

Diagonal braces are the dissipative elements in CBFs. They are expected to yield under moderate- to high-magnitude earthquake ground motions. Alternate stress reversals can cause buckling of compressed braces, thus inhibiting large energy absorption. In steel and composite CBFs and/or RC frames retrofitted with steel diagonals, the amount of dissipated seismic energy is significantly reduced by the onset of buckling, local and global. Braced frames can be employed in areas of high seismicity provided that they employ high ductile details, especially for braces and their connections with beams and columns.

Inelastic seismic performance of CBFs is considered fairly poor because of proneness to buckling of strut components along with softening due to the Bauschinger effect. The global translation ductility

Table A.3 Comparison between response characteristics of framed and braced systems (*relative measures*).

Frame	Stiffness	Strength	Ductility
MRF	L	H	H
CBF	H	H	L
EBF	M / H	M / H	M / H
KBF	M / H	M / H	M / H

Key: H = high; L = low; M = moderate; CBF = concentrically braced frame; EBF = eccentrically braced frame; KBF = knee-braced frame; MRF = moment resisting frame.

μ_Δ of CBFs is generally small, especially when compared with MRFs. Buckling restrained braces can be used to enhance the inelastic deformation capacity and energy absorption of braced frames (e.g. Inoue *et al.*, 2001; Bozorgnia and Bertero, 2004, among many others). Improved brace configuration layouts have been proposed. Thus, for large bay widths, KBFs are attractive lateral resisting systems because of the shortening of the length of the braces (Balendra, 1993; Sam *et al.*, 1995). In KBFs, one end of the brace is connected to a short knee element instead of beam-column joint. The brace provides the required lateral stiffness, whereas the ductility is obtained through shear yielding of the knee element. KBFs are suitable for the seismic retrofitting of steel, composite and RC moment frames: they enhance lateral stiffness without endangering the ductility of MRFs.

Eccentrically braced frames employ bracing members with axis offsets to deliberately transmit forces by combined bending and shear. Adequate lateral stiffness and ductility may be achieved by means of link beams. For steel and composite structures, experimental and numerical tests have shown that active links should yield in shear (known as 'short link') rather than in bending (known as 'long link') to dissipate a larger amount of energy (Hjelmstadt and Popov, 1984; Kasai and Popov, 1986; Qi *et al.*, 1997). Comparisons between response characteristics of MRFs, CBFs, EBFs and KBFs are outlined in Table A.3. Both EBFs and KBFs exhibit enhanced seismic performance. These systems can be reliably employed in medium- to high-seismicity regions.

EBFs and KBFs can accommodate architectural features such as door and window openings with less intrusion; this is not the case for CBFs. Viable locations for braces are around cores and elevators, where frame diagonals may be enclosed within walls. The braces can be joined together, thus behaving as closed or partially closed spatial cells that may withstand torsional effects. Braced frames are cost-effective for medium- to high-rise buildings, up to 30 storeys (Di Sarno, 2002). These systems are widely used for several other types of constructions, such as towers, bridges and tanks, because high lateral stiffness can be achieved with great economy of materials.

For high-rise building structures, e.g. up to 50–60 storeys, internal braced cores (either CBFs or EBFs) are often connected to exterior columns of frames through deep and rigid truss beams (termed 'outriggers'). The resulting structural systems are known as 'outrigger-braced frames' (OBFs). These systems, which are suitable for steel, composite and RC tall buildings (Di Sarno, 2002), consist of four components: (braced) core, outriggers, columns and beams. Typical layouts for OBFs are displayed in Figure A.15. It is observed that outriggers, which are generally truss deep beams, can be as high as two to three storeys and are generally located at the top and/or mid-height in the structure.

Under horizontal seismic actions, the core behaves in a flexural mode. It exhibits high uplift forces and base overturning moments because of its slenderness. Therefore, its efficiency, as a free-standing structure, is reduced as its height increases: this reduction can be expressed as a cubic function of the height (Taranath, 1998). Exterior columns, which are offset with respect to the core (Figure A.15), act as stays, increasing the lever arm to resist overturning moments. Elongation and shortening of these columns may be advantageously used to prevent uplift in foundations (Balendra, 1993). Deep cap trusses (also known as 'roof outriggers'), which connect cores to exterior columns, limit the curvature in the core, thus reducing lateral deformations of the system (high stiffness) and bending moments (low

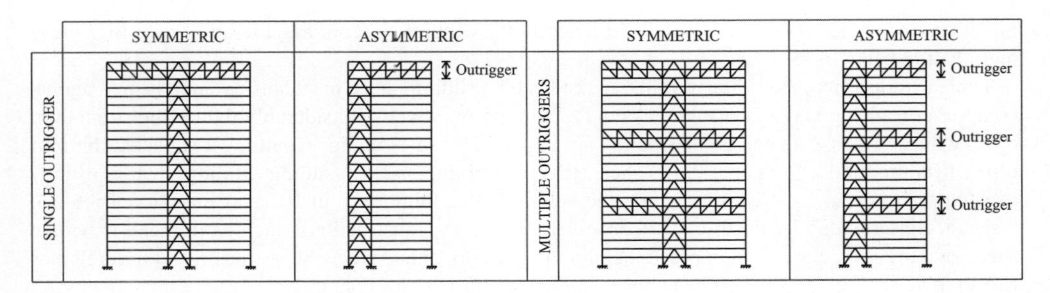

Figure A.15 Typical layouts of outrigger braced frames with central and offset cores

Table A.4 Behaviour of walls for different height-to-width ratios (*see* also Figure 2.12).

Height-to-width ratios	Contribution to response	
	Shear	Bending
1	✓	
3	✓	
4	✓	✓
6		✓

overturning moments). Lateral load resistance of OBFs is primarily provided by core, outrigger and exterior columns. The dynamic behaviour of OBFs is significantly affected by the location and the number of outriggers along the structure height. Relative location of outriggers considerably influences storey drift response (Stafford-Smith and Salim, 1981; Hoenderkamp and Snijder, 2003). Their optimum location along the building height is a trade-off between the (flexural) stiffness of the deep truss beam and the rotation of the core section where it is connected to the outrigger. Parametric analyses have shown that systems with outriggers at intermediate frame heights are very effective for earthquake response (Taranath, 1998). Multi-truss systems are preferable to single-outrigger configurations because they increase structural redundancy and enhance global strength, overstrength and ductility.

(iii) Structural Walls

Structural walls (SWs) are vertical systems which are frequently combined with RC, steel and composite framed structures to control lateral deflections. These systems are often classified according to their height-to-width (H/L) ratio (also known as vertical aspect ratio) in 'squat' and 'slender' (or 'cantilever') walls. Squat walls have low slenderness: their H/L ratios vary between 1 and 3. Slender or cantilever walls are those with $H/L > 6$. Under horizontal loads, the ratio of bending-to-shear deflections of structural walls increases with the system aspect ratio H/L. Consequently, squat and slender walls are governed by shear and flexural modes, respectively. Relationships between horizontal forces and corresponding deformations are provided in Section 2.3.1.1. Rough estimates of the structural behaviour of SWs can be obtained from Table A.4.

Squat and cantilever walls have high in-plane stiffness and strength (also known as 'membrane action'). Bending is resisted in wall systems through chord effects at the edges. This load mechanism is similar to diaphragm actions, which have been discussed for horizontal systems in Section A.2.1. Lateral stiffness and strength of structural walls are increased by using cross sections with I-shape rather than narrow rectangular shapes, as discussed in Section 2.3.2.2. The former layout is a viable solution in RC and composite systems to confine efficiently the concrete in compression and hence to achieve high ductile behaviour.

Typical failure modes of squat walls are caused by sliding shear mechanisms and shear diagonal compressions. Diagonal X-shaped cracks in RC and masonry walls considerably reduce the strength of SWs, while stiffness and energy dissipation are impaired by shear sliding. Cantilever walls exhibit four distinct failure modes: (i) flexural, (ii) shear, (iii) overturning and (iv) sliding. Sliding shear is usually resisted by friction. In structures where the self-weight is high, e.g. in RC and masonry, frictional resistance is provided by dead loads. Conversely, in lighter constructions, such as in metal or wood, shear anchorages are required to prevent sliding under horizontal forces. Shear (or brittle) failure gives rise to lower energy dissipation capacity than flexural (or ductile) response.

Consideration of out-of-plane deflections is also important to prevent brittle failure of SWs. Limitations on wall slenderness are usually employed to prevent out-of-plane buckling, caused by diagonal compression effects. Analytical and laboratory tests have demonstrated that steel and composite walls employing compact sections exhibit excellent ductile response under earthquake loads (Astaneh, 2001; Bruneau and Bhagwagar, 2002).

Lateral stiffness, strength and ductility of structural walls, either squat or slender, are significantly affected by the type and seismic detailing of the joint between superstructure and foundation system. Whereas for MRFs it is relatively easy to ensure fixity at column bases, this is not the case for slender structural walls. Lateral stiffness of walls is high; it is thus either impractical or uneconomic to have fixed-base walls. A degree of flexibility is generally present at the base of wall structures. Additionally, because walls under horizontal actions behave like free-standing columns, the connection between superstructure and foundation systems should possess high curvature ductility μ_{χ}; defined in Section 2.3.3. Inelastic deformations are, in fact, concentrated at the base of SWs, i.e. at plastic hinge location. High values of μ_{χ} of plastic hinges lead to adequate displacement ductilities μ_{Δ}. To accommodate windows and doors, 'pierced walls' are often utilized for multi-storey buildings. The seismic behaviour of SWs is influenced by the presence of openings along the height. These openings may be either small or large as shown, for example, in Figure A.16. The size of windows and doors also considerably affects the structural response of SWs. Gross reductions of area jeopardize stiffness, strength and ductility of the wall, especially if localized at corners.

For small openings, SWs behave like monolithic cantilever columns; the influence of windows and doors is negligible. It is, however, difficult to define small openings quantitatively. As a rough estimate, it can be assumed that small holes for windows and doors are those with width (l_o) less than 10–15% of the wall length (L_w), $l_o/L_w < 0.10$–0.15, as displayed in Figure A.16. On the other hand, SWs with large openings behave as coupled walls under horizontal forces. These wall systems are connected by stiff floors or deep beams (or spandrels) at each storey. The lateral deformability of coupled walls depends on the stiffness of connecting members. For example, each wall bends independently about its own axis for very flexible spandrels. Conversely, if rigid connecting members are present, coupled walls behave as a cantilever bending about a common centroid axis. Coupled shear walls can thus deflect in either flexural, shear, or a combination of flexural and shear modes. In building structures, coupled walls deflect generally in a shear-flexure mode. The flexural mode dominates in the lower storeys while the shear mode is prevalent in the upper storeys.

In RC and masonry structures under seismic loads, spandrel panels beneath windows exhibit X-shaped cracks because of high vertical shear effects (also known as 'shearing effects'). The latter are similar to the slippage, which occurs in laminated (or composite) beams under bending actions. Shearing effects increase with the height of the walls and significantly erode the lateral global resistance of medium- to high-rise structures employing coupled SWs.

Comprehensive experimental and numerical tests have demonstrated that RC coupled walls can dissipate a large amount of energy provided that spandrels are very ductile (Paulay and Priestley, 1992). Studies on a full-scale specimen carried out by Astaneh (2001) have also shown that steel and composite SWs (with and without openings) exhibit high deformation capacity and energy dissipation, provided that seismic details with high ductility are employed.

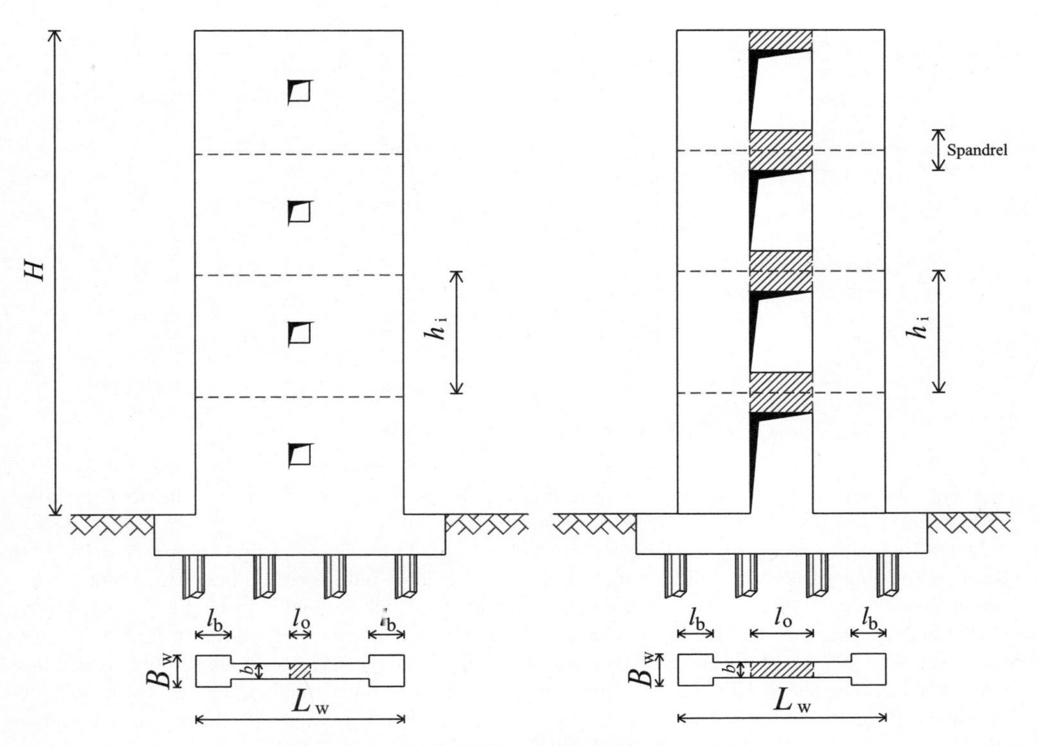

Figure A.16 Walls with small (*left*) and large (*right*) openings

Favourable locations in plan of SWs, with or without openings, are along the perimeter of the plan layout of the structure, as discussed in Section A.1.1. Arrangements of SWs in plan, which lead to closed cross sections (or thin-walled sections) for wall systems, possess high torsional rigidity and are desirable for adequate earthquake response. Cores employing SWs are also suitable for services, such as stairwells and lifts. Lateral force-resisting systems employing SWs are generally cost-effective for RC, steel and composite buildings with a number of storeys up to 25–30 (Di Sarno, 2002).

(iv) Hybrid Frames

Rigid moment-resisting frames are ductile systems with high resistance, but their lateral stiffness is often inadequate to prevent large drifts under earthquake forces. To reduce storey and roof drifts, MRFs are often connected to bracing systems or structural walls (also known as 'hybrid frames or dual systems'). It is generally cost-effective for hybrid frames (HFs) to employ frames that are designed for gravity loads only, while horizontal forces are resisted by bracing systems, e.g. braced frames, or structural walls. However, under lateral earthquake loads, frames and bracing systems and frames and walls interact to withstand seismic actions (Ghoubhir, 1984). This interaction varies along the height of the structure; it also depends upon the type and the stiffness of structural components used to connect the two components of the HFs, e.g. struts or beam with rigid or semi-rigid connections. For example, if beams with rigid joints are used, bending moments, shear and axial loads can be transferred from the bracing system or wall to the frame and vice versa. Under lateral forces, beams and joints connecting the components of the HFs undergo large deformations: high ductile details are required to achieve satisfactory seismic performance, especially in areas of high seismicity. Columns of systems employing moment-resisting and braced frames are subjected to high axial loads, which are caused by overturning

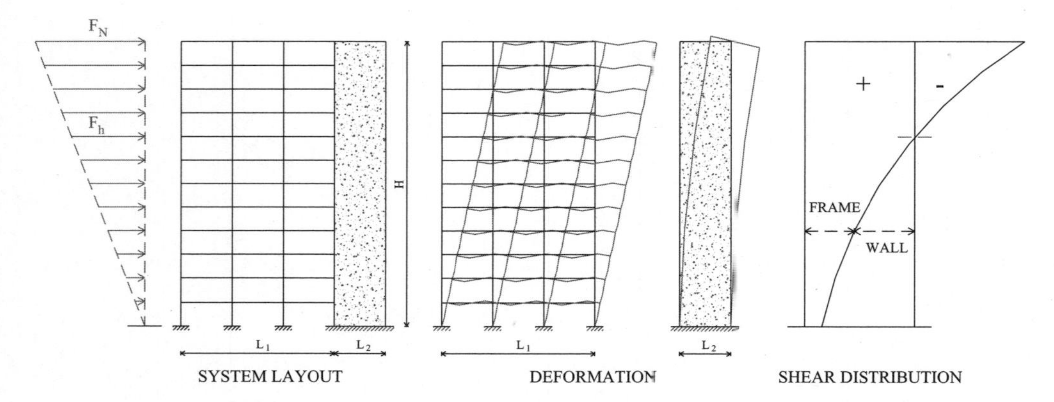

Figure A.17 Interaction between frame and structural wall

moments and layout of the diagonals. To prevent net tensile actions, it is important to distribute bracing systems in HFs in both principal directions of the structure.

Overall lateral deformations of HFs are primarily generated by shear racking (frame) and flexural bending (bracing system or wall). Frame lateral displacements reduce as the height increases; conversely, lateral deflections of braced frames and structural walls increase with the height. The net effect is that at lower storeys, bracing systems and walls are stiffer than frames, while, in turn, the latter possess higher stiffness at upper floors. This difference in lateral stiffness along the height between the structural components of HFs significantly affects the distribution of seismic actions as shown, for example, in Figure A.17. The shear resisted by the frame in HFs increases with the height, vice versa for the interacting wall. It is also observed that the total shear carried by the MRF at top storeys can exceed the applied seismic action (also referred to as 'negative storey shear share'). Effects of negative storey shear share in HFs are exacerbated if rotation of the wall anchors is allowed (Wakabayashi, 1986).

Several in-plan configurations for HFs exist. They consist basically of a braced core or structural wall and a frame system, frequently MRF, which can be arranged in-plan in different relative positions, e.g. at interior and exterior locations in high-rise structures, thus reducing conflicts between structural and architectural requirements. Hybrid frames are generally efficient for steel, RC and composite buildings with 30–40 storeys.

(v) Tube Systems

Tube systems (TSs) are structural systems in which lateral stiffness and strength are provided by MRFs, BFs, SWs or hybrid systems that form either a single tube around the perimeter of the structure, or nested tubes around the perimeter and core of the structure. Tube systems are frequently used for high-rise structures; they include the following:

- Framed tubes;
- Trussed tubes;
- Tube-in-tube;
- Bundled tubes.

Framed tubes are lateral force-resisting systems that combine the efficiency of MRFs and core systems, e.g. SWs or BFs. These TSs consist of closely spaced perimeter columns, with spacing ranging between 2 and 4 m, and short-span girders (or spandrels), about 1 to 1.5 m deep. This layout of columns and beams gives rise to stiff perimeter tubes like that displayed, for example, in Figure A.18.

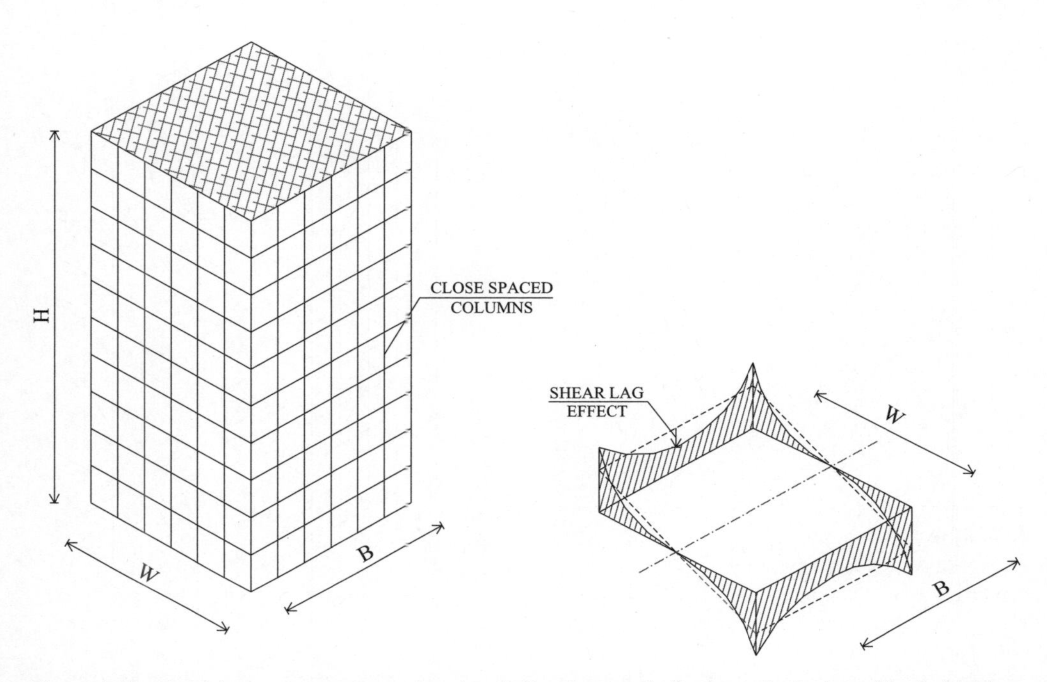

Figure A.18 Framed tube system: perspective view (*left*) and stress distribution under horizontal loads (*right*)

The structural response of TSs under gravity and seismic loads is similar to cantilever hollow box columns or tubes. The perimeter of the tube may be decomposed into two webs and two flanges. Webs and flanges are parallel and orthogonal to horizontal forces, respectively. Webs are characterized by in-plane frame behaviour, thus exhibiting flexural and racking deformations. Conversely, the two flanges are in tension and compression, respectively. Webs and flanges of the tube system have wide sizes, thus shear lag is generated. The latter leads to action distributions within the tube components (webs and flanges) violating Bernoulli's hypothesis for plane sections. Distributions of actions and deformations in TSs should be derived on the basis of principles of structural components with thin-walled cross sections, because traditional engineering theory for beams and columns (*De Saint Venant's theory*) is no longer valid. In particular, shear stresses and strains generated by flexural actions are higher than the values provided by De Saint Venant's theory. As a result, stresses at the corner of the framed tube lag the stress distribution towards the centre of webs and flanges of the perimeter cell (also known as 'shear lag effects'), as shown, for example, in Figure A.18. Shear lag effects can be significantly reduced by selecting system configurations with adequate column spacing and spandrel spans. Spacing and size of columns and girders, width-to-height ratios of the plan layout, and height-to-base slenderness ratios affect both the lateral stiffness and strength of TSs. Requirements for system slenderness ratios to achieve satisfactory seismic performance have been discussed in Section A.1. Deformation capacity and energy dissipation of TSs are strongly dependent on the details used to connect vertical and horizontal framing systems, e.g. ductility of the connections between framed tubes and horizontal diaphragms.

Tube framed systems may, however, show some drawbacks. These include relatively high flexibility of spandrels, significant contributions of shear deformations (or raking) to the total lateral deflection of the structural system and limited spacing between columns. Trussed tubes are a viable alternative to framed tubes. In this alternative tube configuration, large diagonal members are located within perimeter

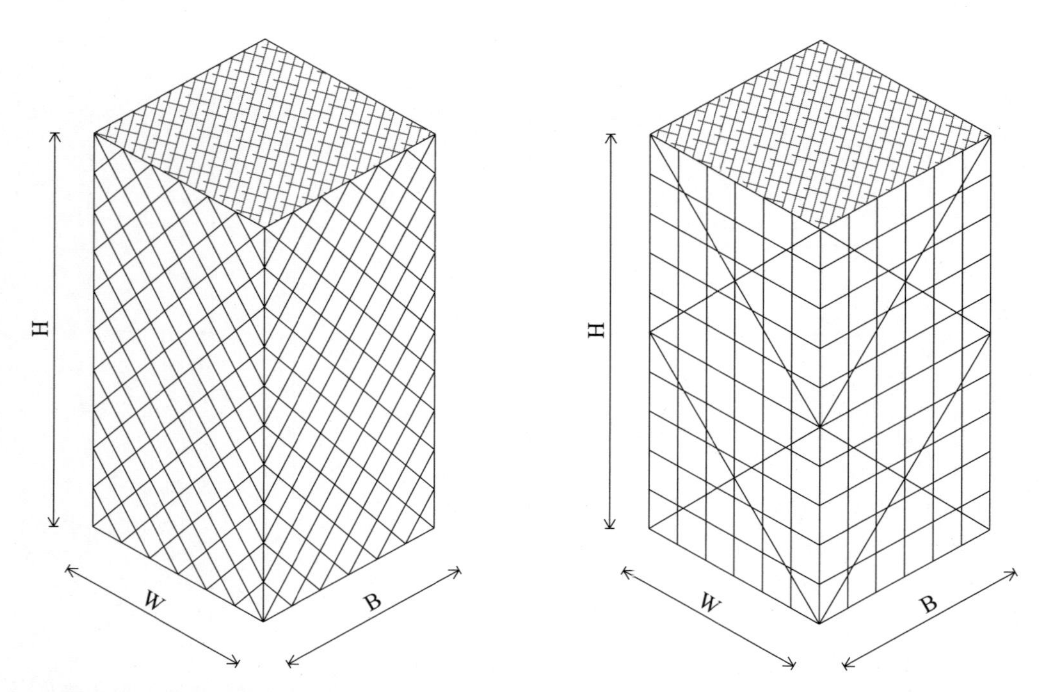

Figure A.19 Trussed tube systems: optimal configuration with inclined perimeter columns (*left*) and configuration with mega-bracings (*right*)

frames and the spacing between columns can be very much larger than in framed tubes, as shown in Figure A.19. The function of diagonal members in trussed tubes is twofold. They withstand shear actions generated by horizontal seismic forces and transfer gravity loads to the ground acting like inclined columns. Diagonal members are tied together with spandrel beams along the perimeter of the system, thus exhibiting a pure flexural mode under horizontal forces. Optimal configurations for trussed tube systems require closely spaced diagonal braces in both directions of the tube. Nevertheless, this layout is impractical due to the interaction with architectural elements in the facade, e.g. claddings and openings for windows. Multi-storey diagonal braces (also known as 'mega-braces') are often utilized for high-rise structures: in this configuration, diagonals intersect peripheral columns at tube corners (Figure A.19). Mega-brace members can also help to reduce shear lag effects either in webs and flanges of tube walls.

Trussed tube systems may exhibit large inelastic deformations and energy dissipation, provided that buckling of diagonal braces is prevented and base columns employ seismic details with high ductility.

Tube-in-tube and bundled tube systems possess higher lateral stiffness and strength than both framed and trussed TSs. Tube-in-tube structures resist earthquake-induced horizontal forces through an interior and an exterior framed tube (Figure A.20).

Floor slabs, acting as horizontal rigid diaphragms, tie exterior and interior tubes together so that they interact under horizontal loads. The structural interaction between perimeter and interior tubes is similar to that discussed above for HFs. The exterior tube resists most lateral loads in the upper floors, while the interior tube carries most lateral loads at lower storeys. The lateral strength of tube-in-tube systems is superior to that of HFs (Balendra, 1993). Similarly, the ductility and energy dissipation capacity of high-rise with tube-in-tube structures are higher than those of HFs. The former possess, in fact, higher redundancy and can give rise to more uniform action redistributions.

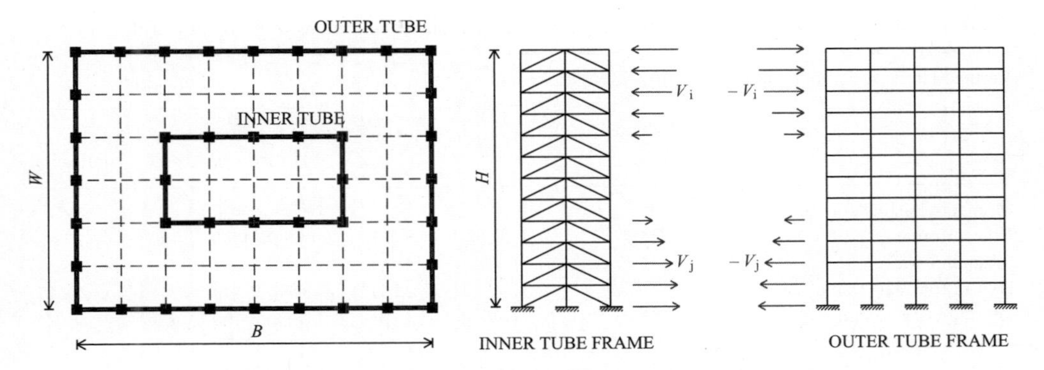

Figure A.20 Typical tube-in-tube system: layout (*left*) and action distribution (*right*)

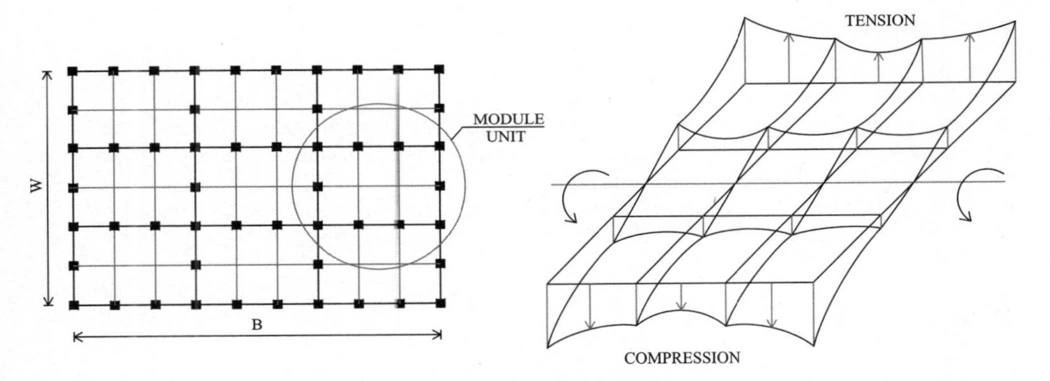

Figure A.21 Bundled tube systems: typical layout (*left*) and stress distribution due to horizontal forces (*left*)

Bundled tube systems are lateral force-resisting tube structures with reduced shear lag effects. Their typical configuration consists of nested framed tubes. The layout is generally modular as displayed, for example, in Figure A.21. Exterior framed tubes are stiffened by interior vertical diaphragms, thus forming a modular layout with several tubes. Interior vertical diaphragms consist of closely spaced columns tied by spandrel beams. Each tube comprises two webs and two flanges, as for framed TSs. Internal webs, in the direction of the applied horizontal loads, reduce shear lag effects. Stress distributions in central columns along the perimeter in bundled systems are more uniform than in other tube structures. This effect is also due to orthogonal internal diaphragms, which tend to distribute axial stresses equally along the flange frames (Taranath, 1998). As a result, perimeter columns more adequately resist overturning moments caused by lateral loads, although shear lag effects may still be present. In addition, bending and torsional warping response is greatly improved with respect to tube frames. Similarly, strength and energy dissipation are also augmented.

Bundled tubes employ wider column spacing than in a single tube, and each modular tube may be terminated at different heights without any loss of efficiency, e.g. lateral stiffness and reduction of shear lag effects. Squared shapes behave better than triangular, hence the former should be used as sub-elements between internal diaphragms (Figure A.21).

Earthquake structural response characteristics of MRFs, BFs, SWs, HFs and TSs have been summarized in Table A.5. The suitability of each system for seismic applications has also been provided along

Table A.5 Efficiency of vertical lateral resisting systems for seismic applications (*relative measure*).

Vertical lateral resisting system	Stiffness	Strength	Ductility	Suitability	
				Max. number of storeys	Seismic application
Moment-resisting frame	L	H	H	15–20	✓✓
Braced frame	H	H	L-M	20–30	✓
Structural wall	H	H	L-M	25–30	✓
Hybrid (or dual) frame	H	H	M-H	30–40	✓✓
Outrigger-braced frame	H	H	L-M	50–60	✓
Framed tube system	H	H	M-H	60–70	✓✓
Tube-in-tube system	H	H	M-H	70–80	✓✓
Trussed tube system	H	H	M-H	80–100	✓✓
Bundled tube system	H	H	M-H	120–150	✓✓

Key: H = high; M = moderate; L = low; ✓ = suitable; ✓✓ = very suitable.

with the maximum number of storeys for which they are cost-effective. As the height increases, the systems exhibit high lateral stiffness and strength. In high-rise structures, e.g. with number of storeys greater than 30–40, the design is often governed by drift limitations under wind loading rather than earthquakes. The ductility for such structures can vary between moderate and high. The lateral force-resisting systems summarized in Table A.5 are extensively utilized for RC, steel and composite structures (CTBUH, 1995).

References

Ambrose, J. and Vergun, D. (1999). *Design for Earthquakes*. John Wiley & Sons, New York, NY, USA.

Aoyama, H. (2001). *Design of Modern High Rise Reinforced Concrete Structures*. Imperial College Press, London, UK.

Arnold, C. and Reitherman, R. (1982). *Building Configuration and Seismic Design*. John Wiley & Sons, New York, NY, USA.

Astaneh, A.A. (2001). Seismic behaviour and design of steel shear walls. *Structural Tips*, Structural Steel Education Council, Technical Information and Product Service, July, Berkeley, CA, USA.

Balendra, T. (1993). *Vibration of Buildings to Wind and Earthquake Loads*. Springer-Verlag, London, UK.

Balendra, T., Sam, M.T., Liaw, C.Y. and Lee, S.L. (1991). Preliminary studies into the behaviour of knee braced frames subjected to seismic loading. *Engineering Structures*, 13 (1), 67–74.

Balendra, T., Lim, E.L. and Liaw, C.Y. (1997). Large-scale seismic testing of knee-brace-frame. *Journal of Structural Engineering, ASCE*, 123(1), 11–19.

Bertero, R.D. and Bertero, V.V. (1999). Redundancy in earthquake-resistant design. *Journal of Structural Engineering, ASCE*, 125(1), 81–88.

Bozorgnia, Y. and Bertero, V.V. (2004). *Earthquake Engineering. From Engineering Seismology to Performance-Based Engineering*. CRC Press, Boca Raton, FL, USA.

Bruneau, M. and Bhagwagar, T. (2002). Seismic retrofit of flexible steel frames using thin infill panels. *Engineering Structures*, 24(4), 443–453.

Bruneau, M., Uang, C.M. and Whittaker, A. (1998). *Ductile Design of Steel Structures*. McGraw-Hill, New York, NY, USA.

Comite Euro-Internationale du Beton (CEB) (1996). *RC Frames Under Earthquake Loading. State of The Art Report*. Thomas Telford, London, UK.

Council on Tall Buildings and Urban Habitat (CTBUH) (1995). *Structural Systems For Tall Buildings. Tall Buildings and Urban Environment Series*. McGraw-Hill, New York, NY, USA.

DesRoches, R. and Fenves, G.L. (1997). New design and analysis procedures for intermediate hinges in multiple-frame bridges. Report No. UCB/EERC-97/12, Earthquake Engineering Research Center, University of California, Berkeley, CA, USA.

DesRoches, R. and Muthukumar, S. (2002). Effect of pounding and restrainers on seismic response of multi-frame bridges. *Journal of Structural Engineering, ASCE*, 128(7), 860–869.

Di Sarno, L. (2002). Dynamic Response of Tall Framed Buildings Under Environmental Actions. PhD Thesis, University of Salerno, Salerno, Italy.

Dowrick, D.J. (1987). *Earthquake Resistant Design for Engineers and Architects*. 2nd Edition, John Wiley & Sons, New York, NY, USA.

Elnashai, A.S. and Dowling, P.J. (1991). Seismic design of steel structures. In *Steel Subjected to Dynamic Loading. Stability and Strength*, R. Narayanan and T.M. Roberts, Eds., Elsevier Applied Science, London, England.

Eurocode 8 (2004). Design provisions for earthquake resistance of structures. Part 1.3: General rules. Specific rules for various materials and elements. European Communities for Standardisation, Brussels, Belgium.

Federal Emergency Management Agency (FEMA) (2004). NEHRP recommended provisions for seismic regulations for new buildings and other structures, Part 2 – Commentary. Report No. FEMA 450, Washington, DC, USA.

Foliente, G.C. (1997). *Earthquake Performance and Safety of Timber Structures*. Forest Products Society, Madison, WI, USA.

Ghoubhir, M.L. (1984). Earthquake resistance of structural systems for tall buildings. Proceedings of the 8th World Conference on Earthquake Engineering, Vol. V, San Francisco, CA, USA, pp. 491–498.

Gupta, A. and Krawinkler, H. (2000). Estimation of seismic drift demands for frame structures. *Earthquake Engineering and Structural Dynamics*, 29(9), 1287–1305.

Hassan, O.F. and Goel, S.C. (1991). Modeling of bracing members and seismic behaviour of concentric braced steel structures. Report No. UMCE 91-1, Department of Civil Engineering, University of Michigan, Ann Arbor, MI, USA.

Hjelmstadt, K.D. and Popov, E.P. (1984). Characteristics of eccentrically braced frames. *Journal of Structural Engineering, ASCE*, 110(2), 340–353.

Hoenderkamp, J.C.D. and Snijder, H.H. (2003). Preliminary analysis of high-rise braced frames with façade riggers. *Journal of Structural Engineering, ASCE*, 129(5), 640–647.

Inoue, K. Sawaizumi, S. and Higashibata, Y. (2001). Stiffening requirements for unbounded braces encased in concrete panels. *Journal of Structural Engineering, ASCE*, 127(6), 712–719.

Kasai, K. and Popov, E.P. (1986). A study on the seismically resistant eccentrically steel frames. Earthquake Engineering Research Center, Report No. UCB/EERC-86/01, University of California, Berkeley, CA, USA.

Kasai, K., Jagiasi, A.R. and Jeng, V. (1996). Inelastic vibration phase theory for seismic pounding mitigation. *Journal of Structural Engineering, ASCE*, 112(10), 1136–1146.

Khatib, I., Mahin, S.A. and Pister, K.S. (1988). Seismic behavior of concentrically braced steel frames. Earthquake Engineering Research Center, Report No.UCB/EERC-8-01, Berkeley, CA, USA.

Kim, S.H., Lee, S.W., Won, J.H. and Mha, H.S. (2000). Dynamic behaviours of bridges under seismic excitations with pounding between adjacent girders. Proceedings of the 12th World Conference on Earthquake Engineering, Auckland, New Zealand, Paper N.1815.

Laogan, B.T. and Elnashai, A.S. (1999). Structural performance and economics of tall high strength RC buildings in seismic regions. *The Structural Design of Tall Buildings*, 8(3), 171–204.

Nowak, A.S. and Collins, K.R. (2000). *Reliability of Structures*. McGraw-Hill Higher Education, New York, USA.

Paulay, T. and Priestley, M.N.J. (1992). *Seismic Design of Reinforced Concrete and Masonry Buildings*. John Wiley & Sons, New York, NY, USA.

Priestley, M.J.N., Seible, F. and Calvi, G.M. (1996). *Seismic Design and Retrofit of Bridges*. John Wiley & Sons, New York, NY, USA.

Qi, X., Chang, K.L. and Tsai, K.C. (1997). Seismic design of eccentrically braced space frames. *Journal of Structural Engineering, ASCE*, 123(8), 977–985.

Sam, M.T., Balendra, T. and Liaw, C.Y. (1995). Earthquake resistant steel frames with energy dissipating knee. *Engineering Structures*, 17(5), 334–343.

Stafford-Smith, B. and Salim, I. (1981). Parameter study of outrigger-braced tall building structures. *Journal of Structural Engineering, ASCE*, 107(10), 2001–2014.

Taranath, B.S. (1998). *Steel, Concrete and Composite Design of Tall Buildings*. McGraw-Hill, New York, NY, USA.

Wakabayashi, M. (1986). *Design of Earthquake-Resistant Buildings*. McGraw-Hill, New York, NY, USA.

Wen, Y.K. and Song, S.H. (2003). Structural reliability/redundancy under earthquakes. *Journal of Structural Engineering, ASCE*, 129(1), 56–67.

Yang, C.S., Leon, R.T. and DesRoches, R. (2006). On the development of zipper frames by pushover testing. Proceedings of the 5th International Conference on the Behaviour of Steel Structures in Seismic Areas, Yokohama, Japan, pp. 555–561.

Appendix B

Damage to Structures

B.1 Structural Deficiencies

Failure modes observed in existing structures during past earthquakes worldwide were caused by a number of member, connection and system deficiencies. Some of these defects are summarized below for buildings and bridges, respectively.

B.1.1 Buildings

Generally, deficiencies in building structures are classified as structural and non-structural. The former refers to: (i) sections, (ii) members, e.g. beam-columns and braces, (iii) connections, (iv) diaphragms, (v) foundations and (vi) structural systems. Non-structural deficiencies comprise: (i) suspended ceilings, (ii) exterior ornamentation, (iii) mechanical and electrical utilities, (iv) poor construction quality and (v) deterioration. This section chiefly focuses on design defects of structural components.

Common structural deficiencies and design defects in reinforced concrete (RC) buildings include:

(i) Poor quality and inadequate detailing;
(ii) Excessive and unexpected member overstrength, especially for dissipative components;
(iii) Change of material and detailing at intermediate floors. In some cases, the bottom storeys may be constructed from composite (steel/composite) changing to RC at an upper level. Deformation demand may be concentrated at the floor, where the change occurs;
(iv) Reduction in column dimensions due to high overstrength if uniform sections are used at higher storeys. An abrupt change in stiffness and strength may lead to failure at the level of change, since the floor load above and below is similar;
(v) Inadequate storey shear strength caused by an insufficient number of columns and walls;
(vi) Irregularities of mass, stiffness and strength distribution in plan and elevation (*see* Sections A.1.1 and A.1.2). Torsional effects may be caused by non-coincidence on the floor plan of the centre of gravity and the centre of stiffness as discussed in Sections 2.3.1.2 and A.1;
(vii) Low structural redundancy, e.g. insufficient number of lateral resisting systems;
(viii) Large openings in floor diaphragms due, for example, to the presence of stairwells and lifts as illustrated in Section 2.3.2.2;
(ix) Inadequate separation joints between adjacent buildings, especially for buildings with different heights and different materials of construction;
(x) Large differential displacements due to settlement of the foundation system.

Fundamentals of Earthquake Engineering Amr S. Elnashai and Luigi Di Sarno

The main factors affecting the level of damage in masonry buildings can be summarized as follows:

(i) Problems relative to the structural configuration, especially asymmetry and inadequate arrangement of openings;
(ii) Weakness in walls, such as low tensile/shear capacity, weak mortar, inadequate connections between intersecting walls;
(iii) Lack of interconnection between masonry structure and roof or floors, especially at upper levels as also discussed in Section A.1.2;
(iv) Poor quality of construction, e.g. materials, workmanship, absence of cross stones or bonding units;
(v) Foundation soil problems, which include liquefaction, settlement, weathering effects.

Steel and composite building structures exhibit structural deficiencies similar to those cited above for RC and masonry structures. Other deficiencies specific to metal constructions include:

(i) Slender sections and members and inadequate lateral supports, especially for bare steel components;
(ii) Inadequate bracing layouts;
(iii) Incompatible deformations between joined parts;
(iv) Excessive column panel flexibility and inadequate resistance capacity;
(v) Inadequate diaphragm strength and rigidity;
(vi) Poor connectivity of diaphragm to vertical elements of the lateral resisting systems, particularly when structural walls are present;
(vii) Incomplete and inadequate lateral force-resisting system as discussed in Appendix A;
(viii) Uplift and high overturning moments in foundation systems.

Examples of damage observed during past earthquakes worldwide abound in reconnaissance reports; significant examples are discussed in Section B.2.

B.1.2 Bridges

The causes of failure for RC, steel and composite bridges are numerous and difficult to categorize; generalizations by definition are fraught with omissions. However, most of the cases of damage and collapse in recent earthquakes may be attributed to the following design defects:

(i) The earthquake-induced deformations were underestimated because gross sections were considered in the computation of displacements instead of cracked sections.
(ii) Serious underestimation of the combined effects of seismic and gravity loads as further illustrated in Section 2.3.2.2. Bridges with few or no seismic requirements are unlikely to survive seismic loading.
(iii) Foundation movements due to local soil conditions. Potential liquefaction and differential settlements may undermine the global stability of the bridge or impair its functionality.
(iv) The requirements of ductility in the plastic hinge area (also known as 'dissipative zone') were not satisfied. The ductility capacity is of primary importance if structures are to survive high levels of inelastic deformation demands as discussed in Section 2.3.3.

Some of the above-cited common causes of damage patterns and collapse for bridges with RC lateral resisting structural systems, are also applicable to steel and composite bridges. These causes include, for example, combined effects of vertical (gravity) and horizontal (earthquake) loads, geotechnical-induced failures, and the detailing for dissipative zones. Other causes of damage specific to design of metal structures include:

(i) Slender sections used for column piers and upper decks, which are inadequate to prevent the occurrence of local and global buckling as presented in Section 2.3.3;
(ii) Partial infills, typically used in large hollow sections of piers, which may generate high stress concentration and squashing;
(iii) Inadequate toughness of welded components can lead to brittle failure modes, such as fracture. Members in tension and bolts with inadequate cross-section areas may also lead to tearing out and breakage.

Examples of bridge failure during recent earthquakes are presented in Section B.3. The observed damage or failure of structural elements can often be related to one or more of the design deficiencies mentioned above.

B.2 Examples of Damage to Buildings

Earthquake damage to building structures is generally assessed according to the type of construction material. The following section provides an overview of types of failures commonly observed in RC, masonry, steel and composite buildings.

B.2.1 RC Buildings

Buildings with structures in RC can experience several types of failure during earthquake loads; these include primarily:

(i) Brittle shear failure of columns and beams;
(ii) Buckling of longitudinal bars in beam-columns due to inadequate spacing or lack of transverse stirrups;
(iii) Shear failure of columns which were shortened by the supporting effect of non-structural elements;
(iv) Brittle failure in corner columns caused by torsion and biaxial bending effects;
(v) Shear cracking in beam-to-column connections, e.g. in panel zones;
(vi) Bond failure, particularly in zones where there are high cyclic stresses in the concrete;
(vii) Brittle failure of single or coupled structural walls, particularly walls with openings;
(viii) Tearing of slabs at discontinuities and junctions with very stiff elements;
(ix) Excessive damage to infills and other non-structural components;
(x) Concentration of damage at a given storey level (also referred to as a 'soft storey');
(xi) Pounding between adjacent buildings;
(xii) Overturning and uplift.

The above failure modes are discussed in the following sections as a function of structural member, joints and global system.

(i) Beams
The ductile design of beams presupposes the formation of plastic hinges at their ends (also called 'dissipative zones') to comply with the requirements of the capacity design philosophy as illustrated in Section 2.3.3. The dissipation of energy through stable hysteresis loops, i.e. without significant degradation of stiffness and strength, plays a significant role in the seismic response of structures.

Under cyclic loading, mechanisms of shear resistance, e.g. compression zone, aggregate interlock, dowel action, truss action, tend to deteriorate as inelasticity is increased. A very undesirable behaviour, known as 'sliding shear', may occur. In members with high shear stress, open cracks in the tension and compression zone remain open, shear behaviour is governed by a vertical (full depth) crack, which does not intersect the hoops, however close they are. Typical damage observed in beams after devastating earthquakes is depicted in Figure B.1. It is important to recognize that the examples shown refer to members characterized by poor detailing and insufficient strength, and which do not comply with current seismic codes; however, they represent an important number of existing buildings. The first damage pattern is flexural cracking in beam span (Figure B.1a). Such cracks would have been pre-existing due to gravity loads and may have opened further because of the effects of the vertical component of the earthquake. The overall safety of the building is not compromised. The second damage pattern is shear cracking (Figure B.1b). These cracks are attributed mainly to inadequate shear reinforcement. They are more hazardous than flexural cracks but in general are not critical with regard to the overall safety of the building. The third damage pattern is flexural cracking in beam supports (Figure B.1c). The quantity and anchorage of bottom reinforcement at supports are critical parameters, which define the extent and severity of the damage. The fourth damage pattern is cracking in beam span at indirect support (Figure B.1d). These cracks are mainly due to the vertical component of the earthquake. They may be prevented if suspension reinforcement is installed.

Examples of failure due to cracking at beam supports are given in Figure B.2; these include shear and flexural cracking, respectively.

The potential strength and ductility of many RC beams is often reduced because of the absence of proper detailing. For example, the lack of proper stirrup spacing (nearly equal to the beam depth!) and

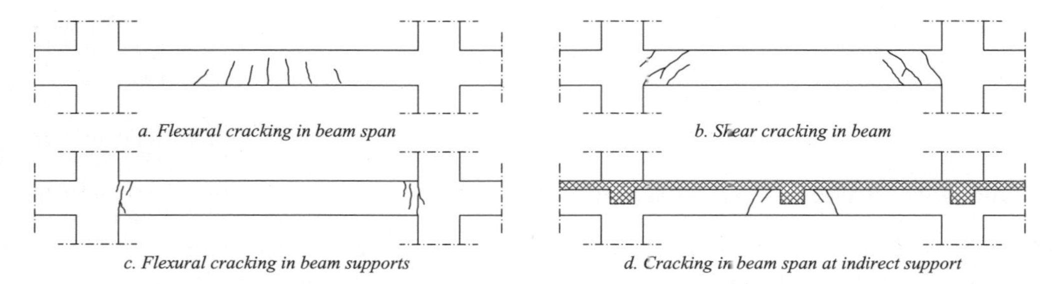

a. Flexural cracking in beam span

b. Shear cracking in beam

c. Flexural cracking in beam supports

d. Cracking in beam span at indirect support

Figure B.1 Typical damage patterns in beam elements: sketches of failure mechanisms (*adapted from* Penelis and Kappos, 1997)

Figure B.2 Failure due to shear (*left*) and flexural (*right*) cracking at beam supports

use of smooth longitudinal steel bars have generated the formation of shear cracks shown in Figure B.2. These cracks have, in turn, led to reductions in both flexural and shear strength. Shear mechanisms should always be avoided in RC members because they are associated with low energy dissipation and sudden failure (also known as 'brittle failure').

(ii) Columns

Columns play an important role in the stability of framed systems. To achieve ductile seismic response, plastic hinges should not form in columns with the exception of base of ground storey columns and top storey columns as discussed in Sections 2.3.3.2 and 2.3.3.3. To prevent the possibility of plastic hinging in columns, beam overstrength factors are often utilized in seismic design. These factors are generally either insufficient or unfeasible since the confinement reinforcement requirements in the ensuing heavily reinforced columns could not be achieved. Hence, there is need for ductility in most columns of the structure.

Compressive axial loading influences response under cyclic loading. The effects are either favourable or unfavourable, as shown for example in Table B.1. Tensile loads, as a result of high overturning moments, although not harmful from the ductility point of view, may cause significant degradation and risk of sliding shear.

The mode of failure in columns depends on the shear span ratio $\alpha_s = M / (VH)$. Short columns ($\alpha_s < 2$) present a brittle failure (or shear type). Figure B.3 summarizes the different failure modes for such types of columns as a function of the steel reinforcement layout. The first type, which has conventional reinforcement (hoops and longitudinal bars) and high axial load, when subjected to cyclic loading results in cross-inclined shear cracks. This behaviour may be improved if cross-inclined reinforcement is utilized, and particularly if multiple cross-inclined reinforcement (forming a truss) is used.

Examples of short-column effects are provided in Figure B.4. The members employ conventional steel reinforcement consisting of ribbed longitudinal bars and rectangular stirrups. The increased relative stiffness of these short columns, as discussed in Section 2.3.1.1, attracts high lateral loads. In turn, shear demand in these structural members is extremely high; even adequate seismic detailing is usually ineffective to prevent the occurrence of shear failure.

On the other hand, columns of medium and high slenderness ($\alpha_s > 3.5$) are characterized by a flexural type of failure (Figure B.5). This type of damage consists of spalling of the concrete cover and then crushing of the compression zone, buckling of longitudinal bars and possible fracture of hoops due to the expansion of the core. Columns of low to medium slenderness ($2.0 \leq \alpha_s \leq 3.5$) and with insufficient shear reinforcement present a mixed (failure/shear) type of failure. The critical parameter is the amount of transverse reinforcement.

Another type of failure may be caused by interaction with masonry infills as also shown in Figure B.5. Infills are present in one side of the column only. The height of the column is a critical region and

Table B.1 Effects of axial loading on column response.

Effects of axial loading	
Favourable	Unfavourable
• Control of flexure/shear crack opening – Elongation of member prevented • Increased stiffness and width of hysteresis loops • Premature failure due to sliding shear or failure of bond anchorages prevented	• Available deformation of compression zone is reached much sooner than in beams, hence early spalling and strength drop and risk of premature buckling of longitudinal bars • Significant P-Δ effects for relatively high levels of inelasticity (say $\mu_\delta > 4$) – effective stiffness reduced, hence risk of collapse

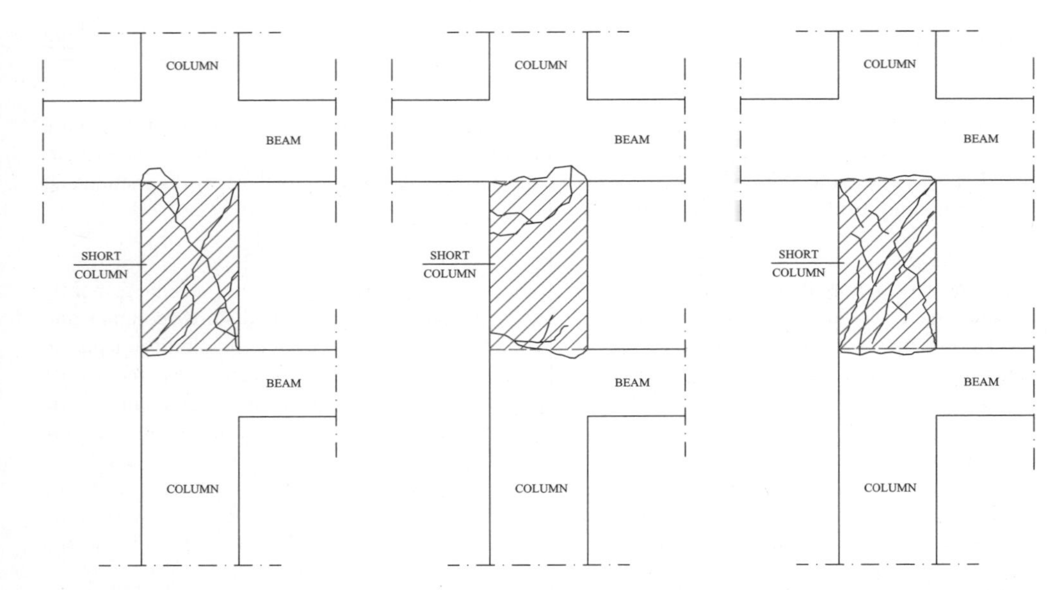

Figure B.3 Failure modes in short columns depending on the reinforcement pattern: conventional hoops and longitudinal bars (*left*), cross-inclined steel reinforcement (*middle*) and multiple cross-inclined (truss) reinforcement (*right*) (*adapted from* Penelis and Kappos, 1997)

Figure B.4 Shear failures in short columns observed in the 1999 Kocaeli (Turkey) earthquake

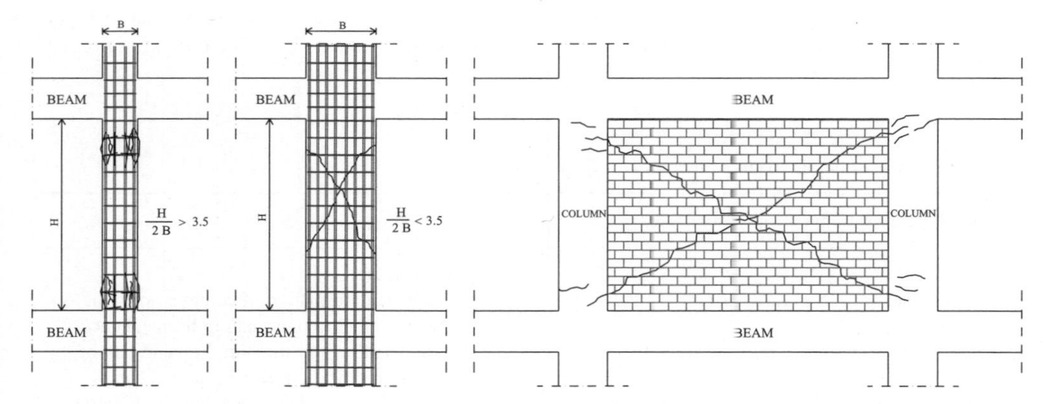

Figure B.5 Typical damage in columns: flexural failure (*left*), mixed (flexural-shear) failure (*middle*) and shear failure due to interaction with masonry infills (*right*) (*adapted from* Penelis and Kappos, 1997)

a large amount of transverse steel reinforcement needs to be employed. Examples of shear failure modes in RC columns are provided in Figures B.6 and B.7. These failures demonstrate that inadequate seismic detailing, especially in the critical zones at the ends of the members, and interaction with masonry infills can cause extensive damage in columns during earthquakes. The Van Nuys Holiday Inn was a seven-storey instrumented building located approximately 7.0 km east of the epicentre of the 1994 Northridge earthquake. This building experienced a peak horizontal acceleration of 0.47 g at base and 0.59 g at the roof. It suffered serious structural damage in all columns of the third floor where signs of a shear-bond splitting type of failure were observed (inset of Figure B.6). The Holiday Inn structure had already suffered extensive non-structural damage during the 1971 San Fernando earthquake.

A sealed separation gap between the brick wall and the columns of the frame in Figure B.7 could have avoided the structural damage experienced by the RC structure.

Part of the damage in the corner columns of adjacent frames in Figure B.7 may also be attributable to pounding, i.e. mutual impact due to out-phase motion of the two structures, which is discussed in Appendix A.

Figure B.6 Shear failure in columns during the 1994 Northridge (California) earthquake in the Van Nuys Holiday Inn (*left*) and in the 1998 Adana–Ceyhan (Turkey) earthquake (*right*)

Figure B.7 Shear failures in columns due to interaction with masonry infills in the 1998 Adana–Ceyhan earthquake: short-column effects and shear failure due to the presence of a masonry wall on a single side of the column

(iii) Beam-to-Column Joints

The design philosophy of RC beam-column joints requires first of all that the strength of the joint shall not be inferior to that of the weakest member framing into it. This is a fundamental requirement and results from the need to avoid seismic energy dissipation through mechanisms characterized by strength and stiffness degradation under cyclic loading conditions, as well as from the fact that the core region is difficult to repair. Secondly, the resistance capacity of a column should not be jeopardized by possible strength degradation of the joint core. During an earthquake of moderate intensity, it is preferable that beam-to-column joints remain in the elastic range, so that no repair is required. Under cycling loading, shear transfer in joint cores takes place mainly through the development of strut-and-tie mechanisms. Contribution of aggregate interlock and dowel action is practicably negligible, since the shear deformation in the joint core is not large enough to activate these mechanisms. It is important to note that adjacent beams restrain joint core expansion, and develop axial forces, which contribute to joint confinement. Floor slabs tend equally to increase both stiffness and strength of joints. However, slab reinforcement increases negative moment capacity of beams (which might cause column hinging, especially at high deformation levels) and, at exterior joints, torsion induced by the slab causes torsional cracking of transverse beams, which tend to become ineffective. Typical mechanisms of damage patterns at interior and exterior beam column joint are sketched in Figure B.8.

As far as interior joints are concerned, the worst type of failure corresponds to the onset of yield penetration at both sides of the joint. Bond conditions may be improved if the diameter of beam bars passing through the joint is limited to minimize slippage, if axial compression from the column is present and, in addition, if gravity loading prevails. In the case of exterior and corner joints, unfavourable bond conditions may also develop. Splitting cracks along the beam bars affect the efficiency of part of the anchorage before the hook. Moreover, column bars at the exterior face are in compression at one end and tension at the other, while being affected by radial forces at hooks; this leads to large splitting cracks, and extensive spalling at the exterior face. Inadequate number of stirrups at beam-to-columns has caused extensive damage in previous earthquakes, especially in exterior connections of framed structures. Figure B.9 provides examples of RC joints with poor seismic detailing.

Smooth longitudinal bars without any confinement at the beam-to-column connections have frequently been found during post-earthquake surveys, particularly in countries in the Mediterranean basin, such as Greece, Italy, Turkey and North Africa (Ambraseys *et al.*, 1990, 1992; Elnashai, 1998, 1999).

Failure of beam-to-column joints has caused several collapses of multi-storey RC frames for buildings as seen during the 1994 Northridge (California) and 1995 Kobe (Japan) earthquakes. Figure B.10 displays, for example, the collapse of the five-storey Kaiser Permanente office block on Balboa Boulevard in Northridge. Most damage occurred in insufficiently ductile members and connections. Shear compression failure of beam-to-column joints led to significant reductions of the capacity of the entire lateral resisting system. The formation of a soft storey was observed at the second floor of the building structure.

Surveys have shown that many residential, commercial and office building structures, like that in Figure B.10, employ typical details for low to moderate seismic hazard levels, although located in regions of high seismicity (e.g. Broderick *et al.*, 1994; Goltz, 1994; Elnashai, 1998, 1999). These details are clearly inadequate to provide sufficient strength and ductility under severe earthquakes, as demonstrated by the observed damage.

(iv) Frames

Framed structures are designed to dissipate energy in beams and at base columns. This global mechanism requires that dissipative zones employ adequate seismic detailing (also known as ductile response as discussed in Section 2.3.3). Several existing buildings were, however, built to resist only gravity and wind loads. Under moderate to large earthquakes, lateral resisting systems, if any, are inadequate to limit storey drifts and resist the additional demand due to earthquake loading. Large horizontal

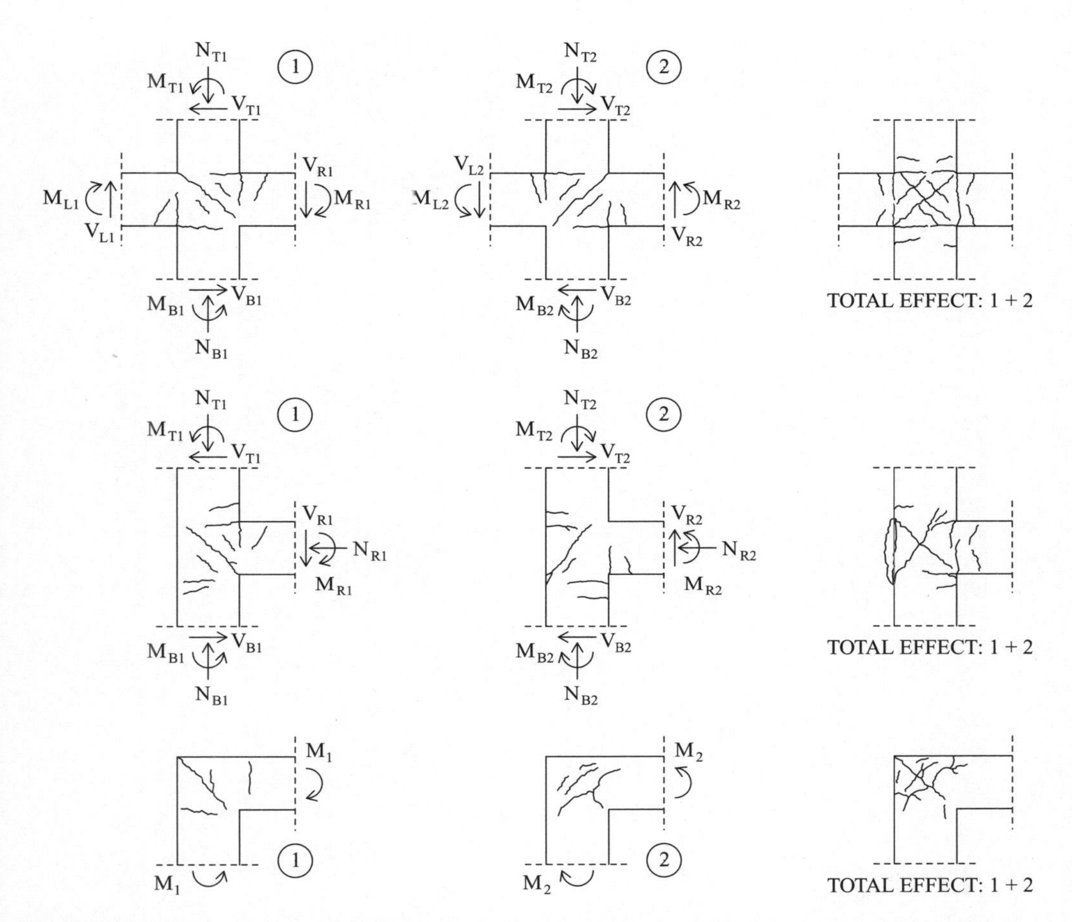

Figure B.8 Seismic damage patterns of beam-to-column joints: interior joints (*top*), exterior T-joints (*middle*) and corner or knee joints (*bottom*) (*adapted from* Penelis and Kappos, 1997)

displacements generate P-Δ effects, thus impairing the global stability of the frame. Overturning moments and foundation sliding are also caused by horizontal forces. Consequently, several damage patterns may be found in framed structures. Some of them are summarized below:

- Extensive damage to exterior walls (masonry panels) caused by large storey drifts. Diagonal cracks may, for example, occur in masonry infills as shown in Figure B.11. The infills provide significant reserve strength to the structures beyond that assumed in the structural design. These panels contribute to the individual storey shear resistances by acting as bracing struts. Consequent damage to masonry panels is evidenced by shear cracking, often in an X-shape. Typically, diagonal cracks originate at the corners of openings for windows (Figure B.11).
- Soft storeys at first and intermediate floors in multi-storey buildings. There are several causes for the occurrence of this type of failure. Generally, soft storeys at ground levels are generated by the absence of infills at this location. Heavy and relatively stiff upper storeys impose large inelastic demand at the bottom levels where, for instance, shop windows or opening for garages replace solid walls. Crushing of intermediate storeys has also been observed in buildings with large windows on the

a. Inadequate number of stirrups at the beam-to-column connection

b. Lack of stirrups at the beam-to-column connection

Figure B.9 Spalling (*top*) and severe damage (*bottom*) of beam-to-column connections with poor detailing in the 1999 Kocaeli (Turkey) earthquake

facade. Figure B.12 shows two examples of soft storeys located at ground and intermediate floors, respectively, in multi-storey RC buildings during the 1995 Kobe earthquake in Japan.

Soft storeys also represent indirect effects caused by dynamic amplifications generated by resonance between the superstructure and the underlying soil layers, which is illustrated in Section 1.3.2. Several 10- to 14-storey-high MRFs, with fundamental periods between about 1.0 and 2.0 seconds, resonated with the soft soil layer (period equal to 2.0 seconds) during the 1985 Michoacan (Mexico City) earthquake. Most of the damage occurred in the transition zone between the firm soil and the

Figure B.10 Collapse of the Kaiser Permanente office building (*left*) and close-up of the joint failures (*right*) during the 1994 Northridge earthquake in California (*courtesy of* National Information Service for Earthquake Engineering, University of California, Berkeley) Reproduced from Matej Fischinger – CD ROM, with permission.

Figure B.11 Diagonal cracks in the infills of an apartment building on Wilshire (Santa Monica, California) in the 1994 Northridge earthquake (*courtesy of* National Information Service for Earthquake Engineering, University of California, Berkeley) Reproduced from: Left: Image: Northridge Collection: NR158 Right: Image: Northridge Collection: NR156 URL: http://nisee.berkeley.edu/elibrary/

Figure B.12 Soft storeys in RC structures in the 1995 Kobe (Japan) earthquake: collapse of the first (*left*) and intermediate (*right*) storey (*courtesy of* Dr. Matej Fishinger) Reproduced from Matej Fischinger – CD ROM, with permission.

Figure B.13 Soft storeys in multi-storey RC frame caused by resonance in the 1985 Michoacan (Mexico City) earthquake: collapsed frames and inset with close-up of the crushed columns (*courtesy of* Dr. Matej Fishinger)

lake zone with soft sedimentary deposit. Soft storeys due to column failures at the base of MRFs caused catastrophic collapses as shown in Figure B.13.

- Torsional effects may impose high inelastic demands upon perimeter columns. The rotation of building floors due to eccentricity between centre of mass C_M and centre of rigidity C_R increases the shear on exterior columns of the frame when subjected to horizontal loads. The further the distance of the column from C_R, the higher the shear is. Additional demand is generated by biaxial bending due to direction of earthquake ground motions. Collapse due to torsional effects is very common especially in multi-storey buildings with weak/soft storeys at ground floor. Torsional forces rotate floor diaphragms and displace columns sideways. Large lateral displacements of first-storey columns and irregularities in elevation have led to the collapse of several framed structures during past earthquakes. Figure B.14 shows two examples of collapses of RC multi-storey buildings during the 1999 Kocaeli (Turkey) and the 1999 North Athens (Greece) earthquakes.
- High redundancy is desirable under earthquakes because local failure does not cause structural collapse if sufficient ductility is available. Redundancy primarily arises from the capability of structures to provide alternative load paths after any component failure. Redundancy is chiefly related to the type of structural configuration adopted as the lateral resisting system, as also discussed in Appendix A. Buildings with few lines of moment-resisting frames, e.g. two along each principal direction of the structure, are commonly used in practice, especially in the USA. These frames are located along the perimeter to optimize the use of the internal space from an architectural standpoint. Several perimeter frames experienced severe damage during the 1994 Northridge earthquake. Figure B.15 shows the collapse of the multi-storey RC frame used as a car park at California State University, in Northridge (CSUN).

Total collapse of CSUN parking structures can be attributed to the loss of structural contact between the precast girders and the short corbel seats (i.e. girder unseating), resulting from large lateral move-

Figure B.14 Collapses caused by torsional effects in RC multi-storey buildings during the 1999 Kocaeli (Turkey) (*left*) and the 1999 North Athens (Greece) earthquakes (*right*)

Figure B.15 Collapse of the multi-storey perimeter frame of the California State University Campus during the 1994 Northridge earthquake: global view (*left*) and close-up of the damage in the frame (*right*)

ment, or shear compression failure in the columns. Furthermore, a catenary action was provided by the floor slab, which caused an 'implosion' of part or all of the structure. Typically, precast RC frames for industrial buildings, as shown for example in Figure B.16, possess low redundancy. Collapses for these types of constructions are also attributed to the inadequate seismic detailing of the connections. Smoothed bars with insufficient anchorage lengths are shown in Figure B.16.

- Hammering of two adjacent structural systems during an earthquake causes pounding as illustrated in Section A.1.2. This failure mode is caused by insufficient spacing between adjacent buildings, which should accommodate the relative displacements under earthquake ground motions. Buildings with different configurations and different materials of construction may be either in phase or in opposition of phase when oscillating. If they are sufficiently close to each other, the frames impact and both may suffer significant structural damage. Adequate separation gaps should be used to prevent these failure modes. Historical data from past earthquakes show that pounding of adjacent buildings has caused enormous losses (Bertero, 1996). This is a typical mechanism of failure that happens

Figure B.16 Collapse of precast RC building during the 1999 Kocaeli (Turkey) earthquake: general damage (*top*) and close-up of the details (*bottom*)

frequently in city centres, where due to the high price of land, buildings are constructed close to each other. Figure B.17 provides some examples of pounding in RC buildings of similar and different heights. Major damage is frequently found in adjacent structures of different heights; taller buildings are, in fact, more flexible than their lower counterparts.

Pounding may also occur in suspended elements connecting adjacent structures. Figure B.18 shows the extensive damage localized at the connection of the suspended walkway in the California State University building during the 1994 Northridge earthquake. Separation joints with an adequate gap may be used to prevent damage caused by pounding of adjacent buildings or connecting elements, as occurred in the cases shown in Figures B.17 and B.18.

- Inadequate seismic detailing is perhaps the most common cause of structural damage in RC structures. Lack of stirrups, insufficient concrete confinement, insufficient concrete cover, use of smooth longitudinal steel bars and insufficient anchorage lengths are frequently observed during post-earthquake surveys. Many failure modes are generated by a combination of the above factors. Figure B.19 illustrates the collapse of a multi-storey RC frame with inadequate seismic detailing in the columns during the 1999 Athens earthquake.

Another common failure mode caused by inadequate detailing is the punching shear, which occurs especially in flat slabs. Several examples of this brittle failure mode were observed, for example, during the 1971 San Fernando, the 1989 Loma Prieta and 1994 Northridge earthquakes, in California (Elnashai *et al.*, 1989). High concentrated loads (combined shear and moment) in columns may lead to the perforation of floor diaphragms. This damage can be prevented by locating close-spaced stirrups horizontally around beam-to-column joints. In some cases, punching shear may also occur at the column base.

Figure B.17 Severe damage caused by pounding between adjacent buildings with same height (*top-left*), different height (*top-right*) in the 1999 Kocaeli (Turkey) earthquake and close-up of damage due to hammering between adjacent columns (*bottom*) in the 1999 North Athens (Greece) earthquake

- Global overturning of multi-storey buildings is a typical damage pattern of RC frames. Several causes may generate this failure mode; the latter can be either geotechnical or structural. Liquefaction caused many buildings to tilt during the 1964 Niigata earthquake in Japan. On the other hand, soft storeys, excessive building slenderness and inadequate foundation systems may generate overturning of the structure as a whole, as further illustrated in Section A.1. Figure B.20 provides two collapses due to global overturning observed in the Niigata (Japan) and the 1999 Kocaeli (Turkey) earthquakes.

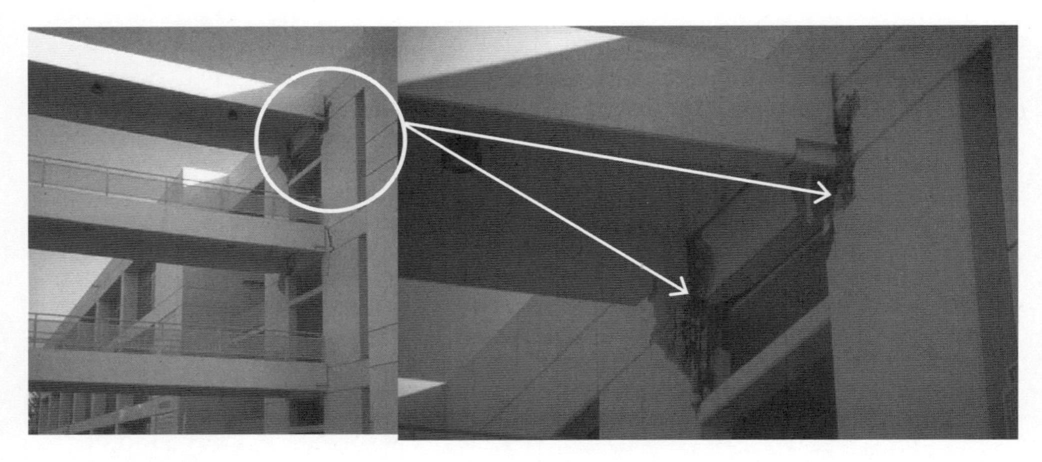

Figure B.18 Damage to suspended walkway in the California State University building during the 1994 Northridge earthquake: damage location (*left*) and close-up (*right*) (*courtesy of* National Information Service for Earthquake Engineering, University of California, Berkeley) Reproduced from: Left: Image: Northridge Collection: NR246 Right: Image: Northridge Collection: NR247 URL: http://nisee.berkeley.edu/elibrary/

Figure B.19 Collapse of framed structure caused by inadequate anchorage of column longitudinal reinforcement bars in the 1999 North Athens (Greece) earthquake: collapsed frames (*top*) and close-up of the pull-out of the steel bars (*bottom*)

(v) Walls

Walls are frequently designed to bear horizontal (seismic) loads. The stiffness of the structure is significantly increased, P-Δ effects are reduced and hence the damage to non-structural components is reduced. Seismic behaviour becomes more predictable compared to that of frames, since formation of unwanted plastic hinges is avoided and the negative influence of asymmetrically arranged infill panels is significantly reduced.

Figure B.20 Global overturning of reinforced concrete buildings caused by liquefaction (*left*) (*courtesy of* National Information Service for Earthquake Engineering, University of California, Berkeley) in the 1964 Niigata (Japan) earthquake and soft storey (*right*) in the 1999 Kocaeli (Turkey) earthquake Reproduced from: Left: Image: Karl V. Steinbrugge Collection: S3161 URL: http://nisee.berkeley.edu/elibrary/

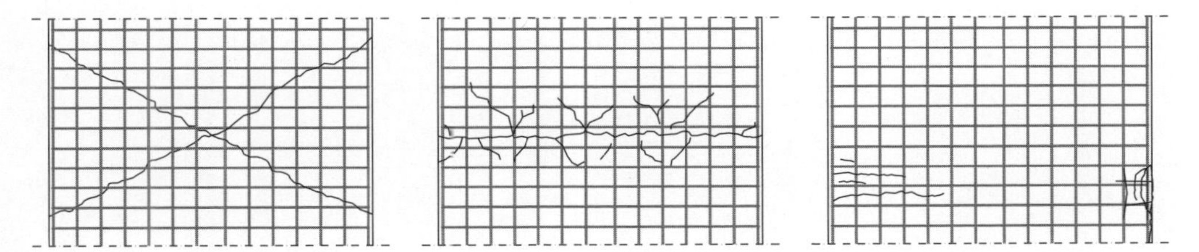

Figure B.21 Typical wall damage types: shear damage (*left*), cracks at construction joints (*middle*) and flexure damage (*right*)

The critical parameter for cyclic shear response is the aspect ratio H/L as illustrated in Sections 2.3.1.1 and A.2.1. If $H/L \geq 2$, the walls are defined as 'slender'. Slender walls are highly ductile and exhibit a flexure-dominated behaviour. If $H/L < 2$, the walls are characterized as 'squat' walls and exhibit a shear (sliding) dominated behaviour. Squat walls are commonly found in low-rise constructions; they have low natural periods and seismic damage is expected to be higher than in slender walls. Typical damage patterns in walls are outlined in Figure B.21.

Shear damage is a typical damage pattern. If boundary elements are not designed and detailed appropriately, then diagonal cracking may lead to failure. Another common damage pattern is that of cracks at construction joints. Possible causes are the poor detailing of the construction joint and insufficient vertical web reinforcement. Additional 'dowel' bars in the construction joint area and fulfilment of the guidelines for construction joints are parameters that should be taken into account. It is important to recognize that this type of damage does not jeopardize the stability of the structure, because the direction (horizontal) of cracks allows the structure to carry the vertical loads. The third damage type is flexure damage. It is rare, although walls in old multi-storey buildings are typically under-designed in flexure. A possible explanation is that the underestimation of actual design moment M_d leads to under-designed foundation members, which rotate and uplift during the earthquake. This results in significant reduction in bending moment M, while the shear V remains almost unchanged. Consequently, shear rather than flexural failure occurs. Examples of shear damage and cracks at construction joints in RC slender walls are given in Figure B.22.

Damage patterns observed in walls of low slenderness (squat walls) are presented in Figure B.23. If horizontal reinforcement is insufficient, diagonal tension failure occurs. This failure mode can appear

Figure B.22 Typical shear damage (*left*) and cracks at construction joints (*right*) in reinforced concrete slender walls (*courtesy of* National Information Service for Earthquake Engineering, University of California, Berkeley) Reproduced from: Left: Image: Karl V. Steinbrugge Collection: S2073 Right: Image: Karl V. Steinbrugge Collection: S2089 URL: http://nisee.berkeley.edu/elibrary/

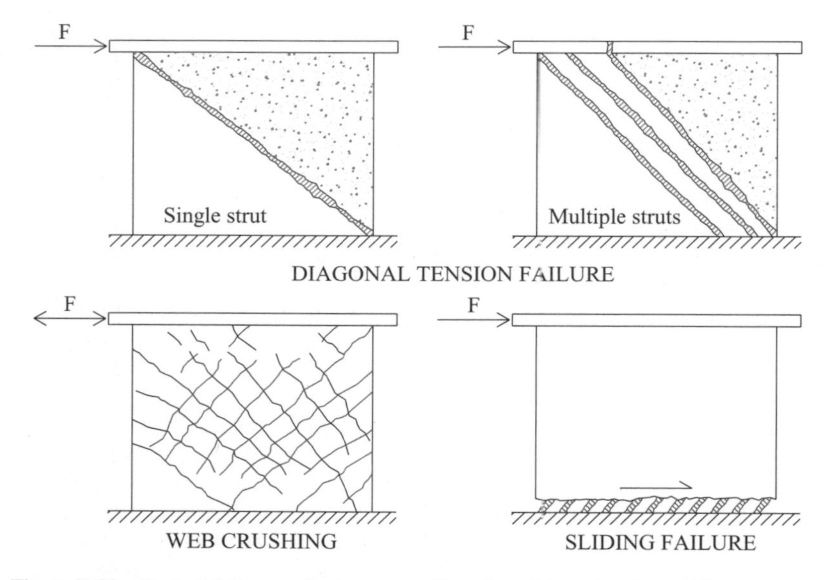

Figure B.23 Typical failure modes in squat walls (*adapted from* Penelis and Kappos, 1997)

as either a single diagonal crack or as a group of inclined cracks. The web crushing shear mode occurs when the diagonal compressive struts of concrete are crushed. They have reduced strength under cyclic loading due to cracks in the other direction. Low slenderness walls fail, as a rule, in sliding shear, except where high axial loading is present. Large displacements are observed along the horizontal crack, which leads to significant strength and energy dissipation capacity reduction.

Figure B.24 Shear cracks along the joint bed (*left*) and through the masonry units (*right*) of load-bearing walls observed during the 1999 North Athens (Greece) earthquake

B.2.2 Masonry Buildings

The majority of existing masonry building stock, especially in Europe, pre-dates the introduction of seismic provisions and hence was designed to resist vertical loads only, leading to heavy damage or collapse under horizontal seismic loading. For example, the 1976 Friuli earthquake in the north-east of Italy ($M_S = 6.5$) caused extensive damage to traditional masonry buildings, historical monuments and churches (Braga *et al.*, 1977). Indeed, the combination of heavy weight and high stiffness along with the low tensile strength of the material renders masonry structures highly vulnerable to earthquakes. Some of the damage patterns relative to RC structures presented in the previous paragraph have also been found for masonry structures, e.g. soft storeys, and pounding. Masonry systems can be either engineered or non-engineered and include primarily unreinforced, confined and reinforced masonry. Systems built through different construction technologies exhibit different seismic response: unreinforced masonry exhibits non-ductile behaviour, while confined and especially reinforced masonry have enhanced strength and ductility. Their typical failure modes as found in previous earthquakes can be classified in a uniform way. In general, three groups are considered:

(i) *Failure in load-bearing walls;*
(ii) *Failure in non-bearing walls;*
(iii) *Failure of wall connections.*

These failure modes are analysed below, together with examples observed during past earthquakes.

(i) Failure in Load-Bearing Walls

The most common damage patterns in load-bearing walls can be summarized as follows:

- Diagonal cracking due to shear, either through the bed joint or through the masonry units (Figure B.24). Diagonal (shear) cracking usually begins at the corners of openings and sometimes at the centre of wall segments (also called 'piers').
- Spandrel beams between adjacent openings may be affected by diagonal shear cracking, usually prior to cracking of piers (Figure B.25). RC bands tying the structure together at floor and door levels can

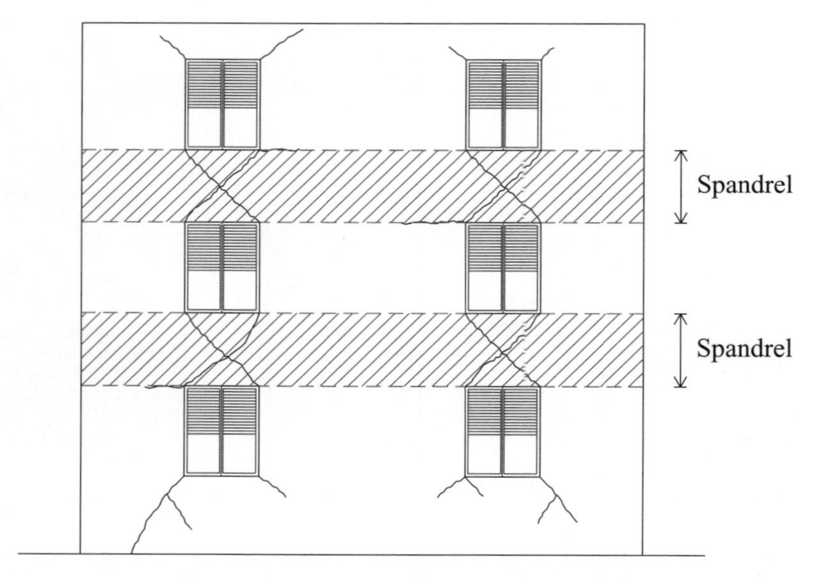

Figure B.25 Cracking of spandrels in masonry walls

prevent premature failure of spandrels and enable distribution of shear among the piers (also known as 'diaphragm action' and illustrated in Section A.1).

- Cracking and collapse due to out-of-plane bending of walls (transverse to the direction of earthquake input). Cracks form vertically close to mid-span or to the corners of walls. Failure of piers usually occurs due to a combination of shear and flexure. Slender walls, especially in unreinforced masonry (URM) buildings, may also buckle under earthquake loads. High incidence of combined in-plane and out-of-plane failure in URM structures was found during the 1994 Northridge earthquake in the Santa Monica area (Broderick *et al.*, 1994). This can be explained both by the low period at which the maximum response amplification occurred in the area and the fact that the city of Santa Monica had not progressed as far with its retrofitting programme as the city of Los Angeles (Goltz, 1994).
- Vertical cracking above door or window openings is also precipitated by the vertical component of earthquakes.
- Corner damage is commonly observed at the intersection of the roof and walls subjected to in-plane and out-of plane demand in moderate earthquakes (Figure B.26).

There are several possible causes of corner failure. It is likely, however, that when a roof diaphragm without shear anchorage moves parallel to the walls subjected to in-plane demands, the walls subjected to out-of plane demands may be pushed outward. The tensile capacity of the wall (from the strength of the bed joints) is exceeded locally, and the wall corner fails.

(ii) Failure in Non-Bearing Walls

Infills or partition walls represent non-structural components of masonry buildings (*non-structural walls*). They suffer damage akin to that of structural walls discussed in the previous paragraph. The occurrence of failure modes in these components seldom affects the global integrity of the structural system. Typical damage patterns observed during earthquakes in non-structural walls include (Figure B.27):

Figure B.26 Corner damage observed during the 1999 North Athens (Greece) earthquake: large vertical separation (*left*) and total collapse (*right*)

Figure B.27 Damage observed in non-structural walls of unreinforced masonry buildings: cracks along joint beds and through the masonry units in the brick facade of the Broadway Store in the 1994 Northridge (California) earthquake (*left*) and global buckling of slender brick wall in a private house during the 1999 North Athens (Greece) earthquake (*right*) (*courtesy of* National Information Service for Earthquake Engineering, University of California, Berkeley) Reproduced from: Left: Image: Karl V. Steinbrugge Collection: S2073 URL: http://nisee.berkeley.edu/elibrary/

- Diagonal cracking through masonry units or the bed joints;
- Sliding along bed joints;
- Global buckling of slender and unrestrained masonry walls;
- Local crushing at corners;
- Overturning due to out-of-plane loading.

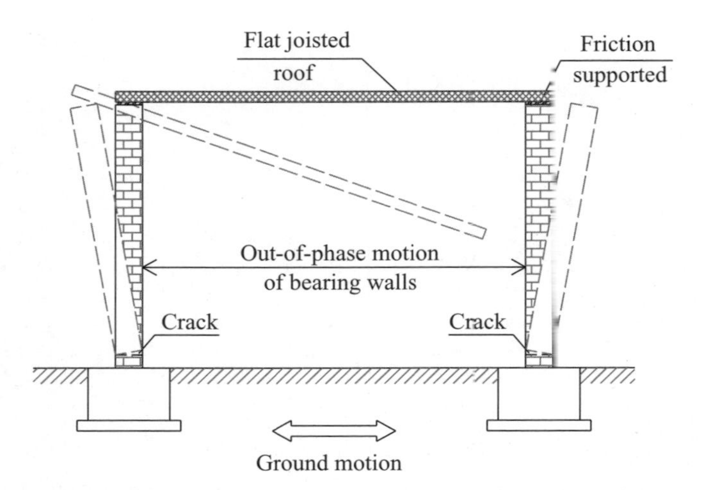

Figure B.28 Fall of roof due to inadequate connection with supporting walls

(iii) Failure of Wall Connections

The strength and energy dissipation capacity of masonry buildings can be enhanced by ensuring a monolithic seismic response of the structure during earthquake-induced vibrations. However, anchorage and tying systems between floor slabs and load-bearing walls have often been proved inadequate. Roofs, in particular flat roofs resting on masonry walls without proper connection, may cause damage to the supporting walls and fall down as shown in Figure B.28.

Inadequate connections between wood floor and vertical walls and between orthogonal walls were the primary causes of damage in several structures during the 1976 Friuli (Italy) earthquake. Failure of connections (near the corners) is quite common in buildings with asymmetric arrangement of walls (torsion).

B.2.3 Steel and Composite Buildings

Damage experienced during past earthquakes worldwide demonstrates that steel and composite building structures generally exhibit adequate dynamic response. This is due to the favourable mass-to-stiffness ratio of base metal (steel) and the enhanced energy absorption of structural systems employed as also discussed in Section A.2.2. Nonetheless, recent earthquakes, e.g. those in 1994 (Northridge, California), 1995 (Kobe, Japan) and 1999 (Chi-Chi, Taiwan), have shown that poor detailing of connections and base columns, and buckling of diagonal braces can undermine the seismic performance of the structure as a whole (*see*, for example, Broderick *et al.*, 1994; Elnashai *et al.*, 1995; Nakashima *et al.*, 1998; Watanabe *et al.*, 1998; Naeim *et al.*, 2000). The occurrence of buckling often in the elastic range in multi-storey buildings, lowers strength capacity and leads to sudden change in the dynamic characteristics of the system. Brittle fractures impair the global ductile response of frames and hence their energy dissipation capacity under earthquake loads as discussed in Section 2.3.1.

Typical failure modes observed in steel and composite buildings can be summarized as follows:

(i) Local and member buckling, particularly for diagonal braces and columns;
(ii) Formation of plastic hinges in columns;
(iii) Fracture of welds and brittle ruptures of bolts at connections, especially beam-to-column, brace-to-beam and brace-to-column;

Figure B.29 Buckling in columns of the Cordova Building during the 1964 Alaska earthquake: local buckling (*left*) and member buckling (*right*) (*courtesy of* National Information Service for Earthquake Engineering, University of California, Berkeley) Reproduced from Left: Image: Karl V. Steinbrugge Collection: S2202 Right: Image: Karl V. Steinbrugge Collection: S2201 URL: http://nisee.berkeley.edu/elibrary/

(iv) Buckling of members and collapse at connections of tubular steel frames;
(v) Severe damage or failure of column bases;
(vi) Fracture at the location of internal stiffeners;
(vii) Fracture and overall buckling of slender bracing members;
(viii) Soft storeys;
(ix) Excessive lateral deformability of framed systems.

Damage in steel and composite members, connections and structural systems may be caused by the effects of a single or combined mode provided above, as discussed below.

(i) Member Failures

Excessive flange and web yielding, brittle fracture, local and global buckling are the most common failure modes for beam-columns and braces in steel and composite buildings. Plastic hinges at the end of structural components can lead to tear-out of steel plates forming the members. Figure B.29 shows severe damage in steel columns. This consists of local buckling at beam-to-column connection and column (global) buckling, which occurred during the 1964 Alaska earthquake in the Cordova Building. The latter was a six-storey office frame with a penthouse; its earthquake-resisting system consisted of a spatial MRF. The main earthquake damage occurred in the first storey and at the penthouse, whose walls collapsed. The local buckling of the south-east corner columns in the Cordova Building was so severe that the flanges tore away from the web and the web crimped. Consequently, columns shortened by about 38 mm.

Inadequate wall thickness-to-width ratios in large boxed columns are one of the causes of the global collapse of the Pino Suarez Complex in Mexico City during the 1985 Michoacan earthquake

Figure B.30 Typical local buckling in steel box columns in the 1985 Michoacan (Mexico City) earthquake (*courtesy of* Dr. Matej Fishinger)

Figure B.31 Typical local (*left*) and global (*right*) buckling of diagonal braces (*courtesy of* National Information Service for Earthquake Engineering, University of California, Berkeley) Reproduced from: Left: Image: Northridge Collection: NR315 Right: Image: KWilliam G.Godden (Vol 4) Collection: GoddenJ38 URL: http://nisee.berkeley.edu/elibrary/

(Rosenblueth *et al.*, 1989). Local buckling and fracture were widespread in this building as shown in Figure B.30. Nonetheless, many other steel buildings, such as the 43-storey Latin American and the 50-storey PeMex Towers, survived the prolonged ground shaking with no damage to structural elements. The large oscillations experienced by these buildings were induced by the resonance with the underlying soil layer, which is illustrated in Section 1.3.2.

Local and global buckling significantly reduces the energy absorption of diagonal braces in multistorey braced systems. This type of failure may occur both in members with thin-walled or open sections (Figure B.31).

Transverse stiffeners along the brace and sections with reduced slenderness can prevent the occurrence of the instability phenomena shown above.

(ii) Connection Failures

Connection failures include excessive plastic deformations and local buckling, brittle fracture and low-cycle fatigue of structural members and welds.

- Excessive yielding and local buckling (flange and web) may occur at both beam-to-column connections and base-columns. Figure B.32 shows common yield and failure modes for welded-flange

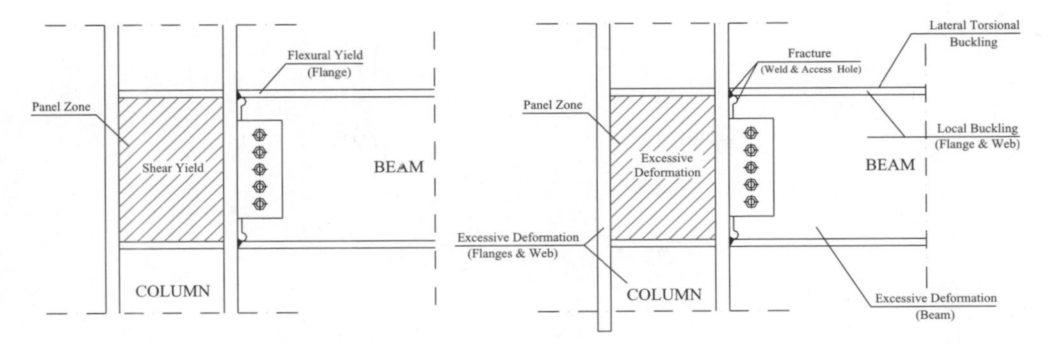

Figure B.32 Yield (*left*) and failure modes (*right*) for typical welded-flange connections used in the USA (*adapted from* Roeder, 2002)

Figure B.33 Buckling in columns of the Cordova Building during the 1964 Alaska earthquake: column top (*left*) and base column (*right*) (*courtesy of* National Information Service for Earthquake Engineering, University of California, Berkeley) Reproduced from: Left: Image: Karl V. Steinbrugge Collection: S2207 Right: Karl V. Steinbrugge Collection: S2206 URL: http://nisee.berkeley.edu/elibrary/

connections observed during the 1994 Northridge earthquake in California. These failure modes are caused by the high inelastic demands either in compression or tension zones of joint systems. Large panel zone yielding may give rise to large storey drifts (often about 20–30% of the total); these drifts can undermine the frame stability. Yielding failure modes are, however, preferable to local buckling because the latter leads to sudden reductions in stiffness, strength and energy dissipation capacity, thus eroding the system seismic performance.

Local buckling of column flanges and column bases are a common damage pattern in multi-storey steel frames. Figure B.33 shows two examples of such damage observed in the Cordova Building during the 1964 Alaska earthquake. The evident overstress at the base of the column indicates that yielding has also occurred at that location (also known as 'plastic hinge').

Figure B.34 Local buckling brace-to-beam connection during the 1995 Kobe (Japan) earthquake (*after* Naeim, 2001)

Connections of diagonal braces with large cross sections can also be affected by severe local buckling. In chevron frames, either concentrically braced frames (CBFs) or eccentrically braced frames (EBFs), local buckling is generally observed at the connection between diagonal brace and deep unstiffened beams. Flange and web stiffeners are very effective means to prevent such phenomena. An example of local buckling in brace-to-beam connection is given in Figure B.34.

- Widespread brittle fracture was found in several connections of steel frames during the 1994 Northridge and 1995 Kobe earthquakes (Elnashai *et al.*, 1995; Youssef *et al.*, 1995). Extensive damage occurred in new and old steel buildings, especially low- to medium-rise, and was primarily localized at beam-to-column connections (Miller, 1998). Predominant failures affected girder groove welds and column flanges (Matos and Dodds, 2002). The damage was typically observed in the lower flange-to-beam portion of the connection, while the top beam flange-to-column flange remained generally intact. This is probably attributable to the presence of composite slabs that, shifting upwards the position of the neutral axis, imposed high strain demands at the bottom of the connection. In some cases, the bolted shear tab – displayed for example in Figure B.32 – experienced shared bolt tears through the tab between the bolt holes. Figure B.35 shows two examples of brittle fracture in beam-to-column connections observed during the 1994 Northridge earthquake.

Brittle failures of welded components, web tear-out for bolted connections and net fracture at bolt holes can take place in ordinary and large braces. Several steel and composite frames for medium- to high-rise buildings in Japan have survived a large number of past earthquakes, e.g. the great Kanto earthquake in 1923 and Tokachi-Oki in 1968, without serious damage. However, during the Miyagiken-Oki earthquake in 1978, some damage was sustained by medium-rise steel buildings, largely due to the fracture of bolted bracing connections. During the 1995 Kobe earthquake, brace members failed mostly at the connection with beams and columns as illustrated in Figure B.36.

Similarly, surveys found brittle fractures localized in braced connections during the 1994 Northridge (California) and 1999 Chi-Chi (Taiwan) earthquakes (FEMA, 2000; Naeim *et al.*, 2000).

Several storey partial collapses and building overturns are likely to have occurred as a consequence of brittle fractures at connections, especially, brace-beams and brace-columns in CBFs.

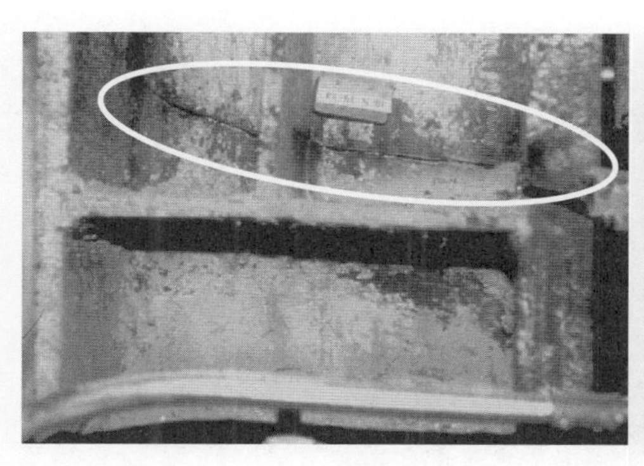
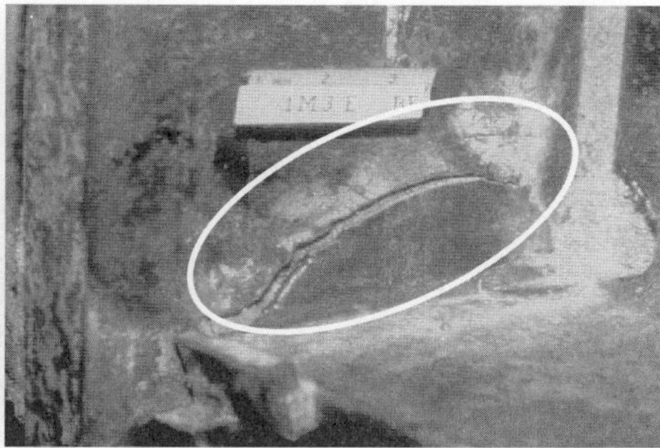

Figure B.35 Damage to beam-to-column connections in the 1994 Northridge (California) earthquake: fracture through web and flange in column (*left*) and causing a column divot fracture (*right*) (*after* Naeim, 2001)

Figure B.36 Fracture in bolted brace connections during the 1995 Kobe (Japan) earthquake: web tear-out (*left*) and net fracture at bolt holes (*right*) (*after* FEMA, 2000)

(iii) System Failures

Steel framed structures typically fail under earthquakes because of their low lateral stiffness as illustrated in Section A.2.2. Inadequate horizontal storey stiffnesses generate extensive damage in non-structural elements (infills) and significantly increase P-Δ effects. The latter may, in turn, cause partial or total collapse of the structural system. For instance, the damage in more than 200 residential steel buildings during the devastating 21 June 1990 Manjil earthquake in Northern Iran ($M_S = 7.7$) was due to excessive lateral deformability (Nateghi, 1995). Slender braces buckled out-of-plane about weak axes, thus causing extensive non-structural damage (Nateghi, 1997).

Failure in columns due to buckling and excessive yielding may give rise to soft storeys, which, as discussed for RC frames, should be avoided. Global mechanisms of failure characterized by formation of plastic hinges in beams are desirable because associated with enhanced energy dissipation capacity of the system as discussed in Sections 2.3.3.2 and 2.3.3.3. Figure B.37 provides two examples of frames with low lateral stiffness surveyed in the 1995 Kobe and the 1999 Chi-Chi earthquakes. These old steel constructions built in the outskirts of large cities did not comply with modern seismic code requirements. Generally, new steel multi-storey frames exhibit adequate structural performance even under intense earthquakes. Figure B.37 demonstrates the inefficacy of very slender diagonal braces (frame on

Figure B.37 Excessive lateral displacement during the 1995 Kobe, Japan, (*left*) and the 1999 Chi-Chi, Taiwan, earthquake (*right*) (*after* FEMA, 2000)

the left-hand side) and the pounding that may occur between adjacent structures. Hammering between buildings employing different materials of constructions, e.g. RC or masonry and steel, is likely to occur because of the difference in relative stiffness of structural systems.

In several cases, damage was found in frames employing different structural materials along the height of the building. This construction technique is very common for high-rise buildings, especially in Japan (Aoyama, 2001). High-strength columns are utilized at the lower storeys, while intermediate and upper floors employ steel members. Abrupt variations of stiffness along the height of structures generate stress concentrations and high inelastic demands, for example, at the connection between steel and RC elements. Lack of maintenance, e.g. evident material corrosion, and inadequate lateral resisting systems caused several partial or total collapses of steel frames in the Kobe region.

B.3 Examples of Damage to Bridges

An efficient transportation system plays a vital role in the development of a modern society, mainly due to the inter-reliance of various industries and the increased trend for outsourcing. Modern transportation networks are referred to as lifelines, the integrity of which has to be protected alongside water, electricity and gas networks. While roads are an important component of transportation networks, bridges are strategic and more sensitive to damage from natural disasters.

Typical earthquake-induced structural failures of bridges are discussed in Section 1.4.1. For RC bridges, observed weaknesses can be summarized as follows:

(i) Abutment backfill settlement and erosion;
(ii) Flexural failures in plastic hinges with inadequate confinement;
(iii) Shear failures in short single columns, piers, multi-column bents, columns with flares and other accidental restraints, and columns in skewed bridges;
(iv) Inappropriate location of lap splices in pier members, causing shear failure;
(v) Compressive failures of columns and piers with corresponding rebar buckling and stirrup openings and ruptures;

(vi) Overstressing of seismic restrainers leading to local failure;
(vii) Uplifting and overturning of bridge foundations and piers with inadequate anchorage at base;
(viii) Pounding and unseating at hinge seats and girder supports;
(ix) Footing failures caused by soil liquefaction and differential settlements.

Some of the above failure modes are similar to those observed for steel and composite steel and concrete bridges. For example, the partial collapse of approach spans due mainly to soil failure and the failure of bearings has occurred in RC and composite bridge structures. Large permanent displacements (about 1.0 m sideways) have been observed in towers of cable-stayed bridges. However, other observed damage modes are more typical of metal constructions such as (Bruneau, 1998):

(i) Local and overall buckling, especially in circular steel columns, either at bottom or intermediate height of the pier;
(ii) Vertical squashing of box sections used for piers;
(iii) Brittle fracture of welds and steel sections prior to yielding, especially in hollow and concrete-filled sections;
(iv) Bolt breakage in tension and shear.

The following sections present the most common damage patterns, illustrate examples of impressive collapses of RC, steel and composite bridge structures worldwide, and discuss their likely causes.

B.3.1 Span Failure

A direct consequence of underestimation of displacements is that the bridge spans may fail due to unseating at the movement joints. This effect is particularly frequent for slender structures. Figure B.38 shows an example of collapse, which occurred at the I-5 (Golden State) and C-14 (Antelope Valley) interchange in the San Fernando Valley in California. The collapse of interior bridge spans was observed in the 1971 San Fernando and 1994 Northridge earthquakes. Extensive damage to lifelines during previous earthquakes, especially in the 1971 San Fernando and 1989 Loma Prieta earthquakes, caused the State Transportation Agency (Caltrans) to initiate and develop a programme to strengthen the vulnerable features of existing structures (Elnashai *et al.*, 1989). While this programme was far from complete at the time of the Northridge earthquake, retrofitted structures, as well as those of more recent construction, displayed superior seismic performance (Broderick *et al.*, 1994). The portion of the highway in Figure B.38, which was designed to pre-1974 seismic standards, had not been retrofitted when the seismic event (Northridge) occurred in 1994.

It is evident that seismic restrainers provided at joints were incapable of resisting the demand imposed. The displacement amplification is aggravated for skewed spans as a consequence of the imposed combination of longitudinal and transversal motion, as shown in Figure B.39.

A further cause that induces unseating is the high displacement amplification due to liquefaction at the foundation. Several collapses of bridge spans were, for example, caused by soil liquefaction during the 1964 Niigata earthquake. Figure B.40 displays the failure of the span of the Showa Bridge due to large movements of the flexible bent piles supporting the deck. Soil liquefaction occurred extensively around the bridge and caused lateral spreading of the surface ground as far as 10 m along the Shinano River.

Spectacular failures due to excessive displacements are also shown in Figure B.41; structural collapses are caused by inadequate support and unseating at the seismic/expansion joints, respectively.

The steel truss Oakland Bay Bridge in California lost a full span during the 1989 Loma Prieta earthquake because of the small size of support angles. The 1995 Kobe earthquake caused the collapse of the simply supported span of the Nishinomiya-ko Bridge due to unseating. Seismic restrainers have

Figure B.38 Span collapses at the Golden State–Antelope Valley interchange collectors during the 1971 San Fernando (*left*) and the 1994 Northridge (*right*) earthquakes (*courtesy of* U.S. Geological Survey)

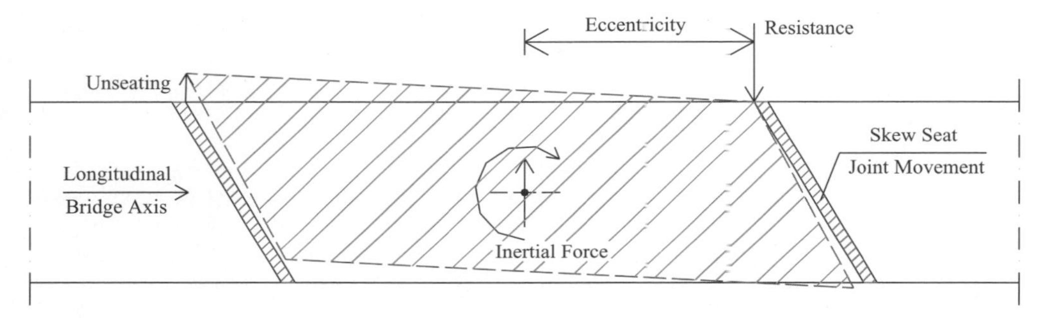

Figure B.39 Unseating due to bridge skew

Figure B.40 Span failure of the Showa bridge due to liquefaction during the 1964 Niigata (Japan) earthquake: aerial view (*left*) and close-up of the deck collapse (*right*) (*courtesy of* Dr. Kazuhiko Kawashima)

Figure B.41 Collapses due to unseating joints: Oakland Bay bridge in the 1989 Loma Prieta (California) and (*left*) and the Nishinomiya-ko bridge in the 1995 Kobe (Japan) earthquakes (*right*) (*courtesy of* U.S. Geological Survey)

Figure B.42 Damage at Higashi–Kobe cable-stayed bridge during the 1995 Kobe (Japan) earthquake: uplift of the deck (*left*) and failure of a viscous damper at the supports (*right*) (*courtesy of* Dr. Kazuhiko Kawashima)

also shown serious weaknesses. High local demands imposed at the restrainer anchorage points damaged the connections with diaphragms. These effects were further aggravated by asymmetric or skew bridge layouts.

Damage at the expansion/seismic joints caused by large displacements of structural systems was observed in many long-span bridges during the 1995 Kobe earthquake. Figure B.42 shows, for example, the 50-cm vertical drop at the expansion joint in the approach span of the Higashi-Kobe bridge, a large modern cable-stayed bridge in Japan with 485-m centre span and 200-m side spans. The damage of the vane-type viscous dampers, used to mitigate wind (transverse) and earthquake (longitudinal) vibrations of the bridge deck, is also displayed in Figure B.42.

During the 1995 Kobe earthquake, the failures were, however, typically more dramatic, as at both sides of a major crossing along the Wangan Expressway. A 200-m-long steel braced-arch bridge failed along this expressway because of lateral displacements of about 30 cm. Vital transportation infrastructures, such as the Shinkanzen line Route 3 of the Hanshin Expressway, were interrupted and this seriously affected all other lines in the Kobe area (Elnashai *et al.*, 1995; Kawashima, 1995).

Figure B.43 Punching of piles through the roadbed of the State Route 1, Watsonville area, span during the 1989 Loma Prieta (California) earthquake (*courtesy of* National Information Service for Earthquake Engineering, University of California, Berkeley) Reproduced from: Left: Image: Loma Prieta Collection: LP0469 Right: Image: Loma Prieta Collection: LP0470 URL: http://nisee.berkeley.edu/elibrary/

Figure B.44 Pounding damage: between adjacent spans at the Interstate-5 at Santa Clara River in Los Angeles County during the 1994 Northridge (California) earthquake (*left*) and at the abutment of a bridge near Nishinomiya Port in the 1995 Kobe (Japan) earthquake (*right*) (*courtesy of* National Information Service for Earthquake Engineering, University of California, Berkeley) Reproduced from: Left: Image: Northridge Collection: NR968 Right: Image: Unknown URL: http://nisee.berkeley.edu/elibrary/

Several examples of punching of piles through the roadbed were observed in RC bridges, especially in the 1989 Loma Prieta earthquake. Bridge spans collapse because of punching of supporting piers as shown in Figure B.43.

Finally, during strong earthquakes, severe pounding damage may also take place at joints between adjacent spans as shown in Figure B.44. This type of damage can be localized both between adjacent bridge spans and at abutments.

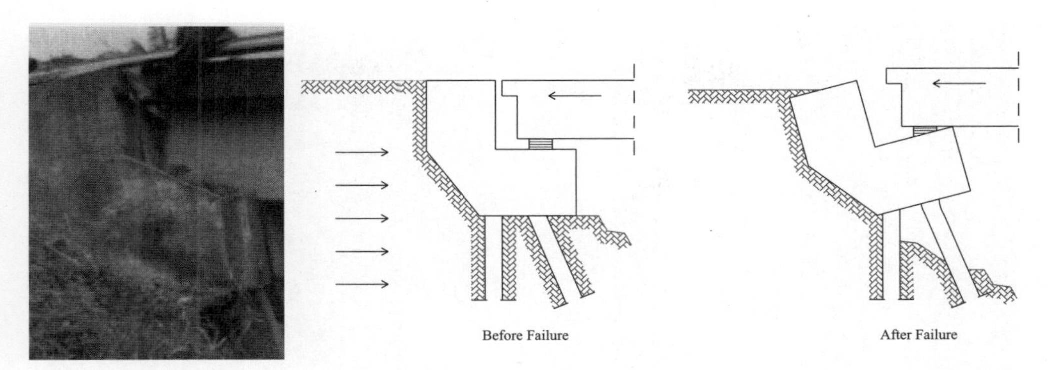

Figure B.45 Abutment slumping and rotation failure of the Rio Bananito Bridge during the 1990 Costa Rica earthquake: post-earthquake site observation (*left*) and sketch of failure mechanism (*right*) (*adapted from* Priestley *et al.*, 1996)

B.3.2 Abutment Failure

The failure of abutments is typically due to slumping of the soil, which produces a global rotation of the structure. This is due to a pressure increase in the infill soil as a consequence of longitudinal response. The sketch and the photo given in Figure B.45 show a failure mechanism of this type that caused the abutment collapse of the Rio Bananito Bridge during the 1990 Costa Rica earthquake.

The rotation of the abutment produced shear collapse of the foundation piles. Furthermore, the pounding of span and back walls may induce damage to the back wall itself. Nevertheless, while abutment failure carries heavy consequences for bridges, this is not a commonly observed mechanism because these components are usually over-dimensioned.

B.3.3 Pier Failure

Failures of RC piers during past earthquakes have often been a consequence of using elastic design (force as opposed to displacements). Strength design may be successful if the demand is estimated accurately, which has been repeatedly shown to be an onerous requirement. The strength is frequently insufficient to guarantee the elastic response of the bridge even though the real resistance is higher than the design value, as a consequence of overstrength. Hence, to survive intense shaking, structures must exhibit an adequate ductility capacity. Additionally, inadequate seismic detailing and slender members have caused severe fractures in steel piers, especially in Japan. The most common damage patterns for bridge piers are listed below.

(i) Column Flexural Failure

The lack of ductility in the flexural failure mechanism is due to inadequate confinement of the plastic hinge zone. Unless the concrete is well confined by closed transverse stirrups, crushing rapidly extends into the core, buckling of longitudinal reinforcement occurs and loss of strength is observed. In extreme conditions, the columns become unable to sustain gravity loads. There are several examples of failure in plastic hinge zones, such as top column failure, as shown in Figure B.46.

Non-ductile response occurs due to conditions not considered in the design phase that lead to the formation of plastic hinges out of the confined areas. A typical example is the Creek Canyon Channel Bridge (Figure B.47), which collapsed during the 1994 Northridge earthquake. A connection wall forced the opening of plastic hinges above the confined area.

Figure B.46 Confinement failure at a bridge pier top during the 1994 Northridge earthquake (*courtesy of* National Information Service for Earthquake Engineering, University of California, Berkeley) Reproduced from: Left: Image: Northridge (California) Collection: NR904 Right: Unknown URL: http://nisee.berkeley.edu/elibrary/

Figure B.47 Flexural plastic hinges in columns connected by a wall, Bull Creek Canyon Channel bridge, 1994 Northridge (California) earthquake (*courtesy of* National Information Service for Earthquake Engineering, University of California, Berkeley) Reproduced from: Left: Image: Northridge Collection: NR902 URL: http://nisee.berkeley.edu/elibrary/

Another common design deficiency is highlighted by discontinuity of longitudinal reinforcement, leading to weak sections at which unexpected inelastic deformations are imposed. The above design deficiency caused spectacular cases of collapse during the 1995 Kobe earthquake, as for example, in the case of the Hanshin expressway shown in Figure B.48.

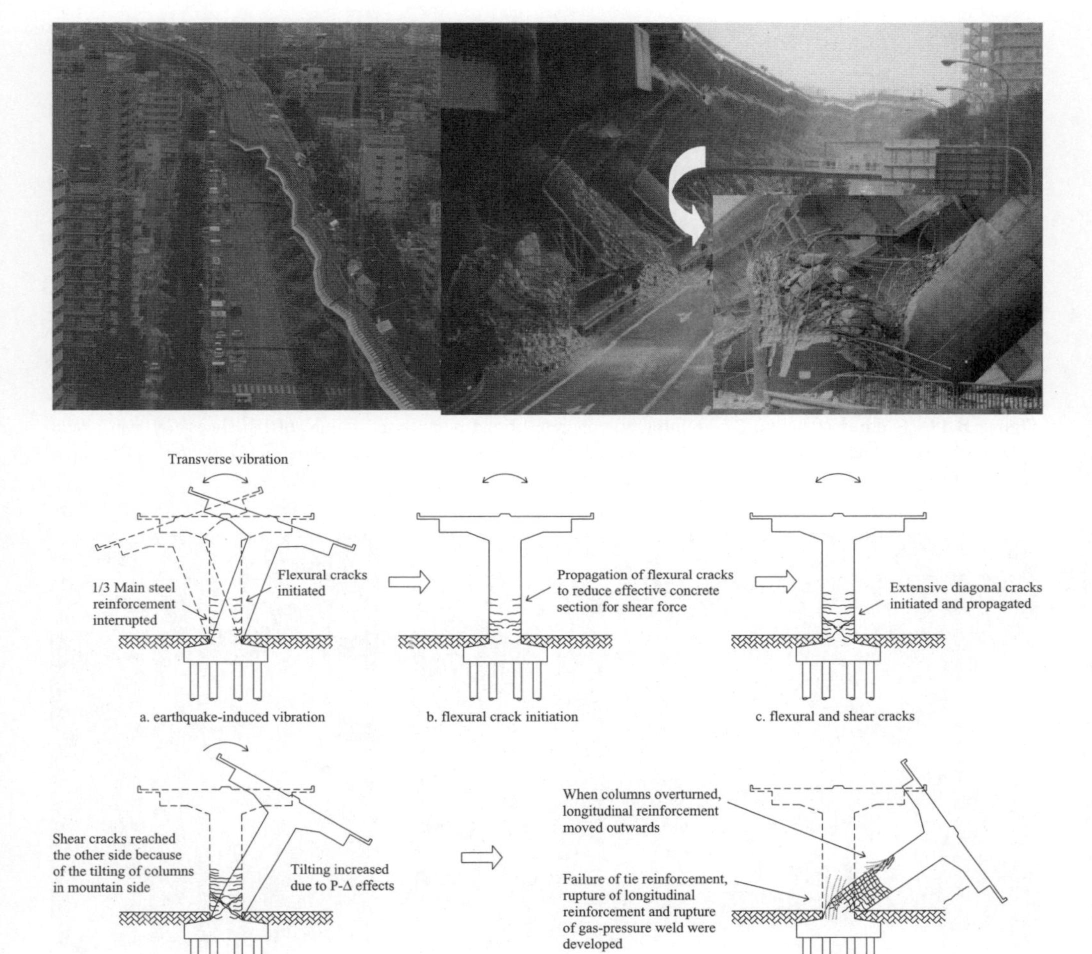

Figure B.48 Flexural failure above column base of columns of the Hanshin expressway, due to premature termination of longitudinal reinforcement and inadequate confinement in the 1995 Kobe (Japan) earthquake: observed failure (*top*) and mechanism of failure (*bottom*) (*courtesy of* Dr. Kazuhiko Kawashima)

Failure may also occur without yielding of vertical reinforcement, due to an inadequate lap-splice length or failure in welded bars as displayed in Figure B.49.

(ii) Column Shear Failure

Elastically designed structures may suffer failure by shear, since the shear strength corresponding to the maximum flexural strength would not have been considered. Shear failure mechanisms are not usually suitable for ductile seismic response, because of the low levels of deformation corresponding to failure. Short columns are particularly susceptible to such effects. A high percentage of bridges lane collapsed during recent earthquakes because of shear failure. Two cases are shown in Figure B.50.

Figure B.49 Failures at the base of reinforced concrete bridge piers: bond failure of lap slices (*left*) and weld failure of longitudinal reinforcement (*right*) in the 1995 Kobe (Japan) earthquake (*courtesy of* Dr. Kazuhiko Kawashima)

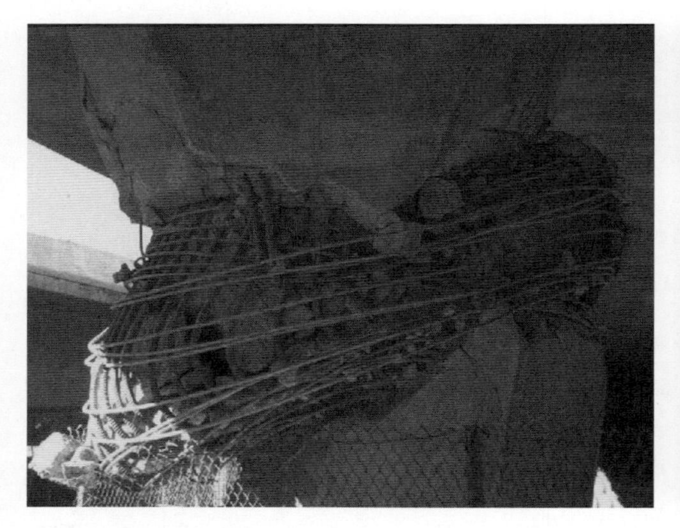

Figure B.50 Shear failure within (*left*) and outside (*right*) the plastic hinge region in the San Fernando Mission Blvd-Gothic Avenue Bridge and I-10 Freeway at Venice Blvd, respectively, during the 1994 Northridge (California) earthquake

In particular, Figure B.50 shows a case in which flexural and shear failure mechanisms were combined. The reduced contribution of concrete to the shear resistance in the plastic hinge area, after the concrete was damaged, led to shear failure.

(iii) Column Buckling and Fractures

A number of steel and composite columns suffered extensive local buckling (also known as 'elephant foot mode') during the 1995 Kobe earthquake (Figure B.51). This failure mode occurred at the base of piers with hollow sections infilled with concrete; the transition zone between infilled and unfilled concrete was critical for buckling. In several cases, this coincided with the termination of concrete infilling, used to protect the piers from vehicle impact.

Figure B.51 Elephant foot mode in steel piers of the Collector from Port Island to Kobe during the 1995 Kobe (Japan) earthquake: buckling at the pier base (*left*) and at intermediate height (*right*) (*courtesy of* Dr. Matej Fishinger)

In many steel bridges, unzipping of corner welds in filled/unfilled box piers has caused collapse; the weight of the heavy deck squashes the piers. This type of failure mechanism was observed in the Tateishi Viaduct during the 1995 Kobe earthquake as shown in Figure B.52.

Several cases of symmetric buckling of reinforcement and compressive failure of piers may be, at least in part, attributable to high vertical earthquake forces both in Kobe and Northridge (Broderick *et al.*, 1994; Elnashai *et al.*, 1995). Three out of four RC piers supporting the I10 (Santa Monica freeway) collector-distributor 36 suffered varying degrees of shear failure due to the short shear span that resulted from on-site modification of the original design (Figure B.53).

B.3.4 Joint Failure

Beam-column connections (or pier-cross beam connections) are subjected to high levels of shear. The heavy damage inflicted on several RC bridges in the San Francisco area during the 1989 Loma Prieta earthquake dramatically brought this problem to the fore. Current design philosophy is to attempt to over-design connections in order to force inelastic action in beams and columns. Without adequate transverse reinforcement, concrete diagonal cracks are opened in the joint regions, where shear stresses produce excessive tension cracks, as shown in Figure B.54.

A further factor that may precipitate joint failure is insufficient anchorage of reinforcement in the end regions. Sliding shear at intentional flexural hinges has also been observed, and is possibly the main reason for the collapse of the Cypress Viaduct (Figure B.55).

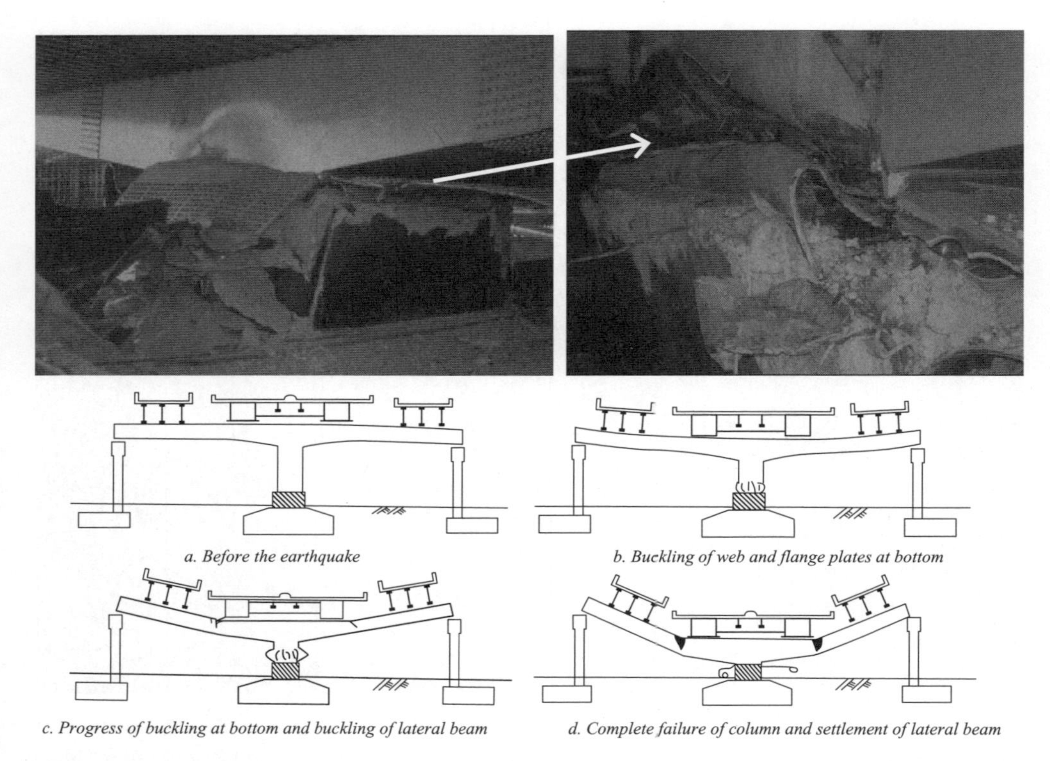

a. Before the earthquake

b. Buckling of web and flange plates at bottom

c. Progress of buckling at bottom and buckling of lateral beam

d. Complete failure of column and settlement of lateral beam

Figure B.52 Failure mechanism of the Tateishi Viaduct during the 1995 Kobe (Japan) earthquake: observed damage (*top*) and failure mechanism (*bottom*) (*courtesy of* Dr. Kazuhiko Kawashima)

Figure B.53 Different shear damage patterns for RC piers at the under-crossing of the Santa Monica Interstate 10 during the 1994 Northridge (California) earthquake: Pier # 5 with inadequate detailing for plastic hinge (*left*), Pier # 6 with symmetric buckling (*middle*) and Pier # 8 with typical shear failure (*right*)

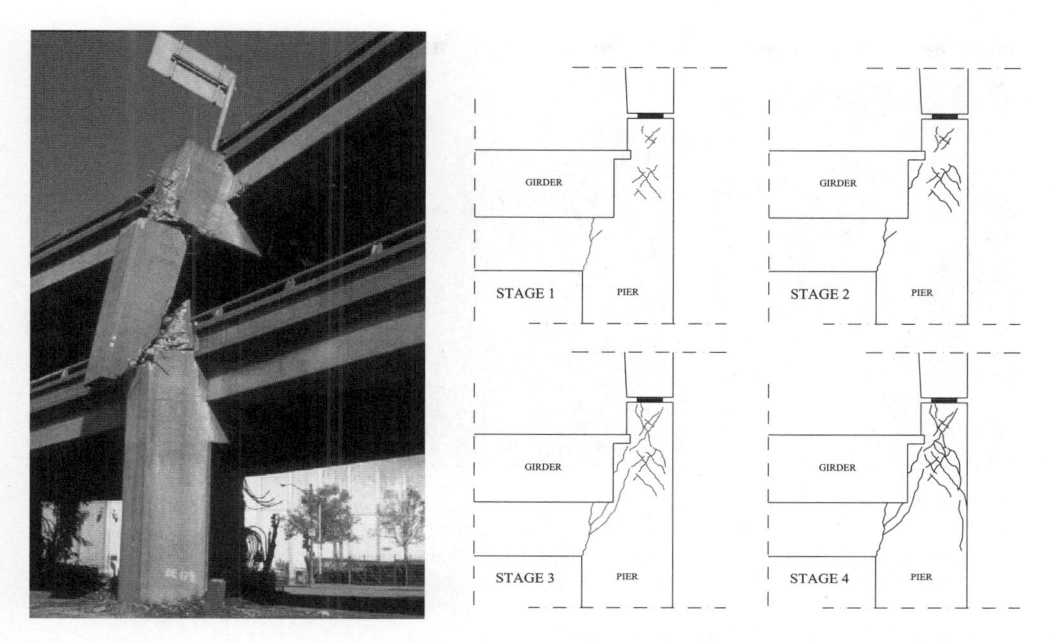

Figure B.54 Joint shear failure of the Cypress Street Viaduct (Interstate 880) during the 1989 Loma Prieta (California) earthquake observed failure (*left*) and sketches of the damage mechanism (*right*) (*courtesy of* National Information Service for Earthquake Engineering, University of California, Berkeley)

Figure B.55 Sliding shear at top columns of the Cypress Viaduct in the 1989 Loma Prieta (California) earthquake (*courtesy of* National Information Service for Earthquake Engineering, University of California, Berkeley) Reproduced from: Left: Image: Karl V. Steinbrugge Collection: S6130 Right: Karl V. Steinbrugge Collection: S6128 URL: http://nisee.berkeley.edu/elibrary/

Figure B.56 Failures of piles supporting RC bridges during the 1989 Loma Prieta (California) (*left*) and 1995 Kobe (Japan) (*right*) earthquakes (*courtesy of* National Information Service for Earthquake Engineering, University of California, Berkeley) Reproduced from: Left: Image: Karl V. Steinbrugge Collection: S6130 Right: Matej Fischinger – CD ROM URL: http://nisee.berkeley.edu/elibrary/

B.3.5 Footing Failure

Compared to other effects, there are few cases of failures caused by footing damage for both RC and steel bridges. Since it is more likely that piers will suffer damage due to inadequate design, actions transmitted to the foundations are limited by the capacity of piers. The rocking of the footing may also have contributed to safeguarding of the foundation system, limiting the level of seismic forces. However, analysis of typical footing detailing points towards several inadequacies, such as:

(i) Footing flexural resistance, mainly due to omission of top reinforcement;
(ii) Footing shear resistance;
(iii) Joint shear resistance;
(iv) Inadequate anchorage of the longitudinal reinforcement of columns;
(v) Inadequate connection between tension piles and footings.

Figure B.56 shows two examples of failure in piles supporting RC bridges; they were observed in the 1989 Loma Prieta and the 1995 Kobe earthquakes, respectively.

In the 1995 Kobe earthquake, a number of investigated cases showed damage to footings, which cracked mainly in shear. Several piles were also damaged. It is relatively difficult to ascertain the cause of failure of sub-grade structures, but it is likely that such failures are due to unconservative estimates of the actions transmitted from the piers to the foundations. Also, the point of contra-flexure of the pile-footing-pier system is often misplaced; hence the critical sections are not treated as such.

B.3.6 Geotechnical Effects

Assessment of geotechnical effects is of great importance for the seismic performance of bridges, of both RC and steel, as discussed in the previous sections. For example, soil lateral spreading or liquefaction, presented in Section 1.4.2, imposes large deformation demands on bridge components, such as piles, abutment walls and simply supported deck spans. Some bridges founded on soft ground in the Kobe area suffered damage to piles due to negative skin friction resulting from soil failure. Approach

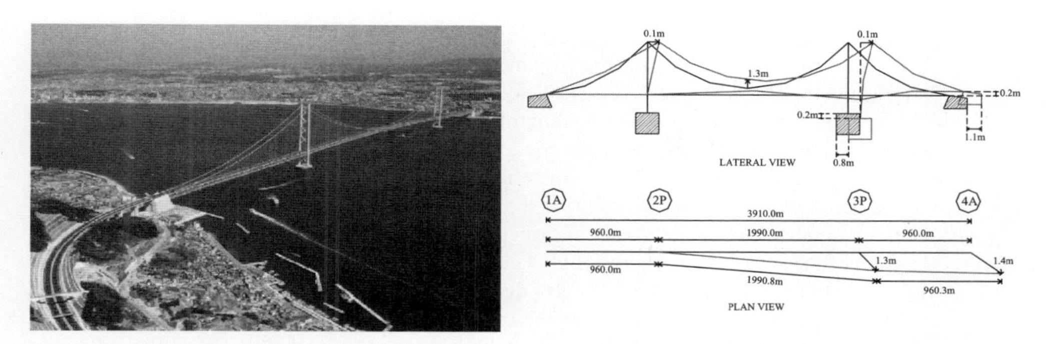

Figure B.57 Damage to the Akashi Kaikyo bridge during the 1995 Kobe (Japan) earthquake: aerial view (*left*) and permanent offset of foundations (*right*)

structures and abutments have suffered substantial movement due to soil slumping. The world's longest suspension bridge (Akashi Kaikyo) was under construction when the 1995 Kobe earthquake occurred. Two abutments and two main towers (2P and 3P, in Figure B.57, respectively) were completed. The fault crossed the bridge between foundations for 2P and 3P. This caused permanent lateral movements and rotations of the anchorage and tower foundations as depicted pictorially in Figure B.57 (Saeki *et al.*, 1997; Yasuda *et al.*, 2000).

Liquefaction was widespread in the 1964 Niigata earthquake, especially in the alluvial plains of the Shinano and Agano rivers. This caused significant damage due to large movements of pier and abutment foundations (*see* for example, Figure B.40). Railway and highway bridges were affected by large ground displacement in the 1990 Costa Rica earthquake, where caisson and pier movements of 2.0 m and 0.8 m, respectively, were observed. Examples of span failures due to liquefaction included the collapse of an exterior and an interior span of the Rio Viscaya and Rio Bananito Bridge during the 1990 Costa Rica earthquake, respectively. The fact that the Rio Bananito Bridge span seat was skewed further amplified the displacement.

Footings and piles are sometimes under-designed for earthquake loading, since the overstrength of the piers they support would not have been taken into account when evaluating actions on the foundations. In the 1923 Kanto (Japan) earthquake, tilting of mass concrete foundations was observed, thus indicating inadequate consideration of overturning. It is likely that such failures are due to unconservative estimates of the actions transmitted from the piers to the foundations.

A detailed description of damage patterns in bridges along with the corresponding structural deficiencies of foundations, sub- and superstructures may be found in specialist textbooks (e.g. Priestley *et al.*, 1996) and reconnaissance reports of recent worldwide earthquakes (*see* Astaneh-Asl *et al.*, 1989; EERI, 1994; Elnashai *et al.*, 1995, among many others).

B.4 Lessons Learnt from Previous Earthquakes

The previous paragraphs (Sections B.1 to B.3) have shed light on the structural deficiencies and relative damage observed in buildings and bridges during past earthquakes. These severe full-scale tests of the seismic performance of structures have taught us important lessons, which are summarized in the sections below.

B.4.1 Requisites of RC Structures

The assessment of the damage discussed in Sections B.2 and B.3 leads to the following conclusions:

(i) Structural members of lateral resisting systems used for buildings and bridges should be detailed so that they exhibit ductile response under severe earthquake ground motions. All other elements should be designed elastically. Dissipative zones, e.g. plastic hinges, require adequate concrete confinements. Buckling of longitudinal steel rebars impairs the anchorage, while splicing of bars should not be carried out in regions of high stress concentration.

(ii) Likely sources of overstrength, e.g. material mechanical properties and presence of slabs, should be accounted for in the design of dissipative elements in ductile systems.

(iii) Values of compressive axial loads in bridge piers and building columns should not exceed 25–30% of the squashing capacity. High values of axial loads significantly reduce the dissipation capacity of piers and columns as also discussed in Section 2.3.3.1. High axial loads lower the maximum plastic rotations and increase the likelihood of buckling of longitudinal steel reinforcement bars. Columns and piers should be designed to exhibit elastic response. Tensile forces should be prevented; the latter give rise to brittle failure modes, e.g. under high vertical components of earthquakes.

(iv) Short-column effects caused by partial infills in framed systems may be prevented by adopting adequate separation gaps. This detail does not increase the shear stiffness of column members.

(v) Failure modes involving shear and bond deteriorations should be avoided. These are brittle failure modes and hence lower the energy dissipation of the structural system. Consequently, flexural failure should anticipate that of shear. Columns with shear span ratios α_s greater than 4.0 are preferable to short columns ($\alpha_s < 2.0$). Close-spaced transverse stirrups or truss reinforcement may be adopted to prevent the degradation of shear resistance.

(vi) Configuration irregularities in plan and elevation should be avoided as also illustrated in Section A.1. Soft-storey mechanisms at the ground floor of buildings are, for example, often caused by infills only in the upper storeys. Structural irregularities may also give rise to significant torsional effects. Eccentricities between centre of mass (point of application of seismic – inertial – forces) and centre of rigidity (point of application of reaction of the structure) should be minimized.

(vii) Continuity in load path is an essential requirement for both gravity (vertical) and earthquake (horizontal and vertical) loads as also shown in Sections 2.3.2.2 and A.1.

(viii) A high degree of structural redundancy should be guaranteed so that as many zones of inelasticity as possible are developed before a failure mechanism is created. Redundant structures can accommodate large plastic redistributions.

(ix) Openings in slabs should be minimized because they detrimentally affect the in-plane strength and rigidity of horizontal diaphragms. To prevent punching, additional steel reinforcement should be located at connection between flat slabs and columns, and between structural walls and slabs.

(x) Joints should be provided at discontinuities between adjacent structures or part of them. Separation gaps should employ adequate provisions for movements so that pounding and unseating, e.g. of bridge spans, is avoided. In multi-span bridges, sufficient gaps should be used both at abutments and between adjacent spans. Overstressing of seismic restrainers should be avoided.

(xi) Uplift and sliding of foundation systems due to high overturning moments and shear forces often have detrimental effects on global structural response.

(xii) Large permanent ground displacements due to soil liquefaction and pile deformations should be accounted for in the design of buildings and bridges.

Several of the above requisites are also applicable to masonry, steel and composite structures. Therefore, the following sections focus only on the design solutions specific to each material.

B.4.2 Requisites of Masonry Structures

Masonry structures exhibit high vulnerability to seismic forces. To prevent the damage patterns outlined in Section B.2.2, it is necessary to account for the following:

(i) Tight quality controls should be performed on construction materials, especially of mortar and masonry units in adobe and stone-masonry buildings.
(ii) Reinforced and confined masonry are preferable to URM. The effect of reinforcement is to limit the amount of diagonal cracks and prevent toppling, particularly in perimeter walls. In systems with confined masonry, this reinforcement should be anchored into the surrounding frame.
(iii) Structural and non-structural walls should possess limited slenderness to prevent global buckling. Connections between orthogonal structural and non-structural walls are adequate to avoid overturning due to out-of-plane seismic forces.
(iv) Adequate connections should be provided between structural walls and slabs at each floor. Slabs act as horizontal diaphragms and distribute horizontal seismic forces among vertical structural walls as illustrated in Section A.1. Diaphragmatic actions should, however, always be checked and are significantly reduced by the presence of large openings. Bond beams should be located at each floor along perimeter walls to achieve monolithic behaviour of masonry structures.
(v) Low values of length-to-width ratios in piers of structural walls should be avoided. This type of geometric layout can give rise to severe brittle shear failures as also discussed in Section A.2.2.
(vi) Large openings should be limited in structural masonry walls. They significantly lower the strength capacity under earthquake loads. Additionally, diagonal cracks often originate at the corner of large openings.
(vii) Adequate building layout is a fundamental requisite to survive moderate and severe earthquakes (*see* Appendix A). Simple and symmetrical configurations along each principal axis with a sufficient number of structural walls, and with approximately the same cross-sectional area and stiffness should be provided in each direction of the building.

B.4.3 Requisites of Steel and Composite Structures

Steel and composite structures have shown generally adequate seismic performance under moderate and severe earthquakes. Their energy dissipation capacity is endangered if the requisites summarized below are not satisfied:

(i) Brittle failure modes, such as weld cracks and fracture, bolt fracture in tension or shear, should be avoided, even in response to a major seismic event.
(ii) Local buckling and global buckling can be avoided by adopting adequate width-to-thickness ratios and member slenderness.
(iii) Excessive column panel zone deformations in beam-to-column connections should be prevented. These deformations may significantly increase lateral drifts of unbraced framed structures and impair their global stability (increased P-Δ effects).
(iv) Overstrength due to the presence of composite slabs should be accounted for in the evaluation of the inelastic seismic demands of capacity-designed components, e.g. beams in MRFs, diagonal braces in CBFs and links in EBFs.

References

Ambraseys, N.N., Elnashai, A.S., Bommer, J.J., Haddar, F., Madas, P., Elghazouli, A.Y. and Vogt, J. (1990). The Chenoua (Algeria) Earthquake of 29 October 1989, Engineering Seismology and Earthquake Engineering, Report No. ESEE/90-4, Imperial College, London, UK.

Ambraseys, N.N., Elnashai, A.S., Broderick, B.M., Salama, A.I and Soliman, M.M. (1992). The Erzincan (Turkey) Earthquake of 13 March 1992, Engineering Seismology and Earthquake Engineering, Report No. ESEE/92-11, Imperial College, London, UK.

Aoyama, H. (2001). *Design of Modern Highrise Reinforced Concrete Structures. Series on Innovation in Structures and Construction*, Vol. 3, A.S. Elnashai and P.J. Dowling, Eds., Imperial College Press, London, UK.

Astaneh-Asl, A., Bertero, V.V., Bolt, B., Mahin, S.A., Moehle, J.P. and Seed, R.B. (1989). Preliminary report on the seismological and engineering aspects of the October 17, 1989 Santa Cruz (Loma Prieta) earthquake. Earthquake Engineering Research Centre, University of Berkeley, Report No. UCB/EERC-89/14, Berkeley, CA, USA.

Bertero, V.V. (1996). Implications of observed pounding of buildings on seismic code regulations. Proceedings of the 11th World Conference on Earthquake Engineering, Disc N.4, Paper N.2102, CD-ROM.

Braga, F., Briseghella, L. and Pinto, P.E. (1977). The Friuli (Italy) May and September 1976 earthquake: A brief survey of the damage. Proceedings of the 6th World Conference on Earthquake Engineering, Sarita Prakashan, Meerut, India, Vol. I, pp. 289–294.

Broderick, B.M., Elnashai, A.S., Ambraseys, N.N., Barr, J.M., Goodfellow, R.G. and Higazy, E.M. (1994). The Northridge (California) Earthquake of 17 January 1994: Observations, Strong Motion and Correlative Response Analysis. Engineering Seismology and Earthquake Engineering, Research Report No. ESEE 94/4, Imperial College, London, UK.

Bruneau, M. (1998). Performance of steel bridges during the 1995 Hyogoken-Nanbu (Kobe, Japan) earthquake – A North American perspective. *Engineering Structures*, 20(12), 1063–1078.

Earthquake Engineering Research Institute (EERI) (1994). Northridge Earthquake January 17, 1994. Preliminary Reconnaissance Report. J.F. Hall, Ed., EERI, CA, USA.

Elnashai, A.S. (1998). Observations on the Effects of the Adana-Ceyhan (Turkey) Earthquake of 27 June 1998. Engineering Seismology and Earthquake Engineering, Report No. ESEE/98-5, Imperial College, London, UK.

Elnashai, A.S. (1999). The Kocaeli (Turkey) Earthquake of 17 August 1999: Assessment of Spectra and Structural Response Analysis. Engineering Seismology and Earthquake Engineering, Report No. ESEE/99-3, Imperial College, London, UK.

Elnashai, A.S., Bommer, J.J. and Elghazouli, A.Y. (1989). The Loma Prieta (Santa Cruz, California) Earthquake of 17 October 1989. Engineering Seismology and Earthquake Engineering, Report No. ESEE/98-9, Imperial College, London, UK.

Elnashai, A.S., Bommer, J.J., Baron, I., Salama, A.I. and Lee, D. (1995). Selected Engineering Seismology and Structural Engineering Studies of the Hyogo-ken Nanbu (Kobe, Japan) Earthquake of 17 January 1995. Engineering Seismology and Earthquake Engineering, Report No. ESEE/95-2, Imperial College, London, UK.

Federal Emergency Management Agency (FEMA) (2000). State of the art report on past performance of steel moment frame buildings in earthquakes. Report No. FEMA 355E, Washington, DC, USA.

Goltz, J.D. (1994). The Northridge, California Earthquake of January 17, 1994: General Reconnaissance Report. National Centre for Earthquake Engineering Research, Report No. NCEER-94-0005, Buffalo, NY, USA.

Kawashima, K. (1995). Seismic design for construction and repair of highway bridges that suffered damage by the Hanshin Awaji great earthquake. In *Building for the 21st Century*, Y.C. Loo, Ed., EASEC-5, Gold Coast, Australia.

Matos, C.G. and Dodds, R.H. (2002). Probabilistic modelling of weld fracture in steel frame connections. Part II: Seismic loadings. *Engineering Structures*, 24(6), 687–705.

Miller, D.K. (1998). Lessons learned from the Northridge earthquake. *Engineering Structures*, 20(4–6), 249–260.

Naeim, F. (2001). *The Seismic Design Handbook*. 2nd Edition, Kluwer Academic Publisher, New York, NY, USA.

Naeim, F., Lew, M., Huang, C.H., Lam, H.K. and Carpenter, L.D. (2000). The performance of tall buildings during the 21 September 1999 Chi-Chi earthquake Taiwan. *The Structural Design of Tall Buildings*, 9(2), 137–160.

Nakashima, M., Inoue, K. and Tada, M. (1998). Classification of damage to steel buildings observed in the 1995 Hyogoken-Nanbu earthquake. *Engineering Structures*, 20(4–6), 271–281.

Nateghi, F.A. (1995). Retrofitting of earthquake damaged steel buildings. *Engineering Structures*, 17(10), 749–755.

Nateghi, F.A. (1997). Seismic upgrade design of a low-rise steel buildings. *Engineering Structures*, 19(11), 954–963.

National Information Service for Earthquake Engineering (NISEE) (2000). Image Database. http://www.nisee.org.

Penelis, G.G. and Kappos, A.J. (1997). *Earthquake Resistant Concrete Structures*. E & FN SPON-Chapman & Hall, London, UK.

Priestley, M.J.N., Seible, F. and Calvi, G.M. (1996). *Seismic Design and Retrofit of Bridges*. John Wiley & Sons, New York, NY, USA.

Roeder, C.W. (2002). Connection performance for seismic design of steel moment frames. *Journal of Structural Engineering, ASCE*, 128(4), 517–525.

Rosenblueth, E., Ruiz, S.E. and Thiel, C.C. (1989). *The Michoacan Earthquake: Collected Papers Published in Earthquake Spectra, Volumes 4 and 5, 1988 and 1989*. Earthquake Engineering Research Institute, El Cerrito, CA, USA.

Saeki, S., Kurihara, T., Toriumi, R. and Nishjtani, M. (1997). Effect of the Hyogoken-Nanbu earthquake on the Akashi Kaikyo bridge. Proceedings of the 2nd Italy-Japan Workshop on Seismic Design and Retrofit of bridges, 27–28 February, Rome, Italy.

Watanabe, E., Sugiura, K., Nagata, K. and Kitane, Y. (1998). Performances and damages to steel structures during 1995 Hyogoken-Nanbu earthquake. *Engineering Structures*, 20(4–6), 282–290.

Yasuda, T., Moritani, T., Fukunaga, S. and Kawabata, A. (2000). Seismic behavior and simulation analysis of Honshu-Shikoku bridges. *Journal of Structural Engineering, JSCE*, 46A, 685–694.
Youssef, N.F.G., Bonowitz, D. and Gross, J.L. (1995). A survey of steel moment-resisting frame buildings affected by the 1994 Northridge earthquake. Report No. NISTR 56254, National Institute for Science and Technology, Gaithersburg, MD, USA.

Index

Fundamentals of Earthquake Engineering Amr S. Elnashai and Luigi Di Sarno